Polish Achievements in Materials Science and Engineering

Polish Achievements in Materials Science and Engineering

Guest Editors

Jakub Zdarta
Agata Zdarta

MDPI

Basel • Beijing • Wuhan • Barcelona • Belgrade • Novi Sad • Cluj • Manchester

Guest Editors
Jakub Zdarta
Institute of Chemical Technology and Engineering
Poznan University of Technology
Poznan
Poland

Agata Zdarta
Department of Organic Chemistry
Poznan University of Technology
Poznan
Poland

Editorial Office
MDPI AG
Grosspeteranlage 5
4052 Basel, Switzerland

This is a reprint of the Special Issue, published open access by the journal *Materials* (ISSN 1996-1944), freely accessible at: https://www.mdpi.com/journal/materials/special_issues/Polish_Achievements_Materials_Science.

For citation purposes, cite each article independently as indicated on the article page online and as indicated below:

Lastname, A.A.; Lastname, B.B. Article Title. *Journal Name* **Year**, *Volume Number*, Page Range.

ISBN 978-3-7258-3467-9 (Hbk)
ISBN 978-3-7258-3468-6 (PDF)
https://doi.org/10.3390/books978-3-7258-3468-6

Contents

About the Editors

Jakub Zdarta

Jakub Zdarta is a Professor at the Poznan University of Technology. He studied General Chemistry (since 2005) at the Adam Mickiewicz University in Poznan (dipl. 2010) and Chemical Technology at the Poznan University of Technology (dipl. 2012), defending his PhD diploma in Chemical Technology in 2017. His career in materials science and biotechnology started in 2008 under the supervision of Prof. Teofil Jesionowski. He continued with a one-year post-doc at the Technical University of Denmark under the mentoring of Prof. Anne S. Meyer, a renowned specialist in enzymology. After that, he returned to the Poznan University of Technology in 2022, where he held a professor position. He now continues his research in enzyme immobilization and application, leading grants from the Polish National Science Centre (Preludium, Sonata) and cooperating with scientists in Denmark (Technical University of Denmark), Australia (University of Technology Sydney), India (IIT Bilai), the United States (MIT), and Spain (University of Almeria). He has received the following awards: Scholarship of the Minister of Science and Higher Education for Outstanding Young Scientists in 2019, Scientific Award of the 4th Division of Technical Sciences of the Polish Academy of Sciences in 2022, Award of the Polish Prime Minister in 2022 for scientific achievements constituting the basis for conferring the degree of habilitated doctors, and the Award of the Ministry of Higher Education and Science in 2023 for outstanding scientific achievements. He is the Chair of the organizing committee of the Green Technology for Sustainable Environment 2024 (GTSE24) and Biotransformations 2026 conferences, co-editor of *Environmental Technology and Innovation*, a Guest Editor in Special Issues for several JCR journals, and he is the author of two books, over 15 book chapters, and more than 120 scientific articles.

Agata Zdarta

Agata Zdarta graduated in Biotechnology from the Poznan University of Life Sciences (2013) and received her PhD in Chemistry in 2020 from the Poznan University of Technology. She works as an Assistant professor at the Poznan University of Technology. She carried out her post-doc research at the Technical University of Denmark under the supervision of Dr. Nanna B. Harmann in microplastic quantification (2021) within the Iwanowska programme (Polish National Agency for Academic Exchange). Currently, her scientific interest focuses on microplastic quantification in environmental samples with a special focus on sample preparation procedures. She is a leader and principal investigator in a Sonata-funded project (Polish National Science Centre) concerning the application of biomembranes to microplastic sample preparation. She is a laureate of the Start scholarship (Foundation for Polish Science, 2021) and the Scholarship of the Polish Minister of Science for Outstanding Young Scientists (2024), as well as a member of the organizing committee of the Green Technology for Sustainable Environment 2024 (GTSE24) international conference and XI Congress of Chemical Technology (XIKTCh) conference and a Guest Editor for international scientific journals.

materials

MDPI

Article

In Vitro Fracture Resistance of Endodontically Treated Premolar Teeth Restored with Prefabricated and Custom-Made Fibre-Reinforced Composite Posts

Michal Bialy [1], Sara Targonska [2,*], Agnieszka Szust [3], Rafal J. Wiglusz [2,*] and Maciej Dobrzynski [1]

1 Department of Pediatric Dentistry and Preclinical Dentistry, Wroclaw Medical University, Krakowska 26, 50-425 Wroclaw, Poland; michal.bialy@umed.wroc.pl (M.B.); maciej.dobrzynski@umed.wroc.pl (M.D.)
2 Institute of Low Temperature and Structure Research, Polish Academy of Sciences, Okolna 2, 50-422 Wroclaw, Poland
3 Faculty of Mechanical Engineering, Wroclaw University of Science and Technology, Wybrzeze Wyspianskiego 27, 50-370 Wroclaw, Poland; agnieszka.szust@pwr.edu.pl
* Correspondence: s.targonska@intibs.pl (S.T.); r.wiglusz@intibs.pl (R.J.W.); Tel.: +48-713-954-274/159 (S.T. & R.J.W.)

Citation: Bialy, M.; Targonska, S.; Szust, A.; Wiglusz, R.J.; Dobrzynski, M. In Vitro Fracture Resistance of Endodontically Treated Premolar Teeth Restored with Prefabricated and Custom-Made Fibre-Reinforced Composite Posts. *Materials* **2021**, *14*, 6214. https://doi.org/10.3390/ma14206214

Academic Editors: Jakub Zdarta and Agata Zdarta

Received: 4 September 2021
Accepted: 13 October 2021
Published: 19 October 2021

Publisher's Note: MDPI stays neutral with regard to jurisdictional claims in published maps and institutional affiliations.

Abstract: (1) Background: The study aimed to compare and analyse the differences between the features of prefabricated fibre-reinforced composite (FRC) posts and custom-made FRC posts in the form of a tape and confirm the necessity of using FRC posts in teeth treated endodontically in comparison to direct reconstruction with a composite material. (2) Methods: Sixty premolars after endodontic treatment were used. The teeth were divided into four groups (n—15). Group 1: teeth with embedded prefabricated posts (Mirafit White); group 2: teeth with embedded prefabricated posts (Rebilda); group 3 teeth with embedded custom-made posts in the form of a tape (EverStick); group 4: teeth without a post restored with composite material. The compressive strength of the teeth was tested using the Instron-5944 testing machine until the sample broke. The crystal structure of the investigated posts was detected with the X-ray diffractometer (3) Results: During the experiment, the maximum values of forces at which the damage of the restored premolar teeth after endodontic treatment occurred were obtained. The best results were obtained for teeth rebuilt with Rebilda Posts (1119 N), while teeth with cemented Mirafit White posts were the weakest (968 N). Teeth without an embedded FRC post, rebuilt only with light-cured composite material, obtained the lowest value—859 N. (4) Conclusions: The use of FRC posts increases the resistance to damage of an endodontically treated tooth when compared to direct restoration with light-cured composite material.

Keywords: premolars; endodontic treated; fibre-reinforced composite post; strength test

1. Introduction

Teeth after endodontic treatment are structurally different from vital teeth and require specialized restorative materials [1,2]. The major difference is a consequence of the loss of dental structures caused by caries, fractures of previous restorations and endodontic procedures. Consequently, the risk of fracture increases after endodontic treatment. [3,4]. Finally, their mechanical and strength properties are directly proportional to the volume of origin tooth tissue as well as the chosen treatment technique [5–11].

The insertion of a post significantly increased the fracture resistance of nonvital premolars [12–14]. One of the main goals of posts' presence is to ensure the dental materials' retention while the lost dental crown is being reconstructed. A post allows for the appropriate stress distribution in the root and may be used to support single crowns and bridges. However, according to numerous researchers, a post does not strengthen the tooth structure, it merely provides the appropriate retention for the material used to restore a crown foundation [15–18]. These findings could be caused by the fact that preparation of space for the crown-root inlay may lead to the diminishing of the remaining part of the tooth structure and increase the risk of the root breaking. This emphasizes the significance of preserving the original anatomy of the root canal and minimizing dentin loss in the reconstruction following the completed endodontic treatment.

There are several factors that could influence the root canal filling related to the mechanical preparation of the post space such as the twisted or vibrated areas [19]. The filled root canals can become contaminated again by microorganisms during the preparation of the canal for the post, when the remaining apical part of the root filling is of an inappropriate density or length. It is a risk factor for the development of periapical inflammation. Santos et al. assess that root canal-filled teeth coronally restored with a rubber dam did not develop severe periapical inflammation but root canal fillings exposed to microbial challenge with exposure to saliva developed severe periapical inflammation [20]. They concluded that the current methods of root canal filling only partially inhibit the penetration of microorganisms into the crown and the development of apical periodontitis after treatment, highlighting the risk of periapical pathology in case of a break in the chain of asepsis during post preparation.

It was stated in several cases that endodontic treatment by using one fibre-reinforced composite (FRC) post may not be enough to ensure suitable mechanical friction in the case of root canals characterized by an unusual and irregular shape. To solve the problems with the unusual shape of root canals, an elastic FRC post (GC Europe, Leuven, Belgium) was introduced [21]. It should be noted, the elastic FRC post can be individually customized and their ability to bond and bend is superior to the prefabricated posts that commonly appear on the market.

The presented study is focused on investigating and comparing the crack resistance of endodontically treated premolar teeth reinforced by different available FRC posts. The null hypotheses were: (1) The application of FRC posts increases the resistance to damage of an endodontically succeeding treated tooth when compared to direct restoration with light-cured composite material and (2) no significant difference was noted in the case of using more elastic fibre posts.

2. Materials and Methods

2.1. Preparation of Materials

In the research, 60 first premolar teeth extracted for orthodontic or periodontal indications were used. The teeth neither showed any signs of caries, root cracks, resorption nor were they treated before. Each of the teeth had two roots and two canals, the buccal and the palatine one. All of the selected teeth had similar dimensions: the length of the crowns was 8.5 mm, the roots' lengths were 14 mm and the diameter of the mesial-distal crowns was 7 mm. They were kept in 0.9% NaCl solution. The root surface was purified of soft tissue by using the hand scaler. Prior to strength tests, they were treated endodontically. The working length was determined using a direct method related to subtracting 1 mm from the actual root length set by introducing a no. 10 K-file (Maillefer-Dentsply, Ballaigues, Switzerland). Irrigation was performed after every change of instrument with 5.25% sodium hypochlorite, citric acid and saline solution. For all of the studied teeth, the step back treating technique was applied, master apical file size 40 and coronal flaring size 70.

The wetness level inside root space was checked by the paper points. After drying, the root canals were filled by an epoxy resin sealer—AH Plus (Dentsply Maillefer, Konstanz, Germany)—and the lateral condensation technique was used. The dental crown was temporarily filled using the self-adhesive posterior restorative system Equia (GC Europe, Leuven, Belgium). The Equia was covered by the apical part of the root to prevent leakage through the apex. The teeth were stored for 7 days in an incubator (at 37 °C, 100% relative humidity). Following this, the palatal cusps of the crowns were removed to the level of the central sulcus, leaving only 2 mm of cingulum above the tooth neck (Figure 1). The palatal canal was prepared using size 3 Pesso drills to obtain a 15 mm length, leaving an apical seal of gutta-percha in the canal (4–6 mm). The teeth were divided into four groups, each group having 15 teeth. Group 1 received prefabricated conventional FRC posts of 1.2 mm diameter (Mirafit White posts, Hager & Werken, Duisburg, Germany). Group 2 received prefabricated conventional FRC posts of 1.2 mm diameter (Rebilda Posts, Voco, Cuxhaven, Germany). Group 3 received one single elastic FRC post of 1.2 mm diameter (EverStick, GC Europe, Leuven, Belgium). Following the manufacturer's recommendations, the elastic post was inserted into the root canal and adjusted to its shape. After adaptation, the post was removed from the root canal with a needle-nose plier and light-cured for 40 s with an Epilar S10 polymerization lamp (3M Espe, Maplewood, NJ, USA). After polymerization, the EverStick posts became stiff. The cavities in group 4 were restored with a microhybrid composite restorative material, Charisma (Heraeus Kulzer, Hanau, Germany).

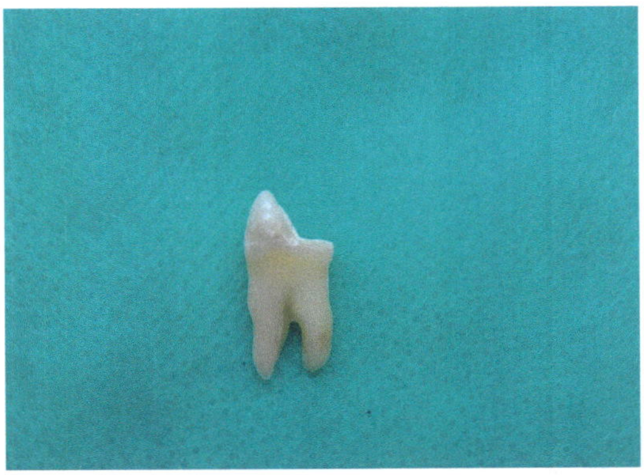

Figure 1. Premolar tooth with removed palatal tubercle.

The posts in groups 1, 2 and 3 were tried and next their lengths were adjusted. The palatal canal and the prepared tooth surface were etched with orthophosphoric acid for 30 s. The dentin surface was dried with filter paper; next, a bond was applied directly in the canal and on the prepared dentin. The dual-cure self-priming dental adhesive system, Prime and Bond NT (Dentsply, Charlotte, NC, USA), was used for bonding, according to the manufacturer's instruction. Luting of the posts was performed with a dual resin composite cement, Core.X Flow (Dentsply, Charlotte, NC, USA). In the following step, the crowns of the teeth were reconstructed with Charisma composite material (Heraeus Kulzer, Hanau, Germany). The cellulite core-forming matrices of the same size were used to ensure the uniformity of the specimens. These matrices were fabricated as vacuum-formed foils on a healthy premolar tooth. The core build-up was polymerized for 40 s, then the cellulite forming matrices were removed. Glycerine gel was used, and a final polymerization was performed from each side for 40 s with a Translux Wave polymerization lamp (Kulzer, Milano, Italy).

After the restorative procedures, the specimens were kept in 0.9% NaCl solution for 1 week. The roots of the prepared teeth were coated with a thin layer of low-density impression material—Oranwash L (Zhermack, Badia Polesine, Italy). It prevented the teeth from rigidly fixing in acrylic blocks and allowed them to maintain their physiological mobility, typical of natural teeth with periodontal fibre (Figure 2). All of the analysed cases were tested in the same way under the same conditions for individual samples. The fixing of the prepared teeth and the method of loading were the same for each case. The teeth were fixed in acrylic blocks into which they were inserted to the level of 2 mm below the tooth neck. The length, width and height of the acrylic blocks were 15 mm for each size. The teeth were fixed at an angle of 90° to the long axis of the tooth in the jaws of a universal testing machine Instron 5944 (Instron, Norwood, MA, USA). A nickel–chrome form reflecting the shape of the chewing surface of a premolar tooth was used as a punch.

Acrylic blocks with fixed teeth were loaded in a way to ensure the optimum occlusive contacts of the upper and lower premolar teeth (Figure 3). The loading rate was 1 mm/min. Strength tests were conducted until the tested sample was damaged, i.e., until the first force decrease was recorded in a universal testing machine graph. The maximum failure load was recorded in Newtons (N) (Table 1).

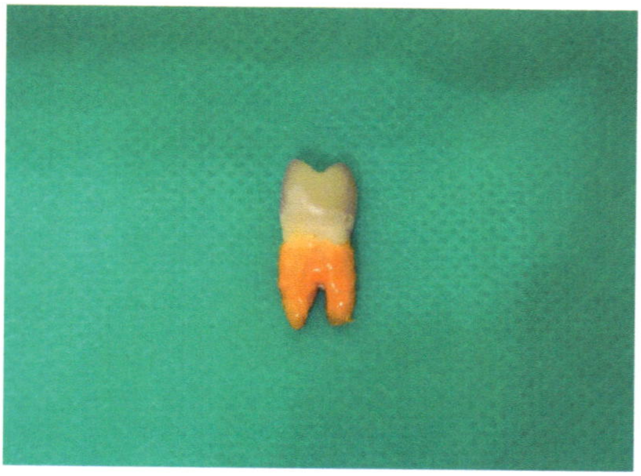

Figure 2. A root covered with a layer of silicone.

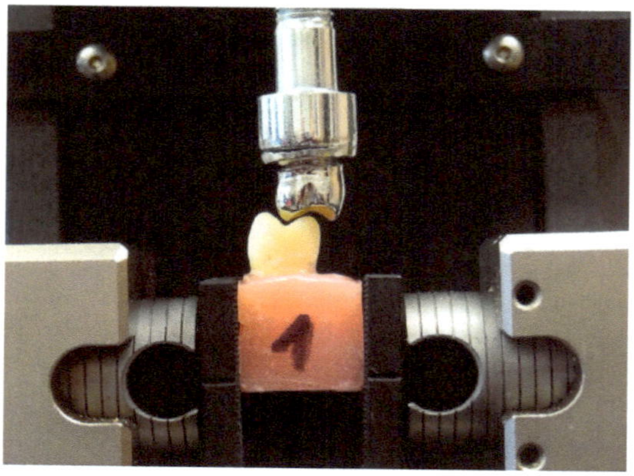

Figure 3. Sample orientation for best occlusal contacts.

Table 1. Fracture thresholds as well as the significance of their difference of studied fibre-reinforced teeth compared to the control group.

Post Type	Valid N	Mean [N]	Min [N]	Max [N]	SD
Mirafit White	15	968	760	1222	149
Rebilda Post	15	1119	973	1323	116
EverStick	15	1000	776	1368	163
No Post	15	859	587	1253	200

After the examination, the number of damaged teeth that could be rebuilt was assessed. There are two groups of damaged teeth proposed. The first group includes teeth able to be rebuilt. The teeth with the experimental restorable fracture above the cementoenamel junction (CEJ) facture were assigned to the first group. Consequently, the result below the CEJ makes a tooth marked as irreparable and it is likely to be extracted [22].

The distribution of the data from the endurance test of endodontically treated teeth did not differ from normal. The significance of the differences in terms of strength was checked using ANOVA (analysis of variance). The significance level of the hypothesis testing was set at $p > 0.05$. The calculations were conducted in the Statistica version 13.3 program (StatSoft, Kraków, Poland).

2.2. Physicochemical Characterization

The crystal structure of the investigated materials was detected using a PANalytical X'Pert Pro X-ray diffractometer (Malvern Panalytical Ltd., Malvern, UK) equipped with Ni-filtered Cu $K\alpha_1$ radiation ($K\alpha_1$ = 1.54060 Å, U = 40 kV, I = 30 mA) in the 2θ range of 5°–70°. X-ray diffraction (XRD) patterns were parsed by Match! software version 3.7.0.124. Energy-dispersive spectroscopy (EDS) analysis was carried out using an FEI Nova NanoSEM 230 scanning electron microscope (SEM, Hillsboro, OR, USA) equipped with an EDS spectrometer (EDAX GenesisXM4) and operating at an acceleration voltage in the range 3.0–15.0 kV and spots at 2.5–3.0 were observed. The Fourier-transformed infrared spectra were carried out by the Thermo Scientific Nicolet iS50 FT-IR spectrometer (Waltham, MA, USA) equipped with an automated beam splitter exchange system (iS50 ABX containing DLaTGS KBr detector), built-in all-reflective diamond ATR module (iS50 ATR), Thermo Scientific Polaris™ and the HeNe laser was used as an infrared radiation source. Absorption spectra were collected by the Agilent Cary 5000 spectrophotometer (Agilent, 5301 Stevens Creek Blvd, Santa Clara, CA 95051, USA), employing a spectral bandwidth (SBW) with a spectral resolution of 0.25 nm in the visible and ultraviolet range and recorded at room temperature.

3. Results

During the experiment, the values of maximum forces, i.e., the forces destroying reconstructed teeth after endodontic treatment, were obtained. The damage could involve dental hard tissues, the post and the light-cured material used in tooth reconstruction. The best results were obtained for teeth reconstructed with Rebilda Posts (mean load 1119 N), while the poorest results were obtained for the teeth in which Mirafit White posts (mean load 968 N) was used (Figure 4). The teeth without any post, reconstructed using only the light-cured material, were damaged by the lowest value of the destruction force (mean 859 N), which confirms the rationale for using posts. Based on the mean values presented in Figure 4, the Rebilda Post group is a more effective treatment than the others. However, in the statistical analysis (*t*-Test ANOVA, $p > 0.05$), no statistically significant differences were found between the various methods. The lack of statistical significance between the investigated treatments may be due to the small number of samples in each group.

After the completion of the strength tests, the assessment of the destructed teeth started. The damage of the supragingival crown, involving natural tissue or composite material, was considered auspicious for further prosthetic reconstruction. Regarding tooth destruction, subgingival damage is considered as the most harmful, especially when it exceeds the biological width. In such cases, obtaining a good impression, removing cement remains or maintaining dryness of the field is required. Any imprecision in the above may lead to periodontitis. Sometimes, even orthodontic extrusion or surgical crown lengthening are necessary to prosthetically restore the lost tooth tissue. Teeth with a broken root are frequently qualified for extraction.

The largest number of supragingival crown fractures in teeth undergoing strength tests were recorded for the samples with Rebilda Posts (60%). It was confirmed that in the case of fibre-reinforced composite posts with good strength parameters, the prognosis for future restoration is better. In the samples with cemented EverStick posts, subgingival fractures were observed more frequently (60%), which made it very difficult or even impossible to prosthetically restore the tooth. The worst prognosis relates to teeth without posts, for 73% of which a subgingival fracture was observed (Tables 1 and 2).

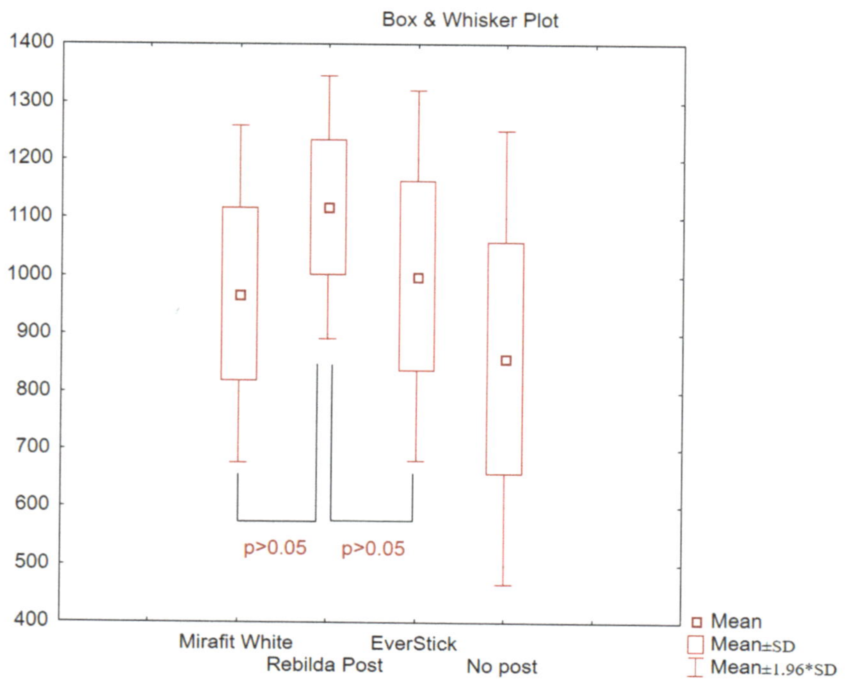

Figure 4. Mean fracture damage values, standard deviations (SD).

Table 2. Prognosis of tooth restoration possibilities after conducted strength test.

Type of FRC Post	Good Prognosis (Supragingival Fracture)	Poor Prognosis (Subgingival Fracture)
Mirafit White	53% (8/15)	47% (7/15)
Rebilda Post	60% (9/15)	40% (6/15)
EverStick	40% (6/15)	60% (9/15)
No Post	27% (4/15)	73% (11/15)

The presented study investigates three types of fibre-reinforced composite posts made of different chemical compounds. The EverStick post is made of poly (methyl methacrylate) (PMMA) polymer and bisphenol A-glycidyl methacrylate (bis-GMA). The neat and UV light-cured EverStick posts were investigated. The manufacturer of Mirafit White informs us that the material is made of E-glass (a type of glass fibre) and epoxy resin with a 65% fibre content (matrix). In the case of the Rebilda Post, information about the composition includes urethane-dimethacrylate (UDMA) 70 wt.%; glass fibre 20 wt.% and glass filling 10 wt.%.

The X-ray diffraction patterns of the EverStick post, UV light-cured EverStick post and Rebilda Post are presented in Figure 5. In the range of 2 theta from 10° to 25°, two broad diffraction lines that were detected corresponded to PMMA (see Figure 5a). The maximum of the intense one is located at 12° and that of the other one at 22°. The typical amorphous phase is confirmed. The UV light-curing did not affect PMMA crystallization. It could be suggested that glass fibres present in the PMMA matrix precludes the transition of the organized polymeric chain into the crystal phase [23]. The series of diffraction line was found in the XRD pattern of the Rebilda Post. The lines corresponded with the theoretical pattern of orthorhombic phase YbF_3 (PDF no. 96-153-2779) [24]. In the range of small 2 theta, a high background was detected. It was found to be associated with the amorphous phase of the polymeric matrix. The XRD pattern of the Mirafit White post does not present diffraction lines.

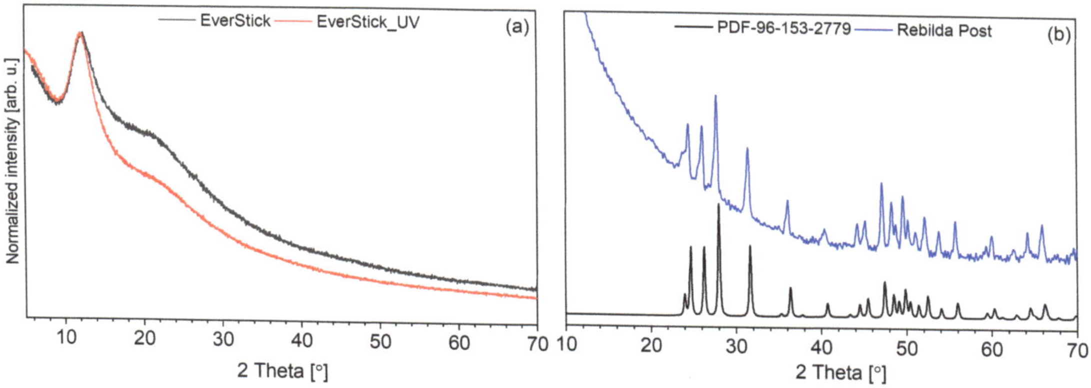

Figure 5. X-ray diffraction pattern of neat EverStick post and UV light-cured EverStick post (**a**) as well as Rebilda Post and PDF theoretical pattern of YbF_3 (PDF-96-153-2779) (**b**).

The presence of the YbF_3 phase was confirmed by EDS measurements. The existence of Yb^{3+} and F^- ions was detected. The results show that the material contains 4.4% of Yb^{3+} and 3.7% of F^-. Small concentrations of sodium, magnesium, potassium and calcium as well as chlorine ions were also observed.

Figure 6 shows the FT-IR spectra of the analysed materials and their main chemical components. In general, the FT-IR spectra are composed of a series of lines that are associated with the absorption of the energy by specific chemical bands. Lines are observed in the case of all of the measured samples corresponding to the PMMA molecules. The line at 1716 cm^{-1} confirms the presence of the ester group (C=O). Lines associated with the ether group O–CH_3 are located at 1163 cm^{-1}. The two lines detected at 2963 cm^{-1} and at 2872 cm^{-1} were ascribed to the methyl group (–CH_3). The methylene group (–CH_2) was located at 1245 cm^{-1} and at 1434 cm^{-1}. The lines at 1510 cm^{-1} were associated with the stretching vibration of an aromatic ring. The stretching vibration of the C–C mode was observed in the range of 1000–800 cm^{-1}. In the range of 1484–1389 cm^{-1}, absorption lines related to the deformation mode of C–H are detected. The lines observed in the range 1243–1272 cm^{-1} are due to the C–O–C vibration modes. On the spectrum of Mirafit White, the line associated with C–NH_2 vibration mode was observed at 1507 cm^{-1}. It suggests that both the Mirafit White and the Rebilda Post are made of UDMA [25–28].

The absorbance spectra of UV-vis radiation of UV light-cured the EverStick post were carried out in the range of 250–800 nm as a function of UV light curing time (see Figure 7). Based on the previous papers, it is known that PMMA does not absorb in the range of UV-vis radiation [29]. Absorption bands in the range of 250–330 nm were associated with the presence of bis-GMA [30]. It was found that the relative absorbance value changed with the UV light curing time. The Lambert–Beer law states that the absorption quantity depends on the concentration of a substance, the path length and the proportionality constant. It could be suggested that during the polymerization process, the thickness of the EverStick post has changed. It was calculated that the thickness decreases by 8% after 60 s of UV light curing.

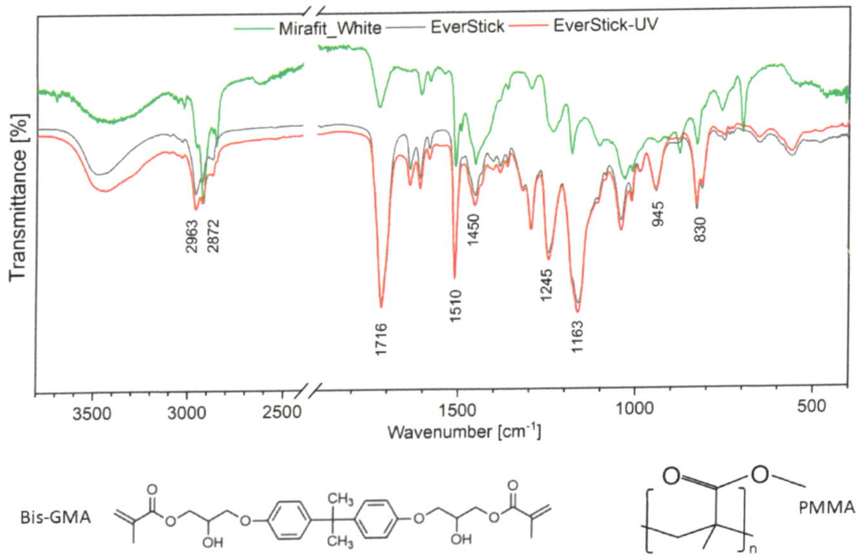

Figure 6. Fourier-transformed mid-infrared spectra of neat EverStick post, UV light-cured EverStick post, Mirafit White and GC Fibre post (**top**) and the structural formula of bisphenol A–glycidyl methacrylate (bis–GMA) (**bottom left**) and poly (methyl methacrylate) PMMA (**bottom right**).

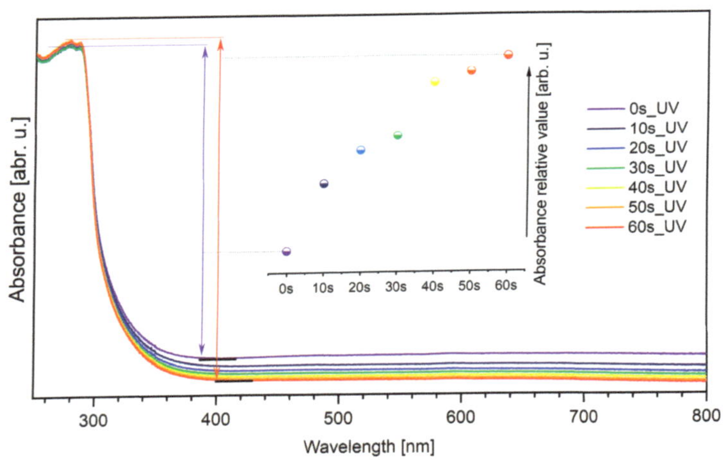

Figure 7. Absorption spectra of neat EverStick post and UV light-cured EverStick post as a function of light-curing time. Inset—The amplitude between the maximum and minimum absorption value.

4. Discussion

Premolars after endodontic treatment were found to be the most frequently fractured teeth [31,32]. A fracture analysis of maxillary premolars detected that teeth with a damage palatal cusp were the most prone to crack under compressive loading [33]. Therefore, in this study, premolars were prepared for strength tests.

Tooth restoration following endodontic treatment is the main objective of dental prosthetics [34]. It is recommended not to insert posts at the cost of the root dentin [35,36]. The research has shown that excessive preparation for a post not only weakens the tooth structure but may also lead to fractures and defects which could result in increasing the probability of tooth fractures or even tooth loss [37,38]. The use of fibre-reinforced composite posts to reduce the risk of tooth fractures has been questioned many times. Phebus et al. demonstrated that the teeth with a cemented fibre-reinforced composite post were significantly stronger than those which were endodontically treated without the use of a post [39]. They used thermal cycles to test a group of endodontically treated incisor teeth with cemented fibre-reinforced composite posts and another group of endodontically treated teeth without such posts. The thermal cycles simulated the changing conditions in the oral cavity. Lassila et al. discovered that significant influence on the fracture load and flexural strength is related to a thermocycling process for FRC posts. [21]. The flexural modulus decrease was estimated about 10% as a result of the thermocycling process. Moreover, the strength and fracture load decreased by about 18%. Hashemikamangar et al. claimed that no significant effect was detected after the thermocycling process on pushing-out bond strength between the fiber posts and resin core [40]. Moreover, the bond strength values in the thermocycled samples were slightly lower than the values in the nonthermocycled samples.

Based on the statistical analyses, the most homogeneous study group was the teeth rebuilt based on Rebilda Posts. In this group, the highest average value of force needed for destroying the treated tooth was obtained. The control group shows the greatest variability of results, where for the mean value of the destructive force of 859 N, SD is 200. The heterogeneity of the results in this group is caused by the absence of a post and the weakening of the crown resulting from endodontic treatment. The results of the strength tests are satisfactory in terms of the value of the standard deviation for the individual test groups. The heterogeneity of the results shows that the tested samples are biological material. The teeth were obtained from different patients of both sexes and different ages. For this reason, it is not possible to obtain perfectly homogeneous samples of a natural origin. The described studies were of a comparative and pilot nature, aimed at justifying the choice of treatment.

In the authors' own research, the teeth with cemented EverStick posts showed lower values of strength parameters than the teeth with standard posts. Similar research was conducted by Cagidiaco et al. who observed the survival rate of endodontically treated premolar teeth for three years [41]. The tested groups encompassed: teeth with cemented fibre-reinforced composite posts—D.T. Light-Post (VDW, Munich, Germany), teeth with individually formed posts (EverStick) and endodontically treated premolar teeth without posts. All of the teeth were covered with a metal ceramic crown. In the examined groups, the 36-month survival rate was 76.7%. The smallest percentage of success was recorded for the teeth without posts (62.5%), while in the case of teeth with the cemented prefabricated D.T. Light-Posts, the success rate was higher (90.9%) than in the case of teeth with the cemented EverStick posts (76.7%).

Ferrari et al. also conducted observations of the survival rate of endodontically treated premolar teeth for six years [42]. The research was conducted on 345 patients with teeth restored using fibre-reinforced composite posts (D.T. Light-Post), individually made posts (EverStick) and the control group were premolar teeth restored using only composite material. All of the teeth were covered with a metal ceramic crown. In the case when all of the walls of the treated teeth were preserved, the success percentage for all of the tested groups was the same and was 100%. This means that no complications were recorded in the teeth restored using the standard and individually made posts, and in the group without posts. The results were different in the teeth with a smaller number of preserved walls. In the case when two walls were preserved, the highest success rate was recorded in the group restored with the standard posts (88.9%), followed by the individually made ones (66.7%) and the lowest success rate was reported for the teeth without posts (52.9%). When only one wall was preserved, the success rate was lower and was 77.8% for teeth with the standard posts, 50% for the individually made posts and 29.4% for teeth without posts. These results are close to the authors' own research results in which the greatest force was needed to destroy crowns in the group of teeth with the standard posts. The higher value of force needed to damage the tooth crown should translate into a higher success rate in treated teeth.

According to the Kivac et al. [43], the presence and the type of the post used did not affect the mechanical strength of the tested teeth, which is a different conclusion from the results obtained in our own research. In the authors' own research, the ANOVA revealed significant differences ($p = 0.05$) in fracture loads, flexural strengths and flexural modulus of the FRC-post systems tested. The teeth with the cemented EverStick posts showed lower values of strength parameters than the teeth with the standard posts. The teeth without any post, reconstructed using only the light-cured material, were damaged by the lowest value of the destruction force.

Fractures of posts or of restored teeth are among the most common failures. A statistical analysis showed that specimens restored without the use of a post were more frequently used for nonrestorable fractures with the fracture line below the CEJ. Prefabricated posts exhibited more favourable fracture patterns of teeth in comparison with the teeth restored using elastic posts. Thus, the null hypothesis concerning fracture patterns was adopted. It is contrary to the studies of Frater et al. [44] who showed that teeth rebuilt with an elastic FRC post had significantly higher fracture resistance than teeth rebuilt with a prefabricated FRC post. Nevertheless, the load tested by Frater and co-workers was applied at 45° to the long axis of the tooth by aligning a stainless-steel ball-shaped stylus with the occlusal surface. It is a completely different than this investigation.

In our own study, elastic FRC posts showed lower flexural properties than the prefabricated FRC posts [45]. There were statistically significant differences in fracture loads, flexural modulus and flexural strengths during the three-point test. This confirms that posts with a good performance during three-point test offer improved damage resistance after endodontic treatment.

Another study examined the flexural properties of different types of prefabricated FRC posts (Snowpost, Carbopost, Parapost, C-post, Glassix, Carbonite) and compared those values with an individually polymerized FRC EverStick post [19]. The highest flexural strength was obtained for EverStick. It was noted that this effect could be explained by the combination of the polymer matrix and fiber properties building the composite. Furthermore, the presence of PMMA chains in the crosslinked polymer matrix is responsible for the differences between everStick and the other investigated FRC posts. PMMA chains plasticize the crosslinked bis-GMA-based matrix of the EverStick FRC, and thus reduce the generation stresses at the fibre–matrix interface during deflection. Consequently, it could be supposed to improve the strength of everStick FRC material.

5. Conclusions

The use of a fibre-reinforced composite post increases the damage resistance in endodontically treated teeth in comparison with teeth restored using light-curing material. Prefabricated Rebilda Posts showed more favourable fracture patterns than the other examined groups.

A statistically significant difference was observed between the use of prefabricated and custom-made FRC posts for multiple restorations. In teeth rebuilt with an elastic FRC post, subgingival fractures were observed more frequently—which made it very difficult or even impossible to prosthetically restore the tooth.

Author Contributions: Conceptualization, M.B. and R.J.W.; methodology, M.B. and A.S.; software, A.S.; validation, M.B. and M.D.; formal analysis, M.B., A.S. and R.J.W.; investigation, M.B.; resources, M.B.; data curation, M.B., A.S., S.T. and R.J.W.; writing—original draft preparation, M.B., S.T. and R.J.W.; writing—review and editing, S.T., A.S. and R.J.W.; visualization, M.B., S.T. and R.J.W.; supervision, M.D. and R.J.W.; project administration, M.D. and R.J.W.; funding acquisition, M.B. and R.J.W. All authors have read and agreed to the published version of the manuscript.

Funding: This research was funded by the Ministry of Health subvention number Pbmn 154 according to the IT Simple system of Wroclaw Medical University, Poland. This work was also financed by a subsidy from Wroclaw Medical University, subsidy number SUB.B180. 21.055.

Institutional Review Board Statement: The study was approved by the Bioethics Committee of Wroclaw Medical University, approval No. 400/2014.

Informed Consent Statement: Not applicable.

Data Availability Statement: Data available on request due to restrictions e.g., privacy or ethical. The data presented in this study are available on request from the corresponding author.

Conflicts of Interest: The authors declare no conflict of interest.

References

1. Cecchin, D.; De Almeida, J.F.A.; Gomes, B.P.F.A.; Zaia, A.A.; Ferraz, C.C.R. Influence of chlorhexidine and ethanol on the bond strength and durability of the adhesion of the fiber posts to root dentin using a total etching adhesive system. *J. Endod.* **2011**, *37*, 1310–1315. [CrossRef]
2. Dietschi, D.; Duc, O.; Krejci, I.; Sadan, A. Biomechanical considerations for the restoration of endodontically treated teeth: A systematic review of the literature—Part 1. Composition and micro- and macrostructure alterations. *Quintessence Int.* **2007**, *38*, 733–743.
3. Taha, N.A.; Palamara, J.E.; Messer, H.H. Fracture strength and fracture patterns of root-filled teeth restored with direct resin composite restorations under static and fatigue loading. *Oper. Dent.* **2014**, *39*, 181–188. [CrossRef]
4. Mohammadi, N.; Kahnamoii, M.A.; Yeganeh, P.K.; Navimipour, E.J. Effect of Fiber Post and Cusp Coverage on Fracture Resistance of Endodontically Treated Maxillary Premolars Directly Restored with Composite Resin. *J. Endod.* **2009**, *35*, 1428–1432. [CrossRef]
5. Balkaya, M.C.; Birdal, I.S. Effect of resin-based materials on fracture resistance of endodontically treated thin-walled teeth. *J. Prosthet. Dent.* **2013**, *109*, 296–303. [CrossRef]
6. Rocca, G.T.; Rizcalla, N.; Krejci, I. Fiber-reinforced resin coating for endocrown preparations: A technical report. *Oper. Dent.* **2013**, *38*, 242–248. [CrossRef] [PubMed]
7. Tay, F.R.; Pashley, D.H. Monoblocks in Root Canals: A Hypothetical or a Tangible Goal. *J. Endod.* **2007**, *33*, 391–398. [CrossRef] [PubMed]
8. Wu, Y.; Cathro, P.; Marino, V. Fracture resistance and pattern of the upper premolars with obturated canals and restored endodontic occlusal access cavities. *J. Biomed. Res.* **2010**, *24*, 474–478. [CrossRef]
9. Soares, P.V.; Santos-Filho, P.C.F.; Martins, L.R.M.; Soares, C.J. Influence of restorative technique on the biomechanical behavior of endodontically treated maxillary premolars. Part I: Fracture resistance and fracture mode. *J. Prosthet. Dent.* **2008**, *99*, 30–37. [CrossRef]
10. Seow, L.L.; Toh, C.G.; Wilson, N.H.F. Strain measurements and fracture resistance of endodontically treated premolars restored with all-ceramic restorations. *J. Dent.* **2015**, *43*, 126–132. [CrossRef]
11. Stappert, C.F.J.; Att, W.; Gerds, T.; Strub, J.R. Fracture resistance of different partial-coverage ceramic molar restorations: An in vitro investigation. *J. Am. Dent. Assoc.* **2006**, *137*, 514–522. [CrossRef]
12. Scotti, N.; Scansetti, M.; Rota, R.; Pera, F.; Pasqualini, D.; Berutti, E. The effect of the post length and cusp coverage on the cycling and static load of endodontically treated maxillary premolars. *Clin. Oral Investig.* **2011**, *15*, 923–929. [CrossRef]
13. Bitter, K.; Noetzel, J.; Stamm, O.; Vaudt, J.; Meyer-Lueckel, H.; Neumann, K.; Kielbassa, A.M. Randomized Clinical Trial Comparing the Effects of Post Placement on Failure Rate of Postendodontic Restorations: Preliminary Results of a Mean Period of 32 Months. *J. Endod.* **2009**, *35*, 1477–1482. [CrossRef]
14. Nothdurft, F.P.; Seidel, E.; Gebhart, F.; Naumann, M.; Motter, P.J.; Pospiech, P.R. The fracture behavior of premolar teeth with class II cavities restored by both direct composite restorations and endodontic post systems. *J. Dent.* **2008**, *36*, 444–449. [CrossRef] [PubMed]
15. Ho, M.H.; Lee, S.Y.; Chen, H.H.; Lee, M.C. Three-dimensional finite element analysis of the effects of posts on stress distribution in dentin. *J. Prosthet. Dent.* **1994**, *72*, 367–372. [CrossRef]
16. Goodacre, C.J.; Spolnik, K.J. The Prosthodontic Management of Endodontically Treated Teeth: A Literature Review. Part III. Tooth Preparation Considerations. *J. Prosthodont.* **1995**, *4*, 122–128. [CrossRef] [PubMed]
17. Maceri, F.; Martignoni, M.; Vairo, G. Mechanical behaviour of endodontic restorations with multiple prefabricated posts: A finite-element approach. *J. Biomech.* **2007**, *40*, 2386–2398. [CrossRef]
18. Zicari, F.; Van Meerbeek, B.; Scotti, R.; Naert, I. Effect of fibre post length and adhesive strategy on fracture resistance of endodontically treated teeth after fatigue loading. *J. Dent.* **2012**, *40*, 312–321. [CrossRef] [PubMed]
19. Saunders, W.P.; Saunders, E.M. Coronal leakage as a cause of failure in root-canal therapy: A review. *Dent. Traumatol.* **1994**, *10*, 105–108. [CrossRef] [PubMed]
20. Santos, J.M.; Palma, P.J.; Ramos, J.C.; Cabrita, A.S.; Friedman, S. Periapical Inflammation Subsequent to Coronal Inoculation of Dog Teeth Root Filled with Resilon/Epiphany in 1 or 2 Treatment Sessions with Chlorhexidine Medication. *J. Endod.* **2014**, *40*, 837–841. [CrossRef]
21. Lassila, L.V.J.; Tanner, J.; Le Bell, A.M.; Narva, K.; Vallittu, P.K. Flexural properties of fiber reinforced root canal posts. *Dent. Mater.* **2004**, *20*, 29–36. [CrossRef]

22. Scotti, N.; Coreo Borga, F.A.; Alovisi, M.; Rota, R.; Pasqualini, D.; Berutti, E. Is fracture resistance of endodontically treated mandibular molars restored with indirect onlay composite restorations influenced by fibre post insertion? *J. Dent.* **2012**, *40*, 814–820. [CrossRef]
23. Hashem, M.; Rez, M.F.; Fouad, H.; Elsarnagawy, T.; Elsharawy, M.; Umar, A.; Assery, M.; Ansari, S.G. Influence of Titanium Oxide Nanoparticles on the Physical and Thermomechanical Behavior of Poly Methyl Methacrylate (PMMA): A Denture Base Resin. *Sci. Adv. Mater.* **2017**, *9*, 938–944. [CrossRef]
24. Jiang, T.; Qin, W.; Zhou, J. Controllable synthesis and crystal structure determined upconversion luminescence properties of Tm3+ (Er3+) ions doped YbF3 and NaYbF4 crystals. *J. Alloys Compd.* **2014**, *593*, 79–86. [CrossRef]
25. Ahmad, S.; Ahmad, S.; Agnihotry, S.A. Synthesis and characterization of in situ prepared poly (methyl methacrylate) nanocomposites. *Bull. Mater. Sci.* **2007**, *30*, 31–35. [CrossRef]
26. Tham, D.Q.; Hoang, T.; Giang, N.V.; Kim Dung, N.T.; Chung, I. Synthesis and characterization of (4-arm-star-PMMA)/PMMA-g-SiO2 hybrid nanocomposites. *Green Process. Synth.* **2018**, *7*, 391–398. [CrossRef]
27. Wang, H.; Wen, Y.; Peng, H.; Zheng, C.; Li, Y.; Wang, S.; Sun, S.; Xie, X.; Zhou, X. Grafting Polytetrafluoroethylene Micropowder via in Situ Electron Beam Irradiation-Induced Polymerization. *Polymers* **2018**, *10*, 503. [CrossRef]
28. Khan, A.S.; Khalid, H.; Sarfraz, Z.; Khan, M.; Iqbal, J.; Muhammad, N.; Fareed, M.A.; Rehman, I.U. Vibrational spectroscopy of selective dental restorative materials. *Appl. Spectrosc. Rev.* **2017**, *52*, 507–540. [CrossRef]
29. Singhal, A.; Dubey, K.A.; Bhardwaj, Y.K.; Jain, D.; Choudhury, S.; Tyagi, A.K. UV-shielding transparent PMMA/In2O3 nanocomposite films based on In2O3 nanoparticles. *RSC Adv.* **2013**, *3*, 20913. [CrossRef]
30. Łagocka, R.; Mazurek-Mochol, M.; Jakubowska, K.; Bendyk-Szeffer, M.; Chlubek, D.; Buczkowska-Radlińska, J. Analysis of Base Monomer Elution from 3 Flowable Bulk-Fill Composite Resins Using High Performance Liquid Chromatography (HPLC). *Med. Sci. Monit.* **2018**, *24*, 4679–4690. [CrossRef]
31. Yamada, Y.; Tsubota, Y.; Fukushima, S. Effect of restoration method on fracture resistance of endodontically treated maxillary premolars. *Int. J. Prosthodont.* **2004**, *17*, 94–98.
32. Tamse, A.; Fuss, Z.; Lustig, J.; Kaplavi, J. An evaluation of endodontically treated vertically fractured teeth. *J. Endod.* **1999**, *25*, 506–508. [CrossRef]
33. St-Georges, A.J.; Sturdevant, J.R.; Swift, E.J.; Thompson, J.Y. Fracture resistance of prepared teeth restored with bonded inlay restorations. *J. Prosthet. Dent.* **2003**, *89*, 551–557. [CrossRef]
34. Dietschi, D.; Argente, A.; Krejci, I.; Mandikos, M. In vitro performance of class I and II composite restorations: A literature review on nondestructive laboratory trials-part II. *Oper. Dent.* **2013**, *38*, 182–200. [CrossRef]
35. Meyenberg, K. The ideal restoration of endodontically treated teeth-structural and esthetic considerations: A review of the literature and clinical guidelines for the restorative clinician. *Eur. J. Esthet. Dent.* **2013**, *8*, 238–268.
36. Paolone, G.; Scolavino, S.; Gherlone, E.; Spagnuolo, G. Direct Esthetic Composite Restorations in Anterior Teeth: Managing Symmetry Strategies. *Symmetry* **2021**, *13*, 797. [CrossRef]
37. Bolhuis, H.P.B.; De Gee, A.J.; Feilzer, A.J.; Davidson, C.L. Fracture strength of different core build-up designs. *Am. J. Dent.* **2001**, *14*, 286–290.
38. Manning, K.E.; Yu, D.C.; Yu, H.C.; Kwan, E.W. Factors to consider for predictable post and core build-ups of endodontically treated teeth. Part I: Basic theoretical concepts. *J. Can. Dent. Assoc.* **1995**, *61*, 685–695. [PubMed]
39. Phebus, J.G.; Owens, B.M.; de Rijk, W.; Davis, A.; Johnson, W.W. Fracture resistance of permanent anterior incisors using fiber-reinforced composite posts. *Gen. Dent.* **2014**, *62*, 37–42. [PubMed]
40. Hashemikamangar, S.S.; Hasanitabatabaee, M.; Kalantari, S.; Gholampourdehaky, M.; Ranjbaromrani, L.; Ebrahimi, H. Bond Strength of Fiber Posts to Composite Core: Effect of Surface Treatment With Er,Cr: YSGG Laser and Thermocycling. *J. Lasers Med. Sci.* **2017**, *9*, 36–42. [CrossRef] [PubMed]
41. Cagidiaco, M.C.; García-Godoy, F.; Vichi, A.; Grandini, S.; Goracci, C.; Ferrari, M. Placement of fiber prefabricated or custom made posts affects the 3-year survival of endodontically treated premolars. *Am. J. Dent.* **2008**, *21*, 179–184. [PubMed]
42. Ferrari, M.; Vichi, A.; Fadda, G.M.; Cagidiaco, M.C.; Tay, F.R.; Breschi, L.; Polimeni, A.; Goracci, C. A Randomized Controlled Trial of Endodontically Treated and Restored Premolars. *J. Dent. Res.* **2012**, *91*, S72–S78. [CrossRef] [PubMed]
43. Kivanç, B.H.; Alaçam, T.; Görgül, G. Fracture resistance of premolars with one remaining cavity wall restored using different techniques. *Dent. Mater. J.* **2010**, *29*, 262–267. [CrossRef] [PubMed]
44. Fráter, M.; Forster, A.; Jantyik, Á.; Braunitzer, G.; Nagy, K.; Grandini, S. In vitro fracture resistance of premolar teeth restored with fibre-reinforced composite posts using a single or a multi-post technique. *Aust. Endod. J.* **2017**, *43*, 16–22. [CrossRef]
45. Biały, M.; Szust, A.; Napadłek, P.; Dobrzyński, M.; Więckiewicz, W. The three-point bending test of fiber-reinforced composite root canal posts. *Adv. Clin. Exp. Med.* **2020**, *29*, 1111–1116. [CrossRef]

materials

Article

Influence of Boundary Conditions on Numerical Homogenization of High Performance Concrete

Arkadiusz Denisiewicz [1], Mieczysław Kuczma [2,*], Krzysztof Kula [1] and Tomasz Socha [1]

1 Division of Structural Mechanics, University of Zielona Góra, ul. Prof. Z. Szafrana 1, 65-246 Zielona Góra, Poland; a.denisiewicz@ib.uz.zgora.pl (A.D.); k.kula@ib.uz.zgora.pl (K.K.); t.socha@ib.uz.zgora.pl (T.S.)
2 Division of Structural Engineering, Poznan University of Technology, ul. Piotrowo 5, 60-965 Poznań, Poland
* Correspondence: mieczyslaw.kuczma@put.poznan.pl

Abstract: Concrete is the most widely used construction material nowadays. We are concerned with the computational modelling and laboratory testing of high-performance concrete (HPC). The idea of HPC is to enhance the functionality and sustainability of normal concrete, especially by its greater ductility as well as higher compressive, tensile, and flexural strengths. In this paper, the influence of three types (linear displacement, uniform traction, and periodic) of boundary conditions used in numerical homogenization on the calculated values of HPC properties is determined and compared with experimental data. We take into account the softening behavior of HPC due to the development of damage (micro-cracks), which finally leads to failure. The results of numerical simulations of the HPC samples were obtained by using the Abaqus package that we supplemented with our in-house finite element method (FEM) computer programs written in Python and the homogenization toolbox Homtools. This has allowed us to better account for the nonlinear response of concrete. In studying the microstructure of HPC, we considered a two-dimensional representative volume element using the finite element method. Because of the random character of the arrangement of concrete's components, we utilized a stochastic method to generate the representative volume element (RVE) structure. Different constitutive models were used for the components of HPC: quartz sand—linear elastic, steel fibers—ideal elastic-plastic, and cement matrix—concrete damage plasticity. The numerical results obtained are compared with our own experimental data and those from the literature, and a good agreement can be observed.

Keywords: boundary conditions; numerical homogenization; RVE; FEM; HPC; concrete

Citation: Denisiewicz, A.; Kuczma, M.; Kula, K.; Socha, T. Influence of Boundary Conditions on Numerical Homogenization of High Performance Concrete. *Materials* **2021**, *14*, 1009. https://doi.org/10.3390/ma14041009

Academic Editor: Nicholas Fantuzzi

Received: 1 January 2021
Accepted: 15 February 2021
Published: 20 February 2021

1. Introduction

Significant progress in developing high-performance concrete (HPC) has been observed since the invention of a concrete from reactive powders (RPC) in the 1990s [1–3]. Concretes of this type are characterized by a much higher compressive strength, flexural tensile strength, ductility, and resilience to an aggressive environment than traditional concrete. The use of such concretes in the construction industry allows one to build challenging engineering structures (bridges and towers) and high-rising buildings. Compressive and tensile strengths and modulus of elasticity are the main mechanical parameters that affect the design properties of HPC. An accurate determination of these parameters is very important from a practical viewpoint of engineering and design for efficient and safe HPC constructions. Due to the complex composition of high-strength concrete and therefore complexity of its microstructure, which is finally formed during the hydration process, determining the abovementioned parameters is not easy [4]. Publications about the properties of cement-based composites, a name that can be also used for HPC, are concerned in most cases with experimental testing. They contain mainly parametric analysis on the influence of the volumetric fraction of individual components on the HPC response, analysis of different orientations and shape factor of fibres [5–11], or analysis of different types and shapes of fibres [12–16]. The material parameters we observe at the macro-scale are a result of phenomena taking place in the microstructure of the material, i.e. on meso-, micro-, and nano-scales. A thorough analysis of these phenomena also gives an opportunity to trace different aspects of the deformation process of HPC. One of the basic phenomena that significantly affect endurance of HPC is the formation and propagation of cracks in the microstructure of material [17–19].

There are many methods used to describe effective properties of heterogeneous materials, of which the most intuitive is direct homogenization [20]. It requires an estimate of the value of the averages within the area occupied by the representative volume element (RVE). The concept of RVE was defined and used by Hashin [21], where both elastic and viscoelastic properties as well as failure criteria for fibre composites where considered; by Hill [22] for elastic media; and more recently by Ostoja-Starzewski [23] in the context of plasticity of random media. The averaging operation is preceded by a solution of the boundary value problem defined for RVE with homogeneous boundary conditions. Unlike direct methods, indirect homogenization does not require estimation of the average values in the RVE cell. Indirect methods should be understood as a whole group of techniques based on Eshelby's theory [24]. They were used in designing composite materials. Many examples of practical applications of these methods can be found in [21], including the self-consistent scheme [25], which is developed for prediction of elastic moduli of two-phase composites assuming that any (ellipsoidal) inclusion is embedded in a homogeneous medium and subject to boundary conditions at infinity. Another way to describe heterogeneity caused by the introduction of a discrete phase into the homogeneous base, i.e., matrix, is characterized by the Mori-Tanaka method [26]. Another technique to be mentioned is differential indirect method [20]. It consists of a volumetric balance matrix and discrete phase participation. A large group of homogenization methods is based on the minimum principle of potential energy, and they belong to the so-called variational homogenization methods. With their help, the lower or upper limits of parameter values are often effectively estimated [20].

Relatively recent, introduced at the turn of the 1970s and the 1980s by Bensoussan et al. [27] and Sanchez–Palencia [28], a method of asymptotic expansion for multiple scales was used [29]. It is based on a mathematical theory of homogenization, in which disturbances of unknown functions are assumed to be so small that they can be presented as an asymptotic expansion with respect to a small parameter, which is a characteristic dimension of the microstructure. This method was used to analyse materials that are physically nonlinear [30,31].

For analyzing inhomogeneous materials where the geometry and distribution of microstructure components cannot be precisely known, stochastic methods [32–34] can be used that also provide an estimation of properties of a surrogate material model on a macro-scale.

The last group of homogenization methods to be mentioned here are numerical methods, presented, e.g., in [35–38] and in particular for cement pastes in [39,40]. They are free from the restrictions that are imposed in strictly analytical techniques, for example, limiting the shape of a particle to the sphere or ellipse or being unable to analyze problems that are geometrically and/or physically nonlinear. In recent years, together with the increase in computing power of computers, the methods of numerical homogenization have flourished. These methods make use of computational algorithms aimed at estimating numerically the relationship between strain and stress at a macroscopic material point based on separate calculations carried out on a representative RVE that is assigned to the point and reflects the material's microstructure there. Calculations can be carried out using various numerical techniques, of which the finite element method is the most popular one.

Modelling and computation of (high performance) concrete, owing to its complex multi-phase microstructure [5,41–43] and exhibiting irreversible and brittle behavior (deterioration and fracture), is an extremely difficult problem. From a computational point of view, the difficulties are manifested when solving a boundary value problem for concrete specimens or concrete structural members in the strain-softening regime, where well-posedness of the problem is lost and the pathological mesh-dependence of a solution by the finite element method occurs. In order to overcome the computational difficulties and to avoid spurious numerical results, some forms of regularization of the boundary value problem corresponding to nonlinear constitutive laws for concrete have been developed. The regularization can be achieved by incorporating into the constitutive law some nonlocal information about the deformation process from a vicinity of the material point, either by accounting for gradients of selected quantities (damage and plastic strain) or by using a phase-field approach; see Bažant et al. [44], de Borst et al. [45], Ramm et al. [46], smf Kaliske et al. [47], for example.

In this paper, we follow the procedure of a numerical two-scale homogenization approach to model the behavior of high-performance concrete, paying special attention to the boundary conditions imposed on the RVE that embodies the concrete's microstructure. In fact, we treat the HPC and its microstructure as a composite consisting of four (or six) phases: cement matrix, fine (and thick) quartz sand, steel micro fibres (and steel fibres), and air voids. The following constitutive models were used for the components of HPC: quartz sand—linear elastic, steel fibers—ideal elastic-plastic, and cement matrix—concrete damage plasticity (CDP). Thus, the interphase and interfacial phenomena are not treated explicitly, being per assumption accounted for implicitly by the CDP model of cement matrix. In two-scale numerical homogenization, the macro-scale and micro-scale are distinguished, and at each of the scales, the corresponding boundary value problem (BVP) is solved using the finite element method. The boundary conditions imposed on the solution of the macro-scale BVP are a direct result of the support constraints of a body (structural element) and therefore are uniquely defined, whereas the boundary conditions imposed on the solution of the micro-scale BVP defined on RVE may be assumed in a number of ways. We solve the micro-scale BVP using three types of boundary conditions imposed on the boundary of RVE: uniform displacement boundary conditions (DBC), traction boundary conditions (TBC), and periodic boundary conditions (PBC).

The influence of boundary conditions used in the numerical homogenization of non-homogeneous media on the obtained values of material parameters and the stability of computational schemes have been studied by many researchers, mainly for elastic composites, e.g., in [48–50]. A mixed formulation of the BVP on the micro-scale was developed in [48] using weakly periodic boundary conditions, which allowed a part of microstructure to be completely cut loose by cracks and in which the idea of an elastic cohesive zone was then used as a regularization method. In order to remove rigid body motions of the elastic RVE with TBC, the concept of semi-Dirichlet boundary conditions was introduced in [49], the latter enforcing non-homogeneous displacements at selected points simultaneously satisfying the Neumann-type conditions. Numerical tests for elastic woven composites have revealed high sensitivities of the in-plane extensional modulus and Poisson's ratio to the type of boundary conditions, and a mix of TBC and PBC proved to be best in representing an experimental strain field [50].

This contribution is a continuation of our papers [51–54], where a linear-elastic two-scale model of HPC and its experimental validation were presented. Herein, we extend our previous analyses to the case of nonlinear elasto-plastic behavior with damage. Our aim is to examine the impact of the particular type of boundary conditions on the response of two-dimensional RVE of high-performance concrete.

2. Materials and Methods

2.1. Research Methodology

The results of the numerical simulations described herein relate to the experimental findings obtained in our own laboratory tests [53] and those contained in [55]. We restrict our numerical computations to the local BVP on the RVE of high-performance concrete, checking the influence of different boundary conditions imposed on the boundary of RVE on the simulated response of the tested HPC. Accounting for randomness of the RVE microstructure, we generated positions of different phases, each in the amount as in the mix recipes in Table 1, by a pseudo-random number generator (for details, see [51]). The solution of the elasto-plastic damage BVP is solved by the finite element software package, Simulia Abaqus [56], supplemented with our in-house written script in Python and the Homtools package [57].

Table 1. Concrete mix proportions [53].

Component	Mixture I [kg/m^3]	wt%	Mixture II [kg/m^3]	wt%
Cement CEM I 42.5R	905	34.2	905	33.2
Silica fume	230	8.7	230	8.4
Quartz sand 0.063–0.4 mm OS 36	702	26.6	330	12.1
Quartz sand 0.04–0.125 mm OS 38	285	10.8	285	10.5
Quartz sand 0.2–0.8 mm OS 30	-	-	335	12.3
Water	260	9.8	260	9.5
Superplasticizer Woerment FM 787 BASF	29.6	1.1	29.6	1.1
Micro steel fibres DM 6/0.17 KrampeHarex	233	8.8	233	8.6
Steel fibres DW 38/1.0 N KrampeHarex	-	-	117	4.3
Density	2645	-	2725	-

2.2. Recipes for Modelled HPC

Two mix proportions of HPC were considered, the ingredients of which are shown in Table 1. The mean compressive and tensile strengths of our two experimentally tested HPC mixtures were 106 MPa and 12.5 MPa for mixture I, and 141 MPa and 18 MPa for mixture II, cf. [53].

2.3. Microstructure of Modeled HPC

The microstructure of high-performance concrete has been studied by many authors, and the topic is well-recognized in the world literature, e.g., [5,41–43,58]. In this contribution, the microstructure of HPC is defined by means of a two-dimensional representative volume element (RVE) and the finite element method is used for modelling. The idea adopted for the numerical finite element analysis of the RVE microstructure is shown in Figure 1. The method with which we generated the considered RVE was described in [51,54]. The actual geometry of the HPC ingredients was approximated by square finite elements of dimensions 0.2 × 0.2 mm of the amount given in the concrete mix proportions table (Table 1). The size of the entire RVE was 10 × 10 mm.

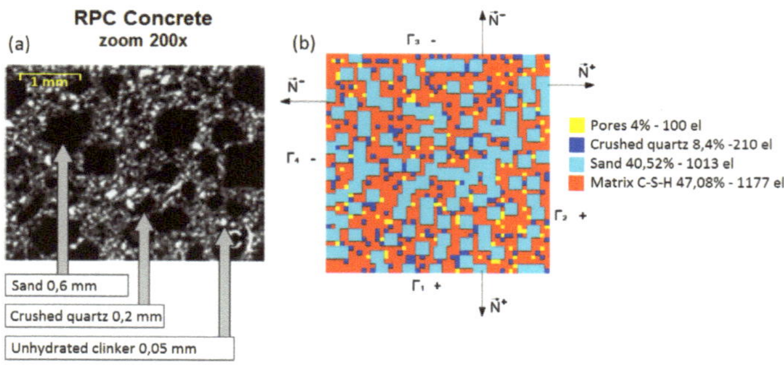

Figure 1. Representative volume element of high-performance concrete (HPC): (**a**) microstructure [58] and (**b**) an exemplary finite element model of representative volume element (RVE) with the given number of finite elements and wt% of components.

The RVE consists of several main micro-ingredients: OS 36 and OS 38-fine quartz sand, OS 30-thick quartz sand, steel (micro) fibres, and cement matrix. The constitutive models used for the components of HPC were quartz sand—linear elastic, steel fibers—ideal elastic-plastic, and cement matrix—concrete damage plasticity (Section 2.4). The RVEs of mixtures I and II were generated for numerical homogenization (Figure 2).

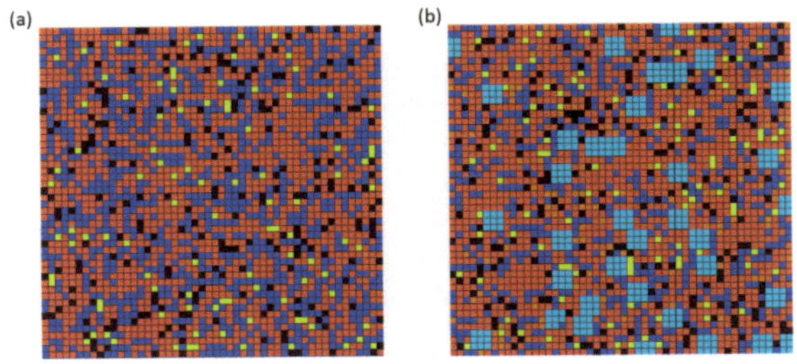

Figure 2. RVE for (**a**) mixture I and (**b**) mixture II. The RVE (10 × 10 mm) is divided into 2500 finite elements, each of dimensions 0.2 × 0.2 mm.

The components (color) of the RVE, the values of the material parameters, and wt% (densities ρ (kg/m^3) for cement—3100, silica fume—2600, sands—2650, water and superplasticizer—1000, and steel fibres—7850) are as follows:

Mixture I (Figure 2a):

1. cement matrix (red): E = 55,000 MPa, ν = 0.17, CDP (see Section 2.4), 49.8 wt%
2. fine quartz sand (dark blue): E = 48,200 MPa, ν = 0.20, 37.4 wt%
3. steel micro fibres (black): E = 210,000 MPa, ν = 0.30, yield stress = 2100 MPa, 8.8 wt%
4. air voids (yellow): empty space (no finite elements), 4 wt%.

Mixture II (Figure 2b):

1. cement matrix (red): E = 55,000 MPa, ν = 0.17, CDP (see Section 2.4), 48.2 wt%
2. fine quartz sand (dark blue): E = 48,200 MPa, ν = 0.20, 22.6 wt%
3. thick quartz sand (sky blue): E = 73,200 MPa, ν = 0.20, 12.3 wt%
4. steel micro fibres (black): E = 210,000 MPa, ν = 0.30, yield stress = 2100 MPa, 8.6 wt%
5. steel fibres (orange): E = 210,000 MPa, ν = 0.30, yield stress = 1100 MPa, 4.3 wt%
6. air voids (yellow): empty space (no finite elements), 4 wt%.

2.4. *Parameters of Concrete Damage Plasticity (CDP) Model*

For modelling the cement-based matrix, i.e., a hardened mixture of cement, silica fume, water, and superplasticizer, we adapted the concrete damage plasticity (CDP) model as described in the influential papers [59–61] and Abaqus documentation [56]. Let us recall that the CDP model captures the nonlinear behavior of concrete by accounting for simultaneous development of permanent (plastic) deformation and elastic stiffness degradation using two damage variables, one for tensile damage (d_t) and the other for compressive damage (d_c), to account for different damage responses of concrete in tension and in compression. The plastic-damage model [56] assumes nonassociated potential flow with the Drucker–Prager hyperbolic function that is continuous and smooth and asymptotically approaches the linear Drucker–Prager yield condition [62]. The evolution of the elastic-plastic-damage deformation process is complex from the numerical viewpoint because it is described by inequality relations and Kuhn–Tucker complementarity conditions (loading/unloading conditions), e.g., [61,63].

The parameters of the CDP model and their values used in the calculations are gathered in Table 2, wherein

Table 2. Mechanical parameters of the cement matrix in the plastic range.

β	m	f_{b0}/f_{c0}	K_c	η
36°	0.1	1.16	0.667	0

β—the internal friction angle of concrete. In the CDP model, β is defined as the inclination angle of the Drucker–Prager surface asymptote to hydrostatic axis of the meridional plane;
m—eccentricity of the surface of the plastic potential. This is the distance measured along the hydrostatic axis between the apex of the Drucker–Prager hyperbola and the intersection of the asymptote of this hyperbola, calculated in practice as a ratio of tensile strength to strength for compression;
f_{b0}/f_{c0}—number specifying the compressive strength ratio in a two-axis state for the strength in a single-axis state;
K_c—parameter defining the shape of the surface of the plastic potential on a deviatoric plane;
η—viscoplasticity parameter, used to regularize the concrete constitutive equations.

The used stress–strain relationships for the cement matrix in compression and in tension are shown in Figures 3 and 4, respectively, with the compressive strength equal to 200 MPa and the tensile strength equal to 20 MPa. These relations are described with the function proposed by Saenz [64] as

$$\sigma_\gamma = \frac{\varepsilon_\gamma}{A + B\varepsilon_\gamma + C{\varepsilon_\gamma}^2 + D{\varepsilon_\gamma}^3} \tag{1}$$

where

$$A = \frac{1}{E}, \quad B = \frac{P_3 + P_4 - 2}{P_3 f_{\gamma m}}, \quad C = -\frac{2P_4 - 1}{P_3 f_{\gamma m}\varepsilon_{\gamma 1}}, \quad D = \frac{P_4}{P_3 f_{\gamma m}{\varepsilon_{\gamma 1}}^2} \tag{2}$$

and

$$P_1 = \frac{\varepsilon_{\gamma u}}{\varepsilon_{\gamma 1}}, \quad P_2 = \frac{f_{\gamma m}}{f_{\gamma u}}, \quad P_3 = \frac{E\varepsilon_{\gamma 1}}{f_{\gamma m}}, \quad P_4 = \frac{P_3(P_2 - 1)}{(P_1 - 1)^2} - \frac{1}{P_1} \tag{3}$$

The lower index $\gamma \in \{c, t\}$ stands for a type of stress, with c standing for compression and t standing for tension, respectively. Furthermore, $\varepsilon_{\gamma u}$ is the ultimate strain in the matrix and $f_{\gamma u}$ is its corresponding stress, whereas $f_{\gamma m}$ is the extremal stress sustained by the matrix with its corresponding strain $\varepsilon_{\gamma 1}$. In the calculations, we assumed for compression $E = 55$ GPa, $f_{cm} = 200$ MPa, $f_{cu} = 196$ MPa, $\varepsilon_{cu} = 0.0028$, and $\varepsilon_{c1} = 0.0025$ and for tension $f_{tm} = 20$ MPa, $f_{tu} = 19$ MPa, $\varepsilon_{tu} = 0.00028$, and $\varepsilon_{t1} = 0.00025$. The values of the parameters of Equation (1) calculated by Equation (2) for compression and tension are collected in Table 3.

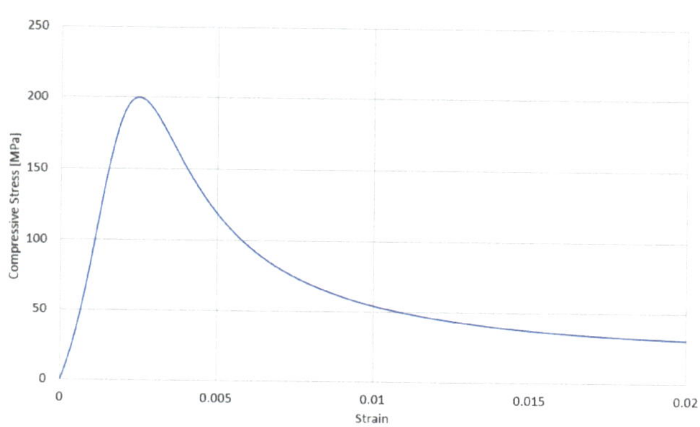

Figure 3. Compressive behavior of the cement matrix (concrete damage plasticity (CDP) model).

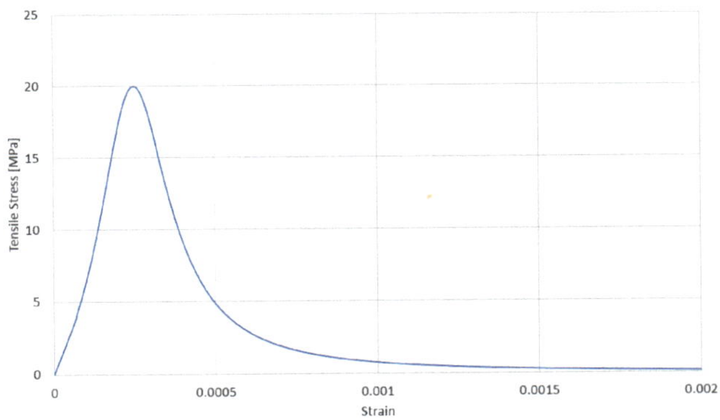

Figure 4. Tensile behavior of the cement matrix (CDP model).

Table 3. The values of the coefficients in Equation (1).

	A	*B*	*C*	*D*
Compression	0.00002	−0.01003	3.25030	−59.29370
Tension	0.00002	0.02236	−651.59869	1885015.6

The evolution of the compression damage scalar d_c for the cement matrix as a function of strain and that of the tension damage scalar d_t are illustrated in Figures 5 and 6 and Equation (4), respectively.

$$d_\gamma(\varepsilon_\gamma) = \frac{f_{\gamma m} - \sigma_\gamma(\varepsilon_\gamma, \ldots)}{f_{\gamma m}}; \quad \gamma \in \{c, t\}; \quad d_\gamma(\varepsilon_\gamma) = 0 \quad for \quad \varepsilon_\gamma \leq \varepsilon_{\gamma 1} \tag{4}$$

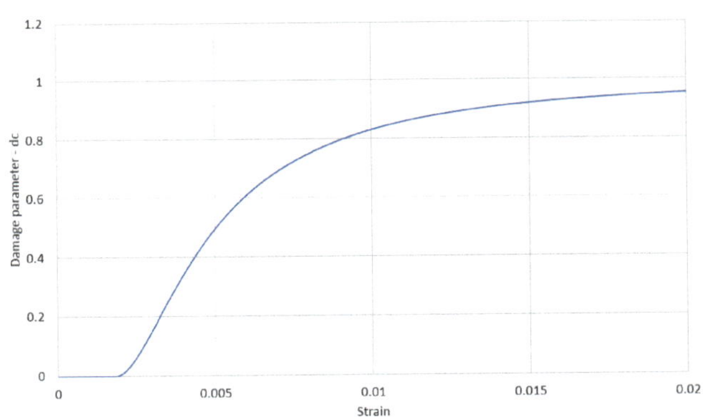

Figure 5. Cement matrix compression damage (CDP model).

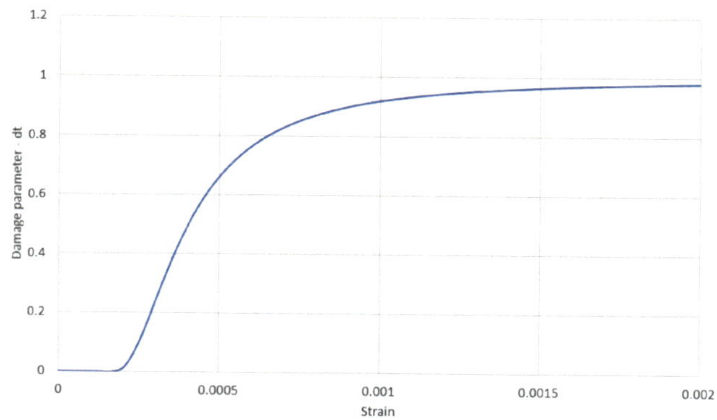

Figure 6. Cement matrix tension damage (CDP model).

2.5. Numerical Homogenization and Boundary Conditions

In the method of two-scale numerical homogenization, the response of a material at the macro-scale is determined by an analysis of the behavior of the material's internal structure at the micro-scale. In the case of nonlinear material behavior, achieving the equilibrium at each scale together with the compatibility of information between the two-scales requires multiple exchanges of needed information between the scales after solving a corresponding nonlinear boundary value problems formulated at each of the scales. On the micro-scale level, the distributions of micro-stresses and micro-strains were calculated, which via homogenization provide the needed information of averaged macroscopic quantities to the macro-scale. In this contribution, our analysis was restricted to the solution of the local nonlinear BVP defined on the two-dimensional representative volume element (RVE). When the characteristic microscopic length was one order smaller than the characteristic macroscopic length, we could take into consideration only effects of the first order.

The analysis was carried out within the realm of linear kinematics. The notation that a bar above a symbol denotes a macroscopic variable or quantity was used. Let $\overline{\boldsymbol{x}}$ denote the vector of coordinates of a material point at the macroscopic level, and let $\boldsymbol{x}$ stand for coordinates of points within the RVE of volume V defined in the material point. Further, let us assume that the microscopic displacement field $\boldsymbol{u}$ can be additively split into a linear part $\overline{\boldsymbol{\varepsilon}}\boldsymbol{x}$ and a fluctuating part $\boldsymbol{r}$:

$$\boldsymbol{u}(\overline{\boldsymbol{x}},\boldsymbol{x},t) = \overline{\boldsymbol{\varepsilon}}(\overline{\boldsymbol{x}},t)\boldsymbol{x} + \boldsymbol{r}(\overline{\boldsymbol{x}},\boldsymbol{x},t) \quad (5)$$

where $\overline{\boldsymbol{\varepsilon}}(\overline{\boldsymbol{x}},t)$ is the macro-strain tensor and t is a time-like parameter. At down-scaling (macro-micro transition) and solving the local BVP at RVE, the strain tensor $\overline{\boldsymbol{\varepsilon}}$ is treated as a known quantity (data). At up-scaling (micro-macro transition), the elements of the macro-strain tensor $\overline{\boldsymbol{\varepsilon}}$ can be defined as mean values of the corresponding micro-strains $\boldsymbol{\varepsilon}$ averaged over the RVE:

$$\overline{\boldsymbol{\varepsilon}} = \frac{1}{|V|}\int_V \boldsymbol{\varepsilon}\, dV = \frac{1}{|V|}\int_\Gamma \frac{1}{2}(\boldsymbol{n}\otimes\boldsymbol{u} + \boldsymbol{u}\otimes\boldsymbol{n})\, d\Gamma \quad (6)$$

where $\boldsymbol{n}$ is the unit outward normal of the boundary of RVE, $\Gamma = \partial V$, and $\otimes$ is a tensor product of vectors. The Formula (6) shows that the macroscopic strain tensor can be expressed by micro-displacements $\boldsymbol{u}$ at the boundary of the RVE. However, Equation (6) is valid provided that the zero gradient condition of the microscopic displacement fluctuation field $\boldsymbol{r}$ is satisfied:

$$\int_V \nabla\boldsymbol{r}\, dV = \int_\Gamma \boldsymbol{n}\otimes\boldsymbol{r}\, d\Gamma = \boldsymbol{0} \quad (7)$$

Fulfilling the above condition ensures that deformation of the RVE boundary in the medium sense is in accordance with the pre-set macro-strain $\overline{\boldsymbol{\varepsilon}}$.

Macro-stresses can be defined, similar to macro-strains, as mean values of the micro-stresses σ:

$$\overline{\sigma} = \frac{1}{|V|} \int_V \sigma \, dV \tag{8}$$

This relationship can be derived from Hill's theorem [25], which says that the work done by macro-stresses on the corresponding macro-strains is equal to the mean value of the work performed by micro-stresses on the corresponding micro-strains:

$$\overline{\sigma} \cdot \overline{\varepsilon} = \langle \sigma \cdot \varepsilon \rangle \tag{9}$$

where the symbol $\langle \bullet \rangle$ stands for averaging over the RVE volume.

$$\langle \bullet \rangle = \frac{1}{|V|} \int_V \bullet \, dV \tag{10}$$

The finite element method is applied for numerical calculations, with the finite element designated as the CPS4R Abaqus system library [56]. The RVE area was discretized with 2500 finite elements, each with dimensions of 0.2×0.2 mm; see Figure 2.

Numerical tests were carried out enforcing three types of boundary conditions on the boundary of RVE: linear displacement boundary conditions (DBC), uniform traction boundary conditions (TBC), and periodic boundary conditions (PBC). Before starting the numerical simulations of the inhomogeneous nonlinear BVP for high-performance concrete, the work and proper interaction between the Homtools and the Abaqus packages were verified with the example of RVE for a homogeneous material with a hole (Section 3.1).

2.5.1. Linear Displacement Boundary Conditions (DBC)

This method consists of applying on the boundary of RVE the displacement field that would occur if the strain were uniform inside the RVE. For the considered linear kinematics, the boundary conditions can be defined by the formulas in Equations (5) and (7):

$$\boldsymbol{u} = \langle \varepsilon(\boldsymbol{u}) \rangle \, \boldsymbol{x} \quad \text{on} \quad \Gamma \tag{11}$$

in which $\varepsilon(\boldsymbol{u})$ is the micro-strain field and $\boldsymbol{x}$ is the position vector of point x on boundary Γ. There is no restriction concerning the use of this method, except that no rigid part must intersect the boundary, and holes are permitted.

2.5.2. Uniform Traction Boundary Conditions (TBC)

This method consists of applying on the boundary of RVE the stress vector field that would occur if the stress were uniform inside the RVE. For the considered linear kinematics, the boundary conditions can be defined as follows:

$$\sigma \, \boldsymbol{n} = \langle \sigma \rangle \, \boldsymbol{n} \quad \text{on} \quad \Gamma \tag{12}$$

where σ is the Cauchy stress tensor and $\boldsymbol{n}$ denotes the outward unit normal. There is no restriction concerning the use of this method, except that no holes must intersect the boundary.

2.5.3. Periodic Boundary Conditions (PBC)

Enforcing the PBC is theoretically relevant for periodic media, which can be defined by a periodicity cell and the associated periodicity vector of translation. The periodic homogenization process consists of assuming that the strains and stresses are periodic at the level of the periodicity cell (which is defined as the RVE). The periodicity of stresses and strains leads to specific periodic boundary conditions for the localization problem on the RVE. In order to introduce periodic boundary conditions, the boundary of RVE is decomposed into two opposing parts, Γ^+ and Γ^-, such that $\Gamma = \Gamma^+ \cup \Gamma^-$. Each point $\boldsymbol{x}^+$

on Γ^+ is associated with a unique point x^- on Γ^-, and the unit normal vectors at these boundaries satisfy $\boldsymbol{n}^- = -\boldsymbol{n}^+$. Then, the PBCs are defined as follows:

$$\boldsymbol{u}^+ - \boldsymbol{u}^- = \langle \boldsymbol{\varepsilon}(\boldsymbol{x}^+ - \boldsymbol{x}^-) \rangle \quad \text{and} \quad \boldsymbol{t}^- = -\boldsymbol{t}^+ \quad \text{on} \quad \Gamma \tag{13}$$

where $\boldsymbol{t}^{\pm} = \sigma \boldsymbol{n}^{\pm}$ is the stress vector. For this type of boundary condition, $\boldsymbol{u}$ is forced to be periodic and $\boldsymbol{t}$ is forced to be antiperiodic. Note that LDBCs satisfy the periodicity of $\boldsymbol{u}$ only whereas TBCs satisfy the antiperiodicity of $\boldsymbol{t}$ only. There is no restriction concerning the use of this method; periodic holes and rigid parts intersecting the boundary are permitted.

3. Results and Discussion

3.1. Test Example: Homogeneous and Linear Elastic RVE with Hole

A shear test of a homogeneous linearly elastic RVE with a hole in the centre was carried out in order to check if the functioning of the Homtools package within the Abaqus calculation environment is correct. This common testing procedure preceded relevant simulations of the complex heterogeneous nonlinear BVP. It allowed us to detect the possible pitfalls in the implementation of the method of numerical homogenization. In the test, the following values of some parameters were assumed: RVE dimensions—10 × 10 mm; hole diameter—5 mm; Young's modulus of linearly elastic material, E = 200 GPa; and Poisson ratio, ν = 0.3.

Figure 7 shows the results of the shear test with the imposed macro-strain $\bar{\varepsilon} = \{0, 0, 1\}$ for different boundary conditions and the corresponding deformation modes of the RVE with the distribution of reduced micro-stresses of Huber-von Mises-Hencky (HMH). The obtained results are in good agreement with similar calculations done by other researchers, e.g., [38]. The solution of homogeneous material does not depend on the boundary condition, so the RVE without a hole was also tested. The homogeneous field of micro-stresses was obtained with the same values for the three different types of the boundary condition. It additionally confirmed the correct implementation of the used numerical homogenization scheme into the Abaqus calculation environment.

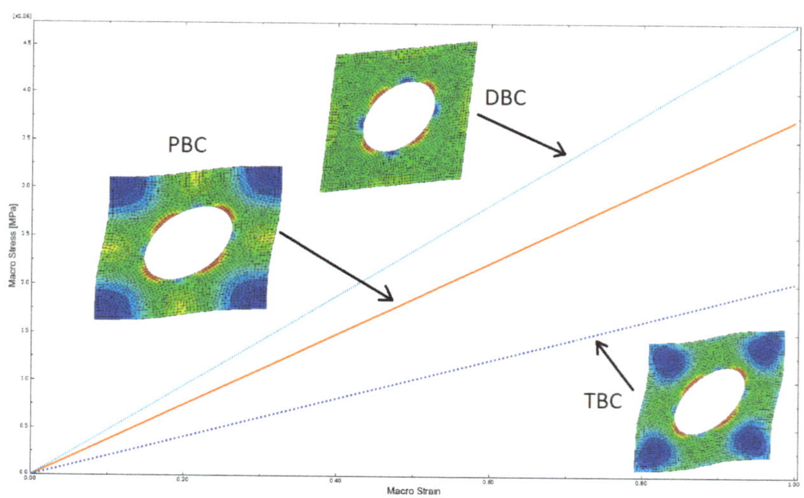

Figure 7. Shear test for homogeneous linear elastic RVE with a hole, demonstrating the influence of the kind of boundary condition (periodic boundary condition (PBC), displacement boundary condition (DBC), and traction boundary conditions (TBC)) on the solution.

3.2. Compression Test for Mixtures I and II

A compression test $\bar{\varepsilon} = \{-1, 0, 0\}$ (Voigt notation) was performed for non-homogeneous RVE with nonlinear materials as a first test. The curves presented in Figure 8 show dependence between macro-strain and macro-stress for the micro-structure model of HPC (mixture I). In all presented results, the following notation was used, Mx_XXX_y, where x indicates the number of mixtures 1 or 2; XXX is the type of boundary condition: DBC, PBC, or TBC; and y is the test class (c—compression, t—tensile, and s—shear).

Analyzing the graphs in Figure 8, one can notice that use of the periodic boundary condition in the deformation range of RVE $\bar{\varepsilon} = (1.75 \div 3.20) \times 10^{-3}$ leads to an upper estimate of the compressive macro-stress. In the case of smaller deformations, the upper estimate was obtained using the displacement boundary condition. The difference between PBC and DBC is not significant; both types of conditions on the RVE boundary give similar results in the analyzed case. However, the traction boundary condition definitely leads to the lowest estimate of the macroscopic value, and under this type of boundary condition, the analyzed RVE of HPC exhibits responses that is characteristic for brittle materials—the plastic part clearly visible for the PBC and DBC does not exist for the TBC. After reaching the deformation of $\bar{\varepsilon} = 1.2 \times 10^{-3}$ with an imposed traction boundary condition, the rapid brittle failure of micro-structures appears. Our own experimental results were compared with the determined numerical ones to validate our computational model. In the Figure 8, the average compressive strength of six concrete specimens 100 × 100 × 100 mm made from mixture I [53] was marked (experiment_M1_c = 106 MPa). The stress–strain path could not be recorded; therefore, in Figures 8–11, only the value of compressive or tensile strength is depicted by a dotted line. Additionally the stress–strain characteristic of concrete made by the authors of the paper [55] is presented. The recipe for that concrete is twin-like in comparison to ours for mixture I. The correlation between the experimental and the numerical results is satisfying.

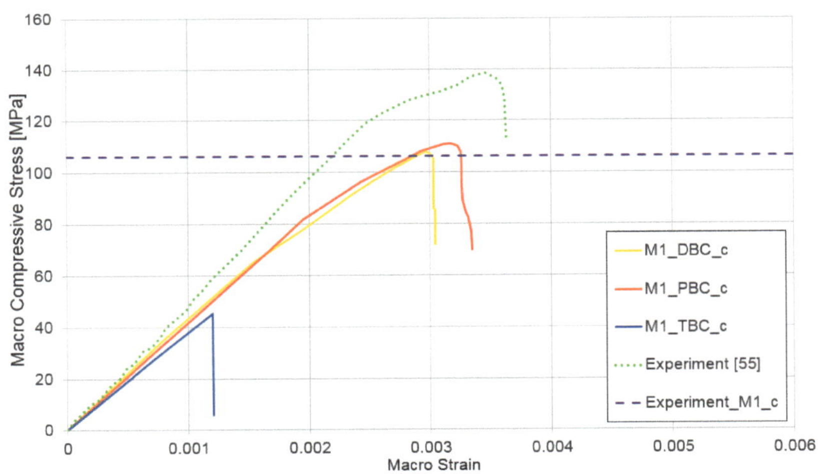

Figure 8. Compression test for mixture I, demonstrating the influence of the type of boundary condition (PBC, DBC, and TBC) on the solution.

A compression test $\bar{\varepsilon} = \{-1, 0, 0\}$ was also performed for the model of micro-structure HPC made from mixture II. The results are shown in Figure 9. In this example, the results obtained with the periodic and the displacement boundary conditions are practically identical to the deformation level of $\bar{\varepsilon} = 2.22 \times 10^{-3}$. The upper estimate of macroscopic values is obtained when using displacement boundary conditions in this case. Use of the traction boundary conditions also leads to an lower estimate of the macro-value. Six concrete specimens 100 × 100 × 100 mm made from mixture II [53] were also examined experimentally. The average compressive strength (marked in Figure 9 as experiment_M2_c) is equal 141 MPa, which is in very good agreement with our own numerical result, similar to the experimental results taken from the paper [55] (marked in Figure 9).

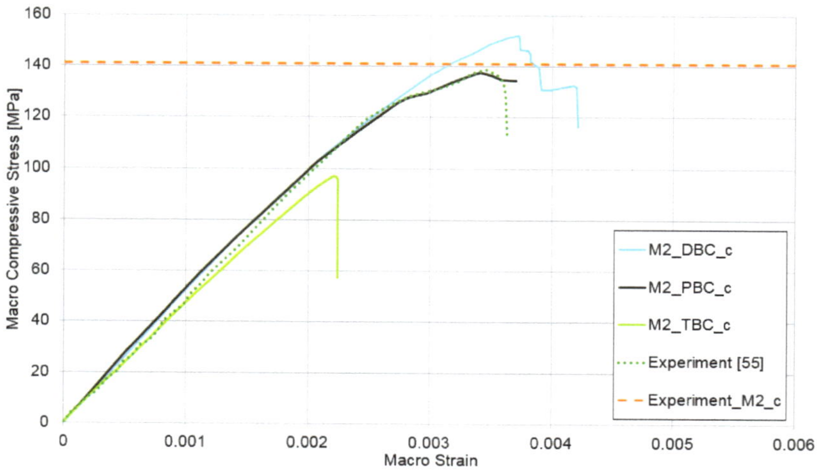

Figure 9. Compression test for mixture II, demonstrating the influence of the type of boundary condition (PBC, DBC, and TBC) on the solution.

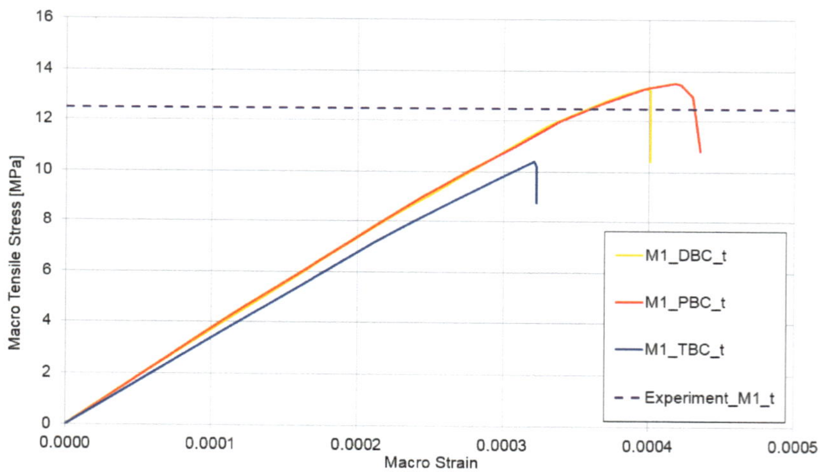

Figure 10. Tensile test for mixture I, demonstrating the influence of the type of boundary condition (PBC, DBC, and TBC) on the solution.

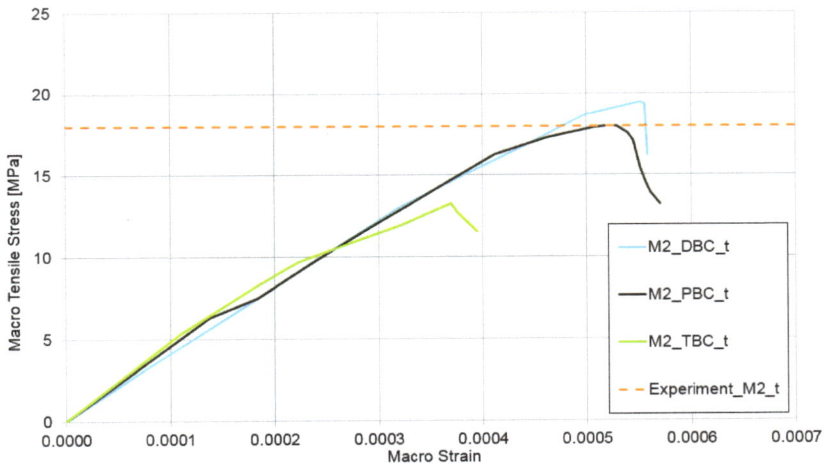

Figure 11. Tensile test for mixture II, demonstrating the influence of the type of boundary condition (PBC, DBC, and TBC) on the solution.

3.3. Tensile Test for Mixture I and II

The next test performed for the non-homogeneous RVE with material nonlinearity was a tensile test $\overline{\varepsilon} = \{1, 0, 0\}$. The curves presented in Figure 10 depict the obtained relation between macro-stress and macro-strain for HPC modelled as RVE-mixture I and, in Figure 11, the similar response for RVE-mixture II. Analogical to that in the compression test, the upper estimate of macro-stresses is provided for mixture I when using the periodic boundary condition and, for mixture II, when using displacement boundary condition.

The RVE deformation range in the tensile test is an order of magnitude lower than that in the compression test, which is an obvious effect of the assumed values of tensile properties for the CDP model of cement matrix. A detailed description of this phenomenon is outside the present research, but it is a very important aspect of HPC, which deserves a separate study. The authors intend to investigate the tensile response of HPC in the future. The average of the experimentally determined tensile strengths [53] is also presented in Figures 10 and 11. The tensile strength was determined in the flexural test for both mixture I—experiment_M1_t = 12.5 MPa—and mixture II—experiment_M2_t = 18.0 MPa. The agreement of the experimental findings with the numerical results is very good.

Figure 12 illustrates some distributions of the compressive and the tensile damage variables. The green areas indicate the places determined numerically where cracks appeared in the HCP micro-structures. A continuity of the micro-structure is breached, and a distribution of the micro-cracks is random. The latter is due to the randomly generated structure of RVE, which reflects the real conditions of HCP manufacturing. The maps of damage variable distribution shown in Figure 12 correspond to the ultimate stress (strength) sustained by the material. This value and that of the corresponding strain are for a map (a) $\overline{\sigma} = 110.91$ MPa, $\overline{\varepsilon} = 3.20 \times 10^{-3}$; for a map (b) $\overline{\sigma} = 13.50$ MPa, $\overline{\varepsilon} = 4.22 \times 10^{-4}$; for a map (c) $\overline{\sigma} = 151.21$ MPa, $\overline{\varepsilon} = 3.70 \times 10^{-3}$; and for a map (d) $\overline{\sigma} = 19.50$ MPa, $\overline{\varepsilon} = 5.52 \times 10^{-4}$.

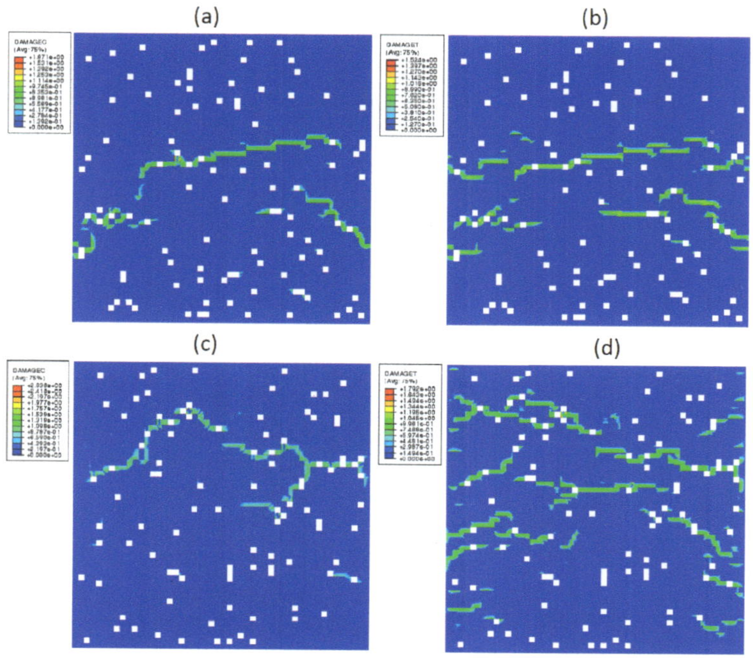

Figure 12. Distribution of damage parameter: (**a**) compression of mixture I with PBC; (**b**) tension of mixture I with PBC; (**c**) compression of mixture II with DBC; and (**d**) tension of mixture II with DBC. The white spots represent pores.

3.4. Shear Test for Mixture I and II

The third performed numerical simulation represents a shear test $\overline{\varepsilon} = \{0, 0, 1\}$. The results of the test are shown in Figures 13 and 14. The upper estimate of macroscopic shear stress in the shear test is provided for both cases of the RVE structure (mixture I and II) by displacement boundary conditions. The lower estimate is reached similar to that in previous cases with traction boundary conditions. The shear strength is not experimentally researched by the authors. Figures 13 and 14 contain this parameter taken from [55], where the shear strength is determined for a similar concrete and on a level of 19.1 MPa, which is depicted in Figures 13 and 14 by a dotted line. In this case, the difference between the experimental and the numerical results is higher by about 25% but could be accepted.

Additionally, the extreme values of the macro-stresses and the corresponding macro-strains determined in the numerical tests are collected in Table 4.

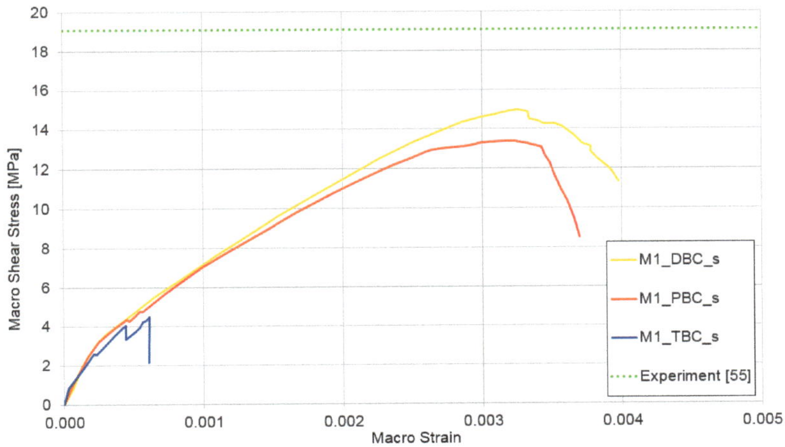

Figure 13. Shear test for mixture I, demonstrating the influence of the type of boundary condition (PBC, DBC, and TBC) on the solution.

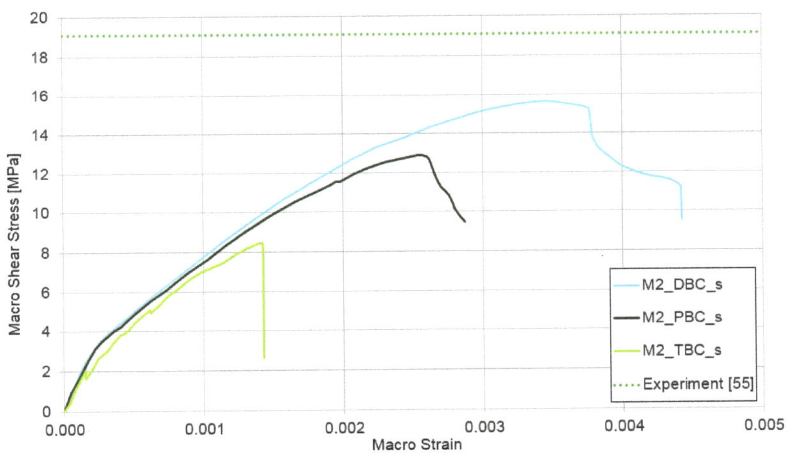

Figure 14. Shear test for mixture II, demonstrating the influence of the type of boundary condition (PBC, DBC, and TBC) on the solution.

Table 4. Extremal macro-stresses and corresponding macro-strains in mixtures I and II (plastic range).

Mixture_BC	Compression		Tensile		Shear	
	$\overline{\sigma}$ [MPa]	$\overline{\varepsilon}$ [-]	$\overline{\sigma}$ [MPa]	$\overline{\varepsilon}$ [-]	$\overline{\sigma}$ [MPa]	$\overline{\varepsilon}$ [-]
M1_PBC	110.91	3.20×10^{-3}	13.50	4.22×10^{-4}	13.40	3.16×10^{-3}
M1_DBC	107.18	3.00×10^{-3}	13.30	3.98×10^{-4}	14.9	3.29×10^{-3}
M1_TBC	45.55	1.20×10^{-3}	10.40	3.21×10^{-4}	4.49	6.14×10^{-4}
M2_PBC	137.77	3.40×10^{-3}	18.00	5.29×10^{-4}	12.90	2.57×10^{-3}
M2_DBC	151.21	3.70×10^{-3}	19.50	5.52×10^{-4}	15.60	3.45×10^{-3}
M2_TBC	96.61	2.20×10^{-3}	13.20	3.70×10^{-4}	8.43	1.42×10^{-3}

4. Concluding Remarks

In this paper, the technique of numerical homogenization was used for determining the macroscopic response of high-performance concrete modelled as a nonhomogeneous multi-phase nonlinear composite. The local plastic-damage boundary value problem defined on a representative volume element of the composite with complex micro-structure was solved by making use of the finite element method. Three types of boundary condition were imposed on the 2-dimensional RVE with nonlinear components: linear displacement boundary conditions (DBC), uniform traction boundary condition (TBC), and periodic boundary condition (PBC). The impact of the boundary conditions on the determined macroscopic response was studied. The obtained results show that the applied boundary conditions exert a significant influence on the calculated values of macro-quantities and that premature numerical stability problems may occur in the case of TBC.

Detailed conclusions concerning the investigated influence of a type of boundary condition on the macro-results could be summarized as follows:

- The periodic boundary conditions (PBC) lead to stable results for the full range of deformations up to failure in all performed numerical calculations.
- In the compression and tensile tests, the upper estimate of values of macro-parameters is reached for mixture I by imposing the PBC whereas for mixture II by imposing the applied displacement boundary conditions (DBC).
- Use of the DBC provides the upper estimate of values of macro-parameters in the shear test for both mixtures I and II.
- Application of the traction boundary conditions (TBC) leads in all analyzed cases to a lower estimate of values of the macro-parameters.

Validation of the numerical results by our own laboratory experiments and the experimental data from the literature shows that the presented model is capable of providing good estimations of macroscopic features for the investigated deformation process. In the first step, we restricted ourselves to an analysis of the micro-scale BVP on the RVE. In the future, it would be useful to enrich the presented analysis by using the concept of modelling error and adaptivity [65,66]. This is especially useful in the circumstance, where the behavior of a structural member as a whole is the objective and then the transition of information between the macro- and micro-scales must be performed iteratively at all integration points of the macro-finite element mesh.

Author Contributions: Conceptualization, A.D. and M.K.; methodology, A.D. and M.K.; software, T.S.; validation, T.S.; formal analysis, M.K.; investigation, A.D.; resources, K.K.; data curation, K.K.; writing—original draft preparation, K.K. and A.D.; writing—review and editing, M.K. and A.D.; visualization, A.D.; supervision, M.K.; project administration, T.S.; funding acquisition, T.S. All authors have read and agreed to the published version of the manuscript.

Funding: This research received no external funding.

Institutional Review Board Statement: Not applicable.

Informed Consent Statement: Not applicable.

Data Availability Statement: Data sharing not available.

Conflicts of Interest: The authors declare no conflict of interest.

References

1. Richard, P.; Cheyrezy, M. Composition of reactive powder concretes. *Cem. Concr. Res.* **1995**, *25*, 1501–1511. [CrossRef]
2. Rahman, S.; Molyneaux, T.; Patnaikuni, I. Ultra high performance concrete: recent applications and research. *Aust. J. Civ. Eng.* **2005**, *2*, 13–20. [CrossRef]
3. Russel, H.G.; Graybeal, B.A. Ultra-High Performance Concrete: A State-of-the-Art Report for the Bridge Community. *Natl. Transp. Libr.* **2013**.
4. Mattei, N.J.; Mehrabadi, M.M.; Zhu, H. A micromechanical constitutive model for the behavior of concrete. *Mech. Mater.* **2007**, *39*, 357–379. [CrossRef]
5. Reda, M.M.; Shrive, N.G.; Gillott, J.E. Microstructural investigation of innovative UHPC. *Cem. Concr. Res.* **1999**, *29*, 323–329. [CrossRef]
6. Okereke, M.I.; Akpoyomare, A.I.; Bingley, M.S. Virtual testing of advanced composites, cellular materials and biomaterials: A review. *Compos. Part B Eng.* **2014**, *60*, 637–662. [CrossRef]

7. Le Hoang, A.; Fehling, E. Influence of steel fiber content and aspect ratio on the uniaxial tensile and compressive behavior of ultra high performance concrete. *Constr. Build. Mater.* **2017**, *153*, 790–806. [CrossRef]
8. Barnett, S.J.; Lataste, J.F.; Parry, T.; Millard, S.G.; Soutsos, M.N. Assessment of fibre orientation in ultra high performance fibre reinforced concrete and its effect on flexural strength. *Mater. Struct.* **2010**, *43*, 1009–1023. [CrossRef]
9. Yoo, D.Y.; Yoon, Y.S. Structural performance of ultra-high-performance concrete beams with different steel fibers. *Eng. Struct.* **2015**, *102*, 409–423. [CrossRef]
10. Denisiewicz, A.; Kula, K.; Socha, T.; Kwiatkowski, G. Influence of silica fume addition on selected properties of fine-grained concrete. *Civ. Environ. Eng. Rep.* **2018**, *3*, 166–176. [CrossRef]
11. Denisiewicz, A.; Śliwa, M.; Kula, K.; Socha, T. Experimental investigation of concrete with recycled aggregates for suitability in concrete structures. *Appl. Sci.* **2019**, *9*, 5010. [CrossRef]
12. He, S.; Qiu, J.; Li, J.; Yang, E.H. Strain hardening ultra-high performance concrete (SHUHPC) incorporating CNF-coated polyethylene fibers. *Cem. Concr. Res.* **2017**, *98*, 50–60. [CrossRef]
13. Kim, D.J.; Park, S.H.; Ryu, G.S.; Koh, K.T. Comparative flexural behavior of Hybrid Ultra High Performance Fiber Reinforced Concrete with different macro fibers. *Constr. Build. Mater.* **2011**, *25*, 4144–4155. [CrossRef]
14. Tai, Y.S.; El-Tawil, S.; Chung, T.H. Performance of deformed steel fibers embedded in ultra-high performance concrete subjected to various pullout rates. *Cem. Concr. Res.* **2016**, *89*, 1–13. [CrossRef]
15. Sorelli, L.; Constantinides, G.; Ulm, F.J.; Toutlemonde, F. The nano-mechanical signature of Ultra High Performance Concrete by statistical nanoindentation techniques. *Cem. Concr. Res.* **2008**, *38*, 1447–1456. [CrossRef]
16. Denisiewicz, A.; Socha, T.; Kula, K.; Pasula, M. Influence of steel and polypropylene fibers addition on selected properties of fine-grained concrete. *Civ. Environ. Eng. Rep.* **2018**, *4*, 138–148. [CrossRef]
17. Kabele, P. Multiscale framework for modeling of fracture in high performance fiber reinforced cementitious composites. *Eng. Fract. Mech.* **2007**, *74*, 194–209. [CrossRef]
18. Kim, S.Y.; Jang, H.G.; Yang, C.M.; Yang, B. Multiscale prediction of thermal conductivity for nanocomposites containing crumpled carbon nanofillers with interfacial characteristics. *Compos. Sci. Technol.* **2018**, *155*, 169–176. [CrossRef]
19. Yang, B.J.; Souri, H.; Kim, S.; Ryu, S.; Lee, H.K. An analytical model to predict curvature effects of the carbon nanotube on the overall behavior of nanocomposites. *J. Appl. Phys.* **2014**, *116*, 033511. [CrossRef]
20. Qin, Q.H.; Yang, Q.S. *Macro-Micro Theory on Multifield Coupling Behavior of Heterogeneous Materials*; Springer: Berlin, Germany, 2008.
21. Hashin, Z. Analysis of composites materials—A survey. *J. Appl. Mech.* **1983**, *50*, 481–505. [CrossRef]
22. Hill, R. Elastic properties of reinforced solids: some theoretical principles. *J. Mech. Phys. Solids* **1963**, *11*, 357–372. [CrossRef]
23. Ostoja-Starzewski, M. Scale effects in plasticity of random media: status and challenges. *Int. J. Plast.* **2005**, *21*, 1119–1160. [CrossRef]
24. Eshelby, J.D. The determination of the elastic field of an ellipsoidal inclusion and related problems. *Proc. R. Soc.* **1957**, *241*, 376–396.
25. Hill, R. A self-consistent mechanics of composites materials. *J. Mech. Phys. Solids* **1965**, *13*, 213–222. [CrossRef]
26. Mori, T.; Tanaka, K. Average stress in matrix and average elastic energy of materials with misfitting inclusions. *Acta Mater.* **1973**, *21*, 571–574. [CrossRef]
27. Bensoussan, A.; Lion, J.L.; Papanicolaou, G. *Asymptotic Analysis for Periodic Structures*; North-Holland: Amsterdam, The Netherlands; New York, NY, USA; Oxford, UK, 1978.
28. Sanchez-Palencia, E. *Non-Homogeneous Media and Vibration Theory*; Lectures notes in Physics; Springer-Verlag: Berlin/Heidelberg, Germany, New York, NY, USA, 1980.
29. Strzelecki, T.; Auriault, J.L.; Bauer, J.; Kostecki, S.; Puła, W. *Mechanics of Heterogeneous Media: Theory of Homogenization*; Lower Silesian Scientific Publisher: Wrocław, Poland, 1996. (In Polish)
30. Fish, J.; Yu, Q. Two-scale damage modeling of brittle composites. *Compos. Sci. Technol.* **2001**, *61*, 2215–2222. [CrossRef]
31. Ghosh, S.; Lee, K.; Moorthy, S. Two scale analysis of heterogeneous elastic-plastic materials with asymptotic homogenization and Voronoi cell finite element model. *Comput. Methods Appl. Mech. Eng.* **1996**, *132*, 63–116. [CrossRef]
32. Fishman, L.; McCoy, J.J. A unified view of bulk property theories for stochastic and periodic media. *J. Appl. Math. Phys.* **1981**, *32*, 45–61. [CrossRef]
33. McCoy, J.J. Macroscopic response of continua with random microstructures. *Mech. Today* **1981**, *6*, 1–40.
34. Guan, X.; Liu, X.; Jia, X.; Yuan, Y.; Cui, J.; Mang, H.A. A stochastic multiscale model for predicting mechanical properties of fiber reinforced concrete. *Int. J. Solids Struct.* **2015**, *56–57*, 280–289. [CrossRef]
35. Geers, M.G.D.; Kouznetsova, V.G.; Brekelmans, W.A.M. Multi-scale computational homogenization: Trends and challenges. *J. Comput. Appl. Math.* **2010**, *234*, 2175–2182. [CrossRef]
36. Kouznetsova, V.G.; Geers, M.G.D.; Brekelmans, W.A.M. Multi-scale second-order computational homogenization of multi-phase materials: a nested finite element solution strategy. *Comput. Methods Appl. Mech. Eng.* **2004**, *193*, 5525–5550. [CrossRef]
37. Schröder, J. A numerical two-scale homogenization scheme: the FE^2-method. *Plasticity and Beyond: Microstructures, Crystal-Plasticity and Phase Transitions*; Schröder, J., Hackl, K., Eds.; CISM Courses and Lectures Vol. 550; Springer: Berlin/Heidelberg, Germany; New York, NY, USA; Dordrecht, The Netherlands; London, UK, 2014; pp. 1–64.
38. Miehe, C.; Koch, A. Computational micro-to-macro tran-sitions of discretized microstructures undergoing small strains. *Arch. Appl. Mech.* **2002**, *72*, 300–317. [CrossRef]
39. Hain, M.; Wriggers, P. Numerical homogenization of hardened cement paste. *Comput. Mech.* **2008**, *42*, 197–212. [CrossRef]

40. Tian, Z.; Bian, C. Numerical modeling of elastic modulus for cement paste using homogenization method. *J. Wuhan Univ. Technol. Mater. Sci. Ed.* **2013**, *28*, 751–760. [CrossRef]
41. Wang, D.; Shi, C.; Wu, Z., Xiao, J.; Huang, Z.; Fang, Z. A review on ultra high performance concrete: Part II. Hydration, microstructure and properties. *Constr. Build. Mater.* **2015**, *96*, 368–377. [CrossRef]
42. Cheyrezy, M.; Maret, V.; Frouin, L. Microstructural analysis of RPC (Reactive Powder Concrete). *Cem. Concr. Res.* **1995**, *25*, 1491–1500. [CrossRef]
43. Wang, C.; Yang, C.; Liu, F.; Wan, C.; Pu, X. Preparation of Ultra-High Performance Concrete with common technology and materials. *Cem. Concr. Compos.* **2012**, *34*, 538–544. [CrossRef]
44. Bažant, Z.P.; Di Luzio, G. Nonlocal microplane model with strain-softening yield limits. *Int. J. Solids Struct.* **2004**, *41*, 7209–7240. [CrossRef]
45. De Borst, R.; Verhoosel, C.V. Gradient damage vs phase-field approaches for fracture: Similarities and differences. *Comput. Methods Appl. Mech. Engrg.* **2016**, *312*, 78–94. [CrossRef]
46. Kuhl, E.; Ramm, E.; de Borst, R. An anisotropic gradient damage model for quasi-brittle materials. *Comput. Methods Appl. Mech. Engrg.* **2000**, *183*, 87–103. [CrossRef]
47. Zreid,I.; Kaliske, M. A gradient enhanced plasticity–damage microplane model for concrete. *Comput. Mech.* **2018**, *62*, 1239–1257. [CrossRef]
48. Svenning, E.; Fagerström, M.; Larsson, F. Computational homogenization of microfractured continua using weakly periodic boundary conditions. *Comput. Methods Appl. Mech. Engrg.* **2016**, *299*, 1–21. [CrossRef]
49. Javili, A.; Saeb, S.; Steinmann, P. Aspects of implementing constant traction boundary conditions in computational homogenization via semi-Dirichlet boundary conditions. *Comput. Mech.* **2017**, *59*, 21–35. [CrossRef]
50. Espadas-Escalante, J.J.; van Dijk, N.P.; Isaksson, P. A study on the influence of boundary conditions in computational homogenization of periodic structures with application to woven composites. *Compos. Struct.* **2017**, *160*, 529–537. [CrossRef]
51. Denisiewicz, A.; Kuczma, M. Two-scale modelling of reactive powder concrete. Part I: representative volume element and solution of the corresponding boundary value problem. *Civ. Environ. Eng. Rep.* **2013**, *10*, 41–61.
52. Denisiewicz, A.; Kuczma, M. Two-scale modelling of reactive powder concrete. Part II: Numerical simulations. *Eng. Trans.* **2015**, *63*, 35–54.
53. Denisiewicz, A.; Kuczma, M. Two-scale modelling of reactive powder concrete. Part III: Experimental tests and validation. *Eng. Trans.* **2015**, *63*, 55–76.
54. Denisiewicz, A.; Kuczma, M. Two-scale numerical homogenization of the constitutive parameters of reactive powder concrete. *Int. J. Multiscale Comput. Eng.* **2014**, *12*, 361–374. [CrossRef]
55. Song, J.; Liu, S. Properties of reactive powder concrete and its application in highway bridge. *Adv. Mater. Sci. Eng.* **2016**, 5460241. [CrossRef]
56. *Abaqus 6.12 Documentation*; Dassault Systèmes Simulia Corp.: Johnston, RI, USA 2012.
57. Lejeunes, S.; Bourgeois, S. Une Toolbox Abaqus pour le Calcul de Propriétés Effectives de Milieux Hétérogènes; CSMA 2011, 10e Colloque National en Calcul des Structures 9-13 Mai 2011, Giens, France. 2011; 8p. Available online: https://hal.archives-ouvertes.fr/hal-00592866 (accessed on 1 January 2021).
58. Zdeb, T.; Śliwiński, J. Reactive powder concrete—Mechanical properties and microstructure. *Build. Ind. Technol. Archit.* **2010**, *51*, 51–55. (in Polish)
59. Hillerborg, A.; Modeer, M.; Petersson, P.E. Analysis of crack formation and crack growth in concrete by means of fracture mechanics and finite elements. *Cem. Concr. Res.* **1976**, *6*, 773–782. [CrossRef]
60. Lubliner, J.; Oliver, J.; Oller, S.; Oñate, E. A plastic-damage model for concrete. *Int. J. Solids Struct.* **1989**, *25*, 299–329. [CrossRef]
61. Lee, J.; Fenves, G.L. Plastic-damage model for cyclic loading of concrete structures. *J. Eng. Mech.* **1998**, *8*, 892–900. [CrossRef]
62. Drucker, D.C.; Prager, W. Soil mechanics and plastic analysis or limit design. *Q. Appl. Math.* **1952**, *10*, 157–165. [CrossRef]
63. Kuczma, M.; Whiteman, J.R. Variational inequality formulation for flow theory plasticity. *Int. J. Eng. Sci.* **1995**, *33*, 1153–1169. [CrossRef]
64. Saenz, L.P. Discussion of paper "Equation for stress-strain curve of concrete" by Desai P. and Krishnan S. *J. Am. Concr. Inst.* **1964**, *61*, 1229–1235.
65. Oden, J.T.; Prudhome, S.; Hammerand, D.C.; Kuczma, M. Modeling error and adaptivity in nonlinear continuum mechanics. *Comput. Methods Appl. Mech. Eng.* **2001**, *190*, 6663–6684. [CrossRef]
66. Jhurani, C.; Demkowicz, L. Multiscale modeling using goal-oriented adaptivity and numerical homogenization.Part I: Mathematical formulation and numerical results & Part II: Algorithms for the Moore-Penrose pseudoinverse. *Comput. Methods Appl. Mech. Eng.* **2012**, *213–216*, 399–417. 418–426.

materials

Article

Resistance of 3D-Printed Components, Test Specimens and Products to Work under Environmental Conditions—Review

Marcin Głowacki [1,*], Adam Mazurkiewicz [1], Małgorzata Słomion [2] and Katarzyna Skórczewska [3]

[1] Department of Mechanical Engineering, Bydgoszcz University of Sciences and Technology, Kaliskiego 7 Street, 85-789 Bydgoszcz, Poland
[2] Department of Management, Bydgoszcz University of Sciences and Technology, Kaliskiego 7 Street, 85-789 Bydgoszcz, Poland
[3] Faculty of Technology and Chemical Engineering, University of Sciences and Technology, Seminaryjna 3, Street, 85-326 Bydgoszcz, Poland
* Correspondence: marcin.glowacki@pbs.edu.pl

Abstract: The development of additive manufacturing methods known as "3D printing" started in the 1980s. In these methods, spatial models are created from a semi-finished product such as a powder, filament or liquid. The model is most often created in layers, which are created from the semi-finished product, which is most often subjected to thermal treatment or using light or ultraviolet rays. The technology of additive manufacturing has both advantages and disadvantages when compared to the traditionally used methods of processing thermoplastic materials, such as, for example, injection or extrusion. The most important advantages are low cost, flexibility and speed of manufacturing of elements with different spatial shapes. From the point of view of the user of the product, the most important disadvantages are the lower mechanical properties and lower resistance to environmental factors that occur during the use of the manufactured products. The purpose of this review is to present current information and a compilation of features in the field of research on the effects of the interactions of different types of environments on the mechanical properties of 3D-manufactured thermoplastic products. Changes in the structure and mechanical properties of the material under the influence of factors such as humidity, salt, temperature, UV rays, gasoline and the environment of the human body are presented. The presented article enables the effects of environmental conditions on common materials used in 3D printing technology to be collated in one place.

Keywords: 3D print; environment conditions; properties; environmental resistance

Citation: Głowacki, M.; Mazurkiewicz, A.; Słomion, M.; Skórczewska, K. Resistance of 3D-Printed Components, Test Specimens and Products to Work under Environmental Conditions—Review. *Materials* **2022**, *15*, 6162. https://doi.org/10.3390/ma15176162

Academic Editors: Jakub Zdarta and Agata Zdarta

Received: 31 July 2022
Accepted: 2 September 2022
Published: 5 September 2022

1. Introduction

The phenomena occurring in nature are the most common inspiration for new solutions. Three-dimensional (3D) printing may seem to be an advanced technology, but many living organisms have engaging in a similar process for a long time. For example, molluscs producing their shells (calcareous exoskeleton) can be considered a natural 3D printer. As they grow, the molluscs add calcium carbonate to their external shell. In this way, more internal space is protected by the skeleton and the growth lines, similar to the layers of printed material, are visible on the outside of the shell [1].

The use of photopolymers to create 3D objects in the 20th century led to the development of fast prototyping capabilities using the Fused Deposition Modeling (FDM) technique [2,3]. Stereolithography, patented in 1986 by Charles Hull, was the first ever method for the automatic manufacturing of three-dimensional models using UV radiation and photosensitive resin [4]. At the same time, in Texas, the concept of creating objects by using a laser beam to selectively sinter powder, a technology currently known as Selective Laser Sintering (SLS), was developed [5]. Fused Deposition Modeling, or FDM, is the most commonly used method. Its applications and modifications are discussed in this review study. The method was patented in 1989 by S. Scot and Lisa Crump [6]. It involves the deposition of a polymer filament via a heated nozzle and allows rapid prototyping. The modern FDM techniques will be discussed in further detail later in the study.

Three-dimensional printing is an additive manufacturing technique, in which the product begins as a single layer applied on the base and is formed by adding subsequent layers. A large number of conventional production technologies involve subtractive manufacturing, meaning that the final product starts as a block of material, e.g., wood or metal, from which excess material is removed to form a desired shape. A large amount of waste material—for example, sawdust or metal chips—is produced in the process. Additive manufacturing techniques, such as laying a wall by adding layers of bricks and mortar according to a formal plan, have been used for many years. All the above methods, from the design to the production of the end product, have developed into modern 3D printing techniques. A key stage is the planning of the layer arrangement process, and 3D printers contain components providing automatic control [7].

Polymers are some of the most commonly used materials in additive techniques. The polymer technology is known as the "innovation of the millennium". They are widely used because they are simple to produce and modify at a relatively low cost compared to other materials, and their wide-ranging properties can easily be adapted to many applications. The polymers are used in many domestic and industrial applications [8]. Due to a single technology, we now have the capability to create, at our own leisure, prototypes or common household items. Three-dimensional (3D) printing, or additive manufacturing (AM), is a technology widely used, not only by the industry or individuals, but also to aid scientific research. This technology allows one to create high-precision, low-cost prototypes, and, with minor hardware requirements and no special qualifications or licenses needed for the operators, it can be used by anyone. The items manufactured with this printing process are robust and lightweight, with a greatly reduced amount of waste product. AM is defined as the process of creating a 3D solid from a digital file. In this process, the building material is applied layer by layer with selective sintering [9]. Depending on the printing technology and materials used, the selective sintering process may differ. The history of 3D printing, its first uses and the available 3D printing techniques are detailed in [10–20].

The following techniques are available.

Fused Deposition Modeling, or FDM, uses thermoplastics for printing. A polymer, in the form of a filament, is fed into a 3D printer extruder and heated to a semi-liquid material. The first layer is applied in the *XY* axis. Next, the extruder moves up and the bed moves down by the thickness of the first layer and the second layer is applied. Hot semi-liquid polymer cools down rapidly and solidifies, forming a designed shape. The solidification rate depends on the position, glass transition temperature, flow temperature and filament thickness [10]. A detailed review of this technology can be found in many publications [19,21–30].

Stereolithography, or SLA, is a printing technique using light-cured resin (cured by a laser beam). A container is filled with light-cured resin, and, during printing, the bed is immersed in the resin to the very bottom. The laser beam traces the shape of the object in the XY plane on the bed, curing the first layer of the resin. After tracing, the bed moves up, and the process is repeated. The disadvantage of this AM technique is the small printing area, and its advantage is that the printed item is not affected by external factors [10]. A detailed description of the method and its applications can be found in [31–35].

Digital Light Processing, or DLP, is a technique similar to SLA, involving the curing of light-sensitive materials using light emitted by a projector. The difference is that DLP creates models by curing the whole layer at one time, and in SLA, the laser beam moves from point to point, tracing the geometry. This technology is used in many areas [36–38].

Jetting Molding, or MjP, is a molding technique using a process similar to a standard jet or laser printing on a piece of paper. The head moves over the table in the *X* axis and applies a light-cured resin, which in turn is cured using UV light. After printing the layer, the table moves down in the *Z* axis. Example applications of the technology can be found in [39–41].

Selective Laser Sintering, or SLS, uses a powdered thermosetting polymer. The applied powder layer is selectively sintered at a high temperature using a laser beam. Similar to CJP or Binder Jetting, the printed models are removed from a block of non-sintered powder [40,42–44].

The above descriptions of selected additive manufacturing techniques present the scale and wide range of applications of 3D printing techniques. The aim was to show the different additive manufacturing methods and the available directions of research to verify the resistance of printed models to environmental conditions.

Fused Deposition Modeling (FDM) is the most commonly mentioned technique in the literature, and the main scope of this study is based on products manufactured using this technique. The Fused Deposition Modeling (FDM) technique was developed in 1988, and the patent protection expired in 2009. It allowed different companies to design, develop and commercialize the technology in a wide range of specifications and components, including domestic, office and industrial applications. The interest in this technique peaked in recent years, particularly in medical applications, but also in the scientific field, which is clearly reflected in the many research studies on this topic [10]. The print parameters are usually recommended by the filament manufacturers, allowing less qualified operators to produce good-quality prints. This makes the technology an excellent solution in many fields. Its many advantages, including its versatility and the ability to adapt it to new materials, allow it to create multi-colored and complex end products at low costs and high speeds, and make the technique very popular. In recent years, it has found its place in medical applications, electronics and various production processes. Thanks to their many interesting properties, including low weight, high rigidity and low density, the layered structures are widely used in many fields of industry [40].

The initial process conditions selected at the planning stage will affect the quality and structure of the product. No information can be found in the literature on how the environment affects the prints using different materials. There are many studies that research the effects of different parameters on the quality and mechanical properties of prints, but no systemic information can be found in the literature on how the prints are affected by environmental factors. The FDM prints can be exposed to many environmental factors that may affect their strength parameters in time.

Depending on the filament and modifiers (fillers, dyes, pigments, etc.) used, many different prints can be produced with different properties depending on the consumer requirements. Filament materials including acrylonitrile butadiene styrene (ABS), acrylonitrile styrene acrylate (ASA), polylactic acid (PLA), polyethylene terephthalate glycol (PETG) or polyamide (PA) are commonly used in different AM processes. The filament used and its modifiers can significantly affect the resistance of the prints to environmental factors.

The properties of 3D prints are determined based on the analysis of basic mechanical properties, including bending, compression, tension or impact. A surface structure is also assessed using scanning electron microscopy (SEM) or computed microtomography (CT) to detect any defects and analyze the structure of materials produced in the AM processes. The testing of additive techniques is also based on the analysis of mechanical properties. Although the polymers are widely used in 3D printing, a small number of researchers deal with the effects of environmental conditions with regard to a specific set of conditions.

The present article includes a current review of the research on the resistance of polymer 3D prints to different environmental factors. The order of the discussed environments with materials is shown in Figure 1. It will allow a systematic review for a better understanding of the effects of environmental conditions on the structure and durability of the prints. It will also allow us to determine further directions of research conducted on this topic. The knowledge will allow us to extend the applicability of the products obtained using AM techniques depending on the material used and its exposure to environmental factors.

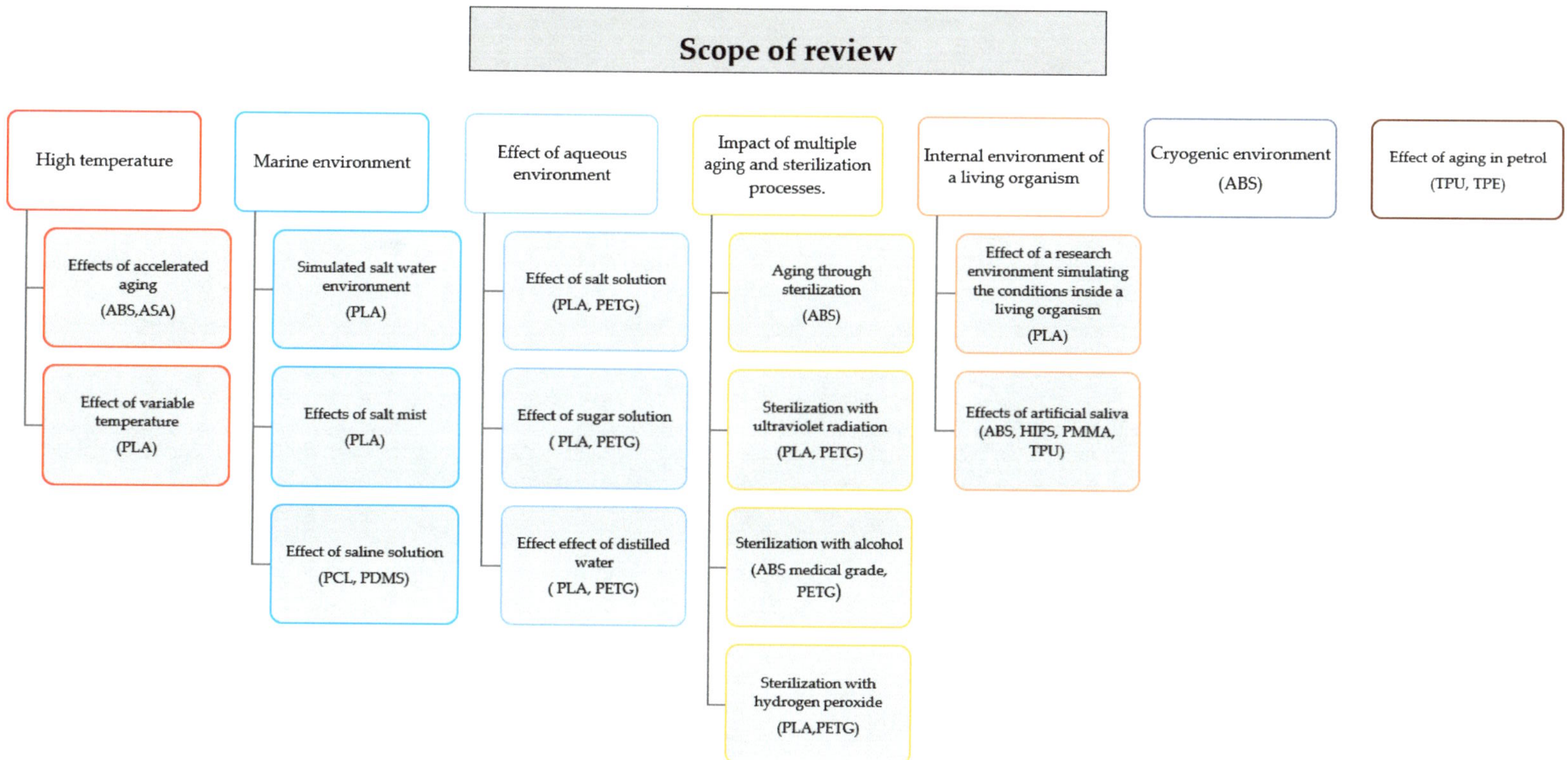

Figure 1. The range of subjects discussed in the review, along with the materials.

2. Effect of Thermal Aging Conditions on the Properties of 3D-Printed Samples

One of the most commonly considered effects of the environment on the print quality is the effect of temperature on the properties of components subject to thermal aging. M. Reza Khosravani et al. [44] analyze the effects of accelerated aging on 3D prints. The products, made of ABS and ASA with a hexagonal and triangular fiber arrangement, were printed with 30% infill density and no outer layer, to allow the analysis of the core structure and its structural evaluation. To prevent defects, including gaps, overlaps or voids, at the production stage, a surface layer made of an epidermal material—a carbon fiber composite—was added. The thermal aging procedure involved placing the test specimens in a climatic chamber at temperatures below the glass transition temperature of the tested material (between 22 °C and 60 °C). The test lasted 240 h. A change in test specimen mass before and after thermal aging was determined. Tensile, bend and compression tests were also carried out [45].

The test specimens, under thermal aging, showed improved rigidity and durability, and the change in mass and dimensions after the aging process was approximately 1%. Despite the 1 wt.%. change in mass, the bending stress in the aged test specimens increased by approximately 15%, which may have been due to annealing as a result of aging, which in turn led to an intermolecular rearrangement [45]. The structure of the aged ASA and ABS components changed, increasing the maximum bending stress, modulus of elasticity and bending resistance. The temperature caused an increase in maximum durability for both fiber arrangements and materials by changing the molecular structure. The ASA component with a honeycomb structure showed the highest breaking strength. The hexagonal core arrangement was the best for both materials, since it resulted in higher strength of both components. The authors of the study highlighted the effect of the working temperature and the core structure as the key parameters in designing components made of ABS and ASA. During the design of the components using 3D techniques, the printing conditions and the printing process must be adapted to the thermal working conditions of the component [45].

The topic of the effect of temperature on the mechanical properties of PLA was discussed by Y. C Niranjan et al. [45]. The authors considered the effects of annealing parameters on the dynamic mechanical properties of PLC components printed using the FDM technique. Six test specimens, 50 mm × 11 mm × 3 mm, were printed and annealed in a convection oven. The test specimens were divided into two groups. The first group of the test specimens was characterized by a variable holding time in the oven at 90 °C. The effect of the holding time (15, 30 and 60 min) on the PLA test specimens at a constant temperature regarding the dynamic mechanical properties was analyzed. The second group of the test specimens was heated for 30 min at different temperatures, namely 80 °C, 90 °C, 100 °C and 110 °C. The PLA heating temperature was limited to 110 °C, since, over 120 °C, the material had deteriorated visibly, affecting its mechanical properties and geometric features. After cooling to room temperature, the test specimens were stored at room temperature for two days to simulate the actual working conditions, followed by dynamic mechanical analysis (DMA).

The results show that both the annealing duration and temperature significantly affect the dynamic mechanical properties of PLA printed using the FDM technique. The heat increased the conservative modulus, the resistance to bending of the PLA components and the glass transition temperature of the tested material. The loss modulus of the annealed test specimens tended to decrease, both at room and increased temperature, due to the increased rigidity of the annealed PLA components compared to the base material. The results showed that the annealing process can also contribute to bonding strength between the layers, and the reduced void content and increased crystallinity can have a positive effect on the stiffness of the viscoelastic polymer.

Annealing processes can be used as a low-cost process to improve the mechanical properties of the material. The interesting effects of thermal aging on the components produced using additive techniques can be found in [46–61].

3. Susceptibility of 3D Prints to Simulated Marine Environment Conditions

Studies discussing the effects of marine environments [62], salt mist [63] and saline solution [64] are the main sources of information on the effects of environmental factors on the mechanical properties of polymer 3D prints. A study discussing the effects of biofouling on 3D-printed components made of PCL immersed in salt water is also available [65].

R. Krishna Upadhyay et al. [62] described a test consisting of immersing the test specimens for 30 days in a simulated salt water environment. The material used for printing was polylactic acid (PLA). The test specimens of different shapes were made using additive manufacturing techniques. These included both standard test specimens for strength testing, propeller blades and plates for the evaluation of the effects of salt water on the material surface structure.

The changes occurring as a result of print storage in a simulated marine environment to ASTM D1141-98 were evaluated. Friction, wear and sliding wear mechanisms, mechanical properties (tensile strength and Vickers hardness) and the surface roughness of the test specimens were analyzed. To determine the changes occurring in the test specimen structure after 30 days in salt water, the test specimens were analyzed using a scanning electron microscope (SEM).

The results of the experiment showed a change in the mechanical properties of the PLA test specimens. The surface roughness measured on both surfaces of the test specimen was much higher than before, which may have an effect on their intended use. After submersion in salt mist for 30 days, the physical properties of the test specimens had changed. The liquid particles that migrate inside the rough structure may degrade the polymer more effectively as a result of a hydrolytic reaction. Low surface roughness significantly affects the tribological properties and ensures problem-free operation and higher resistance to hydrolytic degradation.

The tensile test was conducted at a high load (25 mm/min) to simulate the use of the material in an environment with theoretical conditions present in its actual application. The analysis of the mechanical properties of the primary test specimens and test specimens subjected to the salt mist concluded that the marine environment reduced the PLA test specimens' elongation at rupture, while increasing the yield point and the modulus of elasticity. The analysis of the mechanical properties shows that 3D-printed components can be used for target marine applications, including slide-bearing operations at moderate speed and load.

It is worth noting that the analysis of the external structure of the turbines in [66] after exposure to salt mist showed a deterioration in their surface; however, the observed changes were minor. The XRD spectra of the test specimens before and after salt mist exposure were identical, and the minor changes in 2theta angle within 20–27 2theta indicate the amorphic nature of the PLA used to produce the test specimens. In summary, the 3D-printed PLA components showed satisfactory mechanical strength and the marine environment had a low effect on material deterioration in the analyzed period. This means that the 3D components can be used in slide applications at moderate loads in the analyzed environment.

S. Ambrus et al. presented, in [63], the test results of the analysis of a material aged in salt mist. The components were produced from PLA using the FDM technique. The component production parameters and the crystallinity parameters of the produced components were determined using an XRD technique.

The study was carried out in accordance with the guidelines for component testing in an artificial atmosphere [63]. A Liebisch SI-400 device was used for observation of the mechanical behavior of the test specimens after being held in specific conditions in the experiment involving aging in salt mist. The test specimens were subjected to moisture–thermal testing at 45 °C, with subsequent rinsing with distilled water and 3.5% NaCl solution at a 0.8 m^3s^{-1} flow rate. The test lasted 240 h.

The breaking strength of the test specimens before the aging process ranged from 44.27 MPa to 39.75 MPa, with an average value of 42.14 MPa. Following 240 h testing in sprayed salt mist conditions, the breaking strength of the test specimens showed higher variability, i.e., between 31 MPa and 40 MPa, with the tensile stress ranging from 31 to 34 MPa. The impact test did not yield any significant differences.

The study concluded that no change in test specimen mass was observed, although the strength decreased due to polymer degradation as a result of component aging in the salt chamber.

A non-typical study on the effect of biofouling, i.e., formation of a biofilm as a result of biological contamination in sea water, leading to the biodegradation of plastics, was discussed in the article by Matthew Ryley et. al. [66]. The mechanical degradation of PCL and PDMS glass plates immersed in water was analyzed. The plates were immersed in ultrapure water for 120 h; the water was replaced after the first 24 h to remove potential eluates. The test specimens were rinsed with 70% ethanol solution and immersed in ultrapure water. The printed plates were placed for 12 weeks under a raft (a suspended weight) to prevent them from rising to the surface, at a depth of 6 meters. In this period, the sea water quality parameters, including temperature, conductivity, salinity, pH and dissolved oxygen, were monitored.

After this, an analysis was carried out to determine the differences between the control sample and the aged material. To determine the relationship between the percentage of microfouling coverage and the material properties, statistical analysis (ANOVA or Tukey post hoc tests) was carried out. In the first 4 weeks, in salt water, lower coverage by cyanobacteria was observed on PCL test specimens compared to other materials, and after 12 weeks, all materials were severely fouled by the microorganisms.

PCL was considered the best material due to the lowest overall coverage by microalgae, which, after 12 weeks, was between 86.8% and 98.4%. Among the compared materials, PCL was the least hydrophobic and showed the lowest biofilm coverage of all materials, making it the best choice for marine applications. It is also biodegradable in the marine environment. Moreover, 3D-printed polymers meet the conditions for marine applications; however, they must also be analyzed for degradation by fragmentation, hydrolysis or biodegradation as a result of salt water or microorganisms [66]. The topic was also discussed in [65,67–71].

4. Effect of Aqueous Environment, Salt Solution and Sugar Solution

D. Moreno Nieto et al. [72] presented the results of changes in PLA and PETG properties as a result of interaction with different aqueous environments. The main reason for addressing this issue was to verify the strength of materials used as packaging in the food industry. The test specimens printed for the purpose of the experiment were square-shaped, 30 mm × 30 mm × 3 mm. Sets of 15 test specimens were placed in distilled water and saturated salt and sugar solutions. The test specimens were immersed in metal containers and secured against movement using tin wire. The containers were stored at a constant temperature of 20 °C and 50% humidity. The solution was stirred once a day and replaced once a week.

The experiment involving the degradation tests lasted 10 weeks, whereas the absorption tests were continued until the test specimen mass had stabilized. The measurements were carried out every week at the same time, and if degradation was observed, the absorption was monitored every day for the first week of observation. The method of assessing the degradation involved a weekly sampling of 5 out of 15 test specimens immersed at the beginning of the test, and removing the layer of salt or sugar crystals adhered to their surfaces. The test specimens were dried on absorbent paper and stored in plastic bags. The tests included weighing, measuring and observation under an optical microscope. After this, the samples were immersed in the respective solutions again. The analysis of the experiment's results showed that the test specimens immersed in saturated salt or sugar solutions did not show any changes in color or shape compared to the test specimens immersed in distilled water only.

The absorption tests for the PLA and PETG test specimens were conducted for 10 weeks. Visible absorption was observed in the initial phase of the experiment, i.e., in the first two or three days. An increase in test specimen mass and equilibrium lasting until the end of the experiment was observed. The swelling of the PLA samples had ceased after 8 weeks in the distilled water, 9 weeks in the sugar solution and 3 weeks in the salt solution. For the PETG sample, the equilibrium of swelling in the distilled water was observed after 8 weeks, in the sugar solution after 9 weeks and in the salt solution after 7 weeks.

The degradation effect of these environments was also evaluated. The degradation changes were determined using optical microscopy. The changes occurring in the test specimens and crystalline structures were observed in all three solutions. The observed degradation changes depended on the duration of the experiment. The changes were characterized by dark inclusions and discoloration of the PETG test specimens. Changes in the color of the PLA test specimens were also observed.

To summarize, PETG was the most dimensionally stable among the three solutions, and in a 9-week period, the change in mass was observed to be 0.3%, making it the best option for marine applications. After 8 weeks, the PLA specimens changed in mass by approximately 2.5%.

5. Effect of Multiple Aging and Sterilization Processes

The number of studies on polymers discussing the effects of different types of sterilization procedures has recently increased due to the lack of first aid equipment as a result of limited deliveries, further affected by the COVID-19 pandemic. In [73], Diana Popescu et al. presented an analysis of the effects of multiple sterilization procedures on polymer components and their properties. Catalin Gheorghe Amza [64] analyzed the effect of ultraviolet radiation on test specimen aging. Krzysztof Grzelak et al. [74] discussed the effect of chemical disinfection using alcohol, and in [75], Olivier Oth et al. described the low-temperature sterilization methods using hydrogen peroxide.

The first mentioned article [73] discussed the effects of the process used in a hospital environment to neutralize potential pathogens on medical instruments. The analysis covered ABS. The aim of the experiment was to evaluate the effect of each analyzed factor, i.e., aging only and aging with sterilization, on mechanical properties and structures. This process involved cyclic tests carried out during the first week of exposure to analyzed factors. The tests were carried out for 9 weeks. Mechanical tests, including bending and tensile strength, were also carried out. The test specimens had a 100% infill density and a vertical orientation, which significantly reduced the strength during strength testing due to the opposite distribution of the fibers in relation to the forces acting on them.

The test specimens were sterilized every two weeks, and the process involved exposure to hydrogen peroxide vapors and low-temperature gas plasma. Sterrad equipment was used to expose the test specimens to 45 min sterilization cycles at 134 °C and 0.223 MPa. For the first group of the test specimens, SEM image analysis showed plastic cracking of fibers and delamination between the layers of printed materials that were subjected to identical aging and sterilization conditions. Cracking, with small amounts of material being separated as a result of exfoliation, was observed for the first group aged in the storage process. Cracking along the layer printing direction in both groups was observed at the end of the experiment. Following a tensile test, an analysis of the external structure was carried out for selected series using a micro-CT instrument. The results showed that the ratio of voids to the total volume of the analyzed test specimens was between 6.14% and 7.82% due to the effects of aging and the exfoliation of printed layers.

To summarize, the test specimens were compared at selected intervals, evaluating both their mechanical strength, external and internal structure and change in mass. The results showed no significant changes in mechanical properties regarding rigidity, mass loss and tensile and bending strength, irrespective of the aging technique used.

Catalin Gheorghe Amza et al. [64] discussed topics related to the effects of accelerated aging by exposing 3D-printed test specimens to ultraviolet radiation. The test involved an experimental simulation of the effects of atmospheric conditions and sun on components made of polymers, including PLA and PETG, by irradiation with a UV-A light in a radiation chamber according to ISO 4892-3:2016 [76].

In total, 64 test specimens, in the shape of a dog bone and a 15 mm × 15 mm cube, were printed for the purpose of the test. The test specimens were divided into a refence group and a test group and exposed to 310 nm wavelength radiation. A thermostatic climatic chamber, the Discovery DY110C, was used to experimentally determine the effect of UV radiation. Total UV exposure time was 24 h and included three irradiation and condensation cycles. Each irradiation cycle included an 8 h holding period at 50 °C and 50% humidity with 0.43 $mW^{-2} \times nm^{-1}$ lamps on. The alternating cycle without irradiation included a 4 h rest time with UV lamps off. The total exposure of the material to UV radiation in the chamber corresponded to several months of exposure to the sun in external conditions [64].

The visual inspection showed certain changes in color; however, the dimensional analysis did not show any effect of accelerated aging on changes in the analyzed properties of the components. The analysis of the strength tests (tensile, compression and rigidity) showed a 5.3% decrease in the strength of the PLA test specimens exposed to UV-B radiation. The test specimens were more fragile, losing 6.3% of their compression strength, whereas the rigidity of the irradiated test specimens did not change significantly. For PETG test specimens, a significant 36% strength decrease compared to the control test specimen was observed. Compression strength decreased by 38.6%, whereas the rigidity did not change. The cracking analysis using the SEM technique showed changes in roughness at the surfaces of fibers, which may indicate a degradation process [64].

In PLA test specimens exposed to UV-B radiation, the mechanical strength decreased slightly, whereas significant changes were observed for the PETG test specimens subjected to identical exposure conditions. The analysis of the test results, particularly the mechanical properties of 3D prints, showed a detrimental effect of UV radiation on the components intended for use in direct sunlight [64].

Krzysztof Grzelak et al. analyzed the effects of another sterilization method—alcohol disinfection—on the print properties [74]. The tests included three materials, i.e., PETG filament with color pigment and without additives and ABS for medical purposes, which is a dedicated material for medical applications. The reason for choosing this disinfectant was that it is readily available in most healthcare facilities.

Five test specimens of each material (in the shape of a dog bone) were printed for the purpose of this test. The test specimens were divided into four groups with different exposure times, as well as a control group. Each group was characterized by a different disinfection time, i.e., 0.5 h, 12 h, 24 h and 48 h, respectively. To determine the effect of the disinfectant concentration, selected test specimens were placed in a disinfection container filled with 4% aqueous alcohol solution and selected test specimens were placed in an undiluted disinfectant. After a set disinfection time, the components were dried in a laboratory drier for one hour at 45 °C.

The results showed a minor effect of disinfection on the material structure and its tensile elongation and change in mechanical properties irrespective of the exposure time to disinfectant in the analyzed 48 h period. The slight impact may have been caused by the penetration of liquid molecules into the structure of the samples. The analysis showed that the disinfectant had a minor effect on the PETG prints. Moreover, the type of dye used did not show any effect on the mechanical properties of the material.

Similar effects were observed during polymer tests with different disinfection techniques, e.g., photodynamic [77] or microwave [78]. Among the two types of analyzed materials, the dedicated material—medical-grade ABS—seemed to be a better candidate for component production using AM techniques.

Non-modified PETG (without additives) is at risk of a decrease in tensile strength by 20%, and the addition of a dye slightly increases the chemical resistance of the material [74].

One of the mentioned methods to destroy microorganisms is low-temperature sterilization with hydrogen peroxide. The effect of this technique was analyzed for surgical guides produced using the AM technique. The components described in [75] were designed at the maxillofacial department and made of PETG or PLA. The materials were sterilized in a single 50 min cycle at a temperature of below 55 °C.

Both before and after sterilization, the guides were scanned using a CT scanner and the deviation of starting points of the geometry measurement was compared to the reference guide. A statistical analysis, an ANOVA test with random factors, allowed the authors to determine the morphometric differences between different guides. For PLA, the high temperature used in the conventional sterilization techniques (121 °C) resulted in the material melting after a short cycle (5 min), which was verified by the authors with simultaneous discussion of the results obtained by Boursier et al. [79].

The results of the sterilization with hydrogen peroxide unequivocally showed no effects on the material structure. The components made from these materials can be used for clinical applications. The morphological differences were less than 0.2 mm. The analyzed method can be used as an alternative method to avoid the deformation of 3D prints made of PLA and PETG during sterilization. The results of tests describing the importance and the effects of sterilization are included in [80–85].

6. Internal Environment of a Living Organism

Amirapasha Moetazedian et al. [86] attempted to analyze the effect of a research environment simulating the conditions inside a living organism on the tensile strength of polylactic acid (PLA) components produced using AM techniques. To determine the consequences of placing a plastic implant inside a living organism, the factors present inside the organism must be simulated. The study aimed to determine the effects of three different conditions on the mechanical properties of PLA prints. The effect of an aqueous environment (immersion and increased humidity) and the effect of temperature (room and body temperature (37 °C)) were determined.

To check the component's saturation with water, the absorption was tested within 48 h. At room temperature, after 30 min, the absorption reached 0.561% and did not change until the end of the test. An increase in water temperature increased the absorption to 0.741%. After determining the change in mass, the tensile strength was tested. For the test specimens stored at room temperature, the tensile strength decrease was 0.188%; however, for the second type of test specimens exposed to a physiological environment, a significant weakening of the polymer by 23.4% relative to the reference test specimen was observed. Water absorption by the test specimens before the tensile test significantly affected the elongation at break, whereas changes in the yield point were minor. PLA components fully immersed in water showed a significant change in key properties relative to the control test specimens.

A synergistic effect of high temperature and water absorption resulted in a transition from brittle cracking to intermediate brittle cracking. A significant decrease in the mechanical parameters of 3D-printed test specimens was observed. Immersion in water at 37 °C resulted in a 50% reduction in the mechanical strength of the polymer and a 20% reduction in the tensile modulus of elasticity. The test specimens were deformed at 40% of the reference test specimen.

The test results do not reflect all synergistic environmental conditions present in living organisms due to the lack of use of biomedical materials, i.e., fluids corresponding to those present in the human body.

The second article discussing issues related to the effect of a physiological environment on 3D-printed polymer components was written by Any C. Pinho et al. [87] and describes the effects of artificial saliva on multi-material prints. The experiment covered the properties of test specimens made of a single material or two materials, where TPU was the core and the outer layer used materials including ABS, HIPS and PMMA. The core material was selected based on its ability to disperse impact energy. For each type of configuration and mechanical test, 12 test specimens with 100% infill were printed.

Half of the printed test specimens of each material were aged in artificial saliva before the mechanical tests. The artificial saliva was made by adding 0.426 g (Na_2HPO_4) disodium hydrogen phosphate, 1.68 g ($NaHCO_3$) sodium bicarbonate, 0.147 g ($CaCl_2$) calcium chloride and 2.5 mL (HCl) hydrochloric acid to 800 mL distilled water. Single test specimens were placed in tubes with the prepared solution and left in a shaker for 14 days at a constant temperature of 37 °C at 100 rpm. The procedure aimed to simulate the effect of exposure to saliva on the test specimens for one year.

The test results did not show any significant changes in the dimensions of the components immersed in saliva, which confirmed that the analyzed thermoplastic polymers showed low affinity to aqueous solutions. The same was validated by the results of studies published in [88,89]. The results of the mechanical property tests showed similar trends both for dry test specimens and aged test specimens, except for the ABS and PMMA test specimens, where the maximum bending stress was lower in the case of the aged test specimens. The opposite trend was observed for the HIPS and TPU test specimens.

The information showed that the aging process in a simulated saliva environment did not significantly affect the mechanical behavior of the test specimens, except for the PMMA, which is the only polymer without any aromatic rings in its chemical structure, which may explain its different behavior. The ABS test specimens lost 28.7% elasticity and 34.4% of absorbed energy as a result of the aging processes in the saliva solution.

For the sandwich test specimens, the behavior of all layered structures changed slightly as a result of aging in artificial saliva. Using TPU as a core affected the reduction in the mechanical properties of multi-material test specimens. The highest decrease in strength was observed in the test specimens containing PMMA, probably as a result of lower adhesion between the polymer and TPU, resulting from the difference in chemical composition.

To summarize, the saliva affected some of the multi-material prints. The combination of TPU with other materials affected the bending strength of the test specimens. Among all the material combinations, the best one was ABS-TPU-ABS due to the highest elasticity of the test specimens [87]. The effect of high moisture and variable temperature was discussed in [90–95].

7. Cryogenic Environment

Increased temperature is one of the factors affecting polymers during production or use. In [96], F. Saenz et al. tested the mechanical properties of ABS components produced in additive manufacturing processes, both at room temperature and at −196.15 °C [97]. The manufactured parts were HTS magnetic coils used at low temperatures, expected to transfer moderate mechanical loads at the supports [96]. The effects of cryogenic temperatures on polymers are widely discussed in the literature, and the results may validate the applicability of ABS in the analyzed applications.

For the purpose of an experiment, a cryogenic tank was designed and constructed for use with liquid nitrogen. The test specimens for mechanical tests (32 test specimens) were stored in the cryogenic tank. The test specimens, 115 mm × 19 mm × 6 mm, in the shape of a dog bone, were made with 80–90% infill. The cooling procedure involved placing the test specimen inside the cryogenic container. The cooling time was 3 min due to the insulation added to the test specimens.

Testing of the mechanical properties, both the yield point, Young's modulus and dry strength of the test specimens stored at −196.15 °C, did not show any significant differences for all tensile tests. The maximum strength was reduced by 4–7%, which is a typical effect of material brittleness at low temperatures. For comparison, the strain range for the test specimens stored at room temperature was between 15% and 28%.

The strength tests for test specimens stored at 77 K showed elastic behavior until breaking, a higher Young's modulus and a lower strain rate compared to the test specimens stored at room temperature. A statistical analysis was used to determine the significance of the results. The results showed that the brittleness of ABS test specimens produced using additive techniques increased with a decrease in temperature to 77 K. Based on the analysis of the results, the authors found that the infill density and the pattern affected the change in yield point, Young's modulus and dry strength. The effect of negative/cryogenic temperatures was discussed in [98,99].

8. Effect of Aging in Petrol

Eva Paz et al. [100] discussed issues related to the effect of an abrasive medium—in this case, petrol—on test specimens produced using additive techniques and made of two different materials, TPU and TPE. The study allowed them to determine the effects of a chemical abrasive medium on changes in mechanical properties in time.

The components for the purpose of the study were printed with 5%, 20%, 50% and 80% infill and a rectangular infill pattern. Five test specimens for each infill density, test type and material aging time were printed. The exposure to chemical factors was measured in accordance with ISO 175:2010 regarding the testing of plastics in liquid chemicals [101]. The samples were stored in containers filled with 98 octane petrol, until the samples were fully immersed. The material was stored in a chamber with temperature control and an air exhaust to avoid the possible accumulation of petrol vapors. The test periods were 24 h and 7 days, corresponding to a short-term and long-term test according to the standard. The test specimens were also exposed to petrol for 30 days.

The results for both materials, before and after aging, were compared. TPU components showed higher strength compared to TPE, irrespective of the external factors. For test specimens with over 20% infill density, no significant effects on material hardness were observed at different periods of immersion in petrol. As a result of aging, the TPE test specimen mass increased due to petrol absorption, compared to the TPU test specimens, showing that the TPE test specimens had lower resistance to petrol compared to the TPU test specimens. The test showed that the infill density had a significant effect due to the decrease in petrol absorption rate, resulting in a minimal effect on mechanical properties. The samples with 80% infill density showed surface-only interaction with the solvent.

The results suggest a two-stage interaction between the polymers and the analyzed environment. The first resulted in the breakage of some of the chains forming the polymer, resulting in a loss of mass and the deterioration of its properties. The second resulted in the recovery of the original properties due to the evaporation of the absorbed solvent from the intermolecular lattice of the polymer. For TPE, this process takes less time to complete.

9. Conclusions

The availability and easy application of additive technologies make them popular in many applications. The ease of manufacturing a specific product adapted to individual client needs means that the materials are widely used in many applications. Due to their applications, the components may be exposed to different environmental factors, the effects of which are not fully classified or well known. Another factor affecting the changes in properties is the printing technique used and the choice of materials.

In the discussed experiments involving exposure to different environmental factors, the properties of the components printed using materials most commonly used in additive technologies, including ABS, ASA, PLA, PETG, HIPS, TPU, TPE, PMMA or PCL, were analyzed. The components that can be used as spare parts in many devices were exposed to selected controlled environments, including disinfectants, low and high temperatures, humidity, different liquids and aqueous environments.

The analysis of the results of accelerated aging based on dynamic mechanical characteristics showed that both the annealing time and the temperature caused an increase in the conservative modulus for PLA, resulting in improved mechanical properties of the material [45]. It is worth emphasizing that the infill density is of key significance due to the absorption rate of the medium. The tested specimens with 80% infill density showed surface changes only [10].

PLA test specimens exposed to a simulated marine environment by immersion in salt water and PLC test specimens exposed to salt mist showed only a minor deterioration in the external structure [62,63]. The results do not exclude use of components made of polymers manufactured using additive techniques in the analyzed environment.

The effects of an aqueous environment and a salt and sugar solution on the properties of PLA and PETG prints were evaluated, including absorption and external structure analysis. A change in color in the test specimens immersed in distilled water was the most clear, and the lowest absorption rate and mass stability were observed for PETG test specimens [72].

Sterilization, both using hydrogen peroxide vapors [73] to disinfect ABS, alcohol to disinfect PETG [74] and ultraviolet radiation [75], did not show a significant effect on ABS, PLA and PETG test specimens. The first sterilization method did not affect the component properties, the second showed a minor effect on the material structure, and the third method showed a minimal effect on the mechanical properties of PETG samples only.

A simulation of the internal environment of a living organism was described in [86]. In the mentioned study [86], the environment was simulated without any biomedical fluids, which prevented the simulation of all synergetic conditions present in a living organism. An increase in mechanical properties and 40% deformation compared to the reference sample were observed for PLA test specimens. Another study [87] analyzed the effects of artificial saliva on the properties of multi-material prints made of TPU, ABS, HIPS and PMMA. The chemical structure of PMMA significantly affected the mechanical properties due to the lack of aromatic groups in the macromolecule and thus lower chemical resistance.

Short-term storage of test specimens at −196.15 °C was used to simulate a cryogenic environment and did not show a significant effect on the mechanical properties of ABS test specimens. A reduced maximum strain was observed, which is a phenomenon typical of materials at low temperatures [96]. The following Table 1 presents a summary of the environments along with the research methods considered.

The results, presented in a collective manner, make it possible to discern similarities in terms of the evaluation of the structure or mechanical properties of polymeric materials. The most frequently selected materials for testing were, in the following order, ABS, PLA and PETG. A research gap for further studies may be to investigate the effects of exposure environments on an even wider range of materials and possible strategies to assess changes in mechanical properties—in particular, cyclic fatigue tests.

In the presented review of the literature, components produced using AM techniques were exposed to different environmental factors. The review showed that there is only a small amount of information showing a synergistic effect of different variable environmental factors on the changes in properties of 3D prints; thus, the authors plan to undertake further research on this topic. Experiments should be systematized in terms of the utility properties evaluated, as there are a large number of unknowns and it is impossible to compare relationships. Observation of the behavior of materials exposed to different factors has allowed researchers to draw conclusions on the range of changes in surface structure and variations in strength ranges depending on the 3D printing technique used.

Table 1. Influence of the discussed environments on specific material properties.

	High Temperature	Marine Environment	Aging through Sterilization	Sterilization with Ultraviolet Radiation	Sterilization with Alcohol	Sterilization with Hydrogen Peroxide	Internal Environment of a Living Organism	Cryogenic Environment	Effect of Aqueous Environment, Salt Solution and Sugar Solution	Aging in Gasoline
Static bending	ABS ↑ [44] ASA ↑ [44] PLA ↑ [45]		ABS ↔ [73]				ABS ↓ [87] PMMA ↓ [87] HIPS ↑ [87] TPU ↑ [87]			
Tensile strength		PLA ↑ [62] PLA ↓ [63]	ABS ↔ [73]	PLA ↓ [64] PETG ↓ [64]	PETG ↓ [74] ABS (medical grade) ↔ [74]		PLA ↓ [86]	ABS ↓ [96]		TPE ↓ [100] TPU ↓ [100]
Impact strength		PLA ↔ [63]					ABS ↓ [87] PMMA ↑ [87] HIPS ↑ [87] TPU ↓ [87]			
Rigidity				PLA ↓ [64] PETG ↓ [64]						
Dynamic mechanical analysis	PLA ↑ [45]									
Compression				PLA ↓ [64] PETG ↓ [64]						
Biofilm coverage		PCL ↑ [66] PDMS ↓ [66]								
Degradation changes						PLA ↔ [79] PETG ↔ [79]			PETG ↑ [72] PLA ↓ [72]	

↑ Increase in strength values. ↓ Decrease in strength values. ↔ No impact.

10. Predictions for Further Research

There are also other factors that affect the resistance of 3D-printed elements, which are not described in the paper. Our review may be supplemented with a description of the following issues: the mechanism of damage to the structure of elements produced in 3D technology under the influence of the environment; the effects on the resistance of the parameters of the 3D printer settings; the impact of the modification of the material by adding additional ingredients, e.g., fibers or powders; the influence of the shape of the internal structure of elements resembling natural composites such as bone or wood; and the ways in which post-processing, e.g., annealing or acetone vapor, increases the resistance of components in the work environment [102–113].

All these issues have an impact on the resistance of 3D-printed elements in the work environment. However, it is impossible to discuss them in detail within one work, due to the large scope of each topic. Therefore, we recommend that these topics are considered in future papers.

Author Contributions: Conceptualization, M.G., A.M. and K.S.; methodology, M.G., K.S. and M.S.; validation, M.G., A.M., M.S. and K.S.; investigation, M.G.; data curation, M.G. and K.S.; writing—original draft preparation, M.G.; writing—review and editing, M.G., A.M. and K.S.; visualization, M.G.; supervision, M.G. and K.S. All authors have read and agreed to the published version of the manuscript.

Funding: This research received no external funding.

Institutional Review Board Statement: Not applicable.

Informed Consent Statement: Not applicable.

Data Availability Statement: The data presented in this study are available on request from the corresponding author.

Conflicts of Interest: The authors declare no conflict of interest.

Nomenclature

ABS	Acrylonitrile Butadiene Styrene
AM	Additive Manufacturing
ASA	Acrylonitrile Styrene Acrylate
CT	Computed Microtomography
DMA	Dynamic Mechanical Analysis
FDM	Fused Deposition Modeling
HIPS	High-Impact Polystyrene
PA	Polyamide
PCL	Polycaprolactone
PDMS	Polydimethylsiloxane
PETG	Polyethylene Terephthalate Glycol
PLA	Poly Lactic Acid
PMMA	Poly(methyl methacrylate)
SEM	Scanning Electron Microscopy
Tg	Glass Transition Temperature (°C)
TPE	Thermoplastic Elastomer
TPU	Thermoplastic Polyurethane
UV	Ultraviolet

References

1. Horvath, J. A Brief History of 3D Printing. In *Mastering 3D Printing*; Apress: Berkeley, CA, USA, 2014.
2. Beaman, J.J.; Barlow, J.W.; Bourell, D.L.; Crawford, R.H.; Marcus, H.L.; McAlea, K.P. *Solid Freeform Fabrication: A New Direction in Manufacturing: With Research and Applications in Thermal Laser Processing*; Springer Science & Business Media: Berlin/Heidelberg, Germany, 2013.
3. Hull, C.W. Apparatus for Production of Three-Dimensional Objects by Stereolithography. Patent No. US4575330, 11 March 1986.

4. Deckard, C.R. Apparatus for Producing Parts by Selective Sintering. Patent No. US5597589, 28 January 1997.
5. Wohlers, T.; Gornet, T. *History of Additive Manufacturing*; Wohlers Associates, Inc.: Fort Collins, CO, USA, 2014.
6. Scoot Crump Modeling Apparatus for Three-Dimensional Objectsd. Patent No. US5340433, 23 August 1994.
7. Bourell, D.L.; Beaman, J.; Wohlers, T. *History and Evolution of Additive Manufacturing*; ASM International: Almere, The Netherlands, 2020; Volume 24.
8. Jafferson, J.M.; Debdutta, C.H. A Review on Polymeric Materials in Additive Manufacturing. *Mater. Today Proc.* **2021**, *46*. [CrossRef]
9. Petersmann, S.; Spoerk, M.; Van De Steene, W.; Uecal, M.; Wiener, J.; Pinter, G.; Arbeiter, F. Mechanical properties of polymeric implant materials produced by extrusion-based additive manufacturing. *J. Mech. Behav. Biomed. Mater.* **2020**, *104*, 103611. [CrossRef] [PubMed]
10. Shahrubudin, N.; Lee, T.C.; Ramlan, R. An Overview on 3D Printing Technology: Technological, Materials, and Applications. *Sci. Procedia Manuf.* **2019**, *35*, 1286–1296. [CrossRef]
11. Jain, A.; Bansal, K.K.; Tiwari, A.; Rosling, A.; Rosenholm, J.M. Role of Polymers in 3D Printing Technology for Drug Delivery—An Overview. *Curr. Pharm. Des.* **2018**, *24*, 4979–4990. [CrossRef]
12. Pekkanen, A.M.; Mondschein, R.J.; Williams, C.B.; Long, T.E. 3D Printing Polymers with Supramolecular Functionality for Biological Applications. *Biomacromolecules* **2017**, *18*, 2669–2687. [CrossRef]
13. Daniel, B.; Tillman, A.M. Environmental assessment of additive manufacturing in the automotive industry. *J. Clean. Prod.* **2019**, *226*, 977–987.
14. Daniel, A.; Fentahun, A.K.; Kamal, K. Numerically investigating the AZO/Cu_2O heterojunction solar cell using ZnO/CdS buffer layer. *Optik* **2021**, *228*, 166228.
15. Hofstatter, T.; Pedersen, D.B.; Tosello, G.; Hansen, H.N. State-of-the-art of fiber-reinforced polymers in additive manufacturing technologies. *J. Reinf. Plast. Compos.* **2017**, *36*, 1061–1073. [CrossRef]
16. Lyu, M.Y.; Choi, T.G. Research trends in polymer materials for use in lightweight vehicles. *Int. J. Precis. Eng. Manuf.* **2015**, *16*, 213–220. [CrossRef]
17. Papageorgiou, D.G.; Kinloch, I.A.; Young, R.J. Mechanical properties of graphene and graphene-based nanocomposites. *Prog. Mater. Sci.* **2017**, *90*, 75–127. [CrossRef]
18. MacDiarmid, A.G. Synthetic Metals: A Novel Role for Organic Polymers (Nobel Lecture). *Angew. Chem.* **2001**, *40*, 2581–2590. [CrossRef]
19. Ngo, T.D.; Kashani, A.; Imbalzano, G.; Nguyen, K.T.; Hui, D. Additive manufacturing (3D printing): A review of materials, methods, applications and challenges. *Compos. Part B Eng.* **2018**, *143*, 172–196. [CrossRef]
20. Tofail, S.A.; Koumoulos, E.P.; Bandyopadhyay, A.; Bose, S.; O'Donoghue, L.; Charitidis, C. Additive manufacturing: Scientific and technological challenges, market uptake and opportunities. *Mater. Today* **2018**, *21*, 22–37. [CrossRef]
21. Kristiawan, R.B.; Imaduddin, F.; Ariawan, D.; Zainal, A. A review on the fused deposition modeling (FDM) 3D printing: Filament processing, materials, and printing parameters. *Open Eng.* **2021**, *11*, 639–649. [CrossRef]
22. Spoerk, M.; Savandaiah, C.; Arbeiter, F.; Sapkota, J.; Holzer, C. Optimization of mechanical properties of glass-spheres-filled polypropylene composites for extrusion-based additive manufacturing. *Polym. Compos.* **2019**, *40*, 638–651. [CrossRef]
23. Carneiro, O.S.; Silva, A.F.; Gomes, R. Fused deposition modeling with polypropylene. *Mater. Des.* **2015**, *83*, 768–776. [CrossRef]
24. Dizon, J.R.C.; Espera, A.H.; Chen, Q.; Advincula, R.C. Mechanical characterization of 3D-printed polymers. *Addit. Manuf.* **2018**, *20*, 44–67. [CrossRef]
25. Mitchell, A.; Lafont, U.; Hołyńska, M.; Semprimoschnig, C. Additive manufacturing—A review of 4D printing and future applications. *Addit. Manuf.* **2018**, *24*, 606–626. [CrossRef]
26. Aumnate, C.; Pongwisuthiruchte, A.; Pattananuwat, P.; Potiyaraj, P. Fabrication of ABS/graphene oxide composite filament for fused filament fabrication (FFF)3D printing. *Adv. Mater. Sci. Eng.* **2018**, *2018*, 2830437. [CrossRef]
27. Geng, P.; Zhao, J.; Wu, W.; Ye, W.; Wang, Y.; Wang, S.; Zhang, S. Effects of extrusion speed and printing speed on the 3D printing stability of extruded PEEK filament. *J. Manuf. Process.* **2019**, *37*, 266–273. [CrossRef]
28. Godec, D.; Cano, S.; Holzer, C.; Gonzalez-Gutierrez, J. Optimization of the 3D printing parameters for tensile properties of speci-mens produced by fused filament fabrication of 17-4PH stainless steel. *Materials* **2020**, *202*, 774. [CrossRef]
29. Milosevi, C.M.; Stoof, D.; Pickering, K.L. Characterizing the mechanical properties of fused deposition modelling natural fiber recycled polypropylene composites. *J. Compos. Sci.* **2017**, *1*, 7. [CrossRef]
30. Nguyen, N.A.; Bowland, C.C.; Naskar, A.K. A general method to improve 3D-printability and inter-layer adhesion in lignin-based composites. *Appl. Mater. Today* **2018**, *12*, 138–152. [CrossRef]
31. Huang, J.; Qin, Q.; Jie, W. A Review of Stereolithography: Processes and Systems. *Processes* **2020**, *8*, 1138. [CrossRef]
32. Wang, X.; Jiang, M.; Zhou, Z.; Gou, J.; Hui, D. 3D printing of polymer matrix composites: A review and prospective. *Compos. Part B Eng.* **2017**, *110*, 442–458. [CrossRef]
33. De Beer, M.P.; van der Laan, H.L.; Cole, M.A.; Whelan, R.J.; Burns, M.A.; Scott, T.F. Rapid, continuous additive manufacturing by volumetric polymerization inhibition patterning. *Sci. Adv.* **2019**, *5*, eaau8723. [CrossRef]
34. Van der Laan, H.L.; Burns, M.A.; Scott, T.F. Volumetric Photopolymerization Confinement through Dual-Wavelength Photoinitiation and Photoinhibition. *ACS Macro Lett.* **2019**, *8*, 899–904. [CrossRef] [PubMed]

35. Loterie, D.; Delrot, P.; Moser, C. High-resolution tomographic volumetric additive manufacturing. *Nat. Commun.* **2020**, *11*, 852. [CrossRef] [PubMed]
36. Zhang, J.; Hu, Q.; Wang, S.; Tao, J.; Gou, M. Digital Light Processing Based Three-dimensional Printing for Medical Applications. *Int. J. Bioprinting* **2019**, *6*, 242. [CrossRef]
37. Moin, D.A.; Hassan, B.; Wismeijer, D.H. A Novel Approach for Custom Three-Dimensional Printing of a Zirconia Root Analogue Implant by Digital Light Processing. *Clin. Oral Implant. Res.* **2017**, *28*, 668–670. [CrossRef]
38. Lim, K.S.; Levato, R.; Costa, P.F.; Castilho, M.D.; Alcala-Orozco, C.R.; van Dorenmalen, K.M.A.; Melchels, F.P.W.; Gawlitta, D.; Hooper, G.J.; Malda, J.; et al. Bio-Resin for High Resolution Lithography-Based Biofabrication of Complex Cell-Laden Constructs. *Biofabrication* **2018**, *10*, 034101. [CrossRef] [PubMed]
39. Ziaee, M.; Crane, N.B. *Binder Jetting: A Review of Process, Materials, and Methods*; Elsevier: Amsterdam, The Netherlands, 2019.
40. Shirazi, S.F.S.; Gharehkhani, S.; Mehrali, M.; Yarmand, H.; Metselaar, H.S.C.; Kadri, N.A.; Osman, N.A.A. A review on powder-based additive manufacturing for tissue engineering: Selective laser sintering and inkjet 3D printing. *Sci. Technol. Adv. Mater.* **2015**, *16*, 033502. [CrossRef] [PubMed]
41. Afshar, A.; Mihut, D. Enhancing durability of 3D printed polimer structures by metalization. *J. Mater. Sci. Technol.* **2020**, *53*, 185–191.
42. Le Neel, T.A.; Mognol, P.; Hascoet, J.-Y. A review on additive manufacturing of sand molds by binder jetting and selective laser sintering. *Rapid Prototyp. J.* **2018**, *24*, 1325–1336.
43. Mühler, C.G.T.; Ascheri, M.E.; Nicolaides, D.; Heinrich, J.G.; Günster, J. Slurry-Based Powder Beds for the Selective Laser Sintering of Silicate Ceramics. *J. Ceram. Sci. Technol.* **2015**, *6*, 113–118.
44. Reza Khosravani, M.; Zolfagharian, A.; Jennings, M.; Reinicke, T. Structural performance of 3D-printed composites under various loads and environmental conditions. *Sci. Polym. Test.* **2020**, *91*, 106770. [CrossRef]
45. Anket, J.; Ikshita, C.; Ishika, W.; Ankush, R.; Irfan, U. 3D printing—Review of processes, materials and applications in industry 4.0. *Sci. Sustain. Oper. Comput.* **2022**, *3*, 33–42.
46. Niranjan, Y.C.; Krishnapila, S.; Velmurgan, R.; Sung, K.H. Effect of Annealing Time and Temperature on Dynamic Mechanical Properties of FDM Printed PLA. In *Composite Materials for Extreme Loading*; Springer: Singapore, 2021; pp. 143–160.
47. Bhuvanesh, M.; Kumar, T.; Sathiya, P. Methods and materials for additive manufacturing: A critical review on advancements and challenges. *Sci. Thin Walled Struct.* **2021**, *159*, 107228. [CrossRef]
48. Hart, K.R.; Dunn, R.M.; Sietins, J.M.; Mock, C.M.H.; Mackay, M.E.; Wetzel, E.D. Increased fracture toughness of additively manufactured amorphousthermoplastics via thermal annealing. *Polymers* **2018**, *144*, 192–204. [CrossRef]
49. Joseph, P.; Joseph, K.; Thomas, S.; Pillai, C.; Prasad, V.; Groeninckx, G.; Sarkissova, M. The thermal and crystallisation studies of short sisal fibre reinforced polypropylene composites. *Compos. Part A* **2003**, *34*, 253–266.
50. Vidhate, S.; Ogunsona, E.; Chung, J.; D'Souza, N.A. Rheological, mechanical and thermal properties of PVDF/carbon nanofiber (CNF) composite. In Proceedings of the Regional Technical Conference—Society of Plastics Engineers, Milwaukee, WI, USA, 4–8 May 2008; pp. 74–77.
51. Soltysiak, S.; Selent, M.; Roth, S.; Abendroth, M.; Hoffmann, M.; Biermann, H.; Kuna, M. High-temperature small punch test for mechanical characterizationof a nickel-base super alloy. *Mater. Sci. Eng. A* **2014**, *613*, 259–263. [CrossRef]
52. Koemmling, A.; Jaunich, M.; Wolff, D. Revealing effects of chain scission during ageing of EPDM rubber using relaxation and recovery experiment. *Polym. Test.* **2016**, *56*, 261–268. [CrossRef]
53. Pourmand, P.; Hedenqvist, M.S.; Furó, I.; Gedde, U.W. Deterioration of highly filled EPDM rubber by thermal ageing in air: Kinetics and non-destructive monitoring. *Polym. Test.* **2017**, *64*, 267–276. [CrossRef]
54. Ye, X.; Chen, Z.; Ai, S.; Hou, B.; Zhang, J.; Zhou, Q.; Wang, F.; Liu, H.; Cui, S. Microstructure characterization and thermal performance of reticulated SiC skeleton reinforced silica aerogel composites. *Compos. Part B* **2019**, *177*, 107409. [CrossRef]
55. Khosravani, M.R.; Weinberg, K. Characterization of sandwich 3T-joints under different aging conditions. *Compos. Struct.* **2018**, *197*, 80–88. [CrossRef]
56. Davoudinejad, A.; Khosravani, M.R.; Pedersen, D.B.; Tosello, G. Influence of thermal aging on the fracture and lifetime of additively manufactured ng properties and polymer/fuel interactions of polyamide12 exposed to (bio)diesel at high temperature. *NPJ Mater. Degrad.* **2019**, *115*, 104694.
57. Wach, R.A.; Wolszczak, P.; Adamus-Wlodarczyk, A. Enhancement of mechanical properties of FDM-PLA parts via thermal annealing. *Macromol. Mater. Eng.* **2018**, *303*, 1800169. [CrossRef]
58. Akhoundi, B.; Nabipour, M.; Hajami, F.; Shakoori, D. An experimental study of nozzle temperature and heat treatment (annealing) effects on mechanical properties of high-temperature polylactic acid in fused deposition modeling. *Polym. Eng. Sci.* **2020**, *60*, 979–987. [CrossRef]
59. Park, S.D.; Todo, M.; Arakawa, K. Effect of annealing on fracture mechanism of biodegradable poly (lactic acid). *Key Eng. Mater.* **2004**, *261–263*, 105–110. [CrossRef]
60. Pérez-Fonseca, A.A.; Robledo-Ortíz, J.R.; González-Núñez, R.; Rodrigue, D. Effect of thermal annealing on the mechanical and thermal properties of polylactic acid–cellulosic fiber biocomposites. *J. Appl. Polym. Sci.* **2016**, *133*, 1–10. [CrossRef]
61. Rangisetty, S.; Peel, L.D. *The Effect of Infill Patterns and Annealing on Mechanical Properties of Additively Manufactured Thermoplastic Composites*; ASME International: New York, NY, USA, 2017; Volume 1, pp. 1–12.

62. Krishna Upadhyay, R.; Kumar Mishra, A.; Kumar, A. Mechanical Degradation of 3D Printed PLA in Simulated Marina Environment. *Surf. Interfaces* **2020**, *21*, 100778. [CrossRef]
63. Ambrus, S.; Soporan, R.A.; Kazamer, N.; Pascal, D.T.; Muntean, R.; Dume, A.I.; Mărginean, G.M.; Serban, V.A. Characterization and mechanical properties of fused deposited PLA material. *Mater. Today Proceeding* **2021**, *45*, 4356–4363. [CrossRef]
64. Catalin, G.A.; Zapciu, A.; Baciu, F.; Vasile, M.I.; Nicoara, A.I. Accelerated Aging Effect on Mechanical Properties of Common 3D-Printing Polymers. *Polymers* **2021**, *13*, 4132.
65. Fotopoulou, K.N.; Karapanagioti, H.K. Degradation of Various Plastics in the 603 Environment. Hazardous Chemicals Associated with Plastics in the Marine Environment. In *604 The Handbook of Environmental Chemistry*; Springer: Cham, Switzerland, 2017; Volume 78, pp. 71–92.
66. Ryley, M.; Carve, M.; Piola, R.; Scardino, A.J.; Shimeta, J. Comparison of biofouling on 3D-printing materials in the marine environment. *Int. Biodeterior. Biodegrad.* **2021**, *164*, 105293. [CrossRef]
67. Gebhardt, A.; Hötter, J.-S. *Additive Manufacturing: 3D Printing for Prototyping and Manufacturing*; Hanser Publications: Cincinnati, OH, USA, 2016.
68. Arrieta, C.; Dong, Y.; Lan, A.; Vu-Khanh, T. Outdoor weathering of polyamide and 576 polyester ropes used in fall arrest equipment. *J. Appl. Polym. Sci.* **2013**, *130*, 3058–3065. [CrossRef]
69. Andrady, A. Microplastics in the marine environment. *Mar. Pollut. Bull.* **2011**, *62*, 1596–1605. [CrossRef]
70. Adhikari, D.; Mukai, M.; Kubota, K.; Kai, T.; Kaneko, N.; Araki, K.S.; Kubo, M. Degradation of bioplastics in soil and their degradation effects on environmental 566 microorganisms. *J. Agric. Chem. Environ.* **2016**, *5*, 23–34.
71. Bannister, J.; Sievers, M.; Bush, F.; Bloecher, N. Biofouling in marine aquaculture: A review of recent research and developments. *Biofouling* **2019**, *35*, 631–648. [CrossRef]
72. Nieto, D.M.; Alonso-García, M.; Pardo-Vicente, M.A. Product Design by Additive Manufacturing for Water Environments: Study of Degradation and Absorption Behaviorof PLA and PETG. *Polymers* **2021**, *13*, 1036. [CrossRef]
73. Popescu, D.; Baciu, F.; Vlăsceanu, D.; Mihai Cotruţ, C.; Marinescu, R. Effects of multiple sterilizations and natural aging on the mechanical behavior of 3D-printed ABS. *Mech. Mater.* **2020**, *148*, 103423. [CrossRef]
74. Grzelak, K.; Łaszcz, J.; Polkowski, J.; Mastalski, P.; Kluczyński, J.; Łuszczek, J.; Torzewski, J.; Szachogłuchowicz, I.; Szymaniuk, R. Additive Manufacturing of Plastics Used for Protection against COVID-19—The Influence of Chemical Disinfection by Alcohol on the Properties of ABS and PETG Polymers. *Materials* **2021**, *14*, 4823. [CrossRef] [PubMed]
75. Oth, O.; Dauchot, C.; Orellana, M.; Glineur, R. How to Sterilize 3D Printed Objects for Surgical Use? An Evaluation of the Volumetric Deformation of 3D-Printed Genioplasty Guide in PLA and PETG after Sterilization by Low-Temperature Hydrogen Peroxide Gas Plasma. *Open Dent. J.* **2019**, *13*, 410–417. [CrossRef]
76. *ISO 4892-3*; Plastics—Methods of Exposure to Laboratory Light Sources—Part 3: Fluorescent UV Lamps. 2016. Available online: https://www.iso.org/standard/67793.html (accessed on 30 October 2021).
77. Eldwakhly, E.; Saadaldin, S.; Aldegheishem, A.; Mostafa, M.S.; Soliman, M. Antimicrobial capacity and surface alterations using photodynamic therapy and light activated disinfection on polymer-infiltrated ceramic material contaminated with periodontal bacteria. *Pharmaceuticals* **2020**, *13*, 350. [CrossRef] [PubMed]
78. Basso, M.F.M.; Giampaolo, E.T.; Vergani, C.E.; Machado, A.L.; Pavarina, A.C.; Ribeiro, R.C. Influence of Microwave Disinfection on the Dimensional Stability of Denture Reline Polymers. *J. Prosthodont.* **2010**, *19*, 364–368. [CrossRef] [PubMed]
79. Boursier, J.-F.; Fournet, A.; Bassanino, J.; Manassero, M.; Bedu, A.-S.; Leperlier, D. Reproducibility, accuracy and effect of autoclave sterilization on a thermoplastic three-dimensional model printed by a desktop fused deposition modelling three-dimensional printer. *Vet. Comp. Orthop. Traumatol.* **2018**, *31*, 422–430. [PubMed]
80. Jacobs, P.; Kowatsch, R. Sterrad Sterilization System: A new technology for instrument sterilization. *Endosc. Surg. Allied Technol.* **1993**, *1*, 57–58.
81. Kyi, M.S.; Holton, J.; Ridgway, G.L. Assessment of the efficacy of a low temperature hydrogen peroxide gas plasma sterilization system. *J. Hosp. Infect.* **1995**, *31*, 275–284. [CrossRef]
82. Bagsik, A.; Schöppner, V.; Klemp, E. *Long-Term Aging Effects on Fused Deposition Modeling Parts manufactured with ULTEM*9085*; International Solid Freeform Fabrication Symposium: Downtown Austin, TX, USA, 2012; pp. 629–640.
83. George, M.; Aroom, K.R.; Hawes, H.G.; Gill, B.S.; Love, J. 3D printed surgical instruments: The design and fabrication process. *World J. Surg.* **2017**, *41*, 314–319. [CrossRef]
84. Goodridge, R.D.; Hague, R.J.M.; Tuck, C.J. Effect of long-term aging on the tensile properties of a polyamide 12 laser sintering material. *Polym. Test.* **2010**, *29*, 483–493. [CrossRef]
85. Madej, J.; Śliwka, M.; Trzcionka, W. Influence of UV Radiation on the Mechanical Properties of Specimens Printed with the Use of the FDM Technique at Different Density Levels. *Fibres Text. East. Eur.* **2017**, *27*, 78–83. [CrossRef]
86. Moetazedian, A.; Gleadall, A.; Han, X.; Silberschmidt, V.V. Effect of environment on mechanical properties of 3D printed polylactide for biomedical applications. *Sci. Mater.* **2020**, *102*, 103510. [CrossRef] [PubMed]
87. Pinho, A.C.; Piedade, A.P. Sandwich Multi-Material 3D-Printed Polymers: Influence of Aging on the Impact and Flexure Resistances. *Polymers* **2021**, *13*, 4030. [CrossRef] [PubMed]
88. Brenken, B.; Barocio, E.; Favaloro, A.J.; Pipes, R.B. Simulation of semi-crystalline composites in the extrusion deposition additive manufacturing process. In Proceedings of the American Society for Composites, West Lafayette, IN, USA, 23–25 October 2017; pp. 90–102.

89. Daniel, I.M.; Abot, J.L. Fabrication, testing and analysis of composite sandwich beams. *Compos. Sci. Technol.* **2000**, *60*, 2455–2463. [CrossRef]
90. Andric, T.; Taylor, B.L.; Whittington, A.R. Fabrication and characterization of three-dimensional electrospun scaffolds for bone tissue engineering. *Regen. Eng. Transl. Med.* **2015**, *1*, 32–41. [CrossRef]
91. Bosio, V.E.; Brown, J.; Rodriguez, M.J.; Kaplan, D.L. Biodegradable porous silk microtubes for tissue vascularization. *J. Mater. Chem. B* **2017**, *5*, 1227–1235. [CrossRef] [PubMed]
92. Sonseca, Á.; Menes, O.; Giménez, E. A comparative study of the mechanical, shape-memory, and degradation properties of poly(lactic acid) nanofiber and cellulose nanocrystal reinforced poly(mannitol sebacate) nanocomposites. *RSC Adv.* **2017**, *7*, 21869–21882. [CrossRef]
93. Kim, E.; Shin, Y.J.; Ahn, S.H. The effects of moisture and temperature on the mechanical properties of additive manufacturing components: Fused deposition modeling. *Rapid Prototyp. J.* **2016**, *22*, 887–894. [CrossRef]
94. Vyavahare, O.; Ng, D.; Hsu, S.L. Analysis of structural rearrangements of poly(lactic acid) in the presence of water. *J. Phys. Chem. B* **2014**, *118*, 4185–4193. [CrossRef]
95. Wang, P.J.; Nezami, F.R.; Gorji, M.B.; Berti, F.; Petrini, L.; Wierzbicki, T.; Migliavacca, F.; Edelman, E.R. Effect of working environment and procedural strategies on mechanical performance of bioresorbable vascular scaffolds. *Acta Biomater.* **2018**, *82*, 34–43. [CrossRef]
96. Saenz, F.; Otarola, C.; Valladares, K. Influence of 3D printing settings on mechanical properties of ABS at room temperature and 77 K. *Sci. Addit. Manuf.* **2021**, *39*, 101841. [CrossRef]
97. Drevlak, M.; Brochard, F.; Helander, P.; Kisslinger, J.; Mikhailov, M.; Nührenberg, C.; Turkin, Y.; Nührenberg, J. A quasi-toroidally symmetric stellarator. *Contrib. Plasma Phys.* **2013**, *53*, 459–468. [CrossRef]
98. Weiss, K.-P.; Bagrets, N.; Lange, C.; Goldacker, W.; Wohlgemuth, J. Thermaland mechanical properties of selected 3D printed thermoplastics in the cryogenictemperature regime. *IOP Conf. Ser. Mater. Sci. Eng.* **2015**, *102*, 12–22. [CrossRef]
99. Bartolom, E.E.; Bozzo, B.; Sevilla, P.; Martínez-Pasarell, O.; Puig, T.; Granados, X. ABS 3D printed solutions for cryogenic applications. *Cryogenics* **2017**, *82*, 30–37. [CrossRef]
100. Paz, E.; Jiménez, M.; Romero, L.; del Mar Espinosa, M.; Domínguez, M. Characterization of the resistance to abrasive chemical agents of test specimens of thermoplastic elastomeric polyurethane composite materials produced by additive manufacturing. *Appl. Polym. Sci.* **2020**, *138*, 50791. [CrossRef]
101. *ISO 175*; Plastics—Methods of Test for the Determination of the Effects of Immersion in Liquid Chemicals. ISO: Geneva, Switzerland, 2010.
102. Zaghloul, M.Y.; Zaghloul, M.M.Y.; Zaghloul, M.M.Y. Influence of Stress Level and Fibre Volume Fraction on Fatigue Performance of Glass Fibre-Reinforced Polyester Composites. *Polymers* **2022**, *14*, 2662. [CrossRef]
103. Zaghloul, M.M.Y.; Mohamed, Y.S.; El-Gamal, H. Fatigue and tensile behaviors of fiber-reinforced thermosetting composites embedded with nanoparticles. *J. Compos. Mater.* **2019**, *53*, 709–718. [CrossRef]
104. Głowacki, M.; Mazurkiewicz, A.; Slomion, M. Evaluation of the effectiveness of methods of methods for improving the mechanical properties of plastic products manufactured in the FDM technology—A mini review. In Proceedings of the 27/28th International Conference Engineering Mechanics, Milovy, Czech Republic, 9–12 May 2022.
105. Kocak, H. Surface Modification of a Model Part Produced with 3D Printing from PLA Material by Means of Composite Coating. *J. Mater. Eng. Perform.* **2020**, *30*, 3903–3910. [CrossRef]
106. Zaghloul, M.M.Y.; Zaghloul, M.Y.M.; Zaghloul, M.M.Y. Experimental and modeling of mechanical-electrical behavior of polypropylene composites filled with graphite and MWCNT fillers. *Polym. Test.* **2017**, *63*, 467–474. [CrossRef]
107. Zaghloul, M.Y.M. Mechanical properties of linear low-density polyethylene fire-retarded with melamine polyphosphate. *J. Appl. Polym. Sci.* **2018**, *46*, 46770. [CrossRef]
108. Haihong, J.; Kamdem, P. Development of poly(vinyl chloride)/wood composites. A literature review. *J. Vinyl Addit. Technol.* **2004**, *10*, 59–69.
109. Topolinski, T.; Cichanski, A.; Mazurkiewicz, A.; Nowicki, K. Fatigue Energy Dissipation in Trabecular Bone Samples with Step-wise-Increasing Amplitude Loading. *Mater. Test.* **2011**, *53*, 344–350. [CrossRef]
110. Zaghloul, M.M.Y.; Zaghloul, M.Y.M.; Zaghloul, M.M.Y. Developments in polyester composite materials–An in-depth review on natural fibres and nano fillers. *Compos. Struct.* **2021**, *278*, 114698. [CrossRef]
111. Dányádi, L.; Móczó, J.; Pukánszky, B. Effect of various surface modifications of wood flour on the properties of PP/wood composites. *Compos. Part A Appl. Sci. Manuf.* **2010**, *41*, 199–206. [CrossRef]
112. Topoliński, T.; Mazurkiewicz, A. Relationship between the mineral content of human trabecular bone and selected parameters obtained from its fatigue test with stepwise increasing amplitude. *Acta Bioeng. Biomech.* **2017**, *19*, 19–26.
113. Zaghloul, M.M.Y.; Zaghloul, M.Y.M.; Zaghloul, M.M.Y. Influence of flame retardant magnesium hydroxide on the mechanical properties of high density polyethylene composites. *J. Reinf. Plast. Compos.* **2017**, *36*, 1802–1816. [CrossRef]

materials

MDPI

Article

New Biocomposite Electrospun Fiber/Alginate Hydrogel for Probiotic Bacteria Immobilization

Adam Grzywaczyk [1], Agata Zdarta [1], Katarzyna Jankowska [1], Andrzej Biadasz [2], Jakub Zdarta [1], Teofil Jesionowski [1], Ewa Kaczorek [1] and Wojciech Smułek [1,*]

1 Institute of Chemical Technology and Engineering, Faculty of Chemical Technology, Poznan University of Technology, Berdychowo 4, 60-965 Poznan, Poland; adam.grzywaczyk@doctorate.put.poznan.pl (A.G.); agata.zdarta@put.poznan.pl (A.Z.); katarzyna.jankowska@doctorate.put.poznan.pl (K.J.); jakub.zdarta@put.poznan.pl (J.Z.); teofil.jesionowski@put.poznan.pl (T.J.); ewa.kaczorek@put.poznan.pl (E.K.)

2 Institute of Physics, Faculty of Materials Engineering and Technical Physics, Poznan University of Technology, Piotrowo 3, 60-965 Poznan, Poland; andrzej.biadasz@put.poznan.pl

* Correspondence: wojciech.smulek@put.poznan.pl

Abstract: Biotechnological use of probiotic microorganisms involves providing them with appropriate conditions for growth, but also protection against environmental changes caused by an exchange of the medium, isolation of metabolites, etc. Therefore, the research on effective immobilization of probiotic microorganisms should be focused in this direction. The present study aimed to evaluate the effectiveness of an innovative hybrid immobilization system based on electrospun nanofibers and alginate hydrogel. The analyses carried out included the study of properties of the initial components, the evaluation of the degree and durability of cell immobilization in the final material, and their survival under stress conditions. Effective binding of microorganisms to the hydrogel and nanofibers was confirmed, and the collected results proved that the proposed biocomposite is an efficient method of cell protection. In addition, it was shown that immobilization on electrospun nanofibers leads to the preservation of the highest cell activity and the least cell growth restriction as compared to free or lyophilized cells only. The completed research opens new perspectives for the effective immobilization of microorganisms of significant economic importance.

Keywords: biocomposite; contact angle; electrospun nanofibers; immobilization; *Lactobacillus*; alginate hydrogel

Citation: Grzywaczyk, A.; Zdarta, A.; Jankowska, K.; Biadasz, A.; Zdarta, J.; Jesionowski, T.; et al. New Biocomposite Electrospun Fiber/Alginate Hydrogel for Probiotic Bacteria Immobilization. *Materials* **2021**, *14*, 3861. https://doi.org/10.3390/ma14143861

Academic Editor: Maurice N. Collins

Received: 7 June 2021
Accepted: 8 July 2021
Published: 10 July 2021

1. Introduction

To maintain optimal microorganism cultivation, growth media parameters such as pH, temperature, oxygen level, and nutrition agents must be ensured, depending on the species of microbe (Thomas, 2015 [1]). This causes the use of many different specific and non-specific growth media, such as general-purpose nutrient agar (Smułek et al., 2020 [2]) and enrichment broth (Rahman et al., 2020 [3]), specific for certain fungi Sabouraud's agar (Cultrera et al., 2021 [4]), and Middlebrook and Löwenstein-Jensen media used for *Mycobacterium* spp. culture (Schön et al., 2020 [5]; Xie et al., 2021 [6]) (Mikolajczyk et al., 2016 [7]).

No matter what type of growth media is required, it is crucial is to maintain sterility when working with microorganisms, and prevent microbes from hazardous environmental conditions such as temperature fluctuation, UV radiation, pH changes, organic solvents, and salts. One possible way to do this is the immobilization of microorganisms—chemically bonding or adsorbing microorganisms onto the surface of the carrier, and entrapment in a form of capsule or gel (Drauz and Waldmann, 1995 [8]). Requirements for the appropriate carrier, such as non-toxicity, optimal mechanical properties, chemical resistance, and cost-effectiveness make the entrapment in calcium or sodium alginate hydrogels the most-used technique for cell immobilization (Zhu, 2007 [9]). There are two main types of this process: gel entrapment and encapsulation. Gel entrapment differs from encapsulation in the form of beads. During the gel entrapment, cells are immobilized in a gel matrix, whereas the encapsulation process involves the preparation of porous barrier and entrapment of microorganism in liquid core of such bead (Willaert and Baron, 1996 [10]). Probiotics are the group of microorganism which are most commonly stabilized by the process of encapsulation since their survival in harsh conditions is crucial to trigger the beneficial effect of those microorganisms in human body (Tripathi and Giri, 2014 [11]). Moreover, capsuled microbes have been used in the medical application (Rocha et al., 2021 [12]), the food industry (Penhasi, Reuveni and Baluashvili, 2021 [13]), delivery systems (Cook et al., 2012 [14]), or even in larval rearing system (Ghoname et al., 2020 [15]).

When presenting the process of microbial immobilization, the adsorption process on nanofibers should be mentioned. Nanofibers are one-dimensional structures whose cross-sectional diameter is in the nanometric range (from 1 to 100 nm). Due to their structure and morphology, there can be distinguished porous, ribbon, branched, or hollow nanofibers, and each of them is created under specific conditions of their production process (Stanishevsky, Wetuski, and Yockell-Lelièvre, 2016 [16]; Li et al., 2017 [17]; Ali, Ain, and HuanHe, 2020 [18]). There are several approaches of nanofibers production such as phase separation, consisting of joint gelation of the polymer and solvent (Ma and Zhang, 1999 [19]). Another technique is molecular self-assembly, which combines individual molecules based on the action of intermolecular forces (Xu, Samways, and Dong, 2017 [20]). During template synthesis polymer is pressed into a special corundum template, which allows obtainment of nanofibers of a defined shape (Wang et al., 2010 [21]). The technology, which has been intensively developed in recent years, known as electrospinning, deserves special attention. This method uses an electrostatic field to obtain fibers with a diameter of a few nanometers. Compared to other methods, electrospinning is cheaper and has a wide range of materials from which nanostructures can be spun. Nanofibers prepared by the electrospinning method have been widely used in many different applications, such as tissue engineering, cosmetics industry, filtration processes, and wastewater treatment (Huang et al., 2003 [22]; Bhardwaj and Kundu, 2010 [23]), as well as microbes adsorption process for different usage. For example, Bao et al. used polyethersulfone nanofibers for simultaneous removal of dyes and bacteria (Bao et al., 2021 [24]), and Jayani et al. used bacterial cellulose nanofibers prepared by the electrospinning method as a carrier material for *Lactobacillus acidophilus* 016 (Jayani et al., 2020 [25]). However, so far no report could be found for the usage of nanofibers—alginate hydrogel layered structure, as support for bacteria immobilization.

In this study, the layered structure of alginate hydrogel beads with immobilized bacteria cells and polystyrene nanofibers was prepared. The alginate hydrogel were prepared with gravitational dropping 2% sodium alginate solution from syringe into 1% calcium chloride solution and then sandwiched with two PS mats via electrospinning process. This layer-by-layer immobilization process was used for improving the stability, chemical, and thermal resistance, as well as reduce the leaching of entrapped bacteria. Moreover, the viability of probiotic bacteria was determined, as well as UV resistance.

2. Materials and Methods

2.1. Bacteria Cultivation and Immobilization on Electrospun Material

In the research the bacteria strain *Lactobacillus plantarum* (PCM 2675, Polish Collection of Microorganisms, Wrocław, Poland), which is typically representative of probiotic bacteria. It has been extensively characterized in the literature and has been used in biotechnology. Hence, finding possibilities for its effective immobilization and protection will be very important for producers and also, indirectly, for consumers. The first step in the preparation of the bacterial culture was the establishment of the inoculum of the strain under study. For this purpose, the biomass from Petri dish was transferred from the solid medium to 45 mL of MRS broth (BTL Poland, Łódź, Poland) using a microbiological loop and incubated for 24 h at 30 °C with shaking (120 rpm). Then, all the bacterial inoculum and

1.5 mL of Tween 80 were added to the 1.5 l sterile MRS broth previously prepared in the bioreactor. The bioreactor (Biostat B plus, Sartorius, Germany) operating parameters, maintained throughout the culture, were set as follows: 25 °C, stirring using a Rushton turbine type stirrer 150 rpm, culture running time 6 days. Afterward, the bacterial culture was transferred to bottles and centrifuged (20 min, 4 °C, 4500 rpm). The biomass was resuspended in 100 mL of PBS buffer to the optical density OD_{600} around 2.0, and used for the immobilization procedure. Freeze-dried bacteria and sodium alginate (3%) immobilized bacteria were used as a comparison in the experiments. For freeze-drying, 2 mL of bacterial suspension was placed in a lyophilizer (Alpha 1–2 LD plus, Christ, Germany) for 96 h, −30 °C, 0.36 mbar. For bacteria immobilization in sodium alginate (SA), 2 mL of bacterial suspension was mixed with equal volume of 3% SA. The beads were formed by dropping the mixture to 200 mM $CaCl_2$ solution.

2.2. Production of Electrospun Fibers with Encapsulated Bacteria

Polystyrene (PS, average molecular weight 192,000 g mol^{-1}, Sigma-Aldrich, Warsaw, Poland) was dissolved in *N,N*-dimethylformamide (DMF, Sigma-Aldrich, Warsaw, Poland) to obtain a 32% (w/v) solution and next was mixed for 24 h. After that 500 μL of Pluronic®® P-123 (Sigma-Aldrich, Warsaw, Poland) was added to increase the hydrophilicity of PS electrospun fibers and mixed for another 2 h. At the same time, sodium alginate was dissolved in water (24 h) to obtain a 3% of SA solution. Subsequently, 2 mL of bacteria solution (OD600 ≈ 2.0) was added to the 2 mL of SA solution thoroughly. The prepared PS/DMF and SA/bacteria solutions were placed separately into syringes with needle sizes of 20 G and 22 G, respectively. The production of electrospun fibers from PS layered with bacteria encapsulated into SA was carried out using electrospinning and electrospraying techniques, respectively. Alternately, electrospun polystyrene (PS) fibers and SA/bacteria electrospray were layered to obtain "sandwich-structure" material consists of three layers of PS fibers and two layers of encapsulated bacteria into SA, in three repetitions (Figure 1). Both outer layers of the material were made from polystyrene.

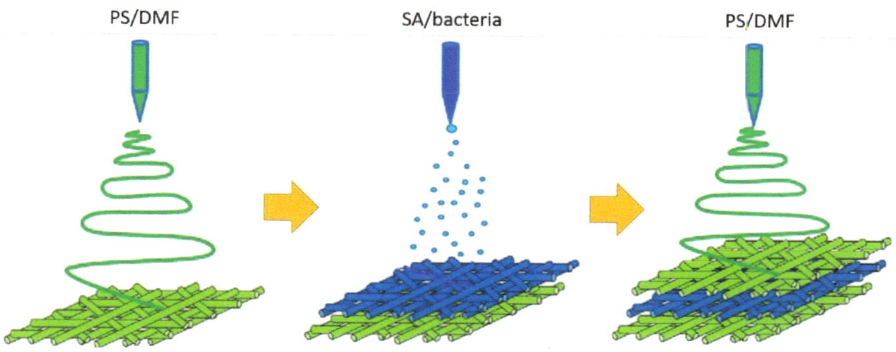

Figure 1. Biocomposite preparation—a workflow scheme.

The electrospinning process of PS was carried out under the following conditions: applied voltage 14 kV, feed rate 1 mL h^{-1}, the distance between nozzle and collector 15 cm, time of process 2 min. The electrospraying conditions were: applied voltage 25 kV, feed rate 0.5 mL h^{-1}, the distance between nozzle and collector 15 cm, time of process 10 min. The fibers with bacteria encapsulated were collected on the aluminum foil, which covered the steel collector and dried for 24 h in a vacuum drier at 23 °C and 50 mbar (Memmert, Schwabach, Germany).

2.3. Material Properties Analysis

Confocal laser scanning microscopy (CLSM) photographs of the PS electrospun material with or without bacteria were obtained with the LSM710 microscope (Zeiss, Jena, Germany) equipped with the argon laser (laser operated at 453 nm for reflected light, and 488 nm for fluorescence mode with the fluorescence observed in the range 510–797 nm).

A scanning electron microscope (SEM) (Hitachi S-3400N, Tokyo, Japan) was applied to observe the surface morphology of the PS/SA/bacteria composite. The average diameter of the PS electrospun fibers was calculated using ImageJ National Institute of Health software (1.53, Bethesda, MD, USA) from 100 points randomly selected from the CLSM images.FT-IR analyses were performed on a Vertex V70 FT-IR spectrometer (Bruker Optik GmbH, Leipzig, Germany) further equipped with a Platinum-ATR-unit (Bruker Optik GmbH, Leipzig, Germany), The material was placed onto the ATR crystal and scanned. To analyze recorded IR spectra the OPUS (7.2, Bruker Optik GmbH, Leipzig, Germany) software was used.

The contact angle was calculated based on measurements made of contact angles of water, LB broth, and milk drops on the surfaces of the materials with the accuracy ±0.01 mN/m using a Drop Shape Analysis System DSA100E (KRÜSS GmbH, Hamburg, Germany).

2.4. Evaluation of Bacterial Metabolic Activity

Immobilized probiotics were subjected to high temperature and UV light tests (UV-30A model UV lamp, Esco, Singapore, Singapore). For this purpose, from each material three 3 cm × 3 cm squares were cut and placed in the incubator at 50 °C for 24 h or under UV light (λ = 253 nm; distance to the light 45 cm; exposition time 30 min). Subsequently, materials were placed in the cell culture plate and flooded with 5 mL of sterile MRS broth. Samples were incubated for 24 h, 30 °C, 120 rpm and subjected to microbial activity analysis: MTT test and alamarBlue test. The control samples were material squares of the same dimensions without temperature or UV light treatment. The results were compared with the lyophilized bacteria and alginate (3%) encapsulated bacteria.

In the 96-well microplate, 200 µL of each tested bacterial suspension and 20 µL of MTT solution (5 mg/mL, Sigma-Aldrich, Warsaw, Poland) were mixed to measure metabolic activity. The plate was incubated for 24 h at 30 °C until formazan crystals formed. The biomass was then centrifuged and 200 µL of isopropanol was added to each well. The plates were left on a shaker with gentle mixing for about 15 min until the formazan crystals were completely dissolved, then briefly centrifuged, and the clear supernatant was transferred to a clean measuring plate and the absorbance was measured at 600 nm.

Analysis of cell metabolic activity using the alamarBlue reagent (Biorad, Warsaw, Poland) involved applying 200 µL of each bacterial culture to a 96-well plate at 3 dilutions with phosphate buffer (1:1; 1:2, 1:4), then 20 µL of alamarBlue reagent was added to each well and incubated at 25 °C with shaking, taking spectrophotometric measurements every hour at 570 nm and 600 nm for 8 h (Multiskan Sky, Thermo Fisher Scientific, Warsaw, Poland). The results were calculated as a percentage difference in the reduction of the reagent in comparison to control samples, according to Equation (1).

$$M = \frac{(O2 * A1) - (O1 * A2)}{(O2 * P1) - (O1 * P2)} * 100 \quad (1)$$

where:

$O1$ = molar extinction coefficient of oxidized alamarBlue at 570 nm (80,586 M^{-1} cm^{-1})
$O2$ = molar extinction coefficient of oxidized alamarBlue at 600 nm (117,216 M^{-1} cm^{-1})
$A1$ = test wells absorbance at 570 nm
$A2$ = test wells absorbance at 600 nm
$P1$ = positive growth control well absorbance at 570 nm
$P2$ = positive growth control well absorbance at 600 nm

Two hundred microliters of distilled water supplemented with alamarBlue or MTT reagent were used as blanks. For each system tested, three replicates of the experiment were performed.

3. Results and Discussion

3.1. Characteristic of the Material

Microscopy

Cross-section morphology of electrospun mats and PS nanofibers with alginate immobilized bacteria (PS/Alg) were characterized based on the CLSM photographs (LSM710, Zeiss, Jena, Germany) in normal and fluorescence mode (Figure 2). Presented results confirm successful immobilization of the probiotic bacteria on the nanofibers. Untreated PS nanofibers display a smooth, uniform surface, with no fluorescence. The fibers morphology became beaded after the immobilization process, comparing it to the pure mat (Figure 2a,c). The beads of oblong shape are distributed along the fibers, with some bigger cluster spots. Moreover, a strong fluorescence of these structures confirms that they are alive probiotic cells. A similar phenomenon has been reported by a previous studies (Mamvura et al., 2011 [26]; Yu et al., 2020 [27]).

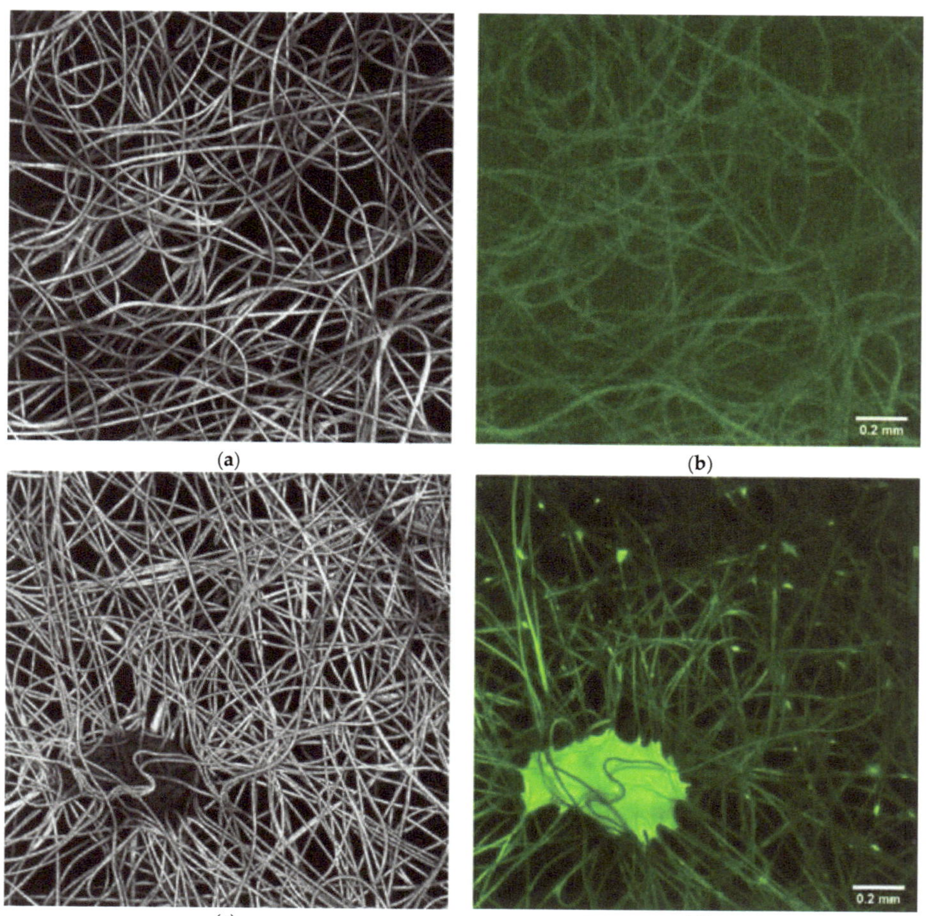

Figure 2. Cross-section CLSM pictures in normal (**a**,**c**) and fluorescence (**b**,**d**) mode of pure PS material (**a**,**b**) and PS material with alginate immobilized bacteria (**c**,**d**).

Furthermore, the electrospun PS nanofibers and PS/Alg material were characterized based on the SEM images (S-3400N, Hitachi, Tokyo, Japan) after gold coating (Figure 3). Moreover, the SEM photographs were used for the calculation of the average diameters of electrospun fibers. Presented photos of the top view confirmed that the electrospinning process was successfully used for nanofibers production. Fibers did not possess any beads or entanglements which proves that and process conditions were chosen correctly. Figure 3b shows that PS/Alg material possesses intercalated alginate hydrogel particles (visible in the background, behind PS fibers). Nanofibers' diameter is constant through the whole membrane and equals ~6 nm for both PS and PS/Alg materials. The same conclusions were drawn by Deng et al., who used electrospun polystyrene nanofibers as a mat for filtering atmospheric air from the heavy metal trace. They used the PS/DMF/THF solution for the production process, with the following electrospinning settings: voltage 20 kV, dosing speed (feeding rate) 1 mL h^{-1}, and needle-collector distance 20 cm. The diameter of obtained nanofibers was 407.6 ± 118.7 nm (Deng et al., 2020 [28]).

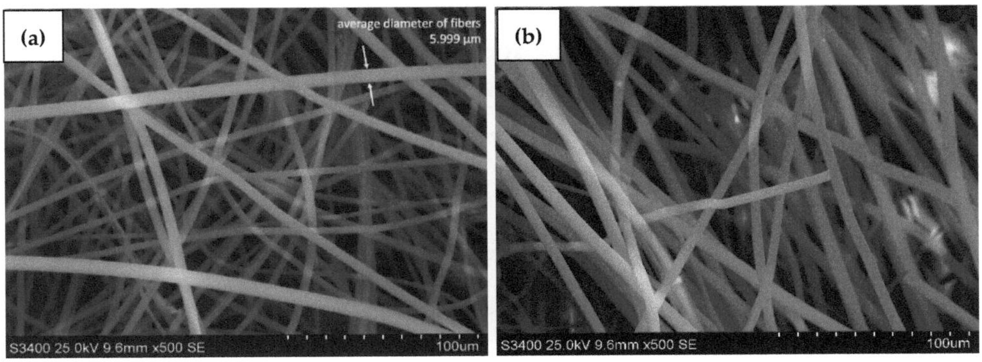

Figure 3. Top-view SEM pictures of PS nanofibers (**a**) and PS nanofibers with alginate immobilized bacteria (**b**) with a magnitude of 500.

To further investigate the tested materials, infrared spectra of the materials were undertaken (Figure 4). The obtained spectra clearly show signals at 1600 and 1497 cm^{-1} as well as at 758 and 699 cm^{-1} corresponding respectively to the signals from chemical bonds: C=C (stretching vibrations of benzene-rings) and the out-of-plane bending of C–H of mono-substituted benzene-ring). Moreover, the absence of signals from O–H bond vibrations is notable. Signals in the region of 3000 cm^{-1} can be attributed to C–H bond dranes in both the alkyl chain and the aromatic ring. The significant similarity of both spectra should be emphasized, which represent mainly the same nanofiber material—polystyrene, which comprises the outer layer of the PS/Alg nanomaterial. Nevertheless, the infrared spectra proved the stability of the PS nanofibers during the preparation of composite material and indicated that the surface properties of PS/Alg material are dominated by PS nanofibers properties.

Successively, the contact angles (CA) measured for electrospun fibers with the immobilized bacteria were analyzed and they are presented in Figure 5. Additionally, calculated values are collected in Table 1. Hydrophobicity of obtained material was characterized based on a measurement of contact angle (DSA100E Drop Shape Analyzer, KRÜSS GmbH, Hamburg, Germany). In general, materials with liquid drop CA on their surface greater than 90° are hydrophobic, and with CA lower than 90° are hydrophilic. PS/Alg nanofibers showed slightly hydrophobic (CA > 90°) properties when a drop of water was applied. This is directly connected to polystyrene hydrophobic properties and was not affected by the presence of sodium alginate hydrogels. Moreover, the presence of alginate decreases the CA in the material, as we can observe higher values of CA for PS than for PS/Alg.

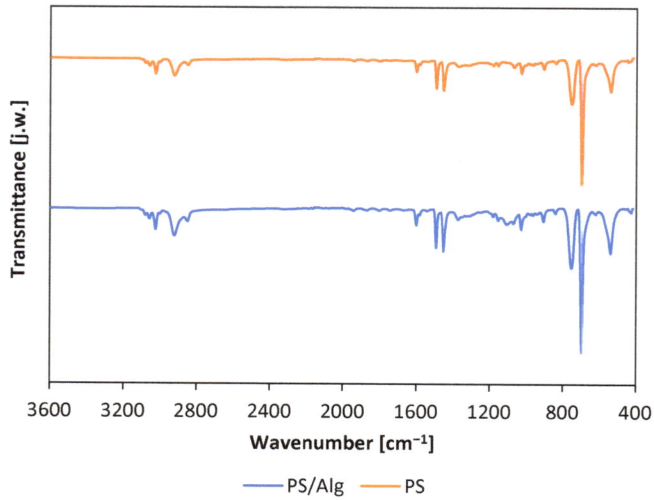

Figure 4. Infrared spectra of polystyrene nanofibers (PS) and polystyrene nanofibers with alginate hydrogel immobilized bacteria (PS/Alg).

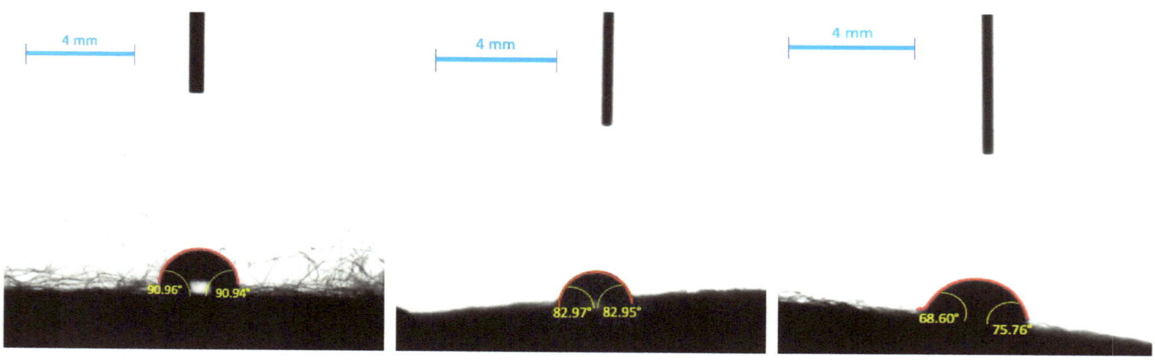

(a) (b) (c)

Figure 5. Contact angles of electrospun nanofibers from polystyrene-alginate-*L. plantarum* 2675 obtained with: (**a**) water; (**b**) MRS broth; (**c**) milk (2% fat).

Table 1. Contact angles values of tested materials with selected dietary liquids.

Material	Medium	Left Angle [°]	Right Angle [°]	Average Angle [°]
PS	Water	97.10 ± 3.6	97.10 ± 3.6	97.10 ± 3.6
	Broth	97.23 ± 1.0	97.23 ± 1.0	97.23 ± 1.0
	Milk (2% fat)	77.24 ± 2.2	77.24 ± 2.2	77.24 ± 2.2
PS/Alg	Water	91.0 ± 2.9	90.9 ± 3.3	91.0 ± 3.1
	Broth	83.0 ± 3.3	83.0 ± 4.1	83.0 ± 3.7
	Milk (2% fat)	68.7 ± 0.3	75.8 ± 5.5	72.2 ± 2.9

These result correspond to results presented by Huan et al. who obtained water CA 106 ± 1.5° and 90 ± 1.2° when electrospinning PS nanofibers from DMF and THF respectively (Huan et al., 2015 [29]). The lower contact angle was obtained for broth and milk and equals 83.0 ± 3.7° and 72.2 ± 2.89°, respectively, which means that both liquids possess the higher wetting ability. This could be caused by the presence of additives like proteins, sugars, and lipids, of which some are surface-active agents responsible for increasing wetting ability (Lorentz, Rogers and Jones, 2007 [30]; Campos Bernardes et al., 2012 [31]). A similar conclusion was drawn by Handojo et al. who reported that CA for water was higher than for whole milk on a glass surface and was equal to 49.5 ± 3.1° and 21.0 ± 4.7°, respectively (Handojo et al., 2009 [32]).

3.2. Bacterial Metabolic Activity

Probiotic bacteria cells' survivability is challenged by heat processing methods, compromising their functionality. To enhance the thermal stability of cells, various immobilization techniques are applied. For this purpose, the analysis of the metabolic activity of bacteria immobilized into the produced polystyrene electrospun nanofibers/alginate hydrogel was performed. The metabolic activity of the cells was measured based on the MTT and alamarBlue assays, comparing the obtained results with traditionally preserved probiotics (lyophilized and encapsulated in sodium alginate). The results collected in Figure 6a present a decrease of cells' metabolic activity upon 24 h exposition to elevated temperature and UV radiation, regardless of preservation technique. However, the electrospun-fibers immobilized cells possessing the highest metabolic activity of the analyzed samples. The type of the factor influencing cells was also meaningless, which proves good protective properties of the produced fibers.

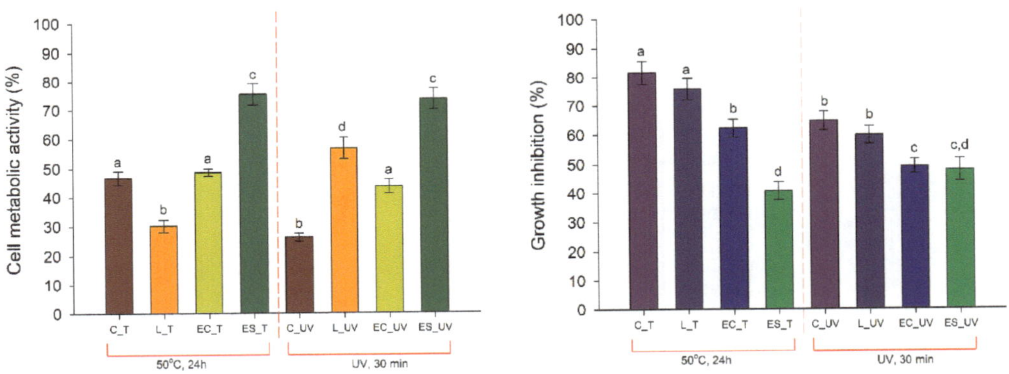

Figure 6. Results of the measurements of: (**a**) cells' metabolic activity, (**b**) growth inhibition of the cells, performed for control (C), lyophilized (L), alginate hydrogel-encapsulated (EC), and electrospun fibers immobilized (ES) cells of *L. plantarum* 2675 exposed to increased temperature (50 °C, 24 h) or UV radiation (λ = 253 nm, 30 min). Error bars represent the standard error of the mean of the three replicates in three independent experiments. Different lowercase letters suggest the significant differences among mean values for each treatment group (*t*-test; $p < 0.05$).

Moreover, these results are also in consistency with the alamarBlue test results which measures cellular respiration during proliferation. In this case, the electrospun fibers immobilized bacteria were characterized with low growth inhibition, equal to 40% and 48% by the applied temperature and UV light, respectively (Figure 6b). At the same time, the growth of free cells was inhibited by 81% (exposure to higher temperature) and 64% (exposure to UV radiation), and a significant decline in growth was observed for other analyzed samples. The protective effect of the matrix surrounding the cells on bacterial proliferation properties was especially visible when the UV light was applied. The free and lyophilized cells showed over a 60% decline in growth, while for immobilized cells it was around 10% lower.

The protection efficiency of *L. plantarum* 2675 cells by the produced membrane was investigated proving increased cell metabolic activity and growth under heat and UV radiation treatment in bath, MTT, and alamarBlue tests. Better thermal stability of the *Lactobacillus* cells encapsulated in the core–shell fiber mat (PVA) was noted by Feng et al. (2020) [33], with the viability loss of 0.32 log(CFU mL^{-1}) after 30 min in 70 °C. As suggested earlier by Feng et al. (2018) [34], the high capacity of the produced PS nanofibers might be related to the shielding effect of the nanomaterial, thus physically inhibiting the hot diffusion into the material. The combination of polymer and hydrogel was successfully applied by Gensheimer et al. (2011) [35]. They immobilized the *E. coli* and *M. luteus* cells in PVA-hydrogel and obtained microparticles that were linked with PLA, PVB, or PCL nanofibers. However, the authors tested only protection against organic solvents. Hence it is hard to compare their results with ours. Moreover, it should be underlined, that the review prepared by Stojanov and Barlec (2020) [36] includes the most frequently used polymers used in nanofibers production for cell immobilization and does not mention broader data about polystyrene. Our study can be considered as a valuable contribution to the present state of knowledge as a proof of usability of the PS nanofibers with alginate hydrogel to immobilize bacteria. The method can be applied mainly for Gram-positive bacteria with cell structure similar to *Lactobacillus* cells. For other groups of bacteria, Gram-negative, the effectiveness of the hybrid material should be similar. However, because of the differences between cell walls of Gram-positive and Gram-negative bacteria, for the second group additional experiments must be conducted.

4. Conclusions

Increasing demand for safe, efficient, and environmentally-safe solutions for pharmaceutical and food industries force researchers to look for new methods of bacteria storage. We propose that composite polystyrene materials/alginate immobilized bacteria meets mentioned expectations, since the research has proved the stability of the hybrid material. Significant reduction of the toxic effect of UV radiation and the higher temperature (50 °C, 24 h) was observed when the bacteria were immobilized in composite in comparison with free cells and cells in alginate only. Good wettability properties make the hybrid material a promising tool for industrial applications.

Author Contributions: Conceptualization, A.Z., W.S., and A.G.; methodology, A.Z., K.J., and A.G.; validation, A.B. and J.Z.; formal analysis, A.B., E.K. and T.J.; investigation, A.Z., K.J. and A.G.; resources, A.Z. and A.G.; data curation, K.J.; writing—original draft preparation, A.Z. and A.G.; writing—review and editing, A.Z. and W.S.; visualization, A.Z. and J.Z.; supervision, E.K. and T.J.; project administration, W.S.; funding acquisition, W.S. All authors have read and agreed to the published version of the manuscript.

Funding: The research was funded by Polish Ministry of Science and Higher Education (Grant No. 0912/SBAD/2015)—a subvention for Poznan University of Technology.

Institutional Review Board Statement: Not applicable.

Informed Consent Statement: Not applicable.

Data Availability Statement: No new data were created or analyzed in this study. Data sharing is not applicable to this article.

Conflicts of Interest: The authors declare no conflict of interest.

References

1. Thomas, E. Microbial growth and physiology: A call for better craftsmanship. *Front. Microbiol.* **2015**, *6*, 1–12.
2. Smułek, W.; Zdarta, A.; Grzywaczyk, A.; Guzik, U.; Siwińska-Ciesielczyk, K.; Ciesielczyk, F.; Strzemiecka, B.; Jesionowski, T.; Voelkel, A.; Kaczorek, E. Evaluation of the physico-chemical properties of hydrocarbons-exposed bacterial biomass. *Colloid Surf. B* **2020**, *196*, 111310. [CrossRef] [PubMed]
3. Rahman, F.U.; Andree, K.B.; Salas-Massó, N.; Fernandez-Tejedor, M.; Sanjuan, A.; Figueras, M.J.; Furones, M.D. Improved culture enrichment broth for isolation of Arcobacter-like species from the marine environment. *Sci. Rep.* **2020**, *10*, 14547. [CrossRef] [PubMed]

4. Cultrera, R.; Torelli, R.; Sarnicola, C.; Segala, D.; Mengoli, A.; Chiaretto, G.; Perri, P.; Sanguinetti, M. Identification and molecular characterization of Subramaniula asteroides causing human fungal keratitis: A case report. *BMC Infect. Dis.* **2021**, *21*, 4–9. [CrossRef] [PubMed]
5. Schön, T.; Werngren, J.; Machado, D.; Borroni, E.; Wijkander, M.; Lina, G.; Mouton, J.; Matuschek, E.; Kahlmeter, G.; Giske, C.; et al. Antimicrobial susceptibility testing of Mycobacterium tuberculosis complex isolates—The EUCAST broth microdilution reference method for MIC determination. *Clin. Microbiol. Inf.* **2020**, *26*, 1488–1492. [CrossRef]
6. Xie, L.; Chen, Z.; Liu, W.; Gu, D.; Yu, Y.; Chen, X.; Wu, Y.; Xu, N.; Xie, J.; Zhao, G.; et al. A sensitive EZMTT method provides microscale, quantitative and high-throughput evaluation of drug efficacy in the treatment of Mycobacterium tuberculosis infectious diseases. *J. Microbiol. Methods* **2021**, *181*, 106136. [CrossRef]
7. Mikolajczyk, A.; Stefaniuk, E.; Bosacka, K.; Hryniewicz, W. Właściwości i zastosowanie podłoży bakteriologicznych. [Properties and uses of bacteriological media. *Postępy Mikrobiol.* **2016**, *55*, 320–329.
8. Drauz, K.; Waldmann, H. Immobilization of Enzymes. In *Enzyme Catalysis in Organic Synthesis*, 1st ed.; Drauz, K., Gröger, H., May, O., Eds.; Wiley-VCH Verlag: Weinheim, Germany, 1995; pp. 73–87.
9. Zhu, Y. Immobilized Cell Fermentation for Production of Chemicals and Fuels. In *Bioprocessing for Value-Added Products from Renewable Resources*, 1st ed.; Yang, S.-T., Ed.; Elsevier Science: New York, NY, USA, 2007; pp. 373–396.
10. Willaert, R.G.; Baron, G.V. Gel entrapment and micro-encapsulation: Methods, applications and engineering principles. *Rev. Chem. Eng.* **1996**, *12*, 160–205. [CrossRef]
11. Tripathi, M.K.; Giri, S.K. Probiotic functional foods: Survival of probiotics during processing and storage. *J. Funct. Food* **2014**, *9*, 225–241. [CrossRef]
12. Rocha, C.E.; Silva, M.F.; Guedes, A.C.; Carvalho, T.P.; Eckstein, C.; Ribeiro, N.Q.; Santos, D.A.; Melo, M.M.; Araújo, M.S.; Martins-Filho, O.A.; et al. Alginate-chitosan microcapsules improve vaccine potential of gamma-irradiated Listeria monocytogenes against listeriosis in murine model. *Int. J. Biol. Macromol.* **2021**, *176*, 567–577. [CrossRef]
13. Penhasi, A.; Reuveni, A.; Baluashvili, I. Microencapsulation May Preserve the Viability of Probiotic Bacteria during a Baking Process and Digestion: A Case Study with Bifidobacterium animalis Subsp. lactis in Bread. *Curr. Microbiol.* **2021**, *78*, 576–589. [CrossRef]
14. Cook, M.; Tzortzis, G.; Charalampopoulos, D.; Khutoryanskiy, V. Microencapsulation of probiotics for gastrointestinal delivery. *J. Control Release* **2012**, *162*, 56–67. [CrossRef]
15. Ghoname, R.M.; El-Sayed, H.S.; Ghozlan, H.A.; Sabry, S.A. Application of probiotic bacteria for the improvement of sea bream (Sparus aurata) larval production. *Egypt. J. Aquat. Biol. Fish.* **2020**, *24*, 371–398. [CrossRef]
16. Stanishevsky, A.V.; Wetuski, J.D.; Yockell-Lelièvre, H. Crystallization and stability of electrospun ribbon- and cylinder-shaped tungsten oxide nanofibers. *Ceram. Int.* **2016**, *42*, 388–395. [CrossRef]
17. Li, L.; Peng, S.; Lee JK, Y.; Ji, D.; Srinivasan, M.; Ramakrishna, S. Electrospun hollow nanofibers for advanced secondary batteries. *Nano Energy* **2017**, *39*, 111–139. [CrossRef]
18. Ali, M.; Ain, Q.T.; Huan, H.J. Branched nanofibers for biodegradable facemasks by double bubble electrospinning. *Acta Chem. Malays.* **2020**, *4*, 40–44. [CrossRef]
19. Ma, P.X.; Zhang, R. Synthetic nanoscale fibrous matrix. *J. Biomed. Mat. Res. A* **1999**, *46*, 60–72. [CrossRef]
20. Xu, D.; Samways, D.S.K.; Dong, H. Fabrication of self-assembling nanofibers with optimal cell uptake and therapeutic delivery efficacy. *Bioact. Mater.* **2017**, *2*, 260–268. [CrossRef] [PubMed]
21. Wang, Y.; Zheng, M.; Lu, H.; Feng, S.; Ji, G.; Cao, J. Template synthesis of carbon nanofibers containing linear mesocage arrays. *Nanoscale Res. Lett.* **2010**, *5*, 913–916. [CrossRef] [PubMed]
22. Huang, Z.-M.; Zhang, Y.; Kotaki, M.; Ramakrishna, S. A review on polymer nanofibers by electrospinning and their applications in nanocomposites. *Compos. Sci. Technol.* **2003**, *63*, 2223–2253. [CrossRef]
23. Bhardwaj, N.; Kundu, S.C. Electrospinning: A fascinating fiber fabrication technique. *Biotechnol. Adv.* **2010**, *28*, 325–347. [CrossRef] [PubMed]
24. Bao, J.; Li, H.; Xu, Y.; Chen, S.; Wang, Z.; Jiang, C.; Li, H.; Wei, Z.; Sun, S.; Zhao, W.; et al. Multi-functional polyethersulfone nanofibrous membranes with ultra-high adsorption capacity and ultra-fast removal rates for dyes and bacteria. *J. Mater. Sci. Technol.* **2021**, *78*, 131–143. [CrossRef]
25. Jayani, T.; Sanjeev, B.; Marimuthu, S.; Uthandi, S. Bacterial Cellulose Nano Fiber (BCNF) as carrier support for the immobilization of probiotic, Lactobacillus acidophilus 016. *Carbohydr. Polym.* **2020**, *250*, 116965. [CrossRef] [PubMed]
26. Mamvura, C.I.; Moolman, F.S.; Kalombo, L.; Hall, A.N.; Thantsha, M.S. Characterisation of the Poly-(Vinylpyrrolidone)-Poly-(Vinylacetate-Co-Crotonic Acid) (PVP:PVAc-CA) Interpolymer Complex Matrix Microparticles Encapsulating a Bifidobacterium lactis Bb12 Probiotic Strain. *Probiot. Antimicrob. Proteins* **2011**, *3*, 97–102. [CrossRef]
27. Yu, H.; Liu, W.; Li, D.; Liu, C.; Feng, Z.; Jiang, B. Targeting Delivery System for Lactobacillus Plantarum Based on Functionalized Electrospun Nanofibers. *Polymers* **2020**, *12*, 1565. [CrossRef]
28. Deng, G.; Deng, X.; Deng, J.; Lu, X.; Kang, X.; Song, Y. Performance Evaluation of an Electrospun Nanofiber Mat as Samplers for the Trap of Trace Heavy Metals in Atmospheric Particles and Its Application. *Anal. Sci.* **2020**, *36*, 1453–1459. [CrossRef] [PubMed]
29. Huan, S.; Liu, G.; Han, G.; Cheng, W.; Fu, Z.; Wu, Q.; Wang, Q. Effect of experimental parameters on morphological, mechanical and hydrophobic properties of electrospun polystyrene fibers. *Materials* **2015**, *8*, 2718–2734. [CrossRef]
30. Lorentz, H.; Rogers, R.; Jones, L. The impact of lipid on contact angle wettability. *Optom. Vis. Sci.* **2007**, *84*, 946–953. [CrossRef]

31. Bernardes, P.C.; Araújo, E.A.; Pires, A.C.; Júnior, J.F.Q.F.; Lelis, C.; De Andrade, N.J. Work of adhesion of dairy products on stainless steel surface. *Braz. J. Microbiol.* **2012**, *43*, 1261–1268. [CrossRef]
32. Handojo, A.; Zhai, Y.; Frankel, G.; Pascall, M.A. Measurement of adhesion strengths between various milk products on glass surfaces using contact angle measurement and atomic force microscopy. *J. Food Eng.* **2009**, *92*, 305–311. [CrossRef]
33. Feng, K.; Huang, R.-M.; Wu, R.-Q.; Wei, Y.-S.; Zong, M.-H.; Linhardt, R.J.; Wu, H. A novel route for double-layered encapsulation of probiotics with improved viability under adverse conditions. *Food Chem.* **2020**, *310*, 125977. [CrossRef] [PubMed]
34. Feng, K.; Zhai, M.Y.; Zhang, Y.; Linhardt, R.J.; Zong, M.H.; Li, L.; Wu, H. Improved Viability and Thermal Stability of the Probiotics Encapsulated in a Novel Electrospun Fiber Mat. *J. Agric. Food Chem.* **2018**, *66*, 10890–10897. [CrossRef] [PubMed]
35. Gensheimer, M.; Brandis-Heep, A.; Agarwal, S.; Thauer, R.K.; Greiner, A. Polymer/Bacteria Composite Nanofiber Nonwovens by Electrospinning of Living Bacteria Protected by Hydrogel Microparticles. *Macromol. Biosci.* **2019**, *11*, 333–337. [CrossRef] [PubMed]
36. Stojanov, S.; Berlec, A. Electrospun Nanofibers as Carriers of Microorganisms, Stem Cells, Proteins, and Nucleic Acids in Therapeutic and Other Applications. *Front. Bioeng. Biotechnol.* **2020**, *8*, 130. [CrossRef] [PubMed]

materials

MDPI

Article

Optical Properties and Light-Induced Charge Transfer in Selected Aromatic C60 Fullerene Derivatives and in Their Bulk Heterojunctions with Poly(3-Hexylthiophene)

Maciej Krajewski [1,*], Piotr Piotrowski [2], Wojciech Mech [1], Krzysztof P. Korona [1], Jacek Wojtkiewicz [1], Marek Pilch [1], Andrzej Kaim [2], Aneta Drabińska [1] and Maria Kamińska [1]

1 Faculty of Physics, University of Warsaw, Pasteura 5, 02-093 Warsaw, Poland
2 Faculty of Chemistry, University of Warsaw, Pasteura 1, 02-093 Warsaw, Poland
* Correspondence: maciej.krajewski@fuw.edu.pl

Abstract: Fullerene derivatives offer great scope for modification of the basic molecule, often called a buckyball. In recent years, they have been the subject of numerous studies, in particular in terms of their applications, including in solar cells. Here, the properties of four recently synthesized fullerene C60 derivatives were examined regarding their optical properties and the efficiency of the charge transfer process, both in fullerene derivatives themselves and in their heterojunctions with poly (3-hexylthiophene). Optical absorption, electron spin resonance (ESR), and time-resolved photoluminescence (TRPL) techniques were applied to study the synthesized molecules. It was shown that the absorption processes in fullerene derivatives are dominated by absorption of the fullerene cage and do not significantly depend on the type of the derivative. It was also found by ESR and TRPL studies that asymmetrical, dipole-like derivatives exhibit stronger light-induced charge transfer properties than their symmetrical counterparts. The observed inhomogeneous broadening of the ESR lines indicated a large disorder of all polymer–fullerene derivative blends. The density functional theory was applied to explain the results of the optical absorption experiments.

Keywords: organic electronics; fullerene derivatives; LESR

Citation: Krajewski, M.; Piotrowski, P.; Mech, W.; Korona, K.P.; Wojtkiewicz, J.; Pilch, M.; Kaim, A.; Drabińska, A.; Kamińska, M. Optical Properties and Light-Induced Charge Transfer in Selected Aromatic C60 Fullerene Derivatives and in Their Bulk Heterojunctions with Poly(3-Hexylthiophene). *Materials* **2022**, *15*, 6908. https://doi.org/10.3390/ma15196908

Academic Editors: Jakub Zdarta and Agata Zdarta

Received: 31 August 2022
Accepted: 28 September 2022
Published: 5 October 2022

1. Introduction

Fullerenes are one of the best known molecules and have been used as building blocks for many various derivatives and larger structures with a wide range of applications, primarily in biomedicine [1], redox flow batteries [2], organic transistors [3], and photovoltaic nanostructures [4,5]. Among their many derivatives, the commercially available PC61BM ([6,6]-phenyl-C61-butyric acid methyl ester) and PC71BM (([6,6]-phenyl-C71-butyric acid methyl ester) are the best known. Organic and perovskite photovoltaics takes advantage of good electron conductivity and acceptor properties of fullerenes, enabling an efficient transport of photo-excited electrons. In organic solar cells, an active layer typically contains a blend with percolating regions of donor (polymer) and acceptor materials that together create the so-called bulk heterojunction (BHJ), schematically shown in Figure 1. Fullerene derivatives are a traditional acceptor component of the active layer. Although they currently face strong competition from non-fullerene organic acceptors, exhibiting higher open-circuit voltages and up to 18% efficiency [6,7], they are still a strong alternative to such non-fullerene acceptors, especially in terms of the charge carrier mobility values. Recently, research has been underway on ternary blends containing both fullerene and non-fullerene acceptors [8] to take advantage of the specific benefits of each. Moreover, the use of fullerene derivatives in layered photovoltaic perovskite structures as electron-transporting and hole-blocking layers is becoming very common. Solution coating technologies of organic and perovskite photovoltaics require fullerene derivatives with such a modification of the fullerene cage as to ensure suitable solubility in typical solvents used in layer processing. Moreover, fullerene derivatives should lead to the proper ordering of the layers during the evaporation of the solvent to ensure high mobility of the charge carriers and meeting the required electrical parameters of the fabricated solar cell. Despite the unwavering interest and wide range of applications of fullerene derivatives, the understanding of the effect of a particular fullerene modification on its solubility, ability to organize the solidifying layer, along with charge transfer capability and absorption spectrum, needs more in-depth research.

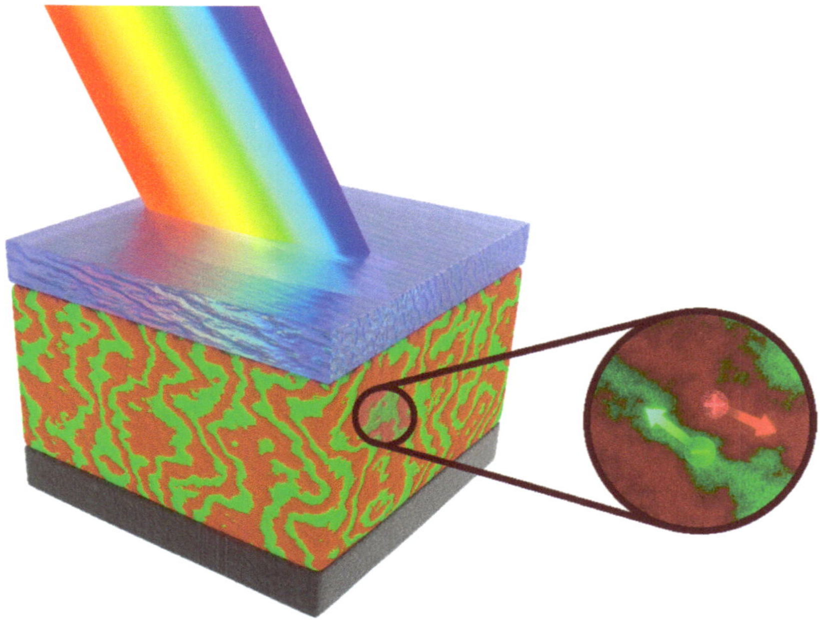

Figure 1. Schematic electron and hole separation in bulk heterojunction photogenerated after illumination of the blend. The red and green arrows on the right scheme show the transport route for photogenerated hole and electron, respectively.

In this study, four recently synthesized fullerene C_{60} derivatives (named [C60] P1, [C60] P2, [C60] T1, [C60] T2) as such and in blends with conductive polymer poly(3-hexylthiophene) (P3HT) were compared regarding their optical properties and light-induced charge transfer ability and their behavior in exemplary solar cells constructed from these mixtures. PC61BM and its blend with P3HT was used as an appropriate reference.

Together with optical absorption measurements, quantum mechanical calculations were performed using the density functional theory (DFT) and time-dependent density functional theory (TDDFT) methods to gain additional insight into the experimental results [9–11]. The energy levels and light absorption probability of the new fullerene derivatives, as well as PC61BM, were determined. For studying the charge transfer process, electron spin resonance (ESR) and time-resolved photoluminescence (TRPL) techniques were applied. As has been shown in previous works, both of these techniques provide crucial information regarding the generation and the separation mechanisms of the charge carriers in polymer–fullerene blends [12]. During the illumination of the blend, the photogeneration of strongly bound (0.3–1.0 eV) exciton (electron–hole pair) takes place. The pair is then spatially separated, resulting in the hole being localized at P3HT and the electron at the fullerene derivative [13–17]. Thanks to the ESR technique, it was possible to compare the mixtures with various fullerene derivatives in terms of their efficiency of formation and then dissociation of excitons. Moreover, the TRPL technique was applied to trace the photo-carriers' behavior in P3HT–fullerene blends. Radiative recombination and charge separation are two competing processes that lead to light emission and photocurrent generation, respectively. Therefore, the observation of the photoluminescence (PL) dumping in the donor–acceptor mixture compared to the PL decay for the individual components of the blend provides information on the effectiveness of charge separation.

2. Materials and Methods

2.1. Samples

Four differently functionalized fullerene derivatives (Figure 2) were chosen for the presented studies. Two of them, 61-etyloxycarbonyl-[61-pyren-1-ylomethylo]-1,2-methano [60]fullerene (called [C60]P1 in the paper) and bis-61-carbonyl [pyren-1-ylomethylo]-1,2-methano [60]fullerene ([C60]P2), were functionalized with either one or two pyrene moieties. Analogically, the other two, 61-etyloxycarbonyl-[61-(2-(thiophen-3-ylo)ethylo)]-1,2-methano [60]fullerene ([C60]T1) and bis-61-carbonyl o [2-(thiophen-3-ylo)ethylo]-1,2-methano [60]fullerene ([C60]T2), were functionalized with either one or two thiophene rings. Additionally, the commercially available PC61BM purchased from Ossila was used as reference material. Fullerene derivatives [C60]P1, [C60]P2, [C60]T1, and [C60]T2 were synthesized using the modified Bingel method [18–20]. Details regarding the synthesis and characterization of the obtained fullerene derivatives are reported for pyrene derivatives in our previous publication [21], and those regarding thiophene functionalized fullerenes can be found in Supplementary Materials. Two types of samples were prepared for the purpose of ESR research: pure fullerene derivatives as such and their blends with P3HT. Functionalized fullerenes were dissolved in chlorobenzene and stirred at a temperature of 50 °C for 3 h under an air atmosphere. The solution was then poured into a 3 mm diameter quartz ESR tube and kept at an air atmosphere of 50 mbar pressure for two days to let the solvent evaporate. The mass of all the studied samples was approximately 3 mg. In the case of fullerene-derivative–P3HT blends, a slightly different procedure was used. Firstly, the commercial regioregular P3HT (purchased at Sigma-Aldrich) and fullerene derivative in the form of powders were mixed at a 3:2 molar ratio. Then, chlorobenzene was applied to create a 25 mg/mL solution. The mixture was stirred at a temperature of 50 °C for 3 h and subsequently poured into an ESR tube. Afterward, the sample was left for two days in an air atmosphere of 50 mbar pressure to let the solvent evaporate to obtain thin and glossy films on the inside wall of the ESR tube. The same solutions were used for the remaining techniques (UV-Vis absorption, PL, and TRPL). However, in these cases, small droplets of the solution were applied onto a glass substrate (UV-Vis) or copper foil (TRPL) and left to dry at 60 °C to create a thin film.

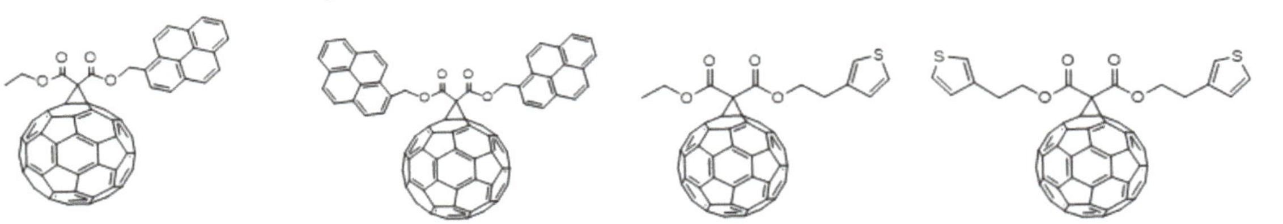

[C60]P1 [C60]P2 [C60]T1 [C60]T2

Figure 2. Chemical structure of the studied fullerene derivatives.

2.2. Methods

UV-Vis absorption spectra were recorded at room temperature by means of a Varian Cary 5000 UV-Vis-NIR spectrophotometer in the range of 300–800 nm at 1 nm resolution. Deuterium and tungsten halogen lamps were used as a light source, and the signal was collected using dual Si diode detectors.

The density functional theory (DFT) and time-dependent density functional theory (TDDFT) calculations were employed using the B3LYP hybrid functional, which is considered a reliable tool for structures with extended conjugated double bonds that provides optimized geometries, corresponding energies, and related properties. The calculations were performed with the implementation of the above methods in the Gaussian 09 package [22]. UV-VIS spectra were calculated by TDDFT, as implemented in the Gaussian package. Computations were performed with the use of the B3LYP functional, in the base 6-31 G(d). The number of excited states was 108. The geometry of all the studied molecules was optimized, so that their energies corresponded to the minimum on the potential energy surface.

Electron spin resonance (ESR) measurements were performed by a Bruker ELEXSYS 580 X-band (f $\approx$ 9.4 GHz) ESR spectrometer with TE102 resonance cavity coupled with an Oxford continuous flow cryostat. During experiments, the modulation frequency and modulation amplitude were 100 kHz and 1 G, respectively. The measurement routine was as follows. The sample was cooled down to 2 K in darkness to avoid any photo-excitation that might affect the intensity of ESR signals. Then, the ESR spectra at a wide range of microwave powers (0.474 µW–0.474 mW) and temperatures up to 80 K were collected. Afterward, the sample was cooled back again to 2 K, and the ESR spectrum was measured under the illumination of the sample with a white LED. Subsequently, the whole series of measurements for different microwave powers and temperatures was repeated for the illuminated sample. In the case of low signals, averaging from 3 to 5 acquisitions was performed to improve the signal-to-noise ratio. Due to residual dark signals of some of the samples, the light-induced ESR (LESR) spectrum will further refer in this paper to the difference in the intensity of the signal with and without illumination. Since the ESR spectrometer measured the derivative of microwave absorption, the signal intensities discussed later relate to a double integral of the measured signals. The fitting of the spectrum was performed using the pepper function (for frozen solution) of the Matlab toolbox–Easyspin [23]. The best fit of the spectra was usually obtained using a convolution of Gaussian and Lorentzian curves. As mentioned before, all ESR experiments were performed for the same amount (mass) of the samples. The ESR spectra of different samples, which were compared for the purposes of the research, were produced under conditions identical to the measurement temperature, microwave attenuation, and field modulation.

Time-resolved photoluminescence (TRPL) was measured at room temperature using Ti:sapphire laser excitation at the third harmonic frequency (300 nm). The obtained spectra were recorded in the wavelength range of 300–900 nm at 1 nm resolution by a Hamamatsu C5680-24C streak camera. To ensure a low signal-to-noise ratio, the final spectra were averaged over 120 measurements. The single transient duration was limited to 12 ns by the laser frequency equal to 80 MHz.

3. Results

3.1. Optical Absorption Spectroscopy

The absorbance spectra of the studied fullerene derivatives were measured in order to find possible differences regarding both their energy range and structure. Figure 3a shows the absorbance normalized to the intensity of the 3.7 eV peak of pure fullerene derivatives (including PC61BM) at room temperature in the energy range of 1.75 to 4.0 eV. The spectra cover the entire measurement region and most likely extend further toward higher energies. The energy range above 4 eV was not available for the measurement due to strong absorption in a glass sample substrate. The absorption spectrum of unmodified C_{60} fullerene molecule measured by other researchers [24] starts with a very weak band (almost invisible in the case of the measured fullerene derivatives), which was identified as a dipole-forbidden transition at a HOMO-LUMO gap of 1.85 eV ($S_0 \rightarrow S_1$ transition). It is followed by the first strong band centered at about 2.7 eV and related to a dipole-allowed transition to the higher excited state ($S_0 \rightarrow S_2$). The third, even stronger peak at 3.6 eV was also ascribed to a dipole-allowed transition, from the lower level in the valence band to the S_1 level. All the studied fullerene derivatives had similar absorption bands to the ones discussed above—however, with some exceptions. An evident difference was observed for the [C60]P2 absorption, where an additional strong line appeared at about 3.4 eV energy.

This line was also present in the [C60]P1 absorption spectrum but as a weaker structure superimposed on the 3.6 eV transition. The line at 3.4 eV may be related to the internal excitation within the pyrene molecule, since its intensity grows with the number of pyrenes attached to the fullerene cage, and it has been found previously that the pyrene monomer exhibits three sharp absorption peaks in the range of 310–340 nm [25], which corresponds well to the energy region of the observed extra line. Further support for such identification comes from the DFT calculations presented in Figure 4 and discussed later. On the other hand, the absorption related to the excitation within the thiophene ring/rings present in the [C60]T1 and [C60]T2 derivatives can be expected in the ultraviolet range, outside of the measurement range [26]. Thus, the results of absorption measurements show only a slight influence of the chemical functionalization of the synthesized fullerenes on the observed spectra, proving the dominance of excitations within the fullerene cage.

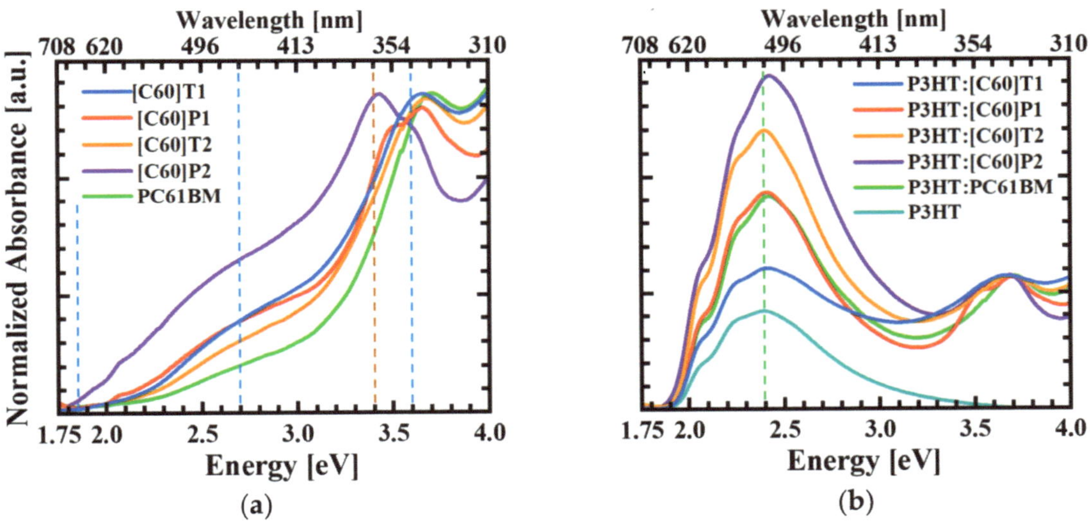

Figure 3. Absorption spectra of (**a**) fullerene derivatives and (**b**) their blends with P3HT. The dashed lines mark the respective transitions: the intra-pyrene transition is marked with the orange line, fullerene transitions are marked with the dark blue lines and P3HT maximum absorption with the green line.

Since fullerene absorption occurs in the area of higher energies than the maximum of the solar spectrum, they are not an absorber in the solar cells; however, they can use the high-energy part of the solar spectrum and enhance the operation of the cells. Among the studied fullerenes, it can be noticed that the absorption capability shifts toward lower energies for fullerenes with pyrene moiety, and the intensity of this absorption increases with the number of pyrenes. For this reason, these fullerene derivatives have the potential to be a better complement to solar cell absorbers than the commonly used PCBM and more efficiently enhance solar energy harvesting by supporting the P3HT donor in the process of carrier photogeneration.

In Figure 3b, the fullerene-derivative–P3HT blends and pure P3HT absorption spectra are shown. As seen, P3HT absorption extended in the energy range of 2.0–2.8 eV, with a maximum at 2.4 eV. The peak at 2.4 eV was present in the spectra of all blends. Interestingly, the intensity of this peak in relation to the fullerene peak at 3.6 eV strongly varied depending on the kind of fullerene derivative present in the blend, despite the preparation of all blends at the same nominal ratio of P3HT to fullerene derivative. This behavior may result from some differences in the solubility of the fullerene derivatives (Table S1).

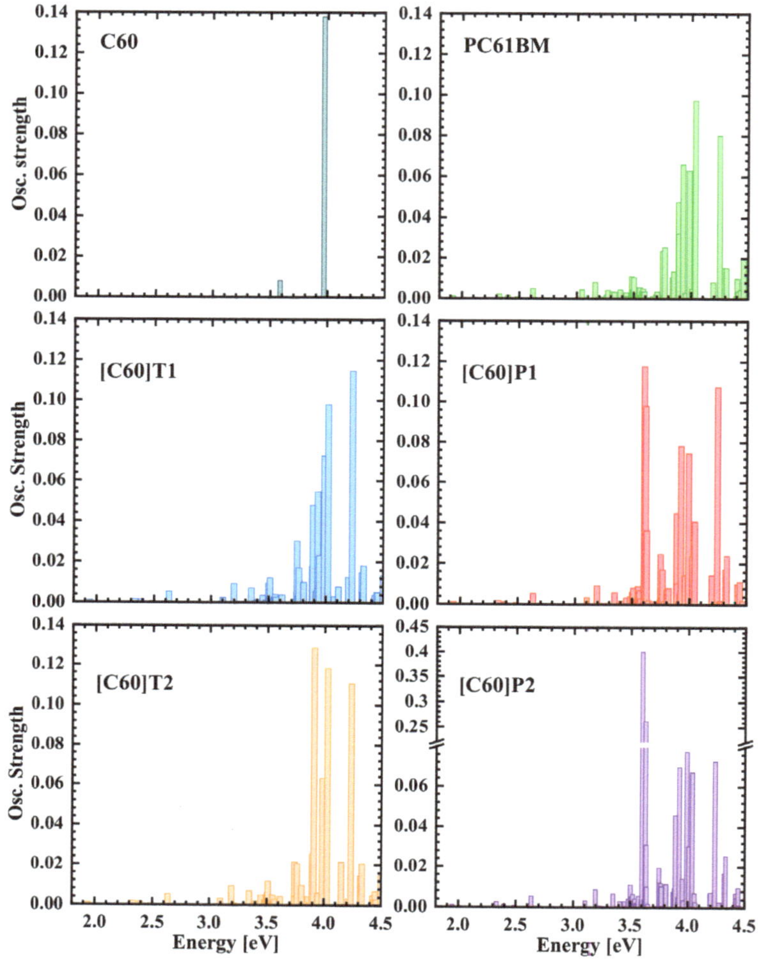

Figure 4. Calculated oscillatory strength of optical transitions for the studied fullerene derivatives and C60 molecule. Note the broken scale in the ordinate axis for the [C60]P2 spectrum (strong intra-pyrene transitions).

3.2. *Density Functional Theory Calculations of Optical Transitions*

For optimized geometries, one-electron energy levels, and among them, HOMO and LUMO energies were calculated for all studied fullerene derivatives, including PC61BM. The results along with the corresponding models of HOMO are presented in Table 1. The already published data for the HOMO/LUMO energy levels of PC61BM using the same basis set are consistent with ours within 0.1–0.2 eV [27–29].

Due to the size of the studied molecules, to optimize their geometry and calculate the corresponding energies, we used a relatively small double-zeta basis sets 6-31G (d). To compare whether the calculation using larger triple-zeta basis sets can qualitatively change the conclusions resulting from HOMO and LUMO energy estimates, single-point calculations for the optimized geometries were performed using triple-zeta basis sets, i.e., 6-311G (d, p). The energy values obtained in this way for HOMO and LUMO are included in Table S4. It can be seen that, in general, there is no change in the sequence of levels for all fullerene derivatives, and the energies of the levels are shifted downward by about the same, −0.4 eV, energy. Therefore, the following discussion and conclusions also remain valid in the larger basis sets.

Table 1. HOMO (also HOMO-1 and HOMO-2 for selected compounds) and LUMO energies of the optimized fullerene derivatives obtained by DFT calculations. Additionally, spatial models of corresponding HOMO and LUMO with yellow and purple lobes corresponding to positive, and green and blue lobes to negative isosurface values.

Material	**PC61BM**	**[C60]T1**	**[C60]P1**	
Spatial distribution of HOMO/HOMO-1 and LUMO isosurfaces	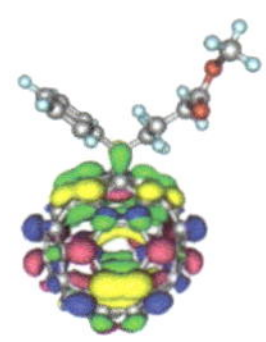	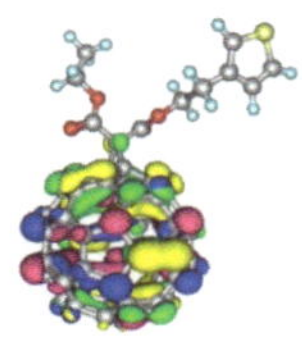	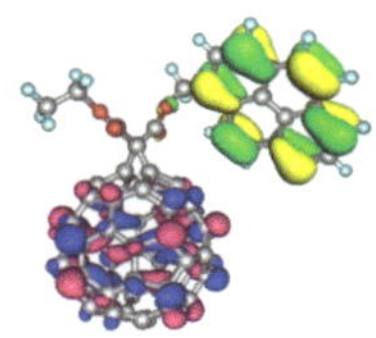	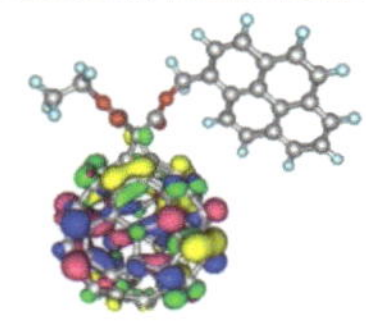 HOMO-1
LUMO [eV]	−3.09	−3.12	−3.10	
HOMO [eV]	−5.66	−5.70	−5.50	−5.68
Frontier Molecular Orbital gaps [eV]	2.57	2.58	2.40	2.58
Material	[C60]T2	[C60]P2		
Spatial distribution of HOMO/HOMO-1/HOMO-2 and LUMO isosurfaces	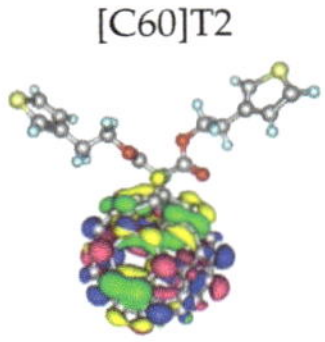	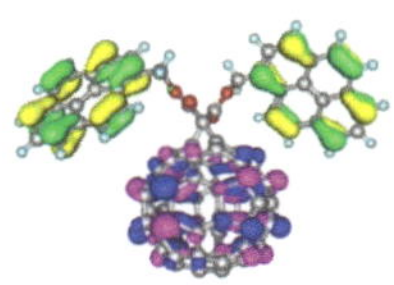	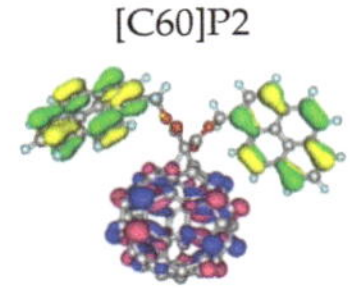 HOMO-1	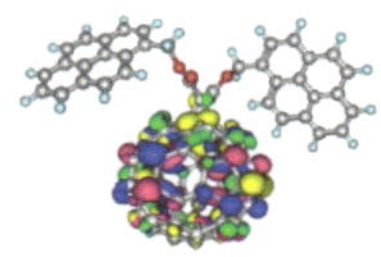 HOMO-2
LUMO [eV]	−3.15		−3.06	
HOMO [eV]	−5.73	−5.48	−5.50	−5.65
Frontier Molecular Orbital gaps [eV]	2.58	2.38	2.44	2.59

It is seen from Table 1 that the chemical modifications of fullerenes by pyrene substituent ([C60]P1 and [C60]P2 derivatives) resulted in significant changes in the energy of HOMO levels. For the rest of the fullerene derivatives, the HOMO level was practically at the same position as for PC61BM. Analysis of the corresponding orbitals revealed that for the pyrene functionalized fullerenes, the HOMO respective orbital was located on the pyrene moieties, whereas for the other studied molecules, it occupied the fullerene cage. The LUMO energies were similar for all fullerene derivatives, and all the LUMO (responsible for accepting electrons in solar cells) were located on the fullerene cage. One can expect optically forbidden transitions between the orbitals localized at the pyrene moiety and the ones localized at the fullerene cage, so the energies of lower levels (below the HOMO level) for the [C60]P1 and [C60]P2 derivatives are also included in Table 1. In the table, the energies of HOMO levels down to the level with the orbital located at the fullerene cage are shown. The highest occupied molecular orbitals located at the fullerene cage were determined as HOMO-1 for [C60]P1 and HOMO-2 for [C60]P2.

The energy positions of these latter levels are very close to the HOMO level of PC61BM as well as fullerenes modified with thiophene rings. The LUMO-HOMO energy gaps of the studied molecules differed by up to 0.21 eV (Table 1). Such a difference does not correspond to the measured absorption spectra, which, according to Figure 3a, were very similar and in the same energy range. To explain this discrepancy, the intensities of optical transitions were calculated in the energy range of 1.5–4.5 eV. The calculated spectra for all the studied fullerene derivatives along with the spectrum for the unmodified C_{60} fullerene are presented in Figure 4. As can be seen, the spectra were dominated by several peaks around 4 eV; those peaks originated from the splitting of a single line of C_{60} absorption observed at 4.0 eV. This line most probably corresponds to the mentioned dipole-allowed transition from the lower level in the valence band to the S1 level (observed in the optical absorption experiment at about 3.6 eV). Noteworthy are the additional structures appearing in the calculated absorption spectra of fullerene derivatives with pyrene moieties, [C60]P1 and [C60]P2, at about 3.6 eV energy, and not visible for the other fullerene derivatives. They are especially strong for [C60]P2, with two pyrene moieties, but are also seen for the calculated absorption spectrum of the [C60]P1 derivative and can therefore be linked to optical intra-pyrene excitations [25]. We relate this theoretical structure to the one observed in the experimental absorption spectra for [C60]P1 and [C60]P2 at 3.4 eV already discussed in the experimental part (Section 2). The calculated absorption threshold and the spectrum at lower energies were much weaker than the structure at about 4 eV for all fullerene derivatives; however, the positions of the lines and their intensities in the 1.9 to 3.5 eV energy range were very similar for all of them. Therefore, it is possible to conclude that, in this energy range, transitions within the fullerene cage were dominant and made the spectra for all the fullerene derivatives very alike. Transitions from HOMO localized at the pyrene moieties for [C60]P1 and [C60]P2 to LUMO localized at the fullerene cage were not visible in the calculated spectrum. The reason for this absence is the weak overlap of HOMO and LUMO wave functions, making the optical excitation very unlikely. Interestingly, optical transitions from orbitals with energy levels below HOMO:HOMO-1 for [C60]P1 and HOMO-2 for [C60]P2 were visible in the spectrum. These orbitals are the highest occupied orbitals localized at the C_{60} fullerene cage and play the HOMO role in terms of optical absorption. As a result, the effective energy gap of fullerene derivatives functionalized with the pyrene is very close to the energy gap of the other functionalized fullerenes. As can be seen from Table 1 (LUMO–HOMO energy differences marked in bold), such effective energy gaps differ only by 0.02 eV.

The calculated transitions for all fullerene derivatives, as shown in Figure 4, seemingly start at a lower energy than the LUMO–HOMO distance. This is related to the DFT calculation method that takes into account the reconfiguration of the charge density distribution after electron excitation (e.g., from HOMO to LUMO). Therefore, the energies corresponding to such reconfigured excited states may differ from those determined for the original shape of the orbital. Additionally, in general, the calculated absorption structures are shifted to higher energies compared to the measured spectra. This, however, is typical for the DFT/B3LYP/6-31G(d) method, mainly due to the overestimation of LUMO energy [28,30]. It is worth noting that only the slightly different values of the HOMO and LUMO levels of the studied fullerene derivatives reveal that all the fullerene derivatives can work with the same donor materials in solar devices.

3.3. Electron Spin Resonance

3.3.1. Signal Analysis

As mentioned above, two kinds of samples were measured using the ESR technique: pure fullerene derivatives and their blends with P3HT (Figure 5). In darkness, most fullerene derivatives did not show any significant ESR signal (data in Supplementary Materials). Some of them exhibited a persistent line of g = 2.0022, which has been attributed earlier to either residual oxygen ions [31] or other intrinsic defects, which may give rise to unpaired spins localized at deep traps [32,33]. Nevertheless, these signals were much weaker than those obtained under illumination and were subtracted during the analysis. As can be seen in Figure 5a, fullerene derivatives exhibited the LESR signal with two lines of various intensities. The line of g = 2.0000 can be related to an electron localized at the fullerene cage due to its similarity to the line identified as such and observed in polymer–fullerene blends [34]. The other line of g = 2.0022 has been usually assigned to a hole localized at the polymer chain in polymer–fullerene blends; however, in this case, the signal might arise from a hole localized at the fullerene functional group. Symmetrical molecules ([C60]P2, [C60]T2) demonstrated a much weaker signal at g = 2.0000 compared to what was observed for asymmetrical ones ([C60]P1, [C60]T1, and also PC61BM). Regarding the line corresponding to the holes, it was weaker for molecules with a pyrene moiety ([C60]P1, [C60]P2), so they seem to be less likely to accumulate holes as compared to those with the thiophene ring ([C60]T1, [C60]T2). The reference sample, PC61BM, exhibited a very similar signal to [C60]T1, yet much more prominent.

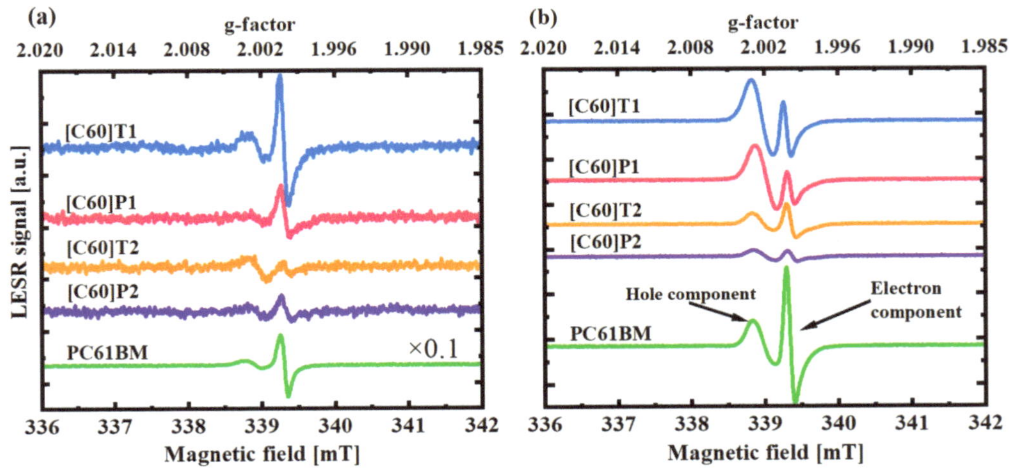

Figure 5. LESR signal at 30 K and for 0.00474 mW of microwave power of (**a**) differently functionalized fullerene derivatives; note that the reference signal of the PC61BM fullerene was divided by 10 due to its high intensity, (**b**) fullerene derivative blends with P3HT. The signals are vertically shifted for better clarity with the scale maintained.

Fullerene derivative blends with P3HT exhibited substantially different ESR spectra than pure fullerene derivatives. In this case, blends with asymmetrical fullerene molecule ([C60]T1:P3HT, [C60]P1:P3HT) gave rise to a strong single line at g = 2.0025 under dark conditions. This signal may be attributed to intrinsic defects inside the molecules [33]. For symmetrical molecule blends ([C60]T2:P3HT, [C60]P2:P3HT), this dark signal was of much lower intensity. Under illumination, two lines appeared in all blends, with g factors as those observed for pure fullerene derivatives. The main noteworthy difference between pure fullerene derivative spectra and those for blends with P3HT was significant signal intensity growth in the case of the blends. Interestingly, for the asymmetric molecule blends, the line intensities of the hole and the electron components gave stronger LESR signals than for the symmetric molecule blends, with a dominant hole component for [C60]T1:P3HT, [C60]P1:P3HT and almost equal intensities of both signals for PC61BM:P3HT. The blends with symmetrical fullerene molecules ([C60]P2:P3HT, [C60]T2:P3HT), on the other hand, exhibited generally lower but similar intensities of the electron and hole components. Thus, the following correlation could be observed: the stronger the LESR signal detected for the pure fullerene derivative, the stronger the LESR signal for its blend with P3HT. To make a more quantitative comparison of the separated electrons and holes in the studied blends, the LESR signals were fitted using the procedure described in Section 2.2.

For accurate fitting, an anisotropic g-tensor, which indicates the existence of some preferential order inside the blends, causing differences in magnetic susceptibility along the respective directions, was necessary. All components of the g-tensor along with the average g-factor attributed to either the hole or the electron signal, (g_{iso}) are listed in Table 2. The components of the g-tensor for a photo-generated hole in all P3HT blends with different fullerene derivatives are consistent within the experimental uncertainty of ± 0.0005 and also agree with the literature values for the PC61BM:P3HT blend [13]. This indicates that a photo-generated hole localized at the polymer backbone does not interact noticeably with electrons localized at the fullerene cage. The absence of such interactions means that photo-induced electron–hole pairs inside these blends are quickly separated [15]. Indeed, fast charge separation has been observed in a PC61BM:P3HT blend, with a reported time of about 40 fs [35,36]. The g-tensor components of the photo-generated electron localized at the fullerene cage in all blends with P3HT were very similar to those in pure fullerene derivatives (data in Supplementary Materials), and all of them were well below the free electron g-factor (g = 2.0023). Such negative deviation most probably results from the spin–orbit coupling with the unoccupied π-orbital, while positive deviation has been considered to arise from the spin–orbit coupling with the occupied orbitals [37,38]. The LESR signals of electrons at the fullerene cage showed only slight differences between the various molecule blends, indicating a rather weak influence of the different fullerene functional groups. However, g-tensors exhibited a different degree of anisotropy. In the case of asymmetric molecules, the highest and uniaxial anisotropy was observed for PC61BM, a little smaller and still uniaxial for [C60]T1, followed by [C60]P1 with no axial symmetry. Interestingly, in the case of symmetric molecules, the g-tensor for [C60]P2 showed very small anisotropy, whereas for [C60]T2, it was completely isotropic. The g-factor anisotropy may indicate a certain ordering of the fullerene molecules; therefore, the obtained results would indicate the best order for the PC61BM:P3HT blend.

Table 2. g-tensor components obtained for all studied fullerene-derivative–P3HT blends.

	g_x	g_y	g_z	g_{iso}
$PC61BM^-$	2.0001	2.0001	1.9992	1.9998
$P3HT^+$	2.0026	2.0023	2.0012	2.0020
$[C60]T1^-$	2.0005	2.0005	1.9998	2.0003
$P3HT^+$	2.0032	2.0022	2.0012	2.0022
$[C60]P1^-$	2.0003	2.0001	1.9998	2.0000
$P3HT^+$	2.0031	2.0022	2.0011	2.0021
$[C60]T2^-$	1.9999	1.9999	1.9999	1.9999
$P3HT^+$	2.0034	2.0022	2.0011	2.0022
$[C60]P2^-$	2.0002	1.9999	1.9999	2.0000
$P3HT^+$	2.0034	2.0022	2.0011	2.0022

3.3.2. LESR Signal Saturation

In order to study the properties of the fullerene derivative with P3HT blends, the LESR spectra as a function of microwave power were measured for the temperature range 5 K–60 K. In Figure 6, for clarity, only the data for the three selected temperatures are shown. As can be seen, the LESR signals tended to saturate, and as the temperature increased, a shift of the microwave power saturation point toward higher power values was observed. At low temperatures, the LESR signals were quickly saturated both for the electrons and the holes, indicating longer relaxation times than for higher temperatures. At 59 K, the electron and hole signals started to saturate at the same microwave power, which was about 0.05 mW.

$$I = A\sqrt{P}\left[1 + \left(2^{\frac{1}{\varepsilon}} - 1\right)\frac{P}{P_{\frac{1}{2}}}\right]^{-\varepsilon} \quad (1)$$

where A is the scaling factor, p is the microwave power, $P_{1/2}$ is the microwave power at which the first derivative amplitude is reduced to half of its unsaturated value, and ε is the homogeneity factor. When the saturation originates from inhomogeneous broadening, $\varepsilon = 0.5$, while for homogeneous broadening, $\varepsilon = 1.5$.

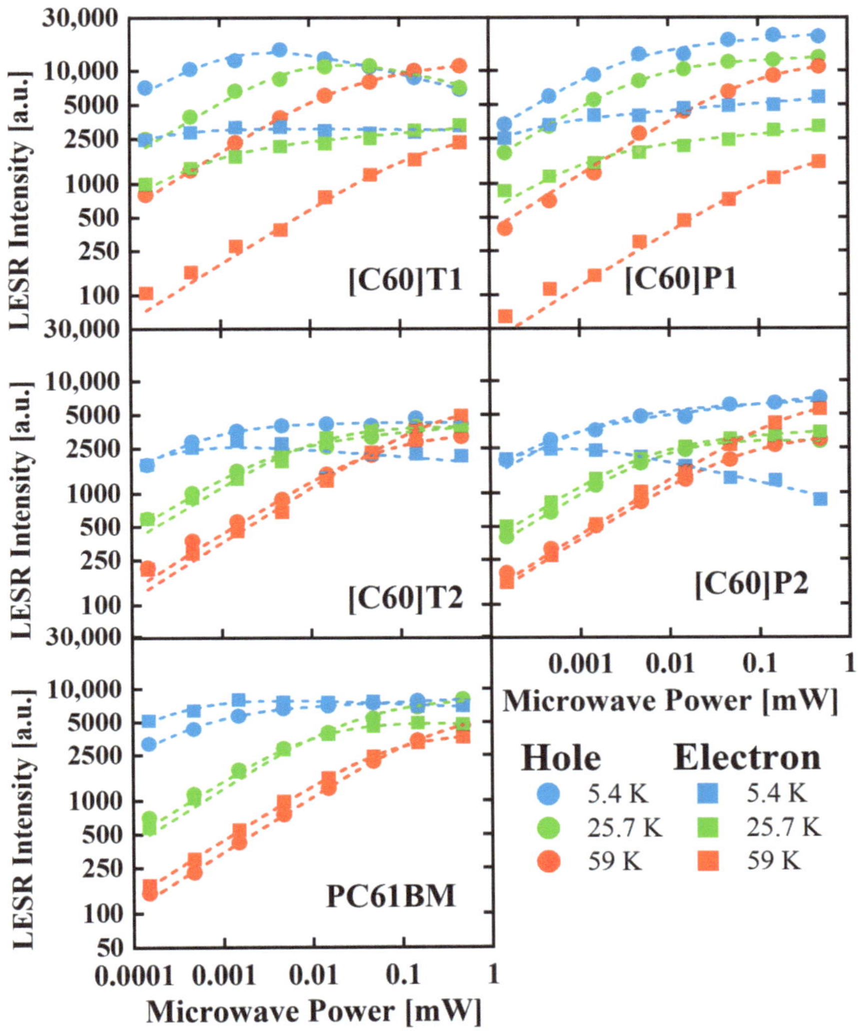

Figure 6. LESR signal intensities as a function of microwave power shown in log–log scale. The indicated fullerene derivatives were measured in blends with P3HT at different temperatures (the results for three selected temperatures are presented). The dashed lines were fitted with the experimental points using Formula (1).

A saturation of the ESR signal is often observed for higher microwave powers due to the finite relaxation time of the excited electrons with spin reversal by microwaves [39]. Generally, spin may relax its polarization but also coherence through the so-called spin–lattice or spin–spin interaction mechanisms. Spin–lattice relaxation involves the exchange of energy with lattice vibrations, causes longitudinal (polarization) relaxation of the spin, and its characteristic time is typically denoted as T_1. On the other hand, spin–spin interaction (with energy conservation) is effective in destroying coherence, and it is characterized by T_2 time. The saturation mechanism includes two extreme cases, called homogeneous and inhomogeneous broadening. The source of homogenous broadening is fluctuating fields, originating mostly from the electron and nuclear spin flips, occurring for an ensemble of spin systems experiencing the same time-averaged local fields. Such situation is typical for paramagnetic centers in single crystals. The width of a homogenous line is given by $2/T_m$, and T_m means the phase relaxation time, and it is often identical to T_2. On the other hand, in an inhomogeneous case, the ESR signal is a superposition of lines from spin ensembles with different Larmor frequencies and typically has a Gaussian shape with a width determined by inhomogeneity. The important sources of inhomogeneous broadening are different orientations of the grains, lack of a long-range order, or changes in the local environment of the spin ensembles in disordered solids (powder, glass, amorphous materials, irradiated or strained crystals), and the inhomogeneous broadening originates from the resulting anisotropy or slight changes of the interaction tensors. Additionally, hyperfine interactions cause inhomogeneous broadening. Whereas the ESR signal intensity is proportional to the square root of microwave power for a low microwave power range in both cases [39], its intensity reaches a maximum and then decreases with increasing microwave power in the case of homogenous broadening. In turn, for inhomogeneous broadening, the ESR signal intensity increases and remains almost at a constant value with increasing microwave power. The following formula describes the changes in ESR signal intensity as a function of microwave power [40,41].

To examine the source of the LESR line broadening, we fitted the experimental intensity of the LESR signal (double integral of the collected signal) as a function of microwave power using Formula (1) with ε as a fitting parameter. The fitted dashed curves are shown in Figure 6, and they reproduce the observed dependencies very well. The obtained values of the ε fitting parameter are presented in Table S3. As can be seen, for most of the curves, the value of the ε parameter is close to 0.5 with an uncertainty of 0.1. Such estimated error of fitting is mostly a result of experimental data processing—primarily the partial overlapping of LESR signals from the electron and the hole, low intensity of the signals in the low microwave powers range, as well as background subtraction. The obtained value of the ε parameter indicates the inhomogeneous origin of the LESR signal broadening. This is due to a significant disorder of the polymer–fullerene blend structure resulting in dispersion of the g-tensor values. A slightly higher value of the ε parameter, reaching 0.7 in a few cases, but still far from 1.5 (characteristic value for homogenous broadening), indicates the dominant contribution to the LESR line width from the disorder of the blend structures, causing local changes in the g-tensor.

The inhomogeneous origin of the LESR signal broadening means that the T_2 relaxation time cannot be derived from the LESR line full width at half maximum (FWHM). The FWHM size is rather a measure of the degree of variation in the local environments of photo-excited carriers [42]. The smallest line widths for both the polymer and the fullerene derivatives were recorded for the mixture of P3HT and PC61BM. This means the smallest degree of differentiation around the corresponding paramagnetic centers and suggests the best ordering of this mixture regarding both of its components. However, generally, the ESR studies indicate a high degree of heterogeneity in the vicinity of the paramagnetic centers, photo-generated electrons, and holes, which proves, in line with the intuition, a high degree of disorder of the polymer–fullerene blends. In agreement with our conclusion about inhomogeneous broadening in polymer–fullerene blends, the authors of Ref. [43] have attributed the wide distribution in longitudinal relaxation times (T_1) to the strong heterogeneity of the blend and the presence of energetically distributed traps.

3.4. Photoluminescence

The kinetics of photoluminescence spectra, TRPL, were measured at room temperature for pure P3HT, pure fullerene derivatives, and blends of P3HT with the fullerene derivatives. Such studies allowed us to observe the influence of the blend components on exciton recombination. The selected TRPL results (a complete set of spectra related to the [C60]P1 fullerene derivative) are shown in Figure 7, while the TRPL spectra for all compounds studied are presented in the Supplementary Materials (Figure S18). Radiative recombination of charge carriers originating from excitonic excitation is mostly governed by first-order kinetics; therefore, the concentration of excitons (n) should decay exponentially with time.

$$n(t) = Aexp\left(-\frac{t}{\tau}\right) \tag{2}$$

where A is a proportionality constant, and τ is the exciton decay time. The recombination rate of the studied materials is dominated by an exponential decay, clearly visible in the long time range (>0.5 ns) (in Figure 7, the contours are then approximately equidistant for a fixed energy). P3HT has quite strong luminescence with a maximum at about 1.8–1.9 eV and a decay time of about 0.2 ns. The [C60]P1 fullerene derivative has about 10 times weaker luminescence intensity. It radiates briefly (0.05 ns) in a broad band of 1.6–2.1 eV and then continues to emit a longer-lasting (ca. 0.5 ns decay time) signal at about 1.7 eV. Figure 8 shows in-sequence time-integrated photoluminescence (PL) spectra of the studied fullerene derivatives in comparison with the spectra of P3HT and the relevant fullerene-derivative–P3HT blend. All fullerene derivatives exhibited a PL peak at 1.7 eV. The P3HT spectrum was about an order of magnitude stronger than those of fullerene derivatives when comparing their maxima (note the logarithmic scale of PL intensity) and consisted of two broad lines at approximately 1.8 eV and 1.9 eV. In the PL spectrum of P3HT blends, it was possible to distinguish structures originating from P3HT and the respective fullerene derivative. In general, all spectra of the blends were attenuated compared to the spectra of the pure components. This was particularly evident for the P3HT contribution, which decayed almost equally strongly in all blends, and proved a fast transfer of electrons from P3HT to the respective fullerene derivative before these electrons could recombine radiatively within the exciton pairs in P3HT. Thus, the blends with all the fabricated fullerene derivatives showed similarly efficient transport of photo-generated electrons from P3HT to fullerenes through the LUMO states. What is worth noticing is that the blend of PC61BM:P3HT exhibited strong and fastest decay of the whole PL spectrum, not only in the part related to recombination within P3HT but also in the part related to PC61BM. The analogous components of the blend spectra with fullerene derivatives fabricated in this work showed generally weaker quenching of the PL spectrum related to fullerene recombination. Such observation can be explained by the weaker transfer of holes from the new fullerene derivatives to P3HT compared to the hole transfer from PC61BM. This may indicate an existence of a barrier between the HOMO levels of the new fullerene derivatives and P3HT, in contrast to the situation in the PC61BM: P3HT blend in which this transport is smooth.

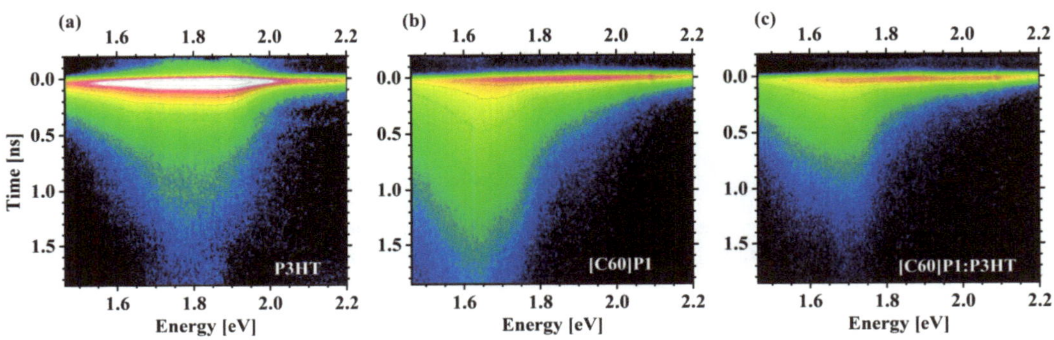

Figure 7. TRPL spectra of (**a**) P3HT, (**b**) [C60]P1, and (**c**) blend of [C60]P1 and P3HT at 3:2 molar ratio. Apparently, the luminescence of the blend is weaker and decays faster than those of the components. The plotted data are contour maps of luminescence intensity as a function of time (vertical scale) and energy (horizontal scale). The intensity is plotted in a logarithmic scale, so the intensity from contour to contour grows e times (where e is the Euler number).

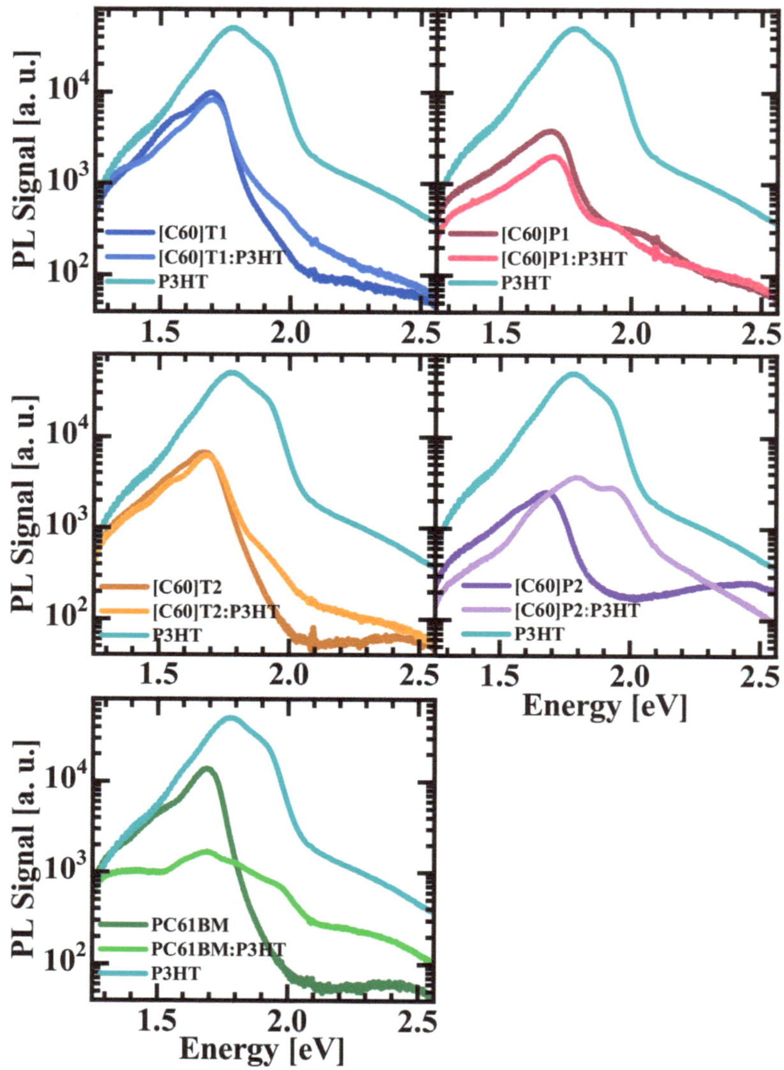

Figure 8. PL spectra of P3HT, fullerene derivative, and their blend.

A rather unusual TRPL spectrum was recorded for the [C60]P2: P3HT blend; it contained a significantly greater P3HT contribution than in the other cases. This may be due to improper phase separation inside the photosensitive layer, mainly due to the low solubility of the fullerene derivative [C60] P2, which excludes part of the derivative [C60] P2 from forming a relatively homogeneous blend phase. This is an issue, which was already discussed above in the context of the absorption spectra for the blends. The solubility table in the Supplementary Materials confirms these assumptions.

To analyze qualitatively the time-resolved luminescence of pure fullerene derivatives and compare it with their luminescence decay in the respective blends with P3HT, all TRPL spectra were integrated in the energy range corresponding to the spectral range of fullerene luminescence (1.54–1.8 eV). The results are shown in Figure 9. In other words, Figure 9 presents luminescence decays over time of the studied fullerene derivatives, both pure and in blends with P3HT. Data for [C60]P2 are not presented here due to the solubility problems of this fullerene derivative, whereby the luminescence of the blend was dominated by the P3HT luminescence (see Figure 8). It can be noticed that the decay of luminescence in blends was faster than that of pure fullerene in all cases except [C60]T2, for which both decays were the same. For pure fullerene derivatives, the TRPL curves were exponential in the long time range (>0.5 ns), which is observed as a straight line in Figure 9, with characteristic times of about 0.4 ns (estimated from fits created in this time range, as shown in Figure 9). This time region can be assigned to radiative recombination with recombination times typical for organic materials. For a short time range, up to about 0.3 ns, the photoluminescence decay occurred faster and was a result of the existence of defect centers in fullerene derivatives trapping the carriers or causing their non-radiative recombination and competing with radiation processes. The TRPL curves for the fullerene derivatives in blends had a similar exponential decay in the long time range, only shifting down in parallel to the curves for pure fullerene derivatives, with similar characteristic times of about 0.4 ns. As mentioned, this long time region can be assigned to the typical radiative recombination. The main difference between the decays of the curves for pure fullerene derivatives and those derivatives in blends occurred in the area of short times, up to about 0.4 ns. In this range, the decay of the photoluminescence occurred faster in the case of blends, which had to be a result of the photo-excited hole transfer from fullerene to P3HT. As can be seen, the hole transfer to P3HT occurred with greater or lesser efficiency in different blends, except for [C60]T2, in which this transfer was imperceptible when it came to TRPL curves. On the other hand, from Figure 9, it is obvious that the hole transfer was most effective in the case of the blend with PC61BM. These results correspond to LESR measurements where the most intensive signals were from the P3HT: PC61BM blend, and symmetrical molecules exhibited the lowest signal intensities, proving a weaker separation of the charges. To quantitatively describe the contribution of various processes to the luminescence decay and to estimate the characteristic times associated with them, we assumed that the decay rate of the exciton in a blend γ, reciprocal to its decay time $\gamma = 1/\tau_{blend}$, is a sum of radiative recombination (γ_{RR}) and carrier escape (γ_E) rates: $\gamma = \gamma_{RR} + \gamma_E$. In the above, we approximated the radiative decay rate by the decay rate for pure fullerene derivatives (and similar rate for fullerenes in blends) $\gamma_{RR} \approx 1/\tau_{pure}$ in the long time range as about 0.4 ns. The blend decay rate should therefore be equal to

$$1/\tau_{blend} = 1/\tau_{pure} + \gamma_E \tag{3}$$

τ_{blend} was estimated from the 0–0.4 ns range of the TRPL transients by the fitting curve

$$I(t) = I_0 e^{-\frac{t}{\tau}} + I_{inf} \tag{4}$$

where I_{inf} represents an emission of lifetime so long, that it is infinite in this time range. The obtained curves are plotted in Figure 9 in the respective fitting ranges.

For all the studied samples, it was about 40–80 ps. The calculated γ_E from Formula (4) exhibited the highest value for PC61BM (22 ns^{-1}), followed by the [C60]P1, [C60]T2, and [C60]T1 fullerenes with values of about 8–17 ns^{-1} (detailed results are presented in Table S4). This result quantitatively shows that the most efficient hole transfer takes place in the PC61BM:P3HT reference sample.

3.5. Preliminary Solar Cells Construction with the Newly Synthesized Fullerenes

Organic solar cells with an active layer in the form of a bulk heterojunction with the P3HT polymer were fabricated from the studied fullerene derivatives. The details regarding the manufacturing process and electrical parameters of the cells are provided in the Supplementary Materials, and they performed significantly worse than the reference cell with the PC61BM acceptor. Among the new fullerene derivatives, the cells with fullerenes functionalized with one aromatic substituent, i.e., [C60]P1 and [C60]T1, were better. However, they were still characterized by substantially lower short circuit current density (J_{SC}) and open circuit voltage (V_{OC}), and as a consequence, lower power conversion efficiency (PCE) than the reference cell with PC61BM. The cells with symmetric fullerene derivatives, functionalized with two aromatic substituents, [C60]T2 and [C60]P2, underperformed even more. Some improvement in cell parameters could probably be achieved through dedicated optimization for newly functionalized fullerenes; yet, this would require separate, in-depth research, which was not our goal at the moment. Secondly, the differences in the solubility of the synthesized fullerene derivatives certainly influenced the final structure of the active layer, and thus, its charge transport properties. Insufficient solubility was observed mainly for [C60] P2, where the PL signal was dominated by P3HT. This may be related to the sediments observed in the active layer. The investigated fullerene derivatives seem to have problems in forming a proper volumetric heterojunction due to their tendency to form thick, impermeable clusters or their inability to easily blend with the conductive P3HT polymer, the crucial factors influencing the final efficiency of the blend [44–46].

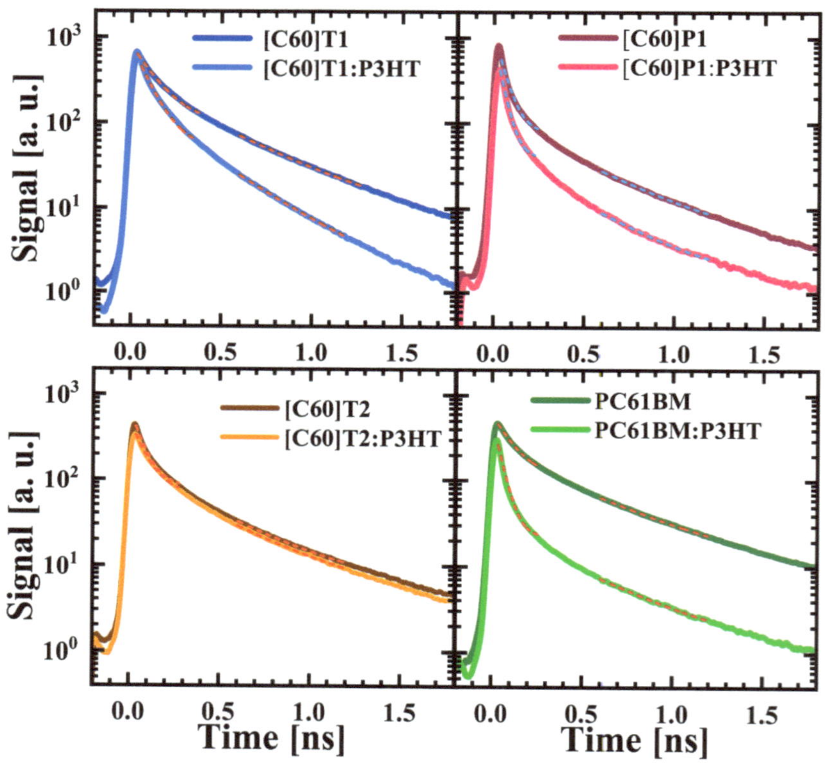

Figure 9. PL signal of the studied fullerene derivatives as integrated in the energy range of 1.54–1.8 eV for pure fullerenes and for fullerene–P3HT blends. The fits with single exponential decays are depicted with dashed lines.

The results of the presented research suggest that asymmetrical, dipole-like structures are more likely to create a desired phase-separated active layer. Additionally, it is important to note that the differences in PCE values were caused not only by the difference in solubility. As shown by the presented research, in blends with the synthesized fullerene derivatives, there was a problem in the transport of holes to P3HT, especially serious for symmetric fullerenes.

4. Conclusions

Four aromatic fullerene C_{60} derivatives were synthesized. They were studied to determine the optical properties and the efficiency of charge separation in blends with P3HT. The research showed that the applied chemical functionalization of the fullerene cage leads to only slight changes in the absorption spectra, since the absorption was dominated by excitations within the fullerene cage. Some additional absorption present in the visible range for fullerene derivatives with pyrene moieties was related to the internal excitations within the pyrene molecules. Its intensity increased with the number of pyrene substituents, which functionalized fullerene. Such excitations seem to be a chance for a modification of fullerene derivative absorption, so that the fullerenes can support the polymer as a sunlight absorber.

Light-induced ESR studies revealed two prominent, overlapping lines—one from a positive charge localized at the P3HT polymer and the other from a negative charge localized at the fullerene cage for all the studied blends. The reference blend with PC61BM showed the most intense LESR signal and PL quenching, which was a sign of efficient charge transfer. PC61BM with its good performance was followed by asymmetrical [C60]P1 and [C60]T1 derivatives, whereas the symmetrical [C60]P2 and [C60]T2 demonstrated a much weaker charge separation in the blends. An interesting result of the research is therefore the finding of stronger charge separation properties of asymmetrical derivatives in blends with P3HT compared to the symmetrical ones. This suggests that the interaction and mixing of materials are better in the case of molecules with a dipole moment. Detailed analysis of the LESR spectra saturation demonstrated inhomogeneity of the environments of paramagnetic centers in blends. However, the PC61BM: P3HT reference blend exhibited the lowest degree of heterogeneity for both the photo-excited electron and the hole surroundings.

The above results were consistent with the performance of the preliminary solar cells. The fullerenes functionalized with one aromatic substituent performed better in the active layer than their symmetrical counterparts bearing two aromatic moieties. Thus, in the case of asymmetric fullerene derivatives, their interfaces with P3HT and the overall arrangement of the blends allowed for better charge transfer. An interesting observation is also the inefficiency of the hole transfer from the synthesized fullerenes to P3HT, which was most severe for the symmetric fullerene derivatives.

Supplementary Materials: The following supporting information can be downloaded at: https://www.mdpi.com/article/10.3390/ma15196908/s1. Figure S1. ESI-MS spectrum of ethyl(2-(thiophen-3-yl)ethyl) malonate as a [M+Na]+ cation. Figure S2. IR spectrum of ethyl(2-(thiophen-3-yl)ethyl) malonate (neat). Figure S3. ^{1}H NMR spectrum of ethyl(2-(thiophen-3-yl)ethyl) malonate in $CDCl_3$. Figure S4. ^{13}C NMR spectrum of ethyl(2-(thiophen-3-yl)ethyl) malonate in $CDCl_3$. Figure S5. ESI-MS spectrum of [C60]T1 as $[M+Na]^+$ anion. Figure S6. IR spectrum of [C60]T1 in KBr disk. Figure S7. ^{1}H NMR spectrum of [C60]T1 in $CDCl_3$. Figure S8. ^{13}C NMR spectrum of [C60]T1 in $CDCl_3$. Figure S9. ESI-MS spectrum of bis(2-(thiophen-3-yl)ethyl) malonate as a $[M+Na]^+$ cation. Figure S10. IR spectrum of bis(2-(thiophen-3-yl)ethyl) malonate (CCl_4). Figure S11. ^{1}H NMR spectrum of bis(2-(thiophen-3-yl)ethyl) malonate in $CDCl_3$. Figure S12. ^{13}C NMR spectrum of bis(2-(thiophen-3-yl)ethyl) malonate in $CDCl_3$. Figure S13. ESI-MS spectrum of [C60]T2 as $[M+Na]^+$ anion. Figure S14. IR spectrum of [C60]T2 in KBr disk. Figure S15. ^{1}H NMR spectrum of [C60]T2 in $CDCl_3$. Figure S16. ^{13}C NMR spectrum of [C60]T2 in $CDCl_3$. Figure S17. ESR signal of C_{60} fullerene derivatives and P3HT blends at 30 K and 0.0047 mW microwaves power. Figure S18. TRPL spectra of (a) P3HT and (b–f) all studied C60 fullerene derivatives and their blend with P3HT. Figure S19. Spectrum of pyrene calculated by DFT. Figure S20. Manufactured organic solar cell scheme. Figure S21. I–V curves of studied solar cells with different active layers. Table S1. Approximate solubility of synthesized fullerene derivatives. Table S2. g-tensor components for negative (−) and positive (+) polarons located at fullerene sphere and functional group respectively during illumination. Table S3. The ε parameter values obtained from fitting of LESR signal intensity as a function of microwave power using Formula (1), and FWHM values of LESR signals. Table S4. Exciton decay time (τ_{blend}), radiative recombination time (τ_{pure}) and escape rate calculated from TRPL spectra presented in Figure 9. Table S5. HOMO (also HOMO-1 and HOMO-2 for selected compounds), LUMO energies, and

HOMO–LUMO energy differences of the studied fullerene derivatives obtained by DFT calculations. Table S6. Electrical parameters of manufactured solar cells.

Author Contributions: Conceptualization, M.K. (Maciej Krajewski), A.D., A.K. and M.K. (Maria Kamińska); methodology, M.K. (Maciej Krajewski), A.D., A.K., P.P. and K.P.K.; investigation, M.K. (Maciej Krajewski), P.P., W.M., K.P.K., J.W. and M.P.; writing—original draft preparation, M.K. (Maciej Krajewski); writing—review and editing, K.P.K., A.D., A.K. and M.K. (Maria Kamińska); visualization, M.K. (Maciej Krajewski), K.P.K.; supervision, A.D., A.K. and M.K. (Maria Kamińska); funding acquisition, M.K. (Maciej Krajewski), J.W. and A.K. All authors have read and agreed to the published version of the manuscript.

Funding: This research was funded by the National Centre for Research and Development (Poland), grant number TECHMATSTRATEG1/347431/14/NCBR/2018, and the Interdisciplinary Centre for Mathematical and Computational Modelling (ICM) at Warsaw University, grant number G87-1101. M. K. thanks the Excellence Initiative-Research University program for the financial support.

Institutional Review Board Statement: Not applicable.

Informed Consent Statement: Not applicable.

Data Availability Statement: The data presented in this study are available on request from the corresponding author.

Conflicts of Interest: The authors declare no conflict of interest.

References

1. Lalwani, G.; Sitharaman, B. Multifunctional Fullerene- and Metallofullerene-Based Nanobiomaterials. *Nano Life* **2013**, *3*, 1342003. [CrossRef]
2. Friedl, J.; Lebedeva, M.A.; Porfyrakis, K.; Stimming, U.; Chamberlain, T.W. All-Fullerene-Based Cells for Nonaqueous Redox Flow Batteries. *J. Am. Chem. Soc.* **2018**, *140*, 401–405. [CrossRef] [PubMed]
3. Bischak, C.G.; Flagg, L.Q.; Yan, K.; Li, C.Z.; Ginger, D.S. Fullerene Active Layers for n-Type Organic Electrochemical Transistors. *ACS Appl. Mater. Interfaces* **2019**, *11*, 28138–28144. [CrossRef]
4. Brabec, C.J.; Sariciftci, N.S.; Hummelen, J.C. Plastic solar cells. *Adv. Funct. Mater.* **2001**, *11*, 15–26. [CrossRef]
5. Deibel, C.; Dyakonov, V. Polymer-fullerene bulk heterojunction solar cells. *Rep. Prog. Phys.* **2010**, *73*, 096401. [CrossRef]
6. Lin, Y.; Firdaus, Y.; Isikgor, F.H.; Nugraha, M.I.; Yengel, E.; Harrison, G.T.; Hallani, R.; El-Labban, A.; Faber, H.; Ma, C.; et al. Self-assembled monolayer enables hole transport layer-free organic solar cells with 18% efficiency and improved operational stability. *ACS Energy Lett.* **2020**, *5*, 2935–2944. [CrossRef]
7. Li, S.; Zhan, L.; Jin, Y.; Zhou, G.; Lau, T.; Qin, R.; Shi, M.; Li, C.; Zhu, H.; Lu, X.; et al. Asymmetric Electron Acceptors for High-Efficiency and Low-Energy-Loss Organic Photovoltaics. *Adv. Mater.* **2020**, *32*, e2001160. [CrossRef] [PubMed]
8. Yu, X.; Lin, H.; Li, M.; Ma, B.; Zhang, R.; Du, X.; Zheng, C.; Yang, G.; Tao, S. Ternary organic solar cells with enhanced charge transfer and stability combining the advantages of polymer acceptors and fullerene acceptors. *Org. Electron.* **2022**, *104*, 106471. [CrossRef]
9. Hohenberg, P.; Kohn, W. Inhomogeneous Electron Gas. *Phys. Rev.* **1964**, *136*, 865–871. [CrossRef]
10. Kohn, W.; Sham, L. Self-Consistent Equations Including Exchange and Correlation Effects. *Phys. Rev.* **1965**, *140*, 1133–1138. [CrossRef]
11. Runge, E.; Gross, E.K.U. Time-dependent density-functional theory for multicomponent systems. *Phys. Rev. Lett.* **1984**, *52*, 997–1000. [CrossRef]
12. Ciechanowicz, S.A.G.; Korona, K.P.; Wolos, A.; Drabinska, A.; Iwan, A.; Tazbir, I.; Wojtkiewicz, J.; Kaminska, M. Toward Better Efficiency of Air-Stable Polyazomethine-Based Organic Solar Cells Using Time-Resolved Photoluminescence and Light-Induced Electron Spin Resonance as Verification Methods. *J. Phys. Chem. C* **2016**, *120*, 11415–11425. [CrossRef]
13. Niklas, J.; Mardis, K.; Banks, B.P.; Grooms, G.M.; Sperlich, A.; Dyakonov, V.; Beaupré, S.; Leclerc, M.; Xu, T.; Yu, L.; et al. Highly-efficient charge separation and polaron delocalization in polymer-fullerene bulk-heterojunctions: A comparative multi-frequency EPR and DFT study. *Phys. Chem. Chem. Phys.* **2013**, *15*, 9562–9574. [CrossRef] [PubMed]
14. Konkin, A.; Sensfuss, S.; Roth, H.-K.; Nazmutdinova, G.; Schroedner, M.; Al-Ibrahim, M.; Egbe, D. LESR study on PPV-PPE/PCBM composites for organic photovoltaics. *Synth. Met.* **2005**, *148*, 199–204. [CrossRef]
15. Zoriniants, G.; Dyakonov, V.; Scharber, M.; Brabec, C.; Janssen, R.; Hummelen, J.; Sariciftci, N. Light-induced ESR studies in conjugated polymer-fullerene composites. *Synth. Met.* **1999**, *102*, 1241–1242. [CrossRef]
16. Watanabe, S.-I.; Tanaka, H.; Kuroda, S.-I.; Toda, A.; Nagano, S.; Seki, T.; Kimoto, A.; Abe, J. Electron spin resonance observation of field-induced charge carriers in ultrathin-film transistors of regioregular poly(3-hexylthiophene) with controlled in-plane chain orientation. *Appl. Phys. Lett.* **2010**, *96*, 3–6. [CrossRef]
17. Poluektov, O.G.; Filippone, S.; Martín, N.; Sperlich, A.; Deibel, C.; Dyakonov, V. Spin signatures of photogenerated radical anions in polymer-[70]fullerene bulk heterojunctions: High frequency pulsed EPR spectroscopy. *J. Phys. Chem. B* **2010**, *114*, 14426–14429. [CrossRef]

18. Bingel, C. Cyclopropanierung von Fullerenen. *Eur. J. Inorg. Chem.* **1993**, *126*, 1957–1959. [CrossRef]
19. Nierengarten, J.; Gramlich, V.; Cardullo, F.; Diederich, F. Regio- and Diastereoselective Bisfunctionalization of C60 and Enantioselective Synthesis of a C60 Derivative with a Chiral Addition Pattern. *Angew. Chem. Int. Ed.* **1996**, *35*, 2101–2103. [CrossRef]
20. Herranz, M.Á.; Cox, C.T.; Echegoyen, L. Retrocyclopropanation reactions of fullerenes: Complete product analyses. *J. Org. Chem.* **2003**, *68*, 5009–5012. [CrossRef]
21. Piotrowski, P.; Mech, W.; Zarębska, K.; Krajewski, M.; Korona, K.; Kamińska, M.; Skompska, M.; Kaim, A. Mono- and Di-Pyrene [60]Fullerene and [70]Fullerene Derivatives as Potential Components for Photovoltaic Devices. *Molecules* **2021**, *26*, 1561. [CrossRef] [PubMed]
22. Frisch, M.J.; Trucks, G.W.; Schlegel, H.B.; Scuseria, G.E.; Robb, M.A.; Cheeseman, J.R.; Scalmani, G.; Barone, V.; Mennucci, B.; Petersson, G.A.; et al. *Gaussian 09, Revision D.01*; Gaussian, Inc.: Wallingford, CT, USA, 2009.
23. Stoll, S.; Schweiger, A. EasySpin, a comprehensive software package for spectral simulation and analysis in EPR. *J. Magn. Reson.* **2006**, *178*, 42–55. [CrossRef]
24. Ren, S.L.; Wang, Y.; Rao, A.M.; McRae, E.; Holden, J.M.; Hager, T.; Wang, K.; Lee, W.; Ni, H.F.; Selegue, J.; et al. Ellipsometric determination of the optical constants of C60 (Buckminsterfullerene) films. *Appl. Phys. Lett.* **1991**, *59*, 2678–2680. [CrossRef]
25. Barry, N.P.E.; Therrien, B. Pyrene: The Guest of Honor. In *Organic Nanoreactors*; Sadjadi, S., Ed.; Elsevier Inc.: Amsterdam, The Netherlands, 2016; pp. 422–461.
26. Holland, D.M.P.; Trofimov, A.B.; Seddon, E.A.; Gromov, E.V.; Korona, T.; de Oliveira, N.; Archer, L.E.; Joyeux, D.; Nahon, L. Excited electronic states of thiophene: High resolution photoabsorption Fourier transform spectroscopy and ab initio calculations. *Phys. Chem. Chem. Phys.* **2014**, *16*, 21629–21644. [CrossRef] [PubMed]
27. Zhang, C.; Chen, H.; Chen, Y.; Wei, Z.; Pu, Z. DFT Study on Methanofullerene Derivative [6,6]-Phenyl-C61 Butyric Acid Methyl Ester. *Acta Phys.-Chim. Sin.* **2008**, *24*, 1353–1358. [CrossRef]
28. Mohajeri, A.; Omidvar, A. Fullerene-based materials for solar cell applications: Design of novel acceptors for efficient polymer solar cells-A DFT study. *Phys. Chem. Chem. Phys.* **2015**, *17*, 22367–22376. [CrossRef]
29. Roy, J.K.; Kar, S.; Leszczynski, J. Optoelectronic properties of C60 and C70 fullerene derivatives: Designing and evaluating novel candidates for efficient P3HT polymer solar cells. *Materials* **2019**, *12*, 2282. [CrossRef]
30. Peverati, R.; Truhlar, D.G. Quest for a universal density functional: The accuracy of density functionals across a broad spectrum of databases in chemistry and physics. *Philos. Trans. R. Soc. A Math. Phys. Eng. Sci.* **2011**, *372*, 1–52. [CrossRef]
31. Sperlich, A.; Kraus, H.; Deibel, C.; Blok, H.; Schmidt, J.; Dyakonov, V. Reversible and irreversible interactions of poly(3-hexylthiophene) with oxygen studied by spin-sensitive methods. *J. Phys. Chem. B* **2011**, *115*, 13513–13518. [CrossRef]
32. Carati, C.; Gasparini, N.; Righi, S.; Tinti, F.; Fattori, V.; Savoini, A.; Cominetti, A.; Po, R.; Bonoldi, L.; Camaioni, N. Pyrene-Fullerene Interaction and Its Effect on the Behavior of Photovoltaic Blends. *J. Phys. Chem. C* **2016**, *120*, 6909–6919. [CrossRef]
33. Krinichnyi, V.I.; Yudanova, E.I.; Denisov, N.N. Light-induced EPR study of charge transfer in poly(3-hexylthiophene)/ fullerene bulk heterojunction. *J. Chem. Phys.* **2009**, *131*, 044515. [CrossRef] [PubMed]
34. Krinichnyi, V.I.; Yudanova, E.I. Light-induced EPR study of charge transfer in P3HT/bis-PCBM bulk heterojunctions. *AIP Adv.* **2011**, *1*, 022131. [CrossRef]
35. Brabec, C.J.; Zerza, G.; Cerullo, G.; De Silvestri, S.; Luzzati, S.; Hummelen, J.C.; Sariciftci, S. Tracing photoinduced electron transfer process in conjugated polymer/fullerene bulk heterojunctions in real time. *Chem. Phys. Lett.* **2001**, *340*, 232–236. [CrossRef]
36. Wudl, F.; Sariciftci, N.; Smilowitz, L.; Heeger, A. Photoinduced Electron Transfer from a Conducting Polymer to Buckminsterfullerene. *Science* **1992**, *258*, 1474–1476. [CrossRef]
37. Stankowski, J.; Piekara-Sady, L.; Kempiński, W. EPR of a fullerene-molecule-derived paramagnetic center as a mesoscopic conducting object. *Appl. Magn. Reson.* **2000**, *19*, 539–546. [CrossRef]
38. Möbius, K. Untersuchung von einfachen pi-Elektronensystem en mit Hilfe von Elektronenspin-Resonanz und Polarographie. *Z. Nat.* **1965**, *20a*, 1093–1102. [CrossRef]
39. Wertz, J.E.; Bolton, J.R. *Electron Spin Resonance—Elementary Theory and Practical Applications*, 1st ed.; Chapman and Hall: London, UK, 1986.
40. Lembicz, F.; Podgórska, D.; Ukielski, R.; Piatek, M. Saturation ESR spectroscopy of PA12 polyamide. *Eur. Polym. J.* **2004**, *40*, 1217–1221. [CrossRef]
41. Altenbach, C.; Greenhalgh, D.A.; Khorana, H.G.; Hubbell, W.L. A collision gradient method to determine the immersion depth of nitroxides in lipid bilayers: Application to spin-labeled mutants of bacteriorhodopsin. *Proc. Natl. Acad. Sci. USA* **1994**, *91*, 1667–1671. [CrossRef]
42. Beletskaya, E.A.; Lukina, E.A.; Uvarov, M.N.; Popov, A.A.; Kulik, L.V. Geminate recombination in organic photovoltaic blend PCDTBT/PC71BM studied by out-of-phase electron spin echo spectroscopy. *J. Chem. Phys.* **2020**, *152*, 044706. [CrossRef]
43. Lukina, E.A.; Popov, A.A.; Uvarov, M.N.; Suturina, E.A.; Reijerse, E.J.; Kulik, L.V. Light-induced charge separation in a P3HT/PC70BM composite as studied by out-of-phase electron spin echo spectroscopy. *Phys. Chem. Chem. Phys.* **2016**, *18*, 28585–28593. [CrossRef]

44. Presselt, M.; Herrmann, F.; Hoppe, H.; Shokhovets, S.; Runge, E.; Gobsch, G. Influence of phonon scattering on exciton and charge diffusion in polymer-fullerene solar cells. *Adv. Energy Mater.* **2012**, *2*, 999–1003. [CrossRef]
45. Laquai, F.; Andrienko, D.; Mauer, R.; Blom, P.W.M. Charge Carrier Transport and Photogeneration in P3HT: PCBM Photovoltaic Blends. *Macromol. Rapid Commun.* **2015**, *36*, 1001–1025. [CrossRef]
46. Campoy-Quiles, M.; Ferenczi, T.A.M.; Agostinelli, T.; Etchegoin, P.G.; Kim, Y.; Anthopoulos, T.D.; Stavrinou, P.N.; Bradley, D.D.C.; Nelson, J. Morphology evolution via self-organization and lateral and vertical diffusion in polymer:fullerene solar cell blends. *Nat. Mater.* **2008**, *7*, 158–164. [CrossRef] [PubMed]

materials

MDPI

Article

Characterization of Synovial Fluid Components: Albumin-Chondroitin Sulfate Interactions Seen through Molecular Dynamics

Natalia Kruszewska [1,*], Adam Mazurkiewicz [2], Grzegorz Szala [2], Małgorzata Słomion [3]

[1] Institute of Mathematics and Physics, Bydgoszcz University of Science and Technology, Kaliskiego 7 Street, 85-796 Bydgoszcz, Poland
[2] Faculty of Mechanical Engineering, Bydgoszcz University of Science and Technology, Kaliskiego 7 Street, 85-796 Bydgoszcz, Poland
[3] Faculty of Management, Bydgoszcz University of Science and Technology, Kaliskiego 7 Street, 85-796 Bydgoszcz, Poland
* Correspondence: nkruszewska@pbs.edu.pl

Abstract: The friction coefficient of articular cartilage (AC) is very low. A method of producing tailor-made materials with even similar lubrication properties is still a challenge. The physicochemical reasons for such excellent lubrication properties of AC are still not fully explained; however, a crucial factor seems to be synergy between synovial fluid (SF) components. As a stepping stone to being able to produce innovative materials characterized by a very low friction coefficient, we studied the interactions between two important components of SF: human serum albumin (HSA) and chondroitin sulfate (CS). The molecular dynamics method, preceded by docking, is used in the study. Interactions of HSA with two types of CS (IV and VI), with the addition of three types of ions often found in physiological solutions: Ca^{2+}, Na^{+}, and Mg^{2+}, are compared. It was found that there were differences in the energy of binding values and interaction maps between CS-4 and CS-6 complexes. HSA:CS-4 complexes were bound stronger than in the case of HSA:CS-6 because more interactions were formed across all types of interactions except one—the only difference was for ionic bridges, which were more often found in HSA:CS-6 complexes. RMSD and RMSF indicated that complexes HSA:CS-4 behave much more stably than HSA:CS-6. The type of ions added to the solution was also very important and changed the interaction map. However, the biggest difference was caused by the addition of Ca^{2+} ions which were prone to form ionic bridges.

Keywords: molecular dynamics simulations; human serum albumin; chondroitin sulfate; synovial fluid; energy of binding; hydrophobic interactions; hydrogen bonds; ionic interactions

Citation: Kruszewska, N.; Mazurkiewicz, A.; Szala, G.; Słomion, M. Characterization of Synovial Fluid Components: Albumin-Chondroitin Sulfate Interactions Seen through Molecular Dynamics. *Materials* **2022**, *15*, 6935. https://doi.org/10.3390/ma15196935

Academic Editor: Antonina Saija

Received: 31 July 2022
Accepted: 1 October 2022
Published: 6 October 2022

1. Introduction

Many novel materials were designed based on ideas that mimic Nature. The approach became so common that a new interdisciplinary branch of science called biomimetics was developed. It is sometimes hard to produce innovative materials with properties nearing or matching the original naturally occurring biological systems. A great example of a natural biosystem characterized by hard-to-mimic properties is articular cartilage (AC). The friction coefficient of AC is very low—it is about ten times lower than ice on ice. Many experimental attempts to characterize the rheological properties of AC were performed. However, they have only been conducted in vitro, never taking into account all the components and features of the system. Physical measurements of the friction coefficient in synovial joints on standard industrial tribometers do not often give satisfactory results. The problems are the complicated shapes of the articular surfaces, load, variable speed, and direction of mutual movement of the surfaces or the variability of the roughness of cartilage [1]. In addition, during movement, the pressure inside the joint system also changes, resulting from changes in the volume of the joint's capsule resulting from the work of the muscles and ligaments surrounding the joint. In addition, during movement, a temporary local load on the cartilage, and consequently, deformation of the articular surfaces appear. Because a layer of AC is very thin, the trabecular bone supporting the cartilage is also elastically deforming during load on bones under motion [2–4]. The mentioned factors make the physical measurements of the friction coefficient very difficult because many different aspects must be considered when planning the experiment. As a consequence, the results of measurements do not always correspond to the real values.

In [5], the authors presented an example of experimental measurement of the friction coefficient of a human shoulder's joint during reciprocal loading in a pendulum testing device at a wide range of sliding speeds. The authors found that the friction coefficient remains very low (0.0015–0.006) for up to 24 hours of continuous loading. They claimed that the low friction coefficients observed in incongruent joints represent rolling rather than sliding friction. A possibility of lowering friction forces in the AC by altering the characteristic action of its components (phospholipid micelles trapped in the network of hyaluronic acid chains) from sliding to rolling was noted in [6]. Another example of friction measurements was reported in [7] for ACs collected from bovine knees. Using sliding pin-on-disc tribotester T-11 under physiological lubrication conditions, the authors measured friction coefficients versus wettability and obtained values in a range of 0.005–0.025.

It is not clearly defined which lubrication model best describes the lubrication mechanism of the AC. The phenomenon has been a subject of many theoretical considerations [8–10]. AC properties depend on the lubrication regime, which depends, in turn, on the amount of load on the system and the health of the joint [11]. Therefore, it is hard to propose an experiment that can imitate the in vivo system, but the puzzle can be solved by analyzing the interactions between system components. Because of all the complexities mentioned above, computer experiment methods appear to be very helpful in explaining many system behaviors [6,12–15].

A synovial fluid (SF) is present between the two opposite cartilages. It is composed mainly of hyaluronic acid (HA), lubricin, phospholipids, and various proteins. SF plays an essential role in synovial joint lubrication [16–18]. Changes in synovial fluid volume and composition reflect changes within the joints [19,20]. This is very important from the medical point of view because diseases can change the balance of the components of synovial fluid. For example, due to various diseases, the concentration of phospholipids and protein can be increased during the concentration, and the molecular weight of the HA can be decreased [21]. These medically important observations point out that the system should be studied as a whole; however, it is very hard due to its complexity. Recently, many research studies focused on subsystems, investigating interactions in pairs [12,13,22–24]. Following this approach, in the present paper, we study interactions between two SF elements: human serum albumin (HSA) and chondroitin sulfate (CS) immersed in a water environment. The importance of ions added to the solution is considered.

The first findings about the binding of CS to HSA were reported in [25]. The Authors performed an experiment using a spectropolarimeter with a UV circular dichroism attachment. Although they found that forces exist between the two molecules, the nature of the forces remained unknown.

In the present research, the intermolecular interaction in the system was studied to describe forces between HSA and CS. A computer simulation method using the molecular dynamics approach was used. A computer model of two complexes was created: HSA:CS-4 and HSA:CS-6. The molecular system was first prepared using the molecular docking (MDoc) method of dry molecules and next studied by molecular dynamics (MD) simulations in an aqueous environment (resembling physiological conditions). The number of intermolecular hydrogen bonds (HBo), hydrophobic–polar interactions (HP), ionic interactions, bridges (water and ionic), and energy of binding (EoB) between HSA and CS-4/CS-6 were calculated to determine the system's dynamics. Moreover, the exact maps of contact were created to show the places of bindings.

2. Materials and Methods

2.1. Characterisation of Simulated Materials

CS stands as a pivotal component of synovial fluid. It is a highly negatively charged unbranched glycosaminoglycan (GAG) compound of N-acetylgalactosamine and glucuronic acid [26]. Two variants of CS are found in human joints: chondroitin sulfate IV and VI (denoted as CS-4 and CS-6, respectively). The difference between both CS types is presented in Figure 1.

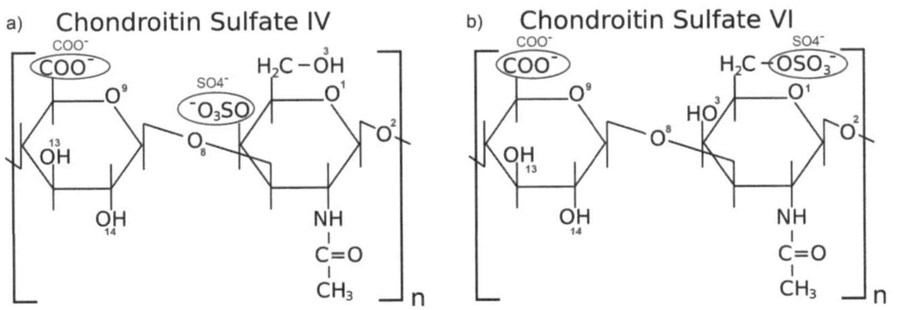

Figure 1. Schematic representation of the CS-4 (**a**) and CS-6 (**b**). The numbers of oxygen atoms and functional groups involved in forming hydrogen bonds are labeled with red letters.

The relative concentration of CS-4 versus CS-6 in joints is associated with cartilage calcification, while the calcification is related to histological degeneration of the joint [27]). In healthy cartilage, the concentration of both components is similar. In contrast, fully calcified cartilages usually have only CS-6 present in the synovial fluid [28]. An assessment of the influence of CS on the frictional–compressive properties of articular cartilage was presented in [29]. The authors deduced that the phenomenon of counter-ion condensation onto the CS chains influences the thermodynamic and frictional-compressive properties of the cartilage system.

This condensation tendency changes with the concentration of ions in the solution. An increase in condensation decreases the electrostatic friction between the chains and their resistance against gliding. Moreover, they demonstrated that the hydration shells of the counter-ions become smaller, which diminishes the resistance of the chains against compression. Their study concluded that at the physiological salt concentration chondroitin sulfate solutions possess optimal frictional–compressive properties. The CS molecules immersed in the SF are exposed to contact not only with ions but also with other molecules, such as proteins [30]. Therefore, the direct non-covalent intermolecular interactions appearing in the system, and indirect ones created by water and ionic bridges, could be of great importance for the rheological and tribological properties of the joint.

Experimental results indicate that CS is an effective lubricant for cartilage under mixed-mode lubrication conditions [31,32]. In [32], the authors speculated that the cartilage tissue might have a specific affinity for lubricants with negatively charged groups and hydroxyl groups on GAG chains, which may help them adsorb better to the cartilage surface, providing effective lubrication. Moreover, in [33], it was shown that HA or CS, when used alone, significantly lower the friction torque and dissipated energy of the fretting interface, which reduces the damage to the articular cartilage. It was also shown that HA and CS combined provided even better cartilage layer protection.

HSA is the most abundant protein in the SF. It is built on a single chain of 585 amino acids. It contains three homologous domains: I, II, and III [34]. Domains include residues (amino acids) as follows: Domain I: 5–197, domain II: 198–382, and domain III: 383–569. Each domain comprises two sub-domains termed A and B (IA: 5–107, IB: 108–197, IIA: 198–296, IIB: 297–382, IIIA: 383–494, and IIIB: 495–569), see Figure 2. The HSA regions responsible for the binding of ligands are known as Sudlow's Site I and II and are located in subdomain IIA and IIIA, respectively. The heme binding site is located in subdomain IB [35–37].

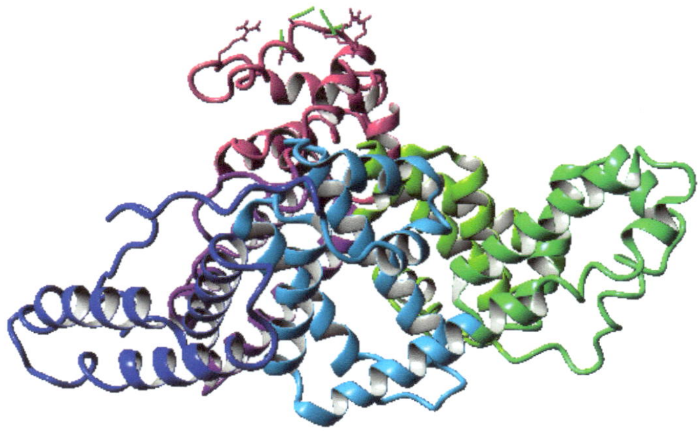

Figure 2. Ribbon representation of albumin in YASARA. Albumin domains are colored as follows: IA-pink, IB-violet, IIA-light green, IIB-green, IIIA-light blue,and IIIB-blue.

HSA shows characteristic binding and transporting properties with fatty acids [38,39], steroids [40], bilirubin [41], ions, and many other molecules. It can also interact with HA. The strength of the interactions depends on, e.g., the amount of ions provided in the system, and affects the rheological and tribological properties of the cartilage [12]. HSA:HA-based complexes are good lubricants, and they considerably lower the friction coefficient [42,43]. New thin-film materials based on albumin and HA have been proposed to be used in biomedicine and cosmetics due to their adhesive properties [44]. Albumin-based nanomaterials are also proposed to be used in drug delivery and many other biomedical applications because, as natural agents, they have high biosafety and biodegradability [45]. Despite HSA and HA having a total negative charge under the physiological conditions, positively charged amino acids in albumin favor interactions with the ionized carboxylic groups in the HA [12]. On the other hand, CS is more negatively charged than HA due to the content of the sulfate group; thus, information about the lubrication properties of HSA:CS complexes could be valuable. Accordingly, to the author's knowledge, no experiments study this complex's friction properties. Moreover, the information about the structural features of HSA:CS molecular complexes and their intermolecular interaction characteristics [46] is still limited.

2.2. Molecular Dynamic Simulation Details

All-atom computer simulations of the model biosystem consisting of HSA and CS-4 or CS-6 molecules were performed. Thus, interactions inside two complexes (HSA:CS-4 and HSA:CS-6) were studied and compared to determine if a place of the sulfate group in the GAGs influences the binding properties.

In a first step, a molecular unbiased docking method was used to find the most energetically optimal places where CS-4 and CS-6 attach to the HSA. Next, 10 from the energetically best-docked structures (sorted from the strongest connection to the weakest connection), with added water solution of chosen ions (Na^+, Mg^{2+}, Ca^{2+} and Cl^-), were subjected to MD simulations. Each realization had a system docked to a different part of HSA; as such, it represented different initial conditions. Both simulations (MDoc and MD) were performed using YASARA molecular modeling software [47].

Chemical structures of single units of CS-4 and CS-6 were obtained from Pubchem and modified to obtain 24 unit chains (around 8 kDa—longest chain allowable to dock with the modeling software used). The HSA structure was taken from the Protein Data Bank

(PDB code: 1e78). Homological modeling using FASTA was performed before docking to fill in atoms missing in the PDB file.

To obtain the most stable complexes of CS ligand docked to HSA, the VINA method [48] was used with their default parameters and point-charge force field [49] initially assigned according to the AMBER14 force field [50]. Next, the system was damped to mimic the less polar Gasteiger charges used to optimize the AutoDock scoring function. A flexible receptor and ligand approach was chosen while docking. In both cases (HSA:CS-4 and HSA:CS-6), 10 of the strongest bound distinctive complexes which differ in the position of GAG docked to HSA with −10 kcal/mol free energy of binding were prepared for MD simulation.

Before MD simulations were conducted, optimization of the hydrogen bonding network was performed to increase the solute stability and pK_a prediction to fine-tune the protonation states of the protein residues at the given pH= 7.4 [51,52]. Optimization was based on three steps as follows: first, pK_a prediction was carried out to consider the influence of the pH on the hydrogen bonding network; next, nonstandard amino acids and ligands were fully accounted for with the use of a chemical knowledge library in SMILES format; finally, the SCWRL algorithm was used to help find the globally optimal solution [53].

Both complexes were immersed in one of the three aqueous 2% salt solutions, NaCl, $CaCl_2$, or $MgCl_2$. After necessary minimization of the model system to remove clashes, the simulation was run for 140 ns using the AMBER14 force field [50] for the HSA, GLYCAM06 [54] for CS-4 and CS-6, and TIP3P for water. The cut-off distance for the van der Waals forces was set to 10 Å[55]. For computing long-range interactions (e.g., electrostatic interactions), the Particle Mesh Ewald algorithm was used [56]. Simulations were performed in a temperature of 310 K and under the pressure of 1 atm (NPT ensemble) [52]. A Berendsen barostat and thermostat were used to maintain constant temperature and pressure (relaxation time of 1 fs) [57]. Periodic boundary conditions were applied to a box of size equal to 120 × 110 × 110 Å^3. The equations of motion were integrated with multiple time steps of 1.25 fs for bonded interactions and 2.5 fs for non-bonded interactions. The time step between stored states of the systems was equal to 100 ps. Thus, the time series for 140 ns of simulations obtained 1400 save points.

In order to characterize the binding between the macromolecules, the energy of binding and the number of intermolecular interactions (direct and via bridges) were computed. The energy of binding, obtained using YASARA's algorithm, is the value (negative in principle) of the change of the energy of the system due to binding between the receptor (HSA) and the ligand (CS). The lower the binding energy, the stronger interaction between the components will be. The binding energy equation, which is the energy needed to disassemble a whole system into separate parts (which is equal to the energy of binding but with a reversed sign), has a form Equation (1)

$$E_{bind} = E_p^r + E_p^l + E_s^r + E_s^l - E_p^c - E_s^c, \tag{1}$$

where E_p^r and E_p^l are potential energies of separated compounds, i.e., receptor: HSA (r) and ligand: CS (l), E_s^r and E_s^l are their solvation energies, E_p^c and E_s^c are the potential and solvation energies of receptor–ligand complex. The energy of binding was computed using YASARA macro md_analyzebindenergy with the assumption that the cost of exposing one Å^2 to the solvent is 0.65 kJ/mol. According to the YASARA Manual, the energy of binding may be shifted by an unknown constant that depends on the receptor (so on the value of the parameter mentioned above); thus, this quantity can only be used to compare various protein–ligand affinities rather than an absolute value.

3. Results and Discussion

In order to check whether the simulation times were long enough, the stability of the simulation based on the root-mean-square deviation (RMSD) of modeled molecules was computed. Exemplary RMSD results (for two best-bound complexes) are presented in Figure 3, and the rest are included in the Supplementary Materials in Figures S1 and S2.

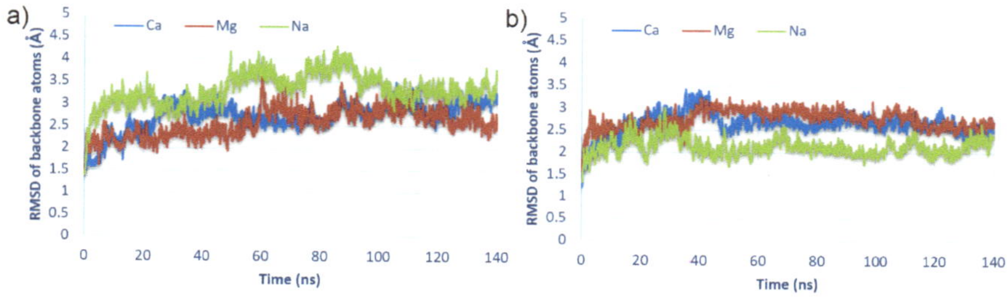

Figure 3. RMSD of backbone atoms of HSA:CS complexes as a function of time for the strongest bound complexes: (**a**) #2 for HSA:CS-6; and (**b**) #3 for HSA:CS-4 (cf., Table 1).

Table 1. Binding ranks of HSA-CS6 (up) and HSA-CS4 (down) complexes. The first column contains two numbers: rank after MD simulations averaged over EoB, and in parenthesis, rank after docking (and before MD simulations). The second column provides values of EoB with STD, averaged over the part of the trajectory after equilibration (and for all ions). The strongest connected domains are marked with bold letters.

HSA-CS6 Rank	EoB (kJ/mol)	HSA Binding Sites
1(2)	−2522 ± 339	IA-IB-IIA-**IIIA**-IIIB
2(6)	−2133± 301	IB-IIIA-**IIIB**
3(9)	−1694± 665	IB-IIIA-**IIIB**
4(7)	−1670± 388	**IA**-IIA-IIB-IIIA
5(1)	−1628± 792	IA-IIA-IIIA-**IIIB**
6(5)	−1542± 665	IA-IIA-IIIA-**IIIB**
7(4)	−1498± 399	IA-IIA-IIIA-**IIIB**
8(3)	−1472± 675	IA-**IIA**
9(10)	−1363± 550	IIA-**IIB**
10(8)	−1033± 453	**IA**-IB
HSA-CS4 Rank	**EoB (kJ/mol)**	**HSA Binding Sites**
1(3)	−2755 ± 624	IB-IIIA-**IIIB**
2(9)	−2737 ± 386	IB-IIA-**IIIA**-IIIB
3(8)	−2194 ± 702	IA-IB-**IIA**
4(10)	−1906 ± 556	**IA**-IB-IIA-IIIA
5(4)	−1904 ± 441	IA-IB-**IIA**-IIB
6(5)	−1641 ± 659	IB-IIA-IIB-**IIIA**-IIIB
7(7)	−1639 ± 374	**IA**-IB-IIA-IIIA-IIIB
8(2)	−1637 ± 674	IA-**IB**-IIIA-IIIB
9(1)	−1613 ± 531	IB-IIA-IIIA-**IIIB**
10(6)	−1516 ± 493	IB-IIIA-**IIIB**

In virtually every case, the RMSD oscillated around a value of 2.5–3 , indicating that the system reached stability in the given simulation time. Furthermore, in most cases, systems stabilized near 40 ns; thus, during intermolecular interactions analysis, the averaged results in the range from 40 ns to 140 ns were taken.

Moreover, the HSA and CS mobility was analyzed by calculating the time-averaged root mean square fluctuation (RMSF) values of HSA:CS-6 and HSA:CS-4 complexes.

The RMSF as a function of a number of the atom is presented in the Supplementary Materials in Figures S3 and S4. In order to compare the stability of HSA:CS-6 and HSA:CS-4 complexes, a sum of the RMSF values over all atoms belonging to the specific parts of the HSA and CS, and over all ten realizations for a specific complex in a specific solution, is presented in Figure S5 of the Supplementary Materials. In the case of HSA, the parts are the protein domains. In the case of CS, the 24 mers were divided into eight groups (each group then has three mers) to simplify the presentation of the results. The behavior of the RMSF maximums seen in Figure S3 of the Supplementary Materials is very similar to those presented for pure HSA computed for the protein amino acids [58]. Comparison of the results for HSA:CS-6 and HSA:CS-4 complexes in Figure S5 of the Supplementary Materials has shown that sum of the RMSF is almost always greater for the case of CS-6 than for CS-4. The greater difference can be seen in the case of the NaCl solution. This suggests that both complexes immersed in NaCl solution are the most unstable because atoms fluctuated more than in all other cases. In general, the peripheral parts of the molecules fluctuated more than the middle ones, but it is mainly seen in the case of NaCl. The difference between the two charts in Figure S5 for HSA atoms was greater for HSA:CS-6 complexes, especially in the case of $MgCl_2$. For CS, the difference in the fluctuation of CS-6 atoms was greater than CS-4, especially in the case of NaCl solution (cf., Figure S5b). The most negligible differences were noticed for $CaCl_2$, which indicates more stable complexes.

In order to check the functionally important region and atoms fluctuations versus ligand contacts, the RMSF of the atoms belonging to the best-bound domains (IIIA for HSA:CS-6; Sudlow's site II, and IIIB for HSA:CS-4) are presented in Figure 4.

In the figure, four of the strongest bound places are marked by arrows. The colors of the arrows show which atom from HSA has bound to the atom coming from CS. In the HSA:CS-6 complex, Pro (carbon atom) made HP interactions with the carbon atom of CS (blue arrow). A single carbon atom from CS (red arrow) created a few HP interactions with HSA bounding two different regions of HSA with CS. Carbon atoms from Glu and Asp created Ca^{2+} bridges with carbon atoms of CS (yellow and orange arrows). The existence of HP and HBo interactions of CS with Lys475 (marked with red letters) which is indicated as the binding site of long-chain fatty acids can be very important [37]. In HSA:CS-4, Lys had a few interactions with the carbon atom of CS (blue arrow). Glu ionic bridged with carbon atoms of CS (orange and yellow arrows; C-Ca-C bridges). Aps ionic bridged with the oxygen atom of CS (green arrow, C-Ca-O bridge). Usually, binding places from CS have lower values of RMSF, but in the case of HSA, it was not a rule. Thus, a stabilizing effect of CS on HSA cannot be reported. Note that in Figure 4, HP interactions, HBo, and ionic bridges are shown, but water bridges have been omitted due to their very short duration. At almost every MD step, other atoms took part in creating these bridges. In Figure S6 of the Supplementary Materials, the RMSF for the IIIA domain for the HSA:CS-4 complex is shown to compare the same fragments between the two best-bound complexes. It can be seen that different parts of HSA interacted with CS-4 in a different place than with CS-6, and mainly the interaction was with Lys and Glu. In general, RMSF values were lower for CS-4 for both HSA and CS molecules.

In the present paper, the focus of the study was on specifying the bonding place of CS to albumin. The changes in albumin conformations were not the subject of the present study as it would need longer simulation times. However, a preliminary analysis of the secondary structure of HSA was performed. Their oscillations as a function of time are presented in Figure S7 of the Supplementary Materials. Moreover, the comparison of the secondary structures at the end of the simulation (at 140 ns) for the two CS isomers in different ionic solutions is shown in Figure S8 of the Supplementary Materials. Only a slight difference can be seen in the percentage content of helices and turns. HSA bound to CS-6 has more turns and fewer helices than in the case of the HSA:CS-4 complex.

Electrostatic interactions are essential for the binding mechanism of HSA and GAG complexes [59]. Electrostatic potential maps of albumin (with and without the addition of ions) are presented in [12] (cf., Figure 3 therein). The authors have shown that the presence of Na^{+}, Mg^{2+}, and Ca^{2+} cations caused a much higher positive charge density that could be observed in the middle of the electrostatic potential map of the HSA. This way, a specific cavity was formed to which GAGs' negatively charged groups have a better chance to bind. This cavity is larger for divalent ions Mg^{2+} and Ca^{2+} than for monovalent Na^{+}. Some HSA domains are more likely to bind to GAGs than others; however, the binding map can be altered under the condition of a disease [59,60]. The binding mechanism is mainly due to ionic bonding, hydrogen bonding, and hydrophobic interactions [59].

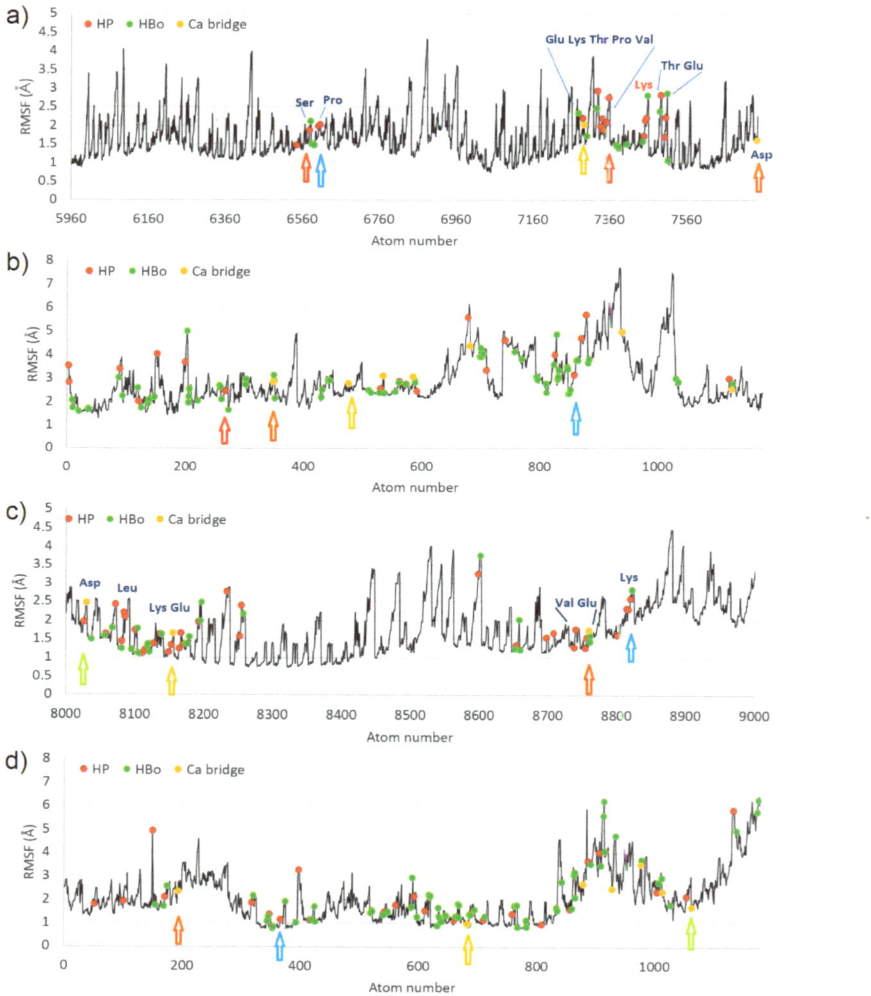

Figure 4. RMSF for best-bound domains of HSA (**a**,**c**) and CS (**b**,**d**) for complexes #2 HSA:CS-6 (**a**,**b**), and #3 HSA:CS-4 (**c**,**d**); the places of specific interactions are marked with dots. In the case of HSA:CS-6, the best-bound domain was IIIA (**a**), and for HSA:CS-4, the best-bound domain was IIIB (**c**).

3.1. Energy of Binding

The binding energies obtained from YASARA VINE for complexes after the MDoc procedure are presented in Tables S1 and S2 of the Supplementary Materials. Based on the table, a list of domains of HSA bound to CS-4 and CS-6 is presented in Table 1. The list of complexes is ranked according to the increasing magnitude of the energies of binding after

MD. Its values were averaged over time from 40 ns to 140 ns (40 ns is when stabilization of the complexes was assumed based on the energy and RMSD of the whole system, cf. Figure 3). While each of the ten best-docked complexes had undergone three separate simulations in different solutions (with the addition of $CaCl_2$, $MgCl_2$, and NaCl) to obtain one value of the energy of binding for comparison and sorting purposes, the three values of energy of binding were averaged. The docking ranks of the complexes showing the binding strength order before MD simulations are presented in parentheses in Table 1.

From the table, it can be concluded that in the case of the HSA:CS-6 complex, the second best-docked structure (#2) turns out to be the best after MD simulations in the solution. The CS-6 docked to a wide range of HSA domains, IA-IB-IIA-IIIA-IIIB, and the greater amount of interactions were with the IIIA domain. In the case of the HSA:CS-4 complex, a third docked structure (#3) was bound strongest after MD, and in this case, the CS-4 built contacts with IB, IIIA, and IIIB (the last one was the strongest). The IA, IB, IIIA, and IIIB subdomains formed the characteristic binding center described in [12]. The albumin domains, IB, IIIA, and IIIB, were reported as very important for the albumin transport function responsible for the heme binding site (IB), Sudlow's site II (IIIA), and the thyroxine-binding site (IIIB) [35]. In addition, all three domains were present in the two first strongest bound complexes for both HSA:CS-4 and HSA:CS-6. A similar feature was reported for HSA:HA complexes [12]. Comparing average MDoc binding energy for all HSA:CS-6 and HSA:CS-4 complexes, the HSA:CS-6 was bound about 23% stronger than HSA:CS-4. However, comparing HSA:CS-6 to HSA:HA, the binding in HSA:CS-6 was about 11% weaker than for HSA:HA [12]. It is consistent with expectations because CS is more negatively charged than HA (CS has two negative groups: COO- and sulfate groups, but HA has only COO-); thus, also binding it with negatively charged albumin is weaker.

Snapshots of the HSA:CS-6 and HSA:CS-4 complexes in $CaCl_2$ solution before and after 140 ns of MD simulation for best-bound complexes (#2 for HSA:CS-6 and #3 for HSA:CS-4) are presented in Figure 5. Similar Figures for $MgCl_2$ and NaCl solutions are presented in Figures S9 and S10 of the Supplementary Materials.

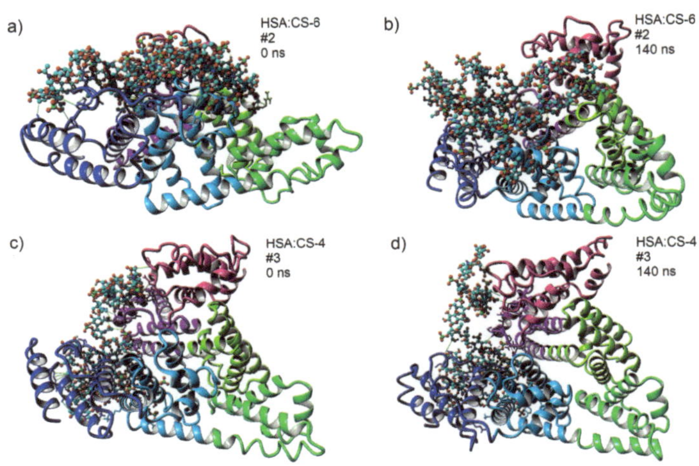

Figure 5. 3D structures of HSA:CS-6 (**a**,**b**) and HSA:CS-4 (**c**,**d**) complexes for the strongest bound complexes after MD in $CaCl_2$ solution (solution is transparent on the picture). HSA domains are colored as follows: IA-pink, IB-violet, IIA-light green, IIB-green, IIIA-light blue, and IIIB-blue. In CS-4 and CS-6, light blue atoms represent carbon; dark, blue nitrogen; red, oxygen; green, sulfur; and white, hydrogen. Snapshots are taken using YASARA software (**a**,**c**) before and (**b**,**d**) after 140 ns of MD simulations [47]. After a closer look at these pictures, green and pink lines can be observed, which show HP and ionic intermolecular interactions, respectively, and also yellow lines, which show intramolecular HBo inside CS.

Generally, best-bound complexes after MDoc are not necessarily best-bound after MD simulations. This statement can be explained by the influence of water solution, which changes both docked molecules' electrostatic map (and conformation). Adding ions into the solution can provide charge inversion and ion-bridge formation [61,62].

In Figure 6, the energy of binding for different complexes is shown. The values are averaged over time from 40 to 140 ns of MD simulation with a doubled standard deviation that reflects fluctuations of the energy values.

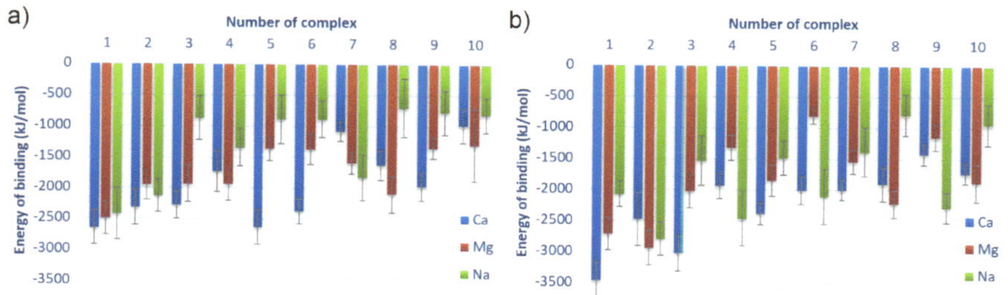

Figure 6. Averaged energy of binding for different complexes for (**a**) HSA:CS-6 and (**b**) HSA:CS-4. Complexes are sorted from lowest to highest averaged energy of binding after MD simulations; thus, the strongest bound are first (cf. Table 1). Error bars present doubled STD.

After MD simulations, the energy of binding for HSA:CS-4 is definitely of lower value than for HSA:CS-6; thus, the binding is more stable for the HSA:CS-4 complexes. Computation of averages for all the energy values over HSA:CS-6 and HSA:CS-4 provides the information that, after MD, complex HSA:CS-4 is about 15% stronger bound than HSA:CS-6, thus the situation is opposite than before MD simulations (cf. Table S1 and S2 in the Supplementary Materials). In about half of the cases (six for CS-6 and four for CS-4), adding $CaCl_2$ into the solution caused the most stable complexes. For 3 out of 10 CS-6 complexes, the highest affinity of CS to HSA was observed in the presence of Mg^{2+} ions and only 1 in the presence of Na^+ ions. In the CS-4 isomer, the energy of binding was the lowest in three complexes for Mg^{2+} ions and in three complexes for Na^+ ions. However, complexes with the addition of NaCl usually created weaker bound systems than $CaCl_2$ and $MgCl_2$. It was also confirmed by RMSF analysis (cf. Figure 4).

The energy of binding as a function of time for best-bound complexes is shown in Figure 7.

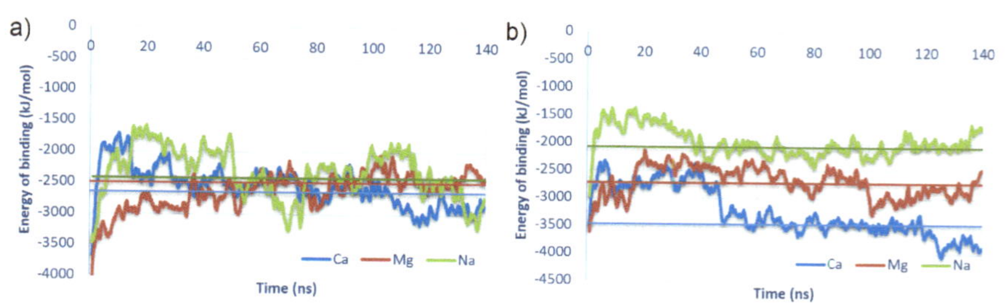

Figure 7. Energy of binding as a function of time for the strongest bound complexes: (**a**) #2 for HSA:CS-6; and (**b**) #3 for HSA:CS-4 (cf., Table 1). The vertical lines represent the average over the last 100 ns of MD.

The charts were plotted for best-bound cases (#2 for HSA:CS-6 and #3 for HSA:CS-4). The vertical lines show values averaged from 40 ns to 140 ns (thus, after stabilization of structures in MD). In both cases, structures with the addition of Ca^{2+} to the solution are bound stronger than in the rest of the cases. It is very interesting that for HSA:CS-6, all averaged energies were similar. The trend is only seen for HSA:CS-4 and accords with the expectation which considers the cavity's role in the electrostatic potential map of the HSA (described earlier in Section 3) and cations bridges formation in the macromolecules binding. This confirms the prominent role of Ca^{2+} in the binding.

The binding of the complexes occurs through intermolecular interactions. The interactions can be direct, when some forces appear between two atoms at a specific, close distance, or indirect, when another atom mediates the binding (creating bridges).

3.2. Intermolecular Interactions

The numbers of HP interactions (between hydrophobic atoms) and HBo were calculated with the algorithm described previously [12,63]. According to the YASARA definition, the HBo is formed when the hydrogen bond energy is greater than 25% of the optimum value for interaction 25 kJ/mol and equals 6.25 kJ/mol. The exact formula is described in the YASARA Manual [51] and previously in [12,63].

The numbers of intermolecular interactions between HSA and CS-6 or CS-4, also averaged over 40–140 ns for complexes sorted by averaged energies of binding after MD are shown in Figure 8. The numbers of the complexes correspond to the ones presented in the first column of Table 1 (before parenthesis).

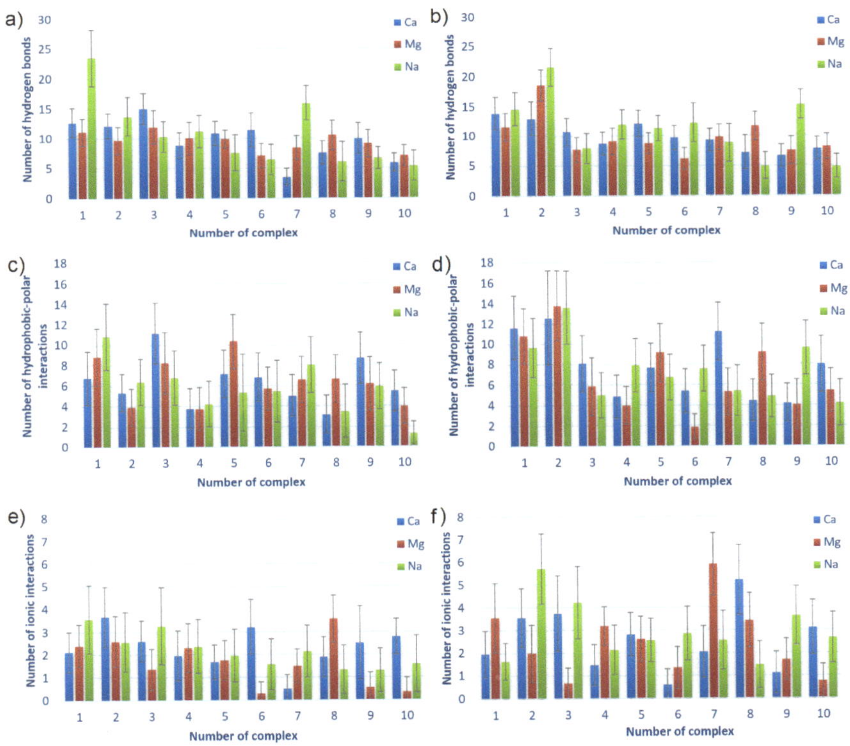

Figure 8. Intermolecular HBo, HP and ionic interactions for HSA:CS-6 (**a**,**c**,**e**) and HSA:CS-4 (**b**,**d**,**f**) complexes. The complexes are sorted from lowest to highest averaged energy of binding after MD simulations; thus, the strongest bound are first (cf., Table 1). Error bars present doubled STD.

The main observation from Figure 8 is that the charts are not much different. It can be seen that the number of HBo is slightly greater for complexes characterized by lower energy of binding, thus a stronger bound. The same can be seen for HP interactions in HSA:CS-4 but not in HSA:CS-6. In general, HSA:CS-4 has created more HBo and HP interactions. It also has a more varied plot of ionic interactions than HSA:CS-6. The number of ionic interactions is the lowest among the others. However, thanks to their electrostatic origin (the electrostatic force of attraction governs them), they are the strongest, so they are also important. HP interactions are also usually stronger than HBo. Thus, it is hard to determine which of the three non-covalent interactions influences most of the HSA:CS binding. An additional important observation is that the number of ionic interactions in most complexes is greater for the NaCl solution (cf., Figure 8e,f). Analyzing the influence of the ions added to the solution, the most visible is the prevalence of the number of HBo in the case of best-bound complexes in NaCl solution. This can be caused by Na^+ having the lowest ionic strength among the three ionic solutions (NaCl, $CaCl_2$, and $MgCl_2$). As a result, it does not have such strong adsorption properties on the surface. As a result, the HBo are formed more often in NaCl than in ionic interactions or ion bridges preferentially formed in divalent ion solutions.

All of the above observations confirm that the binding affinity cannot be related to only one type of interaction but is a result of many different types of interactions.

3.3. Water and Ionic Bridges

The number of bridges created by water molecules and ions between HSA and CS for different complexes HSA:CS-6 and HSA:CS-4 are shown in Figure 9. The water (or ion) bridge is created when one water (or ion) molecule forms an HBo (or ionic interaction) to HSA and another one to CS.

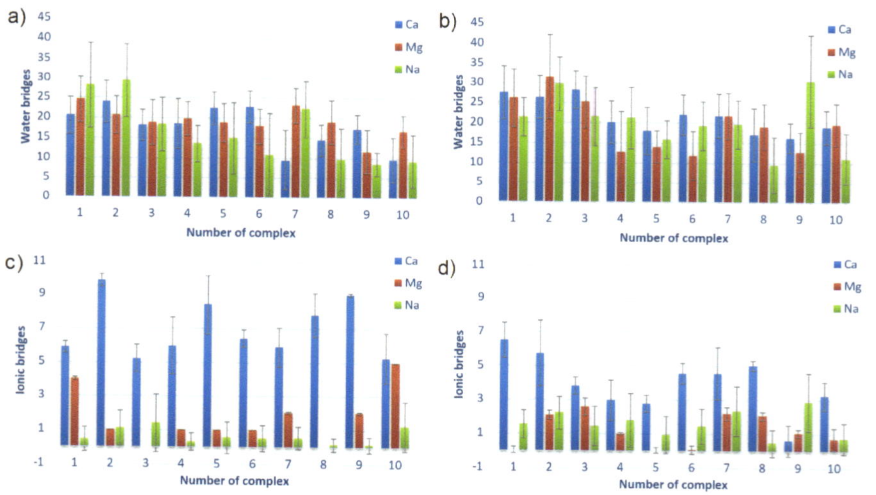

Figure 9. The number of hydrogen bonds mediated by water molecules (water bridges) and the number of ionic interactions mediated by cations (ionic bridges) between HSA and CS-6 (**a**,**c**) and between HSA and CS-4 (**b**,**d**). The complexes are sorted from lowest to highest averaged energy of binding after MD simulations; thus, the first is the strongest bound (cf., Table 1). Error bars present doubled STD.

By analyzing both Figures, it can be seen that HSA:CS-4 complexes characterize a greater number of water bridges, while the HSA:CS-6 complexes have more ionic bridges. Water bridges are very important for energy in binding. Their number usually decreases with the rank of the complex. It can suggest that direct HBo (cf., Figure 8) and indirect ones (i.e., mediated by water bridges) are the most important for HSA:CS binding. HSA:CS-4 has more intermolecular HBo, HP, and ionic interactions than ionic bridges. In the HSA:CS-6 case, in contrast, there are much fewer HBo, HP, and ionic interactions than ionic bridges, even though the greater number of ionic bridges could not make up for energy shortages caused by the lower number of HBo, HP, and ionic interactions. The ionic bridges built by Ca^{2+} are usually created between sulfur and carbon atoms in the case of the HSA:CS-6 complex and between carbon and carbon in the case of the HSA:CS-4 complex.

An important observation should be made for CS-6 complex #1, in which every direct intermolecular interaction type (HBo, HP, ionic) has a greater number of interactions in the case of Na^{+} than Ca^{2+} and Mg^{2+}. It also has slightly more water bridges, but at the same time, it has a much smaller number of ionic bridges. Taking into account that the energy of binding for Na^{+} is greater than for Ca^{2+} and Mg^{2+} and that the complex is weaker bound for Na than for Ca and Mg, it can be stated that the ionic bridges are of great importance for the stability of HSA:CS-6 complexes. The visible difference in the effect of Ca^{2+} and Mg^{2+} for ionic bridge formation is suggested to be due to the lower hydration of Ca^{2+} [12]. The influence of Ca^{2+} and Mg^{2+} on albumin binding can be explained by the fact that albumin interacts with it. HSA has almost no interaction with Na^{+} ions [64].

While hydration properties are crucial for lubrication, a number of HBo created between HSA:CS complexes and the water molecules was analyzed (see Figure S11 in the Supplementary Materials). In nearly every case, HSA:CS-6 complexes created more HBo with water thanHSA:CS-4 did. It confirms better binding between HSA and CS-4 (less space for water molecules to interact) than CS-6. A greater difference can be seen for complexes immersed in $CaCl_2$ solution when water created a lesser number of HBo with both complex types. This, together with observations of a much higher number of ionic bridges and smaller energy of binding in the case of Ca^{2+}, is evidence of stronger binding.

3.4. Maps of Interactions

In Figure 10, the maps of HBo distribution between different groups or atoms are presented. The results were summarized for all HBo interactions found between 40 and 140 ns of simulations.

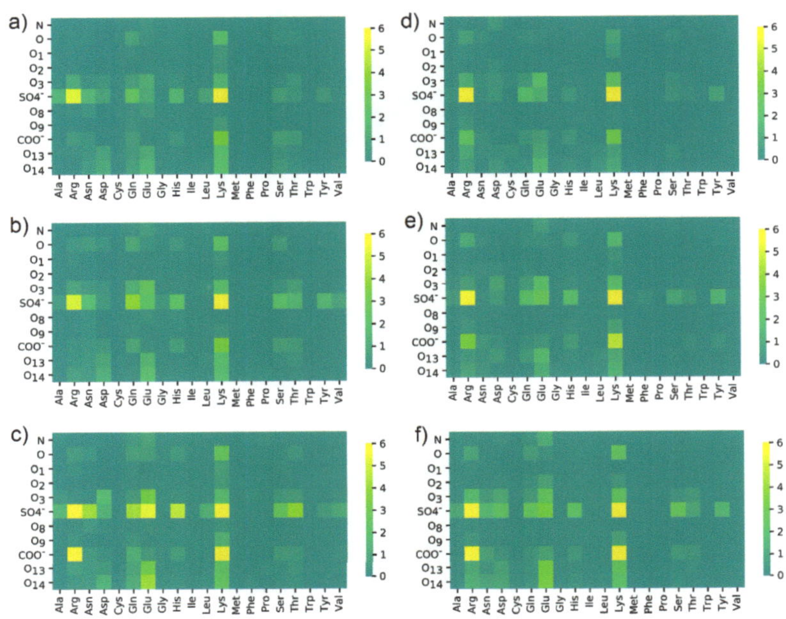

Figure 10. HBo distribution between different oxygen classes in (**a–c**) CS-6 and (**d–f**) CS-4; and different amino acids in HSA. Data were obtained in solutions containing: (**a**,**d**) NaCl; (**b**,**e**) $CaCl_2$; (**c**,**f**) $MgCl_2$. The maps present a number of interactions. The denotation of the atoms and functional groups is presented in Figure 1.

In each case, most HBo interactions were created between SO_4^- and Arg or SO_4^- and Lys, thus positive-charged amino acids. Moreover, the COO^- group has bound to Lys. From CS, most bindings had the group of SO_4^- and for HSA—Lys and Glu. In addition to these expected results, a distinctive impact was noticed for: Gln, Asp, Tyr, and Ser. Moreover, O, O3, and O14 interactions with Glu are also clearly visible in Figure 10. Interestingly, COO^- interacted with Arg using HBo in the case of NaCl water solution; also, O13-O14 with Glu is better visible when NaCl is present (cf., Figure 10c,f). There are no clearly visible differences between CS-6 and CS-4, except in the case of CS-6 + Na^+ where interactions between SO_4^- and Glu and His are noticeable.

In Figure 11, similar maps were created, but for the number of water bridges between HSA and CS.

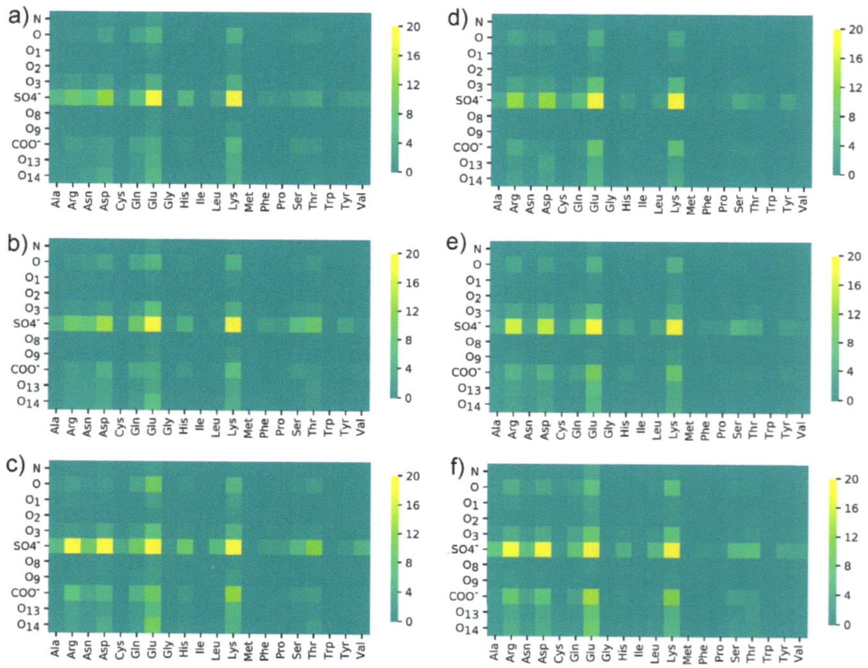

Figure 11. Water bridges distribution between different oxygen classes in (**a**–**c**) CS-6 and ((**d**–**f**) CS-4 and different amino acids in HSA. Data were obtained in solutions containing: (**a**,**d**) NaCl; (**b**,**e**) $CaCl_2$; (**c**,**f**) $MgCl_2$. The maps present a number of interactions. The denotation of the atoms and functional groups is presented in Figure 1.

The water bridges' maps look different from those created for direct HBo interaction. The leading roles of SO_4^- and Lys did not change, but in this case, interactions created by Glu are much more visible. The role of Glu in building water bridges is similar to Lys. SO_4^- made interactions via water bridges with Lys, Glu, Asp, and Arg more frequent. A slight difference can be seen in the case of the NaCl solution, where Glu+O14 is marked more strongly than in the rest of the cases. Moreover, SO_4^- + Arg and Asp interactions are more frequent for NaCl than for the rest. On the other hand, differences between CS-4 and CS-6 are not visible.

4. Conclusions

In the present paper, interactions between HSA and CS-4 and between HSA and CS-6 were studied. In both cases, HSA can form stable complexes, but RMSD and RMSF indicated HSA:CS-4 as behaving much more stably. The binding strength and interaction distribution also differed for HSA:CS-4 and HSA:CS-6 complexes. It can be explained by different intramolecular interactions in the two isomeric forms of CS, which also causes their different conformation in the water solution [65]. MD simulations have shown that CS-4 has a greater affinity for binding to HSA than CS-6 does. Because the percentage content of the two types of CS differs for healthy and ill cartilage, it can be deduced that the lubrication properties of SF containing CS-6 and CS-4 will be different from the ill ones which contain only CS-6. CS-6 is characterized by worse stability when interacting with HSA. Additionally, it can be inferred [32] that the cartilage tissue may have a specific affinity for lubricants with negatively charged groups and hydroxyl groups such as in CS. This may help them adsorb better to the cartilage surface, providing effective lubrication. Thus, the lack of stronger binding provided by the CS-4 type of CS isomer in ill cartilages can explain worse lubrication properties.

The ions contained in the solution are also essential and change the interaction map of the HSA:CS complexes. It is especially seen in the case of complex #1 of HSA:CS-6. Despite having more HP and HBo interactions where a solution of NaCl was concerned, the ionic (Ca) bridges balanced the energies of binding, indicating that HSA:CS-6 with NaCl was weaker bound than with $CaCl_2$. Thus, the availability of Ca^{2+} for ionic interaction formation via bridges seems to be the most important. The presence of Ca^{2+} (and also Mg^{2+}, but less so) amplifies the binding mechanism in the case of HSA:CS, which was mainly associated with the presence of locally positively charged sites (mainly Lys and Glu). The three domains, very important for the albumin transport function, i.e., IB (heme binding site), IIIA (Sudlow's site II), and IIIB (thyroxine-binding site), were presented in the binding in most of the complexes, especially in those characterized by the strongest binding [35].

Analyzing the obtained results, the similarity of HSA:CS binding to HSA:HA binding is evident. The interaction strength was slightly smaller for HSA:CS, but the influence of ions on the binding was similar [12]. Because the addition of HSA and CS separately decreases the friction coefficient [29,42], a complex of HSA:CS could give better results for lubrication properties, similar in the case of HSA:HA [12,42].

To the authors' knowledge, while writing the paper, the findings have not yet been confirmed by experimental data. However, the authors hope that the presented study can inspire other research groups to undertake such an endeavor.

Because understanding the nature of interactions between HSA and CS can be a stepping stone to explaining the lubrication properties of AC, the information contained in this work can be potentially applicable to designing new biomaterials characterized by specific rheological properties.

Supplementary Materials: The following supporting information can be downloaded at: https://www.mdpi.com/article/10.3390/ma15196935/s1, Table S1: Lists of the best binding energy obtained in each cluster after molecular docking of CS-4 to HSA; Table S2: Lists of the best binding energy obtained in each cluster after molecular docking of CS-6 to HSA; Figure S1: RMSD of backbone atoms as a function of time for HSA:CS-6 complexes; Figure S2: RMSD of backbone atoms as a function of time for HSA:CS-4 complexes; Figure S3: RMSF values for each atom of HSA for best-bound a) #2 HSA:CS-6, b) #3 HSA:CS-4 complexes; Figure S4: RMSF values for each atom of CS (1180 atoms) for best-bound a) #2 HSA:CS-6, b) #3 HSA:CS-4 complexes; Figure S5: Sum of RMSF for HSA (top) and CS (bottom) for complexes: a) HSA:CS-6, and b) for HSA:CS-4; Figure S6: RMSF for IIIA domain of HSA (a) and CS-4 (b) for complex #3 HSA:CS-4. Places of specific interactions are marked with dots; Figure S7: Secondary structure of HSA as a function of time for: a) #2 HSA:CS-6, b) #3 HSA:CS-4 complexes in $CaCl_2$ solution; Figure S8: Secondary structure of HSA after 140 ns of MD simulations for best-bound structures; Figure S9: 3D structures of HSA:CS-6 (a and b) and HSA:CS-4 (c and d) for the strongest bound complexes after MD in $MgCl_2$ solution (solution is transparent on the picture). HSA domains are colored as follows: IA-pink, IB-violet, IIA-light green, IIB-green, IIIA-light blue, IIIB-blue. In CS-4 and CS-6, light blue atoms represent carbon, dark blue nitrogen, red oxygen, green sulfur and white hydrogen. Snapshots captured using YASARA software (a and c) before and (b and d) after 140 ns of MD simulations; Figure S10: 3D structures of HSA:CS-6 (a and b) and HSA:CS-4 (c and d) for the strongest bound complexes after MD in NaCl solution (solution is transparent on the picture). HSA domains are colored as follows: IA-pink, IB-violet, IIA-light green, IIB-green, IIIA-light blue, IIIB-blue. In CS-4 and CS-6, light blue atoms represent carbon; dark blue, nitrogen; red, oxygen; green, sulfur; and white, hydrogen. Snapshots captured using YASARA software (a and c) before and (b and d) after 140 ns of MD simulations; Figure S11: Number of HBo between a) HSA:CS-6 or b)HSA:CS-4, and water molecules. Complexes are sorted from lowest to highest energy of binding after MD simulations, thus first are strongest bound (cf. Table 1). Error bars present doubled STD.

Author Contributions: Conceptualization, N.K. and A.M.; methodology, N.K.; software, N.K. and G.S.; validation, N.K., G.S., M.S. and A.M.; formal analysis, N.K.; investigation, N.K. and A.M.; resources, N.K. and G.S.; data curation, G.S. and M.S.; writing—original draft preparation, N.K. and A.M.; writing—review and editing, N.K. and A.M.; visualization, N.K. and M.S.; supervision, N.K.; project administration, N.K.; funding acquisition, A.M. All authors have read and agreed to the published version of the manuscript.

Funding: This research received no external funding.

Institutional Review Board Statement: Not applicable.

Informed Consent Statement: Not applicable.

Data Availability Statement: All data is available in the manuscript or upon request to the corresponding author.

Acknowledgments: The work is supported by BN-WTiICh-11/2022 of the Bydgoszcz University of Science and Technology. Calculations were carried out at the Academic Computer Centre in Gdańsk. The authors would like to thank Piotr Bełdowski from Bydgoszcz University of Science and Technology for providing data from the molecular docking of the albumin-chondroitin sulfate system.

Conflicts of Interest: The authors declare no conflict of interest.

Abbreviations

The following abbreviations are used in this manuscript:

AC	articular cartilage
SF	synovial fluid
HSA	human serum albumin
HA	hyaluronic acid
CS	chondroitin sulfate
CS-4	chondroitin 4 sulfate
CS-6	chondroitin 6 sulfate
GAG	glycosaminoglycan
MD	molecular dynamics
MDoc	molecular docking
PDB	Protein Data Bank
EoB	energy of binding

References

1. Furmann, D.; Nečas, D.; Rebenda, D.; Čipek, P.; Vrbka, M.; Křupka, I.; Hartl, M. The effect of synovial fluid composition, speed and load on frictional behaviour of articular cartilage. *Materials* **2020**, *13*, 1334. [CrossRef]
2. Mazurkiewicz, A. The effect of trabecular bone storage method on its elastic properties. *Acta Bioeng. Biomech.* **2018**, *20*, 21–27. [CrossRef] [PubMed]
3. Mazurkiewicz, A.; Topoliński, T. Relationship between the mineral content of human trabecular bone and selected parameters determined from fatigue test at stepwise-increasing amplitude. *Acta Bioeng. Biomech.* **2017**, *19*, 19–26. [CrossRef] [PubMed]
4. Topoliński, T.; Cichański, A.; Mazurkiewicz, A.; Nowicki, K. Fatigue Energy Dissipation in Trabecular Bone Samples with Stepwise-Increasing Amplitude Loading. *Mater. Test.* **2011**, *53*, 344–350. [CrossRef]
5. Jones, B.K.; Durney, K.M.; Hung, C.T.; Ateshian, G.A. The friction coefficient of shoulder joints remains remarkably low over 24h of loading. *J. Biomech.* **2015**, *48*, 3945–3949. [CrossRef]
6. Bełdowski, P.; Kruszewska, N.; Yuvan, S.; Dendzik, Z.; Goudoulas, T.; Gadomski, A. Capstan-like mechanism in hyaluronan-phospholipid systems. *Chem. Phys. Lipids* **2018**, *216*, 17–24. [CrossRef] [PubMed]
7. Pawlak, Z.; Urbaniak, W.; Hagner-Derengowska, M.; Hagner, W. The Probable Explanation for the Low Friction of Natural Joints. *Cell Biochem. Biophys.* **2015**, *71*, 1615–1621. [CrossRef]

8. Ateshian, G.A. A theoretical formulation for boundary friction in articular cartilage. *J. Biomech. Eng.* **1997**, *119*, 81–86. [CrossRef] [PubMed]
9. Gadomski, A.; Pawlak, Z.; Oloyede, A. Directed ion transport as virtual cause of some facilitated friction–lubrication mechanism prevailing in articular cartilage: A hypothesis. *Tribol. Lett.* **2008**, *30*, 83–90. [CrossRef]
10. Gadomski, A.; Bełdowski, P.; Rubí, J.M.; Urbaniak, W.; Augé, W.K.; Santamaría-Holek, I.; Pawlak, Z. Some conceptual thoughts toward nanoscale oriented friction in a model of articular cartilage. *Math. Biosci.* **2013**, *244*, 188–200. [CrossRef]
11. McNary, S.M.; Athanasiou, K.A.; Reddi, A.H. Engineering lubrication in articular cartilage. *Tiss. Eng. Part B Rev.* **2012**, *18*, 88–100. [CrossRef]
12. Bełdowski, P.; Przybyłek, M.; Raczyński, P.; Dėdinaitė, A.; Górny, K.; Wieland, F.; Dendzik, Z.; Sionkowska, A.; Claesson, P.M. Albumin-Hyaluronan Interactions: Influence of Ionic Composition Probed by Molecular Dynamics. *Int. J. Mol. Sci.* **2021**, *22*, 12360. [CrossRef]
13. Dėdinaitė, A.; Wieland, D.F.; Bełdowski, P.; Claesson, P.M. Biolubrication synergy: Hyaluronan—Phospholipid interactions at interfaces. *Adv. Colloid Interf. Sci.* **2019**, *274*, 102050. [CrossRef]
14. Bełdowski, P.; Mazurkiewicz, A.; Topoliński, T.; Małek, T. Hydrogen and Water Bonding between Glycosaminoglycans and Phospholipids in the Synovial Fluid: Molecular Dynamics Study. *Materials* **2019**, *12*, 2060. [CrossRef]
15. Kruszewska, N.; Domino, K.; Drelich, R.; Urbaniak, W.; Petelska, A.D. Interactions between Beta-2-Glycoprotein-1 and Phospholipid Bilayer-A Molecular Dynamic Study. *Membranes* **2020**, *10*, 396. [CrossRef]
16. Dėdinaitė, A.; Claesson, P.M. Synergies in lubrication. *Phys. Chem. Chem. Phys.* **2017**, *19*, 23677–23689. [CrossRef]
17. Zander, T.; Garamus, V.M.; Dédinaité, A.; Claesson, P.M.; Bełdowski, P.; Górny, K.; Dendzik, Z.; Wieland, D.C.F.; Willumeit-Römer, R. Influence of the Molecular Weight and the Presence of Calcium Ions on the Molecular Interaction of Hyaluronan and DPPC. *Molecules* **2020**, *25*, 3907. [CrossRef]
18. Zander, T.; Wieland, D.F.; Raj, A.; Wang, M.; Nowak, B.; Krywka, C.; Dėdinaitė, A.; Claesson, P.M.; Garamus, V.M.; Schreyer, A.; et al. The influence of hyaluronan on the structure of a DPPC—bilayer under high pressures. *Colloids Surf. B Biointerf.* **2016**, *142*, 230–238. [CrossRef]
19. Nalbant, S.; Martinez, J.; Kitumnuaypong, T.; Clayburne, G.; Sieck, M.; Schumacher, H. Synovial fluid features and their relations to osteoarthritis severity: New findings from sequential studies. *Osteoarthr. Cartil.* **2003**, *11*, 50–54. [CrossRef]
20. Struglics, A.; Larsson, S.; Pratta, M.; Kumar, S.; Lark, M.; Lohmander, L. Human osteoarthritis synovial fluid and joint cartilage contain both aggrecanase- and matrix metalloproteinase-generated aggrecan fragments. *Osteoarthr. Cartil.* **2006**, *14*, 101–113. [CrossRef]
21. Goudoulas, T.B.; Kastrinakis, E.G.; Nychas, S.G.; Papazoglou, L.G.; Kazakos, G.M.; Kosmas, P.V. Rheological study of synovial fluid obtained from dogs: Healthy, pathological, and post-surgery, after spontaneous rupture of cranial cruciate ligament. *Ann. Biomed. Eng.* **2010**, *38*, 57–65. [CrossRef] [PubMed]
22. Lin, W.; Mashiah, R.; Seror, J.; Kadar, A.; Dolkart, O.; Pritsch, T.; Goldberg, R.; Klein, J. Lipid-hyaluronan synergy strongly reduces intrasynovial tissue boundary friction. *Acta Biomater.* **2019**, *83*, 314–321. doi: 10.1016/j.actbio.2018.11.015. [CrossRef] [PubMed]
23. Seror, J.; Zhu, L.; Goldberg, R.; Day, A.J.; Klein, J. Supramolecular synergy in the boundary lubrication of synovial joints. *Nat. Commun.* **2015**, *6*, 6497. [CrossRef] [PubMed]
24. Bełdowski, P.; Yuvan, S.; Dėdinaitė, A.; Claesson, P.M.; Pöschel, T. Interactions of a short hyaluronan chain with a phospholipid membrane. *Colloids Surf. B Biointerf.* **2019**, *184*, 110539. [CrossRef]
25. Gold, E.W. An interaction of albumin with hyaluronic acid and chondroitin sulfate: A study of affinity chromatography and circular dichroism. *Biopolymers* **1980**, *19*, 1407–1414. [CrossRef]
26. Gandhi, N.S.; Mancera, R.L. The Structure of Glycosaminoglycans and their Interactions with Proteins. *Chem. Biol. Drug Des.* **2008**, *72*, 455–482. [CrossRef]
27. Hubert, J.; Beil, F.T.; Rolvien, T.; Butscheidt, S.; Hischke, S.; Püschel, K.; Frosch, S.; Mussawy, H.; Ries, C.; Hawellek, T. Cartilage calcification is associated with histological degeneration of the knee joint: A highly prevalent, age-independent systemic process. *Osteoarthr. Cartil.* **2020**, *28*, 1351–1361. [CrossRef]
28. Mourão, P.A. Distribution of chondroitin 4-sulfate and chondroitin 6-sulfate in human articular and growth cartilage. *Arthritis Rheum.* **1988**, *31*, 1028–1033. [CrossRef]
29. Baeurle, S.; Kiselev, M.; Makarova, E.; Nogovitsin, E. Effect of the counterion behavior on the frictional-compressive properties of chondroitin sulfate solutions. *Polymer* **2009**, *50*, 1805–1813. [CrossRef]
30. Zhang, Z.; Barman, S.; Christopher, G.F. The role of protein content on the steady and oscillatory shear rheology of model synovial fluids. *Soft Matter* **2014**, *10*, 5965–5973. [CrossRef]
31. Katta, J.; Jin, Z.; Ingham, E.; Fisher, J. Chondroitin sulphate: An effective joint lubricant? *Osteoarthr. Cartil.* **2009**, *17*, 1001–1008. [CrossRef]
32. Basalo, I.M.; Chahine, N.O.; Kaplun, M.; Chen, F.H.; Hung, C.T.; Ateshian, G.A. Chondroitin sulfate reduces the friction coefficient of articular cartilage. *J. Biomech.* **2007**, *40*, 1847–1854. [CrossRef]
33. Wu, T.t.; Gan, X.q.; Cai, Z.b.; Zhu, M.h.; Qiao, M.t.; Yu, H.y. The lubrication effect of hyaluronic acid and chondroitin sulfate on the natural temporomandibular cartilage under torsional fretting wear. *Lubr. Sci.* **2014**, *27*, 29–44. [CrossRef]
34. Guizado, T.R.C. Analysis of the structure and dynamics of human serum albumin. *J. Mol. Model.* **2014**, *20*, 2450. [CrossRef] [PubMed]

35. Fasano, M.; Curry, S.; Terreno, E.; Galliano, M.; Fanali, G.; Narciso, P.; Notari, S.; Ascenzi, P. The extraordinary ligand binding properties of human serum albumin. *IUBMB Life Int. Union Biochem. Mol. Biol. Life* **2005**, *57*, 787–796. [CrossRef]
36. De Simone, G.; di Masi, A.; Ascenzi, P. Serum Albumin: A Multifaced Enzyme. *Int. J. Mol. Sci.* **2021**, *22*, 10086. [CrossRef] [PubMed]
37. Sugio, S.; Kashima, A.; Mochizuki, S.; Noda, M.; Kobayashi, K. Crystal structure of human serum albumin at 2.5A resolution. *Protein Eng. Des. Sel.* **1999**, *12*, 439–446. [CrossRef]
38. Spector, A. Fatty acid binding to plasma albumin. *J. Lipid Res.* **1975**, *16*, 165–179. [CrossRef]
39. van der Vusse, G.J. Albumin as Fatty Acid Transporter. *Drug Metab. Pharmacokinet.* **2009**, *24*, 300–307. [CrossRef]
40. Baker, M.E. Albumin's role in steroid hormone action and the origins of vertebrates: Is albumin an essential protein? *FEBS Lett.* **1998**, *439*, 9–12. [CrossRef]
41. Jacobsen, J.; Brodersen, R. Albumin-bilirubin binding mechanism. *J. Biol. Chem.* **1983**, *258*, 6319–26. [CrossRef]
42. Murakami, T.; Yarimitsu, S.; Nakashima, K.; Sawae, Y.; Sakai, N. Influence of synovia constituents on tribological behaviors of articular cartilage. *Friction* **2013**, *1*, 150–162. [CrossRef]
43. Čipek, P.; Vrbka, M.; Rebenda, D.; Nečas, D.; Křupka, I. Biotribology of Synovial Cartilage: A New Method for Visualization of Lubricating Film and Simultaneous Measurement of the Friction Coefficient. *Materials* **2020**, *13*, 2075. [CrossRef]
44. Gadomska, M.; Musiał, K.; Sionkowska, A. New materials based on hyaluronic acid and egg albumin mixture. *Eng. Biomater.* **2021**, *160*, 15–21. [CrossRef]
45. Maurya, P.; Singh, S.; Mishra, N.; Pal, R.; Singh, N.; Parashar, P.; Saraf, S.A. Albumin-based nanomaterials in drug delivery and biomedical applications. In *Biopolymer-Based Nanomaterials in Drug Delivery and Biomedical Applications*; Bera, H., Hossain, C.M., Saha, S., Eds.; Academic Press: Cambridge, MA, USA, 2021; pp. 465–496. [CrossRef]
46. Zewde, B.; Atoyebi, O.; Gugssa, A.; Gaskell, K.J.; Raghavan, D. An Investigation of the Interaction between Bovine Serum Albumin-Conjugated Silver Nanoparticles and the Hydrogel in Hydrogel Nanocomposites. *ACS Omega* **2021**, *6*, 11614–11627. [CrossRef]
47. Krieger, E.; Vriend, G. YASARA View—molecular graphics for all devices—from smartphones to workstations. *Bioinformatics* **2014**, *30*, 2981–2982. [CrossRef]
48. Trott, O.; Olson, A.J. AutoDock Vina: Improving the speed and accuracy of docking with a new scoring function, efficient optimization, and multithreading. *J. Comput. Chem.* **2010**, *31*, 455–461. [CrossRef]
49. Duan, Y.; Wu, C.; Chowdhury, S.; Lee, M.C.; Xiong, G.; Zhang, W.; Yang, R.; Cieplak, P.; Luo, R.; Lee, T.; et al. A point-charge force field for molecular mechanics simulations of proteins based on condensed-phase quantum mechanical calculations. *J. Comput. Chem.* **2003**, *24*, 1999–2012. [CrossRef] [PubMed]
50. Maier, J.A.; Martinez, C.; Kasavajhala, K.; Wickstrom, L.; Hauser, K.E.; Simmerling, C. ff14SB: Improving the Accuracy of Protein Side Chain and Backbone Parameters from ff99SB. *J. Chem. Theory Comput.* **2015**, *11*, 3696–3713. [CrossRef] [PubMed]
51. Krieger, E.; Dunbrack, R.L.; Hooft, R.W.; Krieger, B. Assignment of protonation states in proteins and ligands: Combining PKA prediction with hydrogen bonding network optimization. In *Methods in Molecular Biology*; Springer: New York, NY, USA, 2012; pp. 405–421. [CrossRef]
52. Krieger, E.; Vriend, G. New ways to boost molecular dynamics simulations. *J. Comput. Chem.* **2015**, *36*, 996–1007. [CrossRef] [PubMed]
53. Wang, Q.; Canutescu, A.A.; Roland L. Dunbrack, J. SCWRL and MolIDE: Computer programs for side-chain conformation prediction and homology modeling. *Nat. Protoc.* **2008**, *3*, 1832. [CrossRef]
54. Kirschner, K.N.; Yongye, A.B.; Tschampel, S.M.; González-Outeiriño, J.; Daniels, C.R.; Foley, B.L.; Woods, R.J. GLYCAM06: A generalizable biomolecular force field. Carbohydrates. *J. Comput. Chem.* **2008**, *29*, 622–655. [CrossRef]
55. Hornak, V.; Abel, R.; Okur, A.; Strockbine, B.; Roitberg, A.E.; Simmerling, C. Comparison of multiple Amber force fields and development of improved protein backbone parameters. *Proteins Struct.* **2006**, *65*, 712–725. [CrossRef]
56. Essmann, U.; Perera, L.E.; Berkowitz, M.L.; Darden, T.A.; Lee, H.C.; Pedersen, L.G. A smooth particle mesh Ewald method. *J. Chem. Phys.* **1995**, *103*, 8577–8593. [CrossRef]
57. Berendsen, H.J.C.; Postma, J.P.M.; van Gunsteren, W.F.; Dinola, A.; Haak, J.R. Molecular dynamics with coupling to an external bath. *J. Chem. Phys.* **1984**, *81*, 3684–3690. [CrossRef]
58. Radibratovic, M.; Minic, S.; Stanic-Vucinic, D.; Nikolic, M.; Milcic, M.; Cirkovic Velickovic, T. Stabilization of Human Serum Albumin by the Binding of Phycocyanobilin, a Bioactive Chromophore of Blue-Green Alga Spirulina: Molecular Dynamics and Experimental Study. *PLoS ONE* **2016**, *11*, e0167973. [CrossRef] [PubMed]
59. Shi, D.; Sheng, A.; Chi, L. Glycosaminoglycan-protein interactions and their roles in human disease. *Front. Mol. Biosci.* **2021**, *8*, 639666. [CrossRef] [PubMed]
60. Qiu, H.; Jin, L.; Chen, J.; Shi, M.; Shi, F.; Wang, M.; Li, D.; Xu, X.S.; Su, X.; Yin, X.; et al. Comprehensive Glycomic Analysis Reveals That Human Serum Albumin Glycation Specifically Affects the Pharmacokinetics and Efficacy of Different Anticoagulant Drugs in Diabetes. *Diabetes* **2020**, *69*, 760–770. [CrossRef]
61. Brown, K.L.; Banerjee, S.; Feigley, A.; Abe, H.; Blackwell, T.S.; Pozzi, A.; Hudson, B.G.; Zent, R. Salt-bridge modulates differential calcium-mediated ligand binding to integrin alpha1- and alpha2-I domains. *Sci. Rep.* **2018**, *8*, 1–14. [CrossRef]
62. Matsarskaia, O.; Roosen-Runge, F.; Schreiber, F. Multivalent ions and biomolecules: Attempting a comprehensive perspective. *ChemPhysChem* **2020**, *21*, 1742–1767. [CrossRef] [PubMed]

63. Siódmiak, J.; Bełdowski, P.; Augé, W.; Ledziński, D.; Śmigiel, S.; Gadomski, A. Molecular dynamic analysis of hyaluronic acid and phospholipid interaction in tribological surgical adjuvant design for osteoarthritis. *Molecules* **2017**, *22*, 1436. [CrossRef] [PubMed]
64. van Os, G.A.; Koopman-van Eupen, J.H. The interaction of sodium, potassium, calcium, and magnesium with human serum albumin, studied by means of conductivity measurements. *Recueil Travaux Chimiques Pays-Bas* **2010**, *76*, 390–400. [CrossRef]
65. Yan, L.Y.; Li, W.; Mesgari, S.; Leong, S.S.; Chen, Y.; Loo, L.S.; Mu, Y.; Chan-Park, M.B. Use of a chondroitin sulfate isomer as an effective and removable dispersant of single-walled carbon nanotubes. *Small* **2011**, *7*, 2758–2768. [CrossRef]

materials

Article

Variously Prepared Zeolite Y as a Modifier of ANFO

Łukasz Kuterasiński [1,*], Agnieszka M. Wojtkiewicz [1], Marta Sadowska [1], Paulina Żeliszewska [1], Bogna D. Napruszewska [1], Małgorzata Zimowska [1], Mateusz Pytlik [2,*] and Andrzej Biessikirski [3,*]

1 Jerzy Haber Institute of Catalysis and Surface Chemistry, Polish Academy of Sciences, 8 Niezapominajek Street, 30-239 Krakow, Poland
2 Conformity Assessment Body, Central Mining Institute, 1 Gwarków Square, 40-166 Katowice, Poland
3 Faculty of Civil Engineering and Resource Management, AGH University of Science and Technology, 30 Mickiewicz Avenue, 30-059 Krakow, Poland
* Correspondence: lukasz.kuterasinski@ikifp.edu.pl (Ł.K.); mpytlik@gig.eu (M.P.); abiess@agh.edu.pl (A.B.)

Abstract: In the presented research, we investigated Ammonium Nitrate Fuel Oil (ANFO), with the addition of variously modified zeolite Y as an attractive explosive. Analysis of both blasting tests and thermodynamic models of blasting properties led to the conclusion that the addition of zeolite Y enhanced the detonation properties of such prepared ANFO via the growth of the detonation pressure, temperature, compression energy, and heat of the explosion. Generally, the modification of ANFO with variously prepared zeolite Y also reduced the volume of (CO_x + NO_x) post-blast fumes. Furthermore, it was found that the ANFO's velocity of detonation (VOD) could be controlled by the choice of the way of zeolite Y modification. Namely, for zeolite Y without Mg, as well as Mg-Y prepared via the impregnation method, the VOD rose. The opposite effect was observed when ANFO was modified with Mg-Y, obtained from the deposition of Mg over zeolite Y via the ultrasonic-assisted procedure.

Keywords: zeolite Y; ANFO; modifier; detonation; post-blast fumes

Citation: Kuterasiński, Ł.; Wojtkiewicz, A.M.; Sadowska, M.; Żeliszewska, P.; Napruszewska, B.D.; Zimowska, M.; Pytlik, M.; Biessikirski, A. Variously Prepared Zeolite Y as a Modifier of ANFO. *Materials* **2022**, *15*, 5855. https://doi.org/10.3390/ma15175855

Academic Editor: Lubomira Tosheva

Received: 12 July 2022
Accepted: 22 August 2022
Published: 25 August 2022

1. Introduction

Explosives are currently widely used both in the military industry and in civil works, including mining, demolition works, or macro-leveling [1,2]. One of the most commonly used explosives (especially in mining) is ANFO (Ammonium Nitrate Fuel Oil) due to its easy and cheap procedure of production, as well as its good blasting properties [3,4].

ANFO is manufactured by the blending of ammonium nitrate(V) (AN) (oxygen-bearing component) with fuel oil (FO) (combustible component) in an appropriate mass ratio (usually 94:6). The ratio of 94:6 (AN:FO) provides a zero-oxygen balance, i.e., no excess or deficiency of oxygen in the balance of the composition of the explosive, which is responsible for the formation of maximal detonation energy with simultaneous low toxicity of the post-blast fumes (a minimal content of NO_x and CO), and reduced production of CO_2 being responsible for greenhouse effect [5]. ANFO is considered a non-ideal explosive due to its non-ideal detonation behavior, which can be defined by the impossibility of the reach of its theoretically calculated values, such as the velocity of detonation (VOD) [6].

In mixed-type explosives (like ANFO), the oxygen-bearing component should be characterized by high porosity and low density. That allows to obtain a high value of the adsorption coefficient between the solid phase (AN) and the liquid phase (FO). The presence of a combustible component is necessary for the occurrence of explosive transformation [5–14].

Oxley et al. [15–17] reported a dual mechanism of the AN decomposition, i.e., according to a radical and an ionic pathway. The former type of mechanism is characterized by both a high rate and temperatures, whereas the latter route of reaction takes place at a low temperature with a low rate. In turn, the application of modifiers allows preparing explosives with optimized blasting properties, such as the velocity of detonation or explosion heat and strength. Furthermore, the addition of certain modifiers to ANFO may catalyze the decomposition of explosives, reduce the emission of post-blast fumes, and influence explosive sensitivity. However, the addition of a new component (acting as either a flammable ingredient or modifier) may cause a change in the oxygen balance [15–17].

Apart from FO [18–44], in the available bibliography, many examples of ANFO additives have been reported, such as: charcoal; coal dust; activated carbon [18–26]; 2,4,6-trinitrotoluene (TNT) [19]; nitromethane [27]; iron persulfide (II) [19]; pyrite [28,29]; aluminum, magnesium, copper and zinc dusts [5,20,21,30,31]; sugar [5,20,21]; lubricants [32]; polyolefin wastes [33–36]; polyethylene (PE) [37]; polypropylene (PP) [37]; sodium sulfate [38]; sodium decarbonate [16]; potassium carbonate [16]; ammonium sulfate [16]; calcium oxide [20,21]; dolomite [20,21]; anhydrite [20,21]; aluminum oxides [30,39]; and silica [40,41].

So far, in the present works dedicated to ANFO, only results concerning physicochemical and blasting properties of the systems consisting of AN mixed with oil, as well as with other organic and inorganic modifiers mainly in the form of dust and inorganic salts, were reported. However, the use of additives acting as a carrier for inorganic modifiers has not been published. An example of this type of additive could be zeolite due to the presence of, among others, silicon, and aluminum in the zeolite framework, and due to the possibility of introducing a wide range of other elements to this group of minerals.

In the undertaken research, we investigate zeolite Y due to its low silicon-to-aluminum molar ratio ($2 < Si/Al < 5$), which means that the zeolite with this topology contains a lot of aluminum and therefore is characterized by a high ion exchange capacity, in comparison with the majority of zeolites of other structures. Relative high aluminum content (and automatically introduced metals to the zeolite) allows the use of this mineral as representative support for inorganic ANFO modifiers. The introduced metal into zeolite Y will be magnesium, due to the beneficial effect of this element on the blasting properties of ANFO, which has been published in one of our previous works [31].

We also try to find an answer to the question of whether the choice between (a) ion exchange, (b) impregnation, or (c) ultrasonic irradiation, as methods of the introduction of magnesium to zeolite skeleton, may affect the usefulness of zeolite Y as the carrier of inorganic modifiers in ANFO.

In the present paper, we reported the synthesis of ANFO consisting of AN, FO, and a modifier being the Y-structure type zeolite containing magnesium, which was incorporated to zeolite via three various routes (Mg-Y), or without magnesium. Separate studies are listed for the bare ANFO sample (without zeolite additive), which plays the role of the reference sample. The obtained explosives were subjected to physicochemical characterization including crystallinity/structure (XRD, FT-IR), surface/morphology (AFM/SEM), thermal properties (TG/DSC), and blasting properties (involving velocity of detonation, the heat of the explosion, compression energy, oxygen balance and the analysis of post-blast fumes).

2. Materials and Methods

2.1. Materials and Sample Preparation

Ammonium Nitrate porous prill (UltrAN 70) was produced in 2021 by Yara's International A SA (Szczecin, Poland). UltrAN 70 contained ca. 34.5% nitrogen. The prill diameter and bulk density were in the range of 1.0–2.0 mm and 670–720 kg/m^3, respectively. The moisture did not exceed 0.3% wt. The product was characterized by a purity of 99.4%.

Fuel oil (FO) was manufactured in 2021 by Silesia Oil Sp. z o.o. (Łaziska Górne, Poland) and consisted of C10-C20 hydrocarbons. The bulk density and kinetic viscosity of FO at 40 °C were 800 kg/m^3 and 13.6 mm^2/s, respectively. The detailed characterization of FO can be found in [42]. The reference sample ANFO (designated as "A") was prepared by a blending of UltrAN 70 with FO according to the weight ratio of 94.0:6.0 (Table 1). The procedure was conducted by Yara's International A SA.

Table 1. Chemical composition of non-ideal explosives and their synthesis conditions. The total sample mass was normalized to 5.00 g.

Sample	Chemical Composition [% wt.]			Description
	Ammonium Nitrate	Fuel Oil	Zeolite Y	
A	94.00	6.00	0.00	Commercial ANFO, reference sample (5.00 g)
B1	93.06	5.94	1.00	ANFO (4.95 g) + zeolite Y (0.05 g)
B2	92.12	5.88	2.00	ANFO (4.90 g) + zeolite Y (0.10 g)
C1	93.06	5.94	1.00	ANFO (4.95 g) + Mg-Y (0.05 g). Zeolite Y containing Mg introduced via the impregnation method.
C2	92.12	5.88	2.00	ANFO (4.90 g) + Mg-Y (0.10 g). Zeolite Y containing Mg introduced via the impregnation method.
D1	93.06	5.94	1.00	ANFO (4.95 g) + Mg-Y (0.05 g). Zeolite Y containing Mg introduced via the ion-exchange method.
D2	92.12	5.88	2.00	ANFO (4.90 g) + Mg-Y (0.10 g). Zeolite Y containing Mg introduced via the ion-exchange method.
E1	93.06	5.94	1.00	ANFO (4.95 g) + Mg-Y (0.05 g). Zeolite Y containing Mg introduced via ultrasonic-assisted impregnation method.
E2	92.12	5.88	2.00	ANFO (4.90 g) + Mg-Y (0.10 g). Zeolite Y containing Mg introduced via ultrasonic-assisted impregnation method.

The zeolite with Y-type structure (Si/Al = 2.65) was supplied by Mątwy (Inowrocław, Poland). The parent zeolite was in sodium form. The zeolite Y was added to the ANFO via a simple mixing and played the role of the enhancer of its blasting properties. Depending on the weight loading of the zeolite in the resulting ANFO-type material (1% wt. vs. 2% wt.), the sample was denoted as B1 and B2, respectively (Table 1).

The parent zeolite (Na-Y) was also modified with magnesium. The introduction of Mg into zeolite Y was performed via wet impregnation (Mg-Y_{impr}), ion-exchange (Mg-$_{Yion-exch}$), or ultrasonic-assisted impregnation procedure (Mg-Y_{son}), respectively. A detailed description of the production of Mg-containing zeolites (Mg-Y) was given in Supplementary Materials.

ANFO-based materials, prepared via the addition of Mg-Y_{impr}, Mg-$Y_{ion-exch,}$ and Mg-Y_{son} zeolites to ANFO (via the blending route), were designated as C1 or C2, D1 or D2, and E1 or E2. Digits "1" and "2" refer to the percentage weight contribution of zeolite in the obtained ANFO. Details on the ANFO's synthesis procedure were summarized in Table 1.

2.2. Characterization Methods

The crystallinity of the ANFO samples in the granule form was measured using the X-ray diffraction (XRD) phenomena with a PANalytical X'Pert PRO MPD diffractometer (40 kV, 30 mA), equipped with a CuKα generator (λ = 1.5418 Å). The experiments were conducted for a 2θ angle at 5–50° with a 0.033° step.

Structural analysis of the synthesized materials was performed using a Nicolet iS10 spectrometer (Thermo Scientific, Waltham, MA, USA) equipped with an MCT detector in the Attenuated Total Reflectance (ATR) mode. The FT-IR analysis was performed at 4000–650 cm^{-1}, with a resolution of 4 cm^{-1} and the number of scans during a single measurement was 64.

The morphology of the obtained ANFO was investigated using a Jeol JSM-7500F scanning electron microscope (JEOL Ltd., Tokyo, Japan) equipped with the X-ray energy dispersive (EDS) system INCAPentaFETx3 (JEOL Ltd., Tokyo, Japan). Two detectors were used, and the images were recorded in two modes; the secondary electron detector provided SEI images and the backscattered electron detector provided BSE (COMPO) micrographs. Directly before SEM mapping, the samples were dried for 24 h and coated with chromium (20 nm).

The status of the surface of the studied samples was investigated using an atomic force microscope (AFM) NT-MDT Solver BIO apparatus (Moscow, Russia) equipped with the SMENA SFC050L scanning head. All measurements were conducted in air and in semicontact mode by the application of high-resolution silicon probes (NT-MDT ETALON probes, HA NC series, polysilicon cantilevers with resonance frequencies 140 kHz +/− 10% or, 240 kHz +/− 10% force constants 4.4 N/m +/− 20% or, 9.5 N/m +/− 20%, respectively, a typical curvature radius of the tip was 10 nm and cone angle was less than 20°). Obtained images were referred to the randomly chosen areas over the substrate and within the scan area suitable for the tested samples. All images were flattened and graphically developed using specialized software supplied with the instrument.

Thermal analysis of the prepared samples included Thermogravimetry (TG) and Differential Scanning Calorimetry (DSC). The measurements were performed using NETZSCH STA 409 PC/PG equipment (Netzsch-Gerätebau, Selb, Germany) at 20–700 °C. The temperature ramp was 10 °C/min. All studied samples were kept in airflow (30 mL/min) both in the furnace and the balance chamber. Air functioned as a stimulator of detonation conditions. For all measurements, the prepared sample (20 mg) was placed in the DSC aluminum pan using a spatula. To obtain the correct TG baseline, the same heating profiles were used for both the empty pan and the pan with a measured sample. The TG drift was ca. 5 µg, which corresponds to 0.02 mass%.

The velocity of detonation (VOD) was determined according to the procedure reported in [43]. Briefly, the ANFO charge (600 g) was placed in a glass tube of 46.4 mm inner diameter. The Royal Demolition Explosive (RDX-based) charge (14 g) initiated the detonation. Two short circuit probes were placed throughout the measured ANFO sample. The distance between probes was 150 mm. Moreover, the distance between the primer and the first probe was equal to twice the charge diameter (l). A crucial parameter in this measurement was the time difference (t) between the first recorded signal change in each probe. VOD can be defined according to Equation (1):

$$\mathrm{VOD} = \mathrm{l}/\mathrm{t} \tag{1}$$

Post-blast fumes analysis was conducted according to the procedure described in [44], which was based on regulations reported in [45]. The description of blasting tests has also been reported in our previous works [41,46,47]. In each experiment, 600 g of non-ideal ANFO charge was placed inside the steel mortar locked in the blasting chamber. After the sealing of the blasting chamber, the detonation of the explosive charge was performed. After that, the homogenization of the post-blast fumes took place. The homogenization duration was 3 min. Afterward, post-blast gases were collected in the ventilation system for 20 min. The CO_x ($CO + CO_2$), as well as NO_x volumes, were measured using IR (MIR 25e) and chemiluminescent (TOPAZE 32M) analyzers, respectively. The measurement errors of VOD and post-blast fumes analyses were 2% and 1%, respectively.

Thermodynamic calculations were conducted in the EXPLO5 software produced by OZM Research with a calculation error of 5%. Calculations were made based on Becker-Kwiatkowski-Wilson (BKW) equation state [48]. EXPLO5 software included the calculations of parameters, among others, the explosion pressure and temperature, heat of the explosion, compression energy, and oxygen balance.

3. Results and Discussion

3.1. Structure

From the appearance of XRD images given in Figure 1 for all studied samples, the presence of Ammonium Nitrate in the orthorhombic crystal system and with a Pmmm space group and two molecules per unit cell was confirmed. Based on bibliography, the diffraction peaks at 2θ = 18°, 23°, 29°, 31°, 33°, 36°, 38°, 40°, 43°, and 46° can be assigned to (100), (011), (111), (002), (020), (102), (201), (112), (211), and (210) reflections in ANFO crystallites, respectively [49–52].

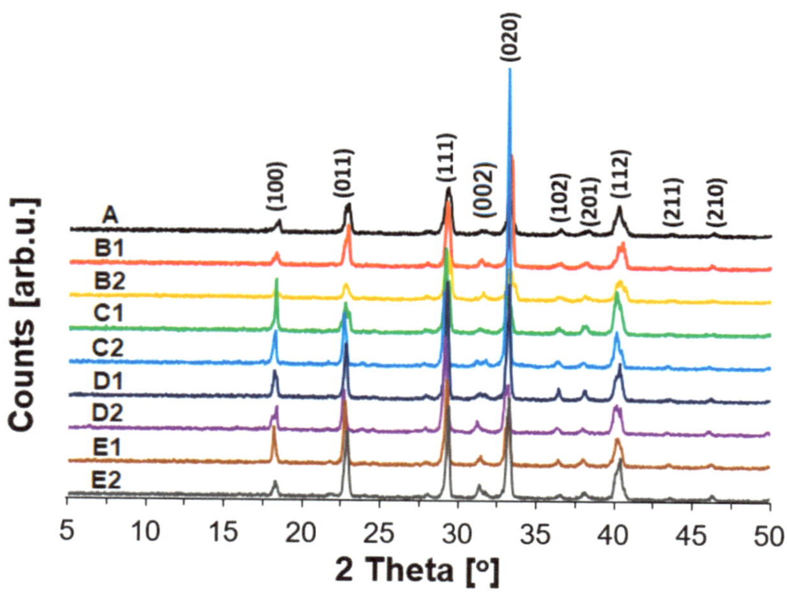

Figure 1. XRD patterns of the studied ANFO samples.

The addition of variously modified zeolite Y to ANFO altered the intensity of reflections coming from AN. Nevertheless, neither the appearance of new reflexes coming from the zeolite phase nor apparent shift was found. Similar effects we observed when we modified ANFO with microstructured charcoal (MC), which was reported in [18]. In turn, other effects were reported by Xu et al. [51], who added organic potassium salts to AN. They showed that the application of this type of additive influenced AN due to the formation of hydrogen bonds by polar groups, which led to the interaction between ammonium and nitrate ions. Based on the results reported by Xu et al. [51], we can expect that the mixing of additives (in this case zeolite Y both in the form of Na-Y and Mg-Y) with AN could result in structural changes in respect to the bare AN. On the other hand, when we compare XRD patterns of variously modified zeolite Y (Figure S1) [53], we can see that some reflexes overlap with the XRD pattern of AN, which can influence the intensity of the reflections assigned to AN. Interestingly, the XRD signals coming from the zeolite phase (Figure S1), and not overlapping with AN reflections, are absent in XRD images obtained for the samples consisting of ANFO mixed with zeolite. At first sight, it could be explained by a low content of zeolite in the prepared materials (not exceeding 2% wt.), but in our previous work [41], we reported that the addition of 1 or 2% wt. of silica resulted in the appearance of a weak signal attributed to this additive.

From FT-IR spectra recorded for all samples (Figure 2), the existence of the bands at 3260–2852 cm^{-1} is due to asymmetric vibrations of ammonium cation both in stretching and deformation modes. The maxima at 1755 cm^{-1} can originate from a stretching vibration or in-plane deformation of nitrate anion, or may result from a combination of an asymmetric deformation of NH_4^+ with a lattice mode. Apparent bands at 1410 and 1290 cm^{-1} can be attributed to the triply degenerated deformation of ammonium cation and the doubly degenerated NO_3^- stretching vibration. In turn, the signals with the maxima at 1041 and 825 cm^{-1} reflect in-plane symmetric stretching and out-of-plane deformation of nitrate ion [49,50,54–56]. The bands at 2951 and 2852 cm^{-1} can be also due to -CH_2- and -CH_3 stretching vibrations from Fuel Oil [57]. Probably, the bands assigned to FO and AN skeletal vibrations overlapped each other.

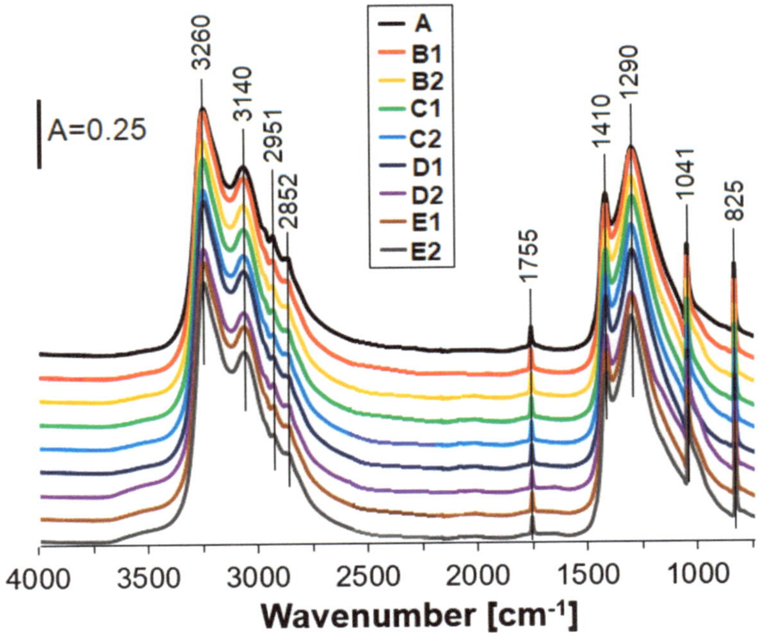

Figure 2. FT-IR spectra of the studied ANFO samples.

The bands attributed to zeolite Y powder (referring to both 1% wt. and 2% wt. of the addition) were not detected. Direct comparison between the appearance of the FT-IR spectra for the prepared ANFO samples depicted in Figure 2, and the FT-IR spectra given in Figure S2 for variously modified zeolite Y samples, leads to the conclusion that the bands typical of AN structure overlapped with the signals coming from zeolite phase. Interestingly, the appearance of a small band at ca 1400 cm^{-1} for all Mg-Y samples can evidence the presence of magnesium in the form of MgO [58,59].

3.2. Morphology and Surface

AFM analysis (Figures 3 and 4) indicated the presence of numerous surface deformations on AN prills in all studied samples. The presence of irregular-shaped AN grains are also apparent. The surface of the tested samples was hilly and was characterized by the existence of bulges. The height of the bulges was the highest for the ANFO without zeolite (for sample A) and reached 5 μm. For the ANFO samples containing 2% wt. of zeolite Y (either in Na-Y or Mg-Y forms)—samples B2, C2, D2, and E2—the occurrence of very small zeolite grains was detected. Another effect was the lowering of the bulge height to 2–3.5 μm.

Zeolite Y was characterized by another character of surface in comparison with ANFO-type materials (Figures S3 and S4). First of all, the surface of zeolite Y seemed to be much more uniform in relation to AN. Furthermore, the variously modified zeolite Y grains were small (with sizes of 1–3 μm and heigh below 1 μm) in respect to AN prills (with diameters in the range of 2–30 μm—Figure 3). The state of the zeolite Y surface depended clearly on the route of zeolite modification. The most uniform surface was found for Mg-Y prepared via the ion-exchange procedure, whereas the most differentiated surface was found for Mg-Y prepared using wet impregnation of zeolite Y with magnesium nitrate.

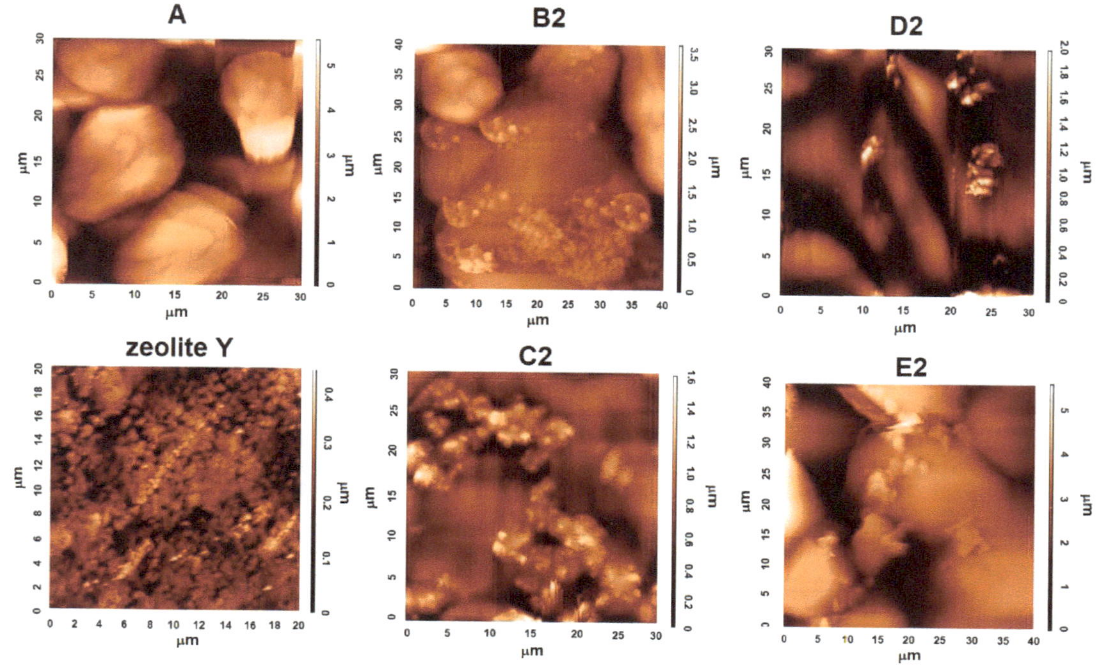

Figure 3. AFM images of the surface of ANFO samples.

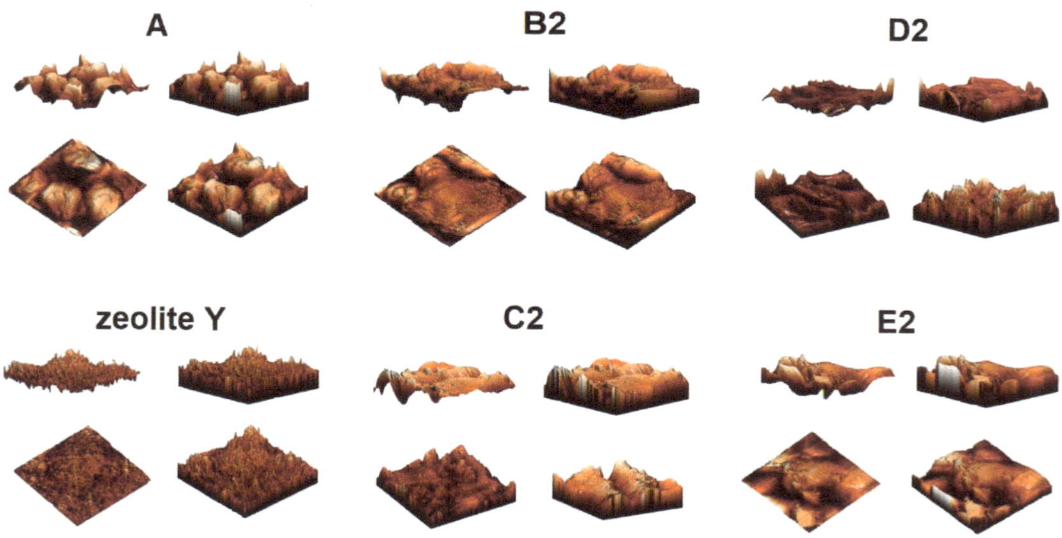

Figure 4. AFM visualization of the surface of ANFO samples in different projections.

SEM images of two magnifications (×1000 and ×10,000) illustrate the influence of the zeolite additive on the morphology of ANFO samples (Figure 5). Ammonium nitrate porous prill crystals mixed with Fuel Oil (sample A) are in the form of wrinkles, surface deformations, and cracks. The morphology of sample A resembles Chinese Cabbage, which is full of holes. The addition of zeolite Y (either with or without Mg) resulted in significant changes in the morphology of the prepared samples. First of all, the appearance of irregular-shaped crystals was found. This new crystalline phase undoubtedly comes from zeolite Y (Samples B2 and C2). Unfortunately, due to the very high sensitivity of the ANFO surface to the electron beam, we were not able to perform the SEM measurements for all prepared samples (D2 and E2). It was caused by the impossibility of achieving a vacuum (probably because of the decomposition of AN) and tearing the surface of the studied samples by the electron beam. Hence, we mainly used the AFM technique to characterize the surface of ANFO-type materials. However, the information obtained from SEM images is in line with the results taken from AFM analysis.

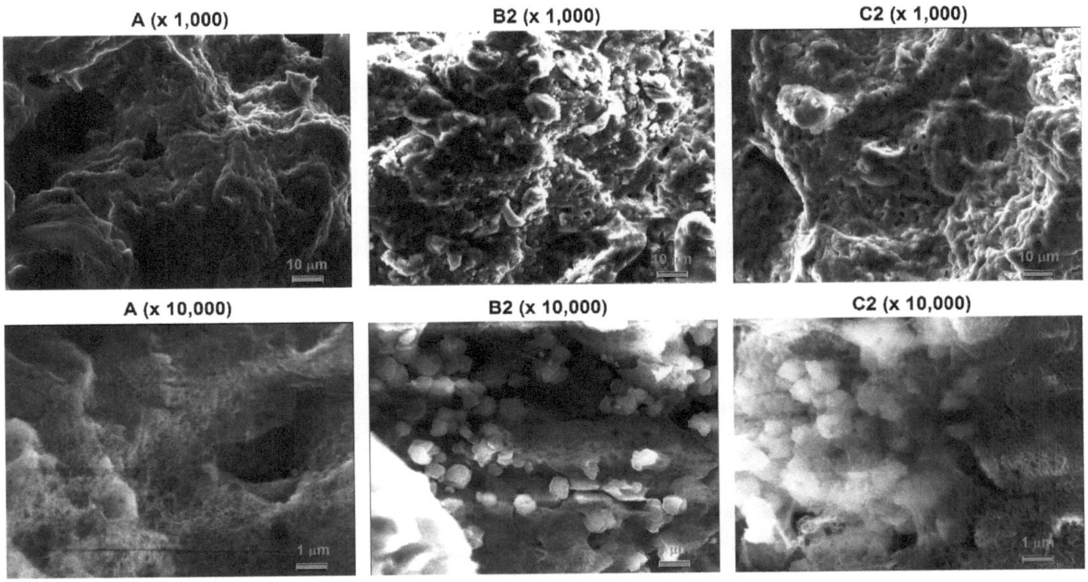

Figure 5. SEM images of ANFO-based samples.

The influence of the way of the zeolite Y modification on its morphology was studied (Figure S5). It was indicated that independently of the zeolite Y modification (both in Na-Y and Mg-Y form), zeolite grains are of irregular shape with sizes smaller than 1 µm. No apparent differences in morphology between variously prepared zeolites were found.

3.3. Thermal Properties

To investigate the influence of the addition of variously modified zeolite Y on the thermal decomposition of ANFO, thermal analysis has been conducted. The TG mass loss and DSC heat effects were measured for ANFO containing 2% wt. of variously prepared zeolite Y (either without or with Mg introduced to zeolite via various techniques). The TG and DSC results are depicted in Figure 6A,B, respectively.

TG data (Figure 6A) showed that for all ANFO samples, a small mass loss (at a temperature lower than 250 °C) was caused by the evaporation of FO from crystalline Ammonium Nitrate. The evaporation of FO took place initially from the surface pores, later from the FO monolayer, and then continued from further FO layers [15].

A further increase of the temperature to the region at 250–300 °C led to the decomposition of AN according to Equation (2):

$$NH_4NO_3 \rightarrow N_2O + 2H_2O \quad (2)$$

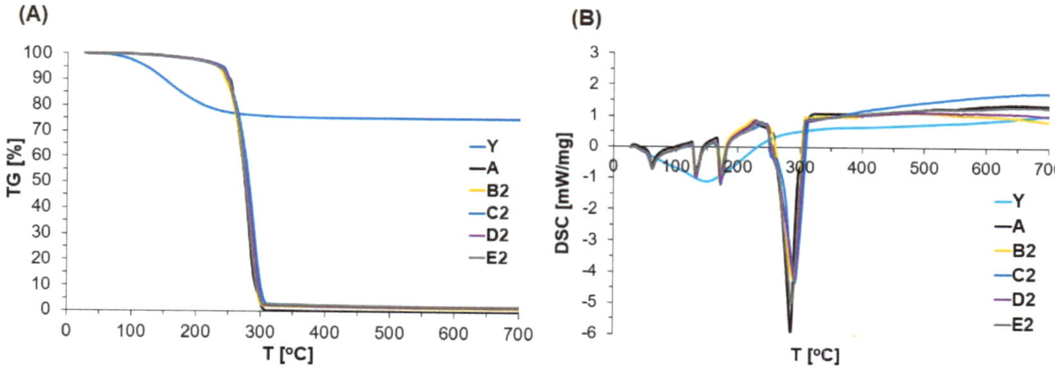

Figure 6. (**A**) TG and (**B**) DSC analysis of the ANFO samples.

That resulted in a total mass decrease of all measured samples. The effect was the same independently of the modification of ANFO with zeolite Y (both in Na-Y form and Mg-Y zeolite prepared via various routes).

For comparison, pure zeolite Y demonstrated high thermal stability. Observed mass loss (22%) resulted from the desorption of water from the surface and then from internal parts of zeolite grains. A relatively high amount of removed water from zeolite can be explained by the high hydrophilicity of zeolite Y, which is typical of zeolites with a low Si/Al molar ratio [60].

DSC curves, obtained at 20–700 °C for all prepared ANFO-type materials, are illustrated in Figure 6B. In order to facilitate interpretation, separated TG/DSC data given for each studied sample were illustrated in Figures S6–S10. The DSC profiles indicate four endothermic peaks at 60, 135, 175, and 287–293 °C. Two signals at ca. 60 and ca. 135 °C originate from the crystallographic transformation of AN III→II and AN II→I, respectively [61]. A third endothermic signal at ca. 175 °C can be attributed to the AN melting point. Furthermore, rising temperature led to the thermal decomposition of AN, which was confirmed by the appearance of the fourth DSC peak with a maximum at ca. 287–293 °C [46,61]. From the analysis of DSC profiles, it may be concluded that the modification route of ANFO with zeolite Y has a small influence on the decomposition temperature of prepared non-ideal explosives. For example, ANFO without zeolite (sample A) undergoes decomposition at 287 °C with a −5.82 mW/mg thermal effect. The addition of zeolite Y containing 2% wt. of Mg introduced via the ultrasonic treatment of aqueous Mg precursor (sample E2) resulted in the decrease of DSC peak to −5.02 mW/mg without the change of peak maximum (at 287 °C). When ANFO was modified with 2% wt. of zeolite without Mg (B2), an energetic effect decreased to −4.25 mW/mg with the simultaneous growth of decomposition temperature to 293 °C. Similar observations took place in the case of the samples prepared via the blending of ANFO with 2% wt. of Mg-Y prepared via wet impregnation (C2) and ion-exchange (D2), for which thermal effects were −4.32 mW/mg and −4.20 mW/mg, respectively. In turn, DCS curve maxima were at 293 °C in both cases.

In our recent work [47], we reported that the maximum of DSC peak responsible for the decomposition of AN can depend on both the size of the microstructured charcoal (MC) grains and the chemical composition of the prepared samples. Generally, higher MC content and smaller MC grains lowered ANFO decomposition temperature (from 292 to 272 °C). In our case, we did not investigate the influence of zeolite additive grains, because we used zeolite in the form of powder.

In another work [62], we reported that the AN provenance (AN as fertilizer: AN-F vs. AN porous prill used in the mining industry: AN-PP) has a profound impact on the temperature of AN decomposition. AN-PP (applied in the undertaken research) is characterized by a lower decomposition temperature.

For comparison with another bibliography, not connected with our previous research, Fedroff et al. [63] reported that AN undergoes decomposition above 230 °C. Nevertheless, AN can deflagrate above 325 °C. Depending on the critical diameter, Ammonium Nitrate can explode at 260–300 °C.

3.4. Detonation Properties

The results of a quantitative analysis of post-blast fumes and velocity of detonation (VOD) measured for the prepared non-ideal explosive samples are summarized in Table 2. Total CO_x and NO_x post-blast volume was equal to 133.4 dm^3 from 1 kg of tested charge (sample A). Generally, the addition of zeolite caused a slight decrease of (CO_x + NO_x) volume to 126.0–131.6 dm^3/kg depending on either the zeolite loading (% wt.) in the ANFO sample or the route of zeolite preparation. One exception was sample E2, for which cumulative (CO_x + NO_x) volume was higher (133.8 dm^3/kg) in respect to the sample without zeolite additive (A).

Table 2. Post-blast fumes, the velocity of detonation (VOD), and density of non-ideal ANFO with the addition of zeolite Y.

Sample	CO_2 [dm^3/kg]	CO [dm^3/kg]	NO_x [dm^3/kg]	CO_x and NO_x Post-Blast Volume [dm^3/kg]	VOD [m/s]	Density [kg/m^3]
A	121.7	4.01	7.64	133.4	2024	695
B1	119.5	4.41	6.83	130.7	2129	683
B2	114.1	4.69	7.24	126.0	2176	672
C1	117.0	4.17	7.96	129.1	2096	676
C2	116.0	4.85	9.00	129.8	2122	671
D1	116.4	4.18	8.73	129.3	1947	684
D2	117.9	3.80	9.94	131.6	2078	676
E1	118.4	3.81	8.73	130.9	1945	686
E2	119.5	4.48	9.83	133.8	1981	686

The most apparent effects were observed for the ANFO containing 1% wt. (B1) and 2% wt. of zeolite Y in sodium form (B2), for which cumulative CO_x and NO_x volumes were equal to 130.7 dm^3/kg and 126.0 dm^3/kg, respectively. In this case, we can see that a higher amount of zeolite additive in the ANFO sample caused a lowering in the formation of CO_x and NO_x. In the case of ANFO modified with the zeolite Y containing Mg, we observed another effect. First of all, when we used the ANFO containing 2% wt. of Mg-Y zeolite (samples C2, D2, and E2), we detected higher total emission of CO_x and NO_x (129.8 dm^3/kg, 131.6 dm^3/kg and 133.8 dm^3/kg, respectively) than for the analogs modified with 1% wt. of Mg-Y (sample C1—129.1 dm^3/kg, sample D1—129.3 dm^3/kg and sample E1—130.9 dm^3/kg). At first sight, the presented results seem to be ambiguous. However, the analysis of EDS results (Table S1) allows us to notice that a rising weight Mg/Na ratio resulted in higher production of CO_x and NO_x from the detonation of 1 kg of explosive. Hence, it seems that the presence of Na in zeolite additive has a positive impact on the reduction of post-blast fumes during detonation of ANFO, whereas Mg evokes an opposite effect. From the analysis of the data summarized in Table S1, we can also establish that the weight ratio between Mg and Na in zeolite additive depends strictly on the way of the preparation of Mg-Y zeolite. The modification of ANFO with Mg-Y, in which Mg was introduced to zeolite via ion-exchange procedure (samples D1 and D2) or sonochemical-assisted technique (samples E1 and E2) resulted in a higher Mg/Na weight ratio followed by higher cumulative production of post-blast fumes.

When we focus only on the emission of CO_2, we can see that for all zeolite-containing samples (B1, B2, C1, C2, D1, D2, E1, and E2), the addition of Na-Y or Mg-Y zeolite resulted in a slight reduction of CO_2 among post-blast gases from 121.7 dm^3/kg (sample A) to 114.1–119.5 dm^3/kg. The lowest emission of CO_2 was found for sample B2 (ANFO modified with 2% wt. of Na-Y zeolite).

Our results on the blasting properties of the studied explosives were compared with the research reported in the available bibliography. However, it was impossible to make a direct comparison between our research described in the current work and other ANFO + zeolite mixtures due to the lack of a bibliography dedicated to the application of zeolites in the manufacturing of ANFO.

Maranda et al. [21] reported that increasing amounts of Al powder caused a distinct reduction of the toxic fumes formed during detonation. In turn, in our recent work [41], we reported that the application of silica affected the oxygen balance of the non-ideal explosive towards positive values, which can be explained by the reduction of FO content and the additional introduction of oxygen (existing in the silica). The shift towards positive oxygen balance caused a high increase in the NO_x and CO_2 volume. It is worth emphasizing that zeolites also belong to oxide materials, and thus should be a source of oxygen. Hence, zeolites can influence the oxygen balance, and we observed an increase in NO_x production among post-blast fumes. On the other hand, minimal reduction of CO_x can be explained by the presence of Al in the structure of the zeolite.

In another previous work [47], we reported the application of microstructure charcoal (MC) powders. In this case, the addition of MC to ANFO also caused the growth of the CO_X and NO_X volume among post-blast gases due to the additional oxygen which was present in the charcoal, and which was reported in the XPS analysis, but which was not included in the calculations. Similar to zeolites and silica, in this case, the shifting oxygen balance toward positive values also takes place. This shift is due to the occurrence of oxygen in the MC chemical composition, which resulted in the MC oxygen balance being lower than assumed at the beginning. That led to the changes in the chemical composition of the studied ANFO + MC systems [47]. Muzyk and Topolnicka [64] indicated that oxygen content decreased with the rank of coal. For example, in flame coal, coke, and anthracite, the oxygen content was at 2–17%. In brown coals, the oxygen amount is in the range of 15–35%.

Our findings concerning the increase of NO_x among post-blasting gases are in agreement with the results reported by Bhattacharyya et al. [65]. Besides the excess of oxygen in the explosive chemical composition or application of the additive which has a positive oxygen balance, the production of NO_x depends on the ANFO's grain size, the amount of FO added to Ammonium Nitrate, as well as the mixing between AN and FO [65–70]. Bhattacharyya et al. [65] indicated that detonation of finer-grained ANFO caused a higher concentration of NO_x in post-blast oxides in relation to the coarser-grained analog, which may be explained by a faster reaction of finer-grained ANFO and a higher oxygen volume available to the chemical reaction leading to a more intensive formation of NO_x. Opposite effects were observed by Sapko et al. [66], who reported that the use of powdered ANFO limited the emission of NO_x fourfold than usual ANFO. Observed phenomena can be explained by a higher mix between fuel and powdered AN prill, leading to a fuller AN decomposition and a higher VOD. On the other hand, it was shown that the application of ANFO of lower FO content caused the growth of NO_x in post-blast fumes, whereas an excess of FO added to AN resulted in less NO_x content among post-blast oxides due to a better stoichiometric blending between AN and FO [67–70].

Experimental data concerning the velocity of detonation (VOD) indicated a clear correlation between VOD values and the type of used ANFO. For ANFO without zeolite (sample A), VOD was 2024 m/s. In the case of ANFO modified with Na-Y zeolite, VOD grew from 2024 m/s to 2129 m/s (for sample B1) and to 2176 m/s (for sample B2). A notable increase in VOD was also observed for ANFO samples modified with Mg-Y prepared via the impregnation method, for which VOD increased to 2096 m/s (for sample C1) and 2122 m/s (for sample C2). An opposite effect was observed for the ANFO modified with Mg-Y via the ultrasonic-assisted procedure. For samples E1 and E2, VOD decreased from 2024 m/s to 1945 m/s and 1981 m/s, respectively. An ambiguous effect was observed for the ANFO modified with Mg-Y zeolite obtained via the ion-exchange procedure. In the case of the sample containing 1% wt. of zeolite (D1), VOD was reduced from 2024 m/s to 1947 m/s, meanwhile, explosive containing 2% wt. of zeolite demonstrated VOD equaling to 2078 m/s. In all cases, VOD increased with the zeolite loading in the tested explosive. Based on the presented results of the blasting properties of the prepared ANFO, it can be concluded that the VOD values of the studied explosives can be easily controlled by both the zeolite Y content and the choice of the zeolite Y modification way. The density of the prepared ANFO samples depended weakly on their route of preparation. For the reference sample (A), density was 695 kg/m^3, while for the ANFO containing zeolite (B1, B2, C1, C2, D1, D2, E1, and E2), density was slightly lower and was in the range of 671–686 kg/m^3.

The results on the VOD values are comparable with our previous studies [71], however, VOD should rise linearly with ANFO's density. In turn, in our undertaken study, we have a similar density of the prepared samples, thus the VOD values are also in a relatively narrow range (1947–2176 m/s). The separate subject for a debate constitutes the influence of the modifier content on the VOD values. As mentioned above, generally, the VOD values increased with the zeolite content in ANFO. When we compared this tendency with the results reported for our previous research [41], where a growing amount of silica in ANFO caused ambiguous changes in VOD values, we stated the opposite effect. This discrepancy can be explained by some fundamental differences between silica and zeolite-type additive. Namely, silica is a typical inert component in ANFO, meanwhile, zeolite Y in the form of both Na-Y and Mg-Y can play a multifunctional role. Zeolite Y can be a source of oxygen (influencing oxygen balance), as well as it supports silicon, sodium, magnesium, and aluminum, being fuel for ANFO. This explanation can be confirmed by Miyake et al. [6], who investigated the velocity of detonation of explosives made of Ammonium Nitrate (AN) blended with activated carbon (AC). They indicated that in AN:AC systems VOD raised with AC content (AN:AC containing 5% wt. of AC was characterized by the VOD value reaching 3400 m/s). Similar observations took place in the case of the mixture of ANFO with microstructured charcoal (ANFO + MC), where the addition of MC to ANFO caused the increase of VOD from 1586 m/s to 1617–2046 m/s depending on both the amount of MC in (ANFO + MC) sample and the size of charcoal enhancer grain [47].

Calculated blasting properties are summarized in Table 3. Performed calculations indicated the potential impact of zeolite additive on the tested ANFO's properties. All investigated parameters (detonation pressure, detonation temperature, heat of explosion, and compression energy defined in [48] being effects of detonation) rose with the loading of ANFO with variously modified zeolite Y. Detonation pressure grew from 3838 MPa for reference AMFO (sample A) to 3844–3966 MPa and to 3954–4126 MPa for samples containing 1% wt. or 2% wt. of zeolite, respectively. For both zeolite contents (1 and 2% wt.), more apparent growths in detonation pressure were indicated for the samples containing Mg-Y zeolite prepared via the ion-exchange or ultrasonic-assisted procedure. The values exceeding 4000 MPa were obtained only for ANFO blended with 2% wt. of zeolite containing magnesium introduced into zeolite Y via ion-exchange process (4008 MPa—sample D2) and using ultrasounds (4126 MPa—sample E2). When we compare our calculated detonation pressure values with EDS results (Table S1), we can see that a rising weight Mg/Na ratio favors a higher detonation pressure. Similar tendencies were found for the heat of the explosion and compression energy. For the heat of the explosion, we obtained rising values from 3913 kJ/kg (sample A) to 4352–4384 kJ/kg (samples B1–E1) and to 4785–4816 kJ/kg (samples B2–E2). In turn, the simulated values of compression energy grew from 795 kJ/kg (sample A) to 819–856 kJ/kg (samples B1–E1) and to 873–889 kJ/kg (samples B2–E2).

In the case of calculated detonation temperature, the addition of zeolite Y (both in the form of Na-Y and Mg-Y) to ANFO resulted in the increase of values from 2970 K (for sample A) to 3204–3224 K and to 3428–3443 K (for ANFO modified with 1% wt. and 2% wt. of zeolite Y, respectively), and was independent of the way of zeolite modification. Only one simulated parameter, which demonstrated such minor changes, was oxygen balance. The addition of variously modified zeolite Y (both in the form of Na-Y and Mg-Y) to ANFO caused very small shifts towards more positive values from −0.99% (for pure ANFO) to −0.98% and −0.97% (when ANFO contained 1% wt. or 2% wt. of zeolite, respectively).

Table 3. Calculated properties of non-ideal ANFO with the addition of variously modified zeolite Y.

Sample	Detonation Pressure [MPa]	Detonation Temperature [K]	Heat of Explosion [kJ/kg]	Compression Energy [kJ/kg]	Oxygen Balance [%]
A	3838	2970	3913	795	−0.99
B1	3871	3204	4352	819	−0.98
B2	3954	3428	4785	873	−0.97
C1	3844	3220	4381	835	−0.98
C2	3961	3443	4811	877	−0.97
D1	3966	3224	4379	856	−0.98
D2	4008	3441	4813	878	−0.97
E1	3900	3211	4384	821	−0.98
E2	4126	3441	4816	889	−0.97

The obtained calculations correspond to our previous research [31], where we reported that the addition of Al or Mg powder to ANFO resulted in significant growth of heat of explosion from 3940 kJ/kg (for pure ANFO) to 4510 J/kg (for ANFO + Al) and to 4500 kJ/kg (for ANFO + Mg), that can be explained by the role of Al and Mg as a fuel being an additional resource of energy [5,20,21,30,31,39]. Another cause of the elevated values of the detonation pressure, temperature, heat of the explosion, and compression energy, can be the afterburning effect, which is a phenomenon resulting from the secondary reaction between unreacted FO (and/or partially oxidized post-blast fumes coming from the primary reactions) with the surrounding air [72,73]. On the other hand, higher values of those parameters are in line with oxygen balance, which also increased and was close to zero.

In a total other view, when we treat zeolite as a system consisting of Si-O and Al-O groups (i.e., semimetal subsystems), we could assume this type of additive introduced to ANFO (being non-ideal composition) combusting behind the reaction front zone. That could lead to the increase in detonation temperature and pressure due to the growth of the surface being responsible for the heat exchange between the zeolite grains and the specific product formed during the reaction of the ANFO decomposition [39].

4. Conclusions

- In the present work, we presented the assessment of the blasting properties of Ammonium Nitrate Fuel Oil (ANFO) with the addition of variously modified zeolite Y.
- The presence of zeolite Y in ANFO did not change the structure, but altered ANFO's surface, morphology, and influenced slightly thermal properties of such synthesized ANFO.
- The addition of zeolite Y to ANFO led to the growth of the detonation pressure, temperature, and heat of the explosion.
- We can control the VOD of ANFO by the choice of the way of the modification of zeolite Y additive. For bare zeolite Y and Mg-Y prepared via the impregnation method, the velocity of detonation (VOD) rose. The opposite effect was observed for ANFO modified with Mg-Y obtained from the deposition of Mg over zeolite Y via the ultrasonic-assisted procedure.
- The utilization of variously modified zeolite Y as an ANFO modifier generally reduced the volume of (CO_x + NO_x) post-blast fumes, which is desired from an ecological point of view.

Supplementary Materials: The following supporting information can be downloaded at: https://www.mdpi.com/article/10.3390/ma15175855/s1, Figure S1: XRD patterns of the variously modified Y-type zeolites; Figure S2: FT-IR spectra of the variously modified Y-type zeolites; Figure S3: AFM images of the surface of variously modified Y-type zeolites; Figure S4: AFM visualization of the surface of variously modified Y-type zeolites in different projections; Figure S5: SEM images of variously modified Y-type zeolites; Figure S6: TG/DSC profiles for sample A; Figure S7: TG/DSC profiles for sample B2; Figure S8: TG/DSC profiles for sample C2; Figure S9: TG/DSC profiles for sample D2; Figure S10: TG/DSC profiles for sample E2; Table S1: EDS (Energy Dispersive Spectroscopy) chemical analysis of variously modified zeolite Y.

Author Contributions: Conceptualization, Ł.K. and A.B.; methodology, Ł.K., A.M.W., M.S., P.Ż., B.D.N., M.Z., M.P. and A.B.; software, Ł.K., M.P. and A.B.; validation, Ł.K., A.M.W., M.S., P.Ż., B.D.N., M.Z., M.P. and A.B.; formal analysis, Ł.K., A.M.W., M.S., P.Ż., B.D.N., M.Z., M.P. and A.B.; investigation, Ł.K., A.M.W., M.S., P.Ż., B.D.N., M.Z., M.P. and A.B.; resources, Ł.K.; data curation, Ł.K.; writing—original draft preparation, Ł.K.; writing—review and editing, Ł.K., A.M.W., M.P. and A.B.; visualization, Ł.K.; supervision, Ł.K.; project administration, Ł.K.; funding acquisition, Ł.K. All authors have read and agreed to the published version of the manuscript.

Funding: The research was financed from ICSC PAS development project "FAU-type zeolites as modifiers of ANFO" no. 2/GR/2020 from the Jerzy Haber Institute of Catalysis and Surface Chemistry, Polish Academy of Sciences.

Institutional Review Board Statement: Not applicable.

Informed Consent Statement: Not applicable.

Data Availability Statement: Not applicable.

Acknowledgments: The authors would like to thank OZM Research s.r.o. and Forcepol Sp. z o.o. for providing the EXPLO5 software for calculating detonation parameters.

Conflicts of Interest: The authors declare no conflict of interest.

References

1. Cook, M.A. *The Science of Industrial Explosives*; IRECO Chemicals: Salt Lake City, UT, USA, 1974.
2. Mathiak, H. *Pothole Blasting for Wildlife*; Wisconsin Conservation Department: Madison, WI, USA, 1965.
3. Biessikirski, A.; Kuterasinski, Ł. *Research on Morphology and Topology of ANFO Based on Various Types of Oxygen Component*; AGH: Kraków, Poland, 2018.
4. Zygmunt, B.; Buczkowski, D. Influence of Ammonium Nitrate Prills' Properties on DetonationVelocity of ANFO. *Propellants Explos. Pyrotech.* **2007**, *32*, 411–414. [CrossRef]
5. Zygmunt, B. Detonation parameters of mixtures containing ammonium nitrate and aluminum. *Cent. Eur. J. Energ. Mater.* **2009**, *6*, 57–66.
6. Miyake, A.; Takahara, K.; Ogawa, T.; Ogata, Y.; Wada, Y.; Arai, H. Influence of physical properties of ammonium nitrate on the detonation behavior of ANFO. *J. Loss Prev. Process Ind.* **2001**, *14*, 533–538. [CrossRef]
7. Cummock, N.R.; Mares, J.O.; Gunduz, I.E.; Son, S.F. Relating a small-scale shock sensitivity experiment to large-scale failure diameter in an aluminized ammonium nitrate non-ideal explosive. *Combust. Flame* **2018**, *194*, 271–277. [CrossRef]
8. Davis, W.; Fauquignon, C. Classical theory of detonation. *J. Phys. IV Colloq.* **1995**, *4*, 3–21. [CrossRef]
9. Smirnov, E.B.; Kostitsin, O.; Koval, A.V.; Akhlyustin, I.A. Model of non-ideal detonation of condensed high explosives. In *Journal of Physics: Conference Series, Proceedings of the XXXI International Conference on Equations of State for Matter, Elbrus, Russia, 1–6 March 2016*; IOP Publishing: Bristol, UK, 2016; Volume 774, p. 012076.
10. Paszula, J.; Kowalewski, E. Study of the detonation development of non-ideal explosive. *High Energy Mater.* **2015**, *7*, 95–105.
11. Clark, G.B. *Principles of Fragmentation*; Wiley-Interscience: New York, NY, USA, 1987.
12. Kwok, Q.S.M.; Jones, D.E.G. Thermodesorption studies of ammonium nitrate prills by high-resolution thermogravimetry. *J. Therm. Anal. Calorim.* **2003**, *74*, 57–63. [CrossRef]
13. Lenspiece, E.; Petr, V. The characterization of ammonium nitrate mini-prills. In *Dynamic Behavior of Materials*; Song, B., Casem, D., Kimberley, J., Eds.; Conference Proceedings of the Society for Experimental Mechanics Series; Springer: Cham, Switzerland, 2015.
14. Biessikirski, A.; Kuterasiński, Ł.; Dworzak, M.; Pyra, J.; Twardosz, M. Comparison of structure, morphology, and topography of fertilizer-based explosives applied in the mining industry. *Microchem. J.* **2019**, *144*, 39–44. [CrossRef]
15. Oxley, J.C.; Smith, J.L.; Rogers, E.; Yu, M. Ammonium nitrate: Thermal stability and explosivity modifiers. *Thermochim. Acta* **2002**, *384*, 23–45. [CrossRef]
16. Oxley, J.C.; Smith, J.L.; Wang, W. Compatibility of Ammonium Nitrate with Monomolecular Explosives 2. Nitroarenes12. *J. Phys. Chem.* **1994**, *98*, 3901. [CrossRef]
17. Brower, K.R.; Oxley, J.C.; Tewari, M. Evidence for Homolytic Decomposition of Ammonium Nitrate at High Temperature. *J. Phys. Chem.* **1989**, *93*, 4029. [CrossRef]
18. Atlagic, S.G.; Biessikirski, A.; Kuterasiński, Ł.; Dworzak, M.; Twardosz, M.; Sorogas, N.; Arvanitidis, J. On the Investigation of Microstructured Charcoal as an ANFO Blasting Enhancer. *Energies* **2020**, *13*, 4681. [CrossRef]
19. Sinditskii, V.P.; Egorshev, V.Y.; Levshenkov, A.I.; Serushkin, V.V. Ammonium nitrate: Combustion mechanism and the role of additives. *Propellants Explos. Pyrotech.* **2005**, *30*, 269–280. [CrossRef]
20. Maranda, A.; Gałęzowski, D.; Papliński, A. Investigation on Detonation and Thermochemical Parameters of Aluminized ANFO. *J. Energ. Mater.* **2003**, *21*, 1–3. [CrossRef]

21. Maranda, A.; Paszula, J.; Zawadzka-Małota, I.; Kuczyńska, B.; Witkowski, W.; Nikolczuk, K.; Wilk, Z. Aluminum powder influence on ANFO detonation parameters. *Cent. Eur. J. Energ. Mater.* **2011**, *8*, 279–293.
22. Izato, Y.; Miyake, A.; Date, S. Combustion characteristics of ammonium nitrate and carbon mixtures based on a thermal decomposition mechanism. *Propellants Explos. Pyrotech.* **2013**, *38*, 129–135. [CrossRef]
23. Hussain, G.; Rees, G. Combustion of NH_4NO_3 and carbon-based mixtures. *Fuel* **1993**, *72*, 1475–1479. [CrossRef]
24. Lurie, B.A.; Lianshen, C. Kinetics and mechanism of thermal decomposition of ammonium nitrate powder under the action of carbon black. *Combust. Explos. Shock Waves* **2000**, *36*, 607–617. [CrossRef]
25. Miyake, A.; Kobayashi, H.; Echigoya, H.; Ogawa, T. Combustion and ignition properties of ammonium nitrate and activated carbon mixtures. *Int. J. Energ. Mater. Chem. Propuls.* **2009**, *8*, 411–419. [CrossRef]
26. Wang, L.; Ago, M.; Borghei, M.; Ishaq, A.; Papageorgiou, A.C.; Lundahl, M.; Rojas, O.J. Conductive Carbon Microfibers Derived from Wet-Spun Lignin/Nanocellulose Hydrogels. *ACS Sustain. Chem. Eng.* **2019**, *7*, 6013–6022. [CrossRef]
27. Nazarian, A.; Presser, C. Forensic methodology for the thermochemical characterization of ANNM and ANFO homemade explosives. *Thermochim. Acta* **2015**, *608*, 65–75. [CrossRef]
28. Gunawan, R.; Zhang, D. Thermal stability and kinetics of decomposition of ammonium nitrate in the presence of pyrite. *J. Hazard. Mater.* **2009**, *165*, 751–758. [CrossRef]
29. Gunawan, R.; Freij, S.; Zhang, D.; Beach, F.; Littlefair, M. A mechanistic study into the reactions of ammonium nitrate with pyrite. *Chem. Eng. Sci.* **2006**, *61*, 5781–5790. [CrossRef]
30. Kramarczyk, B.; Pytlik, M.; Mertuszka, P. Effect of aluminum additives on selected detonation parameters of a bulk emulsion explosive. *High-Energy Mater.* **2020**, *12*, 99–113.
31. Biessikirski, A.; Ziąbka, M.; Dworzak, M.; Kuterasiński, Ł.; Twardosz, M. Effect of metal addition on ANFO's morphology and energy detonation. *Przem. Chem.* **2019**, *98*, 928–931.
32. Gómez-Rico, M.F.; Marty ́ın-Gullón, I.; Fullana, A.; Conesa, J.A.; Font, R. Pyrolysis and combustion kinetics and emissions of waste lube oils. *J. Anal. Appl. Pyrolysis* **2003**, *68–69*, 527–546. [CrossRef]
33. Hakki Metecan, I.; Ozkan, A.R.; Isler, R.; Yanik, J.; Saglam, M.; Yuksel, M. Naphtha derived from polyolefins. *Fuel* **2005**, *84*, 619. [CrossRef]
34. Sriningsih, W.; Garby Saerodji, M.; Trisunaryanti, W.; Triyono; Armunanto, R.; Falah, I.I. Fuel production from LDPE plastic waste over natural zeolite supported Ni, Ni-Mo, Co and Co-Mo metals. *Procedia Environ. Sci.* **2014**, *20*, 215–224. [CrossRef]
35. Al-Salem, S.M.; Antelava, A.; Constantinou, A.; Manos, G.; Dutta, A. A review on thermal and catalytic pyrolysis of plastic solid waste (PSW). *J. Environ. Manag.* **2017**, *197*, 177–198. [CrossRef]
36. Mahari, W.A.W.; Chong, C.T.; Lam, W.H.; Nyuk, T.N.S.T.A.; Moh, L.M.; Ibrahim, D.; Lama, S.S. Microwave co-pyrolysis of waste polyolefins and waste cooking oil: Influence of N_2 atmosphere versus vacuum environment. *Energy Convers. Manag.* **2018**, *171*, 1292–1301. [CrossRef]
37. Owusu, P.A.; Banadda, N.E.; Zziwa, A.; Seay, J.; Kiggundu, N. Reverse engineering of plastic waste into useful fuel products. *J. Anal. Appl. Pyrolysis* **2018**, *130*, 285–293. [CrossRef]
38. Kaniewski, M.; Homann, K.; Homann, J. Influence of selected potassium salts on thermal stability of ammonium nitrate. *Thermochim. Acta* **2019**, *678*, 178313. [CrossRef]
39. Maranda, A. Influence of aluminum chemical activity on detonation parameters of composite explosive containing aluminum powders (CX-Al). *Wiad. Chem.* **2001**, *55*, 353–375.
40. Barański, K. Analysis of the Possibility of Using "Green Pyrotechnics" in the Detonators MW. Ph.D. Thesis, Faculty of Mining and Geoengineering, AGH University of Science and Technology in Cracow, Cracow, Poland, 27 September 2019.
41. Biessikirski, A.; Barański, K.; Pytlik, M.; Kuterasiński, Ł.; Biegańska, J.; Słowiński, K. Application of Silicon Dioxide as the Inert Component or Oxide Component Enhancer in ANFO. *Energies* **2021**, *14*, 2152. [CrossRef]
42. Biessikirski, A.; Wądrzyk, M.; Janus, R.; Biegańska, J.; Jodłowski, G.; Kuterasiński, Ł. Study on fuel oils used in ammonium nitrate-based explosives. *Przem. Chem.* **2018**, *97*, 457–462.
43. *EN 13631-14:2003*; Explosives for Civil Uses. High Explosives. Part 14: Determination of Velocity of Detonation, European Committee for Standardization: Brussels, Belgium, 2003.
44. *EN 13631-16:2004*; Explosives for Civil Uses. High Explosives. Part 16: Detection and Measurement of Toxic Gases, European Committee for Standardization: Brussels, Belgium, 2004.
45. European Union. Council directive 93/15/EEC of 5 April 1993 on the harmonization of the provisions relating to the placing on the market and supervision of explosives for civil uses. *Off. J. Eur. Union* **1993**, *12*, 20–36.
46. Biessikirski, A.; Pytlik, M.; Kuterasiński, Ł.; Dworzak, M.; Twardosz, M.; Napruszewska, B.D. Influence of the Ammonium(V) Nitrate Porous Prill Assortments and Absorption Index on Ammonium Nitrate Fuel Oil Blasting Properties. *Energies* **2020**, *13*, 3763. [CrossRef]
47. Biessikirski, A.; Atlagic, S.G.; Pytlik, M.; Kuterasinski, Ł.; Dworzak, M.; Twardosz, M.; Nowak-Senderowska, D.; Napruszewska, B.D. The Influence of Microstructured Charcoal Additive on ANFO's Properties. *Energies* **2021**, *14*, 4354. [CrossRef]
48. Suceska, M. Calculation of Detonation Parameters by EXPLO5 Computer Program. *Mater. Sci. Forum* **2004**, *465–466*, 325–330. [CrossRef]
49. Hendricks, S.B.; Posnjak, E.; Kracek, F.C. Molecular rotation in the solid state. The variation of the crystal structure of ammonium nitrate with temperature. *J. Am. Chem. Soc.* **1932**, *54*, 2766–2786. [CrossRef]

50. Vargeese, A.A.; Joshi, S.S.; Krishnamurthy, V.N. Effect of the method of crystallization on the IV–III and IV–II polymorphic transitions of ammonium nitrate. *J. Hazard. Mater.* **2009**, *161*, 373–379. [CrossRef]
51. Xu, Z.-X.; Fu, X.-Q.; Wang, Q. Phase stability of ammonium nitrate with organic potassium salts. *Cent. Eur. J. Energ. Mater.* **2016**, *13*, 736–754. [CrossRef]
52. Ferg, E.; Masalova, I. Using PXRD to Investigate the Crystallization of Highly Concentrated Emulsions of NH_4NO_3. *S. Afr. J. Chem.* **2011**, *64*, 7–16.
53. Downs, R.T.; Hall-Wallace, M. The American Mineralogist crystal structure database. *Am. Mineral.* **2003**, *88*, 247–250.
54. National Institute of Standards and Technology: "Ammonium Nitrate". Available online: https://webbook.nist.gov/cgi/cbook.cgi?Name=ammonium+nitrate&Units=SI&cIR=on#Refs (accessed on 1 January 2022).
55. Wu, B.H.; Chan, M.N.; Chan, C.K. FTIR characterization of polymorphic transformation of ammonium nitrate. *Aerosol Sci. Technol.* **2007**, *41*, 581–588. [CrossRef]
56. Steele, B.A.; Oleynik, I.I. New crystal phase ammonium nitrate: First-principles prediction and characterization. In Proceedings of the AIP Conference Proceedings, Tampa Bay, FL, USA, 14–19 June 2015; Volume 1793, p. 130008.
57. Pretsch, E.; Buhlmann, P.; Badertscher, M. *Structure Determination of Organic Compounds. Tables of Spectral Data*; Springer: Berlin, Germany, 2009.
58. Aksay, S. Effects of Al dopant on XRD, FT-IR and UV–vis properties of MgO films. *Phys. B Condens. Matter.* **2019**, *570*, 280–284. [CrossRef]
59. Ansari, A.; Ali, A.; Asifa, M.; Shamsuzzaman. Microwave-assisted MgO NP catalyzed one-pot multicomponent synthesis of polysubstituted steroidal pyridines. *New J. Chem.* **2018**, *42*, 184–197. [CrossRef]
60. Chen, N.Y. Hydrophobic properties of zeolites. *J. Phys. Chem.* **1976**, *80*, 60–64. [CrossRef]
61. Theoret, A.; Sandorey, C. Infrared Spectra and Crystalline Phase Transitions of Ammonium Nitrate. *Can. J. Chem.* **1964**, *42*, 57. [CrossRef]
62. Biessikirski, A.; Kuterasiński, Ł.; Dworzak, M.; Twardosz, M.; Tatko, M.; Napruszewska, B.D. On the Influence of the Ammonium Nitrate(V) Provenance on Its Usefulness for the Manufacture of ANFO Type Explosives. *Energies* **2020**, *13*, 4942. [CrossRef]
63. Fedoro, B.T.; Aaronson, H.A.; Sheeld, O.E.; Reese, E.F.; Clift, G.D.; Dunkle, C.G.; Walter, H.; McLean, D.C. *Encyclopedia of Explosives and Related Items*; US Department of the Army, Picatinny Arsenal: Dover, NJ, USA, 1960; Volume 1.
64. Muzyk, R.; Topolnicka, T. Direct determination of oxygen content in coals by the use of the high-temperature pyrolysis method. *Prz. Gor.* **2013**, *69*, 100–104.
65. Bhattacharyya, M.M.; Singh, P.K.; Singh, R.R. Laboratory methodology for assessment of toxic fumes in post-detonation gases from explosives. *Min. Res. Eng.* **2001**, *10*, 171–183. [CrossRef]
66. Sapko, M.; Rowland, J.; Mainiero, R.; Zlochower, I. Chemical and physical factors that influence NO_x production during blasting–exploratory study. In Proceedings of the 28th Conference on 'Explosives and Blasting Technique', Las Vegas, NV, USA, 10 February 2002; pp. 317–330.
67. Mainiero, R.; Harris, M.; Rowland, J. Dangers of toxic fumes from blasting. In Proceedings of the 33rd Conference on 'Explosives and Blasting Technique', Nashville, TN, USA, 28–31 January 2007; Volume 1, pp. 1–6.
68. Onederra, I.; Bailey, V.; Cavanough, G.; Torrance, A. Understanding main causes of nitrogen oxide fumes in surface blasting. *Min. Technol.* **2012**, *121*, 151–159. [CrossRef]
69. Chaiken, R.F.; Cook, E.B.; Ruhe, T.C. *Toxic Fumes from Explosives: Ammonium Nitrate-Fuel Oil Mixtures*; BuMines RI 7867; US Department of the Interior, Bureau of Mines: Washington, DC, USA, 1974.
70. Rowland, J.H., III; Mainiero, R. Factors Affecting ANFO Fumes Production. In Proceedings of the 26th Annual Conference on Explosives and Blasting Technique, Cleveland, OH, USA, 13–16 February 2000; International Society of Explosives Engineers: Warrensville Heights, OH, USA, 2000; pp. 163–174.
71. Biessikirski, A.; Czerwonka, D.; Biegańska, J.; Kuterasiński, Ł.; Ziąbka, M.; Dworzak, M.; Twardosz, M. Research on the Possible Application of Polyolefin Waste-Derived Pyrolysis Oils for ANFO Manufacturing. *Energies* **2021**, *14*, 172. [CrossRef]
72. Salzano, E.; Basco, A. Comparison of the explosion thermodynamics of TNT and black powder using Le Chatelier diagrams. *Propellants Explos. Pyrotech.* **2012**, *37*, 724–731. [CrossRef]
73. Zawadzka-Małota, I. Testing of mining explosives with regard to the content of carbon oxides and nitrogen oxides in their detonation products. *J. Sustain. Min.* **2015**, *14*, 173–178. [CrossRef]

materials

MDPI

Article

Theoretical and Experimental Analysis of the Hot Torsion Process of the Hardly Deformable 5XXX Series Aluminium Alloy

Konrad Błażej Laber [1,*] and Beata Leszczyńska-Madej [2]

[1] Department of Metallurgy and Metal Technology, Faculty of Production Engineering and Materials Technology, Czestochowa University of Technology, 19 Armii Krajowej Ave., 42-201 Czestochowa, Poland
[2] Faculty of Non-Ferrous Metals, AGH University of Science and Technology, 30-059 Kraków, Poland; bleszcz@agh.edu.pl
* Correspondence: konrad.laber@pcz.pl; Tel.: +48-34-325-0662

Abstract: This work presents the results of the numerical and physical modelling of the hot torsion of a hardly deformable 5XXX series aluminium alloy. Studies were conducted on constrained torsion with the use of the STD 812 torsion plastometer. The main purpose of the numerical tests was to determine the influence of the accuracy of the mathematical model describing the changes in the yield stress of the tested material on the distribution of strain parameters and on the stress intensity. According to the preliminary studies, in the case of numerical modelling of the torsion test, the accuracy of the applied mathematical model describing the changes in the rheological properties of the tested material and the correct definition of the initial and boundary conditions had a particularly significant impact on the correctness of the determination of the strain parameters and the intensity of stresses. As part of the experimental tests, physical modelling of the hot torsion test was conducted. The aim of this part of the work was to determine the influence of the applied strain parameters on the distribution and size of grain as well as the microhardness of the tested aluminium alloy. Metallographic analyses were performed using light microscopy and the electron backscatter diffraction method. Due to the large inhomogeneity of the deformation parameters and the stress intensity in the torsion test, such tests were necessary for the correct determination of the so-called representative area for metallographic analyses. These types of studies are particularly important in the case of the so-called complex deformation patterns. The paper also briefly presents the results of preliminary research and future directions in which it is planned to use complex deformation patterns for physical modelling of selected processes combining various materials.

Keywords: hot torsion test; strain and stress state; numerical modelling; physical modelling; metallographic examinations; EBSD analysis; microhardness; inhomogeneity of microstructure; hardly deformable materials

Citation: Laber, K.B.; Leszczyńska-Madej, B. Theoretical and Experimental Analysis of the Hot Torsion Process of the Hardly Deformable 5XXX Series Aluminium Alloy. *Materials* **2021**, *14*, 3508. https://doi.org/10.3390/ma14133508

Academic Editors: Jakub Zdarta and Agata Zdarta

Received: 19 May 2021
Accepted: 18 June 2021
Published: 23 June 2021

1. Introduction

There are many research methods to determine the value of yield stress as a function of strain parameters and temperature. These methods include tensile, compression, and torsion tests. The methods have been described in detail in other works including [1–5]. According to many authors, the most convenient way of determining the flow curves in high temperature is a hot torsion test. It enables the determination of the yield stress indirectly, using the hypothesis of material effort. This test is widely used in plastometric tests due to the unchanging stress state during the tests, corresponding with high accuracy to pure shear and lack of friction as well as the possibility of obtaining large deformations, significantly exceeding deformations achievable in the tensile or compression test [1,2,6]. The hot torsion method is particularly advantageous when assessing the plasticity of hardly deformable and brittle materials for the conditions of hot plastic working. The advantage of this method is also the fact that during the tests, it is possible to create more secure conditions to obtain a constant strain rate and to model sequential (multistage) deformation patterns in the best possible way [1,7].

The main areas of application of the torsion test are determination of the value of yield stress at a given temperature and for a given stain value and strain rate, assessment of the limit deformability of the tested materials, testing (in conjunction with metallographic analyses) of structural changes in the material caused by the deformation process or physical modelling of plastic working processes [6].

There are many examples of the use of the hot torsion test in basic research in the literature [8–14]. In these works, the authors investigated, for example, the influence of the applied conditions of the deformation process on the yield stress, the development of the microstructure and its gradient in the cross-section, or the influence of the applied conditions on the deformability of the tested materials. While analysing the available literature, numerous papers can also be found in which hot torsion tests have been successfully used to physically copy the actual technological processes of rolling, extrusion or combining materials [15–21].

However, it should be noted that the hot torsion test is marked by an uneven strain distribution, strain rate, and temperature on the cross-section and longitudinal section of the torsion material. In the cross-section, all the parameters mentioned above reach their maximum values on the sample surface, while the lowest values of strain parameters occur in the material axis. The influence of this unevenness should be taken into account through calculating the strain parameters for the so-called representative radius, in which the values of the strain parameters correspond to the average values in the cross-section. The diversity of the strain value and the strain rate depending on the analytical relationships used, linking the equivalent strain with the non-dilatational strain, result in the assumption of different values, the so-called equivalent radius [2]. The projected equivalent radius values are: 0.6 r, 0.67 r, 0.724 r or 0.75 r, where r is the radius of the torsion sample [2,3,22].

Numerical modelling of the torsion test, conducted in accordance with the actual conditions of experimental tests, enables, for example, to determine the representative area for possible metallographic tests. On the basis of the obtained results, it is also possible to analyse the state of strain and stresses in the entire volume of the material. This is especially important for complex deformation patterns (e.g., simultaneous torsion with tension or simultaneous torsion with compression). As the preliminary tests conducted in this area show, the applied strain pattern affects the value of strain parameters (strain intensity and strain rate intensity) and the yield stress of the material. This is especially visible during simultaneous torsion and tension. It was initially determined that this is the result of the necking down formation and, consequently, a strong location (concentration) of the strain in this area. On the basis of the results of many experimental and theoretical studies, it has been proved that the strain rate in the neck during the tensile test is an order higher than the average strain rate calculated on the basis of changes in the measuring length of the samples [1].

Numerical analysis of the torsion test also enables the determination of the degree of unevenness of the strain parameters for different materials and different dimensions of the work part of the torsion samples in a wide range of strain parameters and temperature. However, the necessary condition for the correct determination of the strain and stress parameters during numerical modelling of the hot torsion test is the precise definition of the rheological properties of the tested material and the correct determination of the initial and boundary conditions, in accordance with the experimental tests.

An equally important problem during the theoretical and experimental analysis of the hot torsion test is the determination of the influence of the applied strain parameters, for example, on the distribution and size of the grain as well as the microhardness of the tested material. As shown in the literature, both the grain size and its distribution, as well as the microhardness distribution, depend primarily on the properties of the material itself, as well as on the parameters of the deformation process (caused strain value, strain rate and temperature) and the mechanisms (processes) of rebuilding microstructures [10,11,23–25]. The description of the most common mechanisms of remodelling the microstructure and the behaviour of various materials during hot forming can be found, for example, in the works [26–31]. This applies not only to the torsion processes but also to high-pressure torsion [32–34].

On the basis of the conducted numerical analyses, it was found that the condition for the correct determination of the strain parameters distribution and the stress intensity in the torsion test is high accuracy of the mathematical model describing the rheological properties of the material tested.

No studies have been found in the analysed literature that would comprehensively describe both the influence of the accuracy of the mathematical model of rheological properties on the values and nature of changes in the strain parameters and the yield stress as well as the influence of the applied strain parameters on the nature of the grain size distribution and the microhardness of 5xxx series aluminium alloy. Therefore, according to the authors, the issues raised in the work are topical, especially in the context of complex deformation patterns.

2. Materials and Methods

2.1. Materials and Process Parameters

The tests presented in the paper were conducted for the hardly deformable 5019 aluminium alloy, with the chemical composition compatible with the standard EN 573-3 presentedin details in the paper [35]. Plastometric tests were conducted up to the true strain value equal to 5, for the true strain rate value 0.05, 0.25 and 0.5 s^{-1}, at the temperature 360, 400, 440 480 and 520 °C [35].

2.2. Methods

The methods for the numerical and physical modelling were used for the research. Numerical modelling was performed with the use of commercial FORGE 2011 software [36], while physical modelling was performed with the use of the STD 812 torsion plastometer [14]. Metallographic studies were conducted using light microscopy (Olympus GX 51 microscope, Tokyo, Japan) and electron backscatter diffraction (EBSD) (Hitachi S-3400N microscope, Tokyo, Japan). The microhardness measurements were carried out using the Innovatest Nexus 4000 microhardness tester (INNOVATEST Europe BV, Maastricht, The Netherlands).

2.2.1. Numerical Research: Mathematical Model Using FORGE 2011® Software

The aim of the numerical tests was to determine the influence of the accuracy of defining the rheological properties of the 5019 aluminium alloy on the strain intensity distribution, strain rate intensity, and stress intensity in the hot torsion test. In the first stage, plastometric tests of the analysed aluminium alloy were conducted in constrained torsion tests, the results of which are described closely in the paper [35]. These tests were carried out for conditions characteristic of the extrusion process. The obtained results were approximated with different accuracies (variants no. 1 and 2), determining mathematical models of the rheological properties of the tested material in the analysed range of strain and temperature parameters. In the next stage of the research, numerical modelling of the hot torsion test was conducted with the use of the developed models of changes in the rheological properties of the tested alloy, in accordance with the conditions occurring during experimental tests. These studies were carried out with the use of the FORGE 2011® program [36].

In the FORGE 2011® program [36], the properties of the deformed material were described according to the Norton–Hoff [37–40] conservation law written in the following Equation (1):

$$S_{ij} = 2K(T, \dot{\varepsilon}_i, \varepsilon_i)\left(\sqrt{3}\dot{\varepsilon}_i\right)^{m_m - 1}\dot{\varepsilon}_{ij} \tag{1}$$

where S_{ij} is the deviatoric stress tensor, $\dot{\varepsilon}_i$ is the strain rate intensity $\dot{\varepsilon}_{ij}$ is the strain rate tensor, ε_i is the strain intensity, T is the temperature, K is the consistency depending on the yield stress σ_p, and m_m is the coefficient characterizing hot deformation of the metal ($0 < m_m < 1$).

For the determination of the temperature field, a differential equation is used, which describes variations in temperature with transient heat flow. This is a quasi-harmonic equation in the following Equation (2) [41–44]:

$$\frac{\partial}{\partial x}\left(k_x\frac{\partial T_s}{\partial x}\right)+\frac{\partial}{\partial y}\left(k_y\frac{\partial T_s}{\partial y}\right)+\frac{\partial}{\partial z}\left(k_z\frac{\partial T_s}{\partial z}\right)+\left(Q-c_p\rho\frac{\partial T_s}{\partial t}\right)=0 \quad (2)$$

where k_x,k_y and k_z are functions of the distribution of anisotropic thermal conductivities in the directions x, y and z, T_s is the function that describes temperature in the investigated area,Q is the function of the distribution of deformation heat generation speed, c_p is the function of the distribution of metal-specific heat and ρ is the function of the metal density distribution.

When a die is rotating, the solver computes the torque obtained around the rotation axis. The rolling torque is equivalent to a force multiplied by a distance. So, the solver computes on every surface element the vector force multiplied with the distance to the axis of rotation. The sum of all these local torques provides the total torque as shown in Equation (3) [41,42]:

$$M_w=\int_S r\sigma_y dS=\sum_{e=1}^{N_e}\int_{S_e} r_e\sigma_{ye}l_y dS_e \quad (3)$$

where r is the distance between a point on the material surface and the rotation axis, σ_y is the stress component, S_e is the surface area of a finite element in contact with the tool, e is a finite element in contact with the tool, N_e is the number of finite elements that remain in contact with the tool, σ_{ye} is a component of the stress acting on the element in contact with the rotating tools and l_y is the directional cosine of the normal to the contact area of the current element with the rotating tool.

The initial and boundary conditions for the numerical modelling of the hot torsion test were the same as for the experimental tests using the STD 812 torsion plastometer (Sections 2.1 and 2.2.2). Friction conditions between the sample and the tools were set to bilateral sticking. To obtain a constant temperature distribution on the sample length, the thermal exchange was set as adiabatic.

In order to describe the changes in the value of the yield stress during the approximation of the results of the experimental tests, Equation (4) [43] was adopted:

$$\sigma_p=A\cdot e^{m_1\cdot T}\cdot T^{m_9}\cdot\varepsilon^{m_2}\cdot e^{\frac{m_4}{\varepsilon}}\cdot(1+\varepsilon)^{m_5\cdot T}\cdot e^{m_7\cdot\varepsilon}\cdot\dot{\varepsilon}^{m_3}\cdot\dot{\varepsilon}^{m_8\cdot T} \quad (4)$$

where σ_p is yield stress, T is temperature, ε is true strain, $\dot{\varepsilon}$ is strain rate and A and $m_1 \div m_9$ are function coefficients.

The approximation of the results of the experimental tests was performed with the use of software developed at the Department of Metal Forming and Safety Engineering at the Częstochowa University of Technology.

2.2.2. Experimental Research

As part of the experimental tests, physical modelling of the hot torsion test was conducted. The STD 812 torsion plastometer was used for the tests [14]. The general view of the test chamber and the main parameters of the device are shown in Figure 1.

(**a**)

Heating:	induction
Maximum testing temperature:	1500 °C
Heating and cooling rate:	up to 100 K/s
Minimum time between strains:	60 ms
Medium: Vacuum 10^{-4} mbar, neutral gas, air	
Torsion	
Rotational velocity:	up to 500 rpm
Number of rotations:	up to 30
Torque:	up to 50 Nm
Strain rate:	up to 60 s^{-1}
Tension and compression	
Change in length:	ca. 15 mm
Strain rate:	up to 30 mm/s
Strain force:	up to 25kN
Strain rate:	up to 1.0 s^{-1}
True strain: dependent on specimen dimensions	
The device allows for torsion with tension or compression	

(**b**)

Figure 1. STD 812 torsion plastometer: (**a**) device chamber: 1—specimen, 2—holders, 3—thermocouples type S, 4—induction solenoid, 5—cooling system jets, 6—pyrometer and 7—sensors for laser measurement of specimen diameter; (**b**) basic specification.

Both the experimental and numerical tests were conducted to obtain the true strain value equal to 5, at a constant temperature of the deformed sample and with a constant strain rate. Round samples with a diameter of d = 8 mm and a measuring base length of l = 20 mm were used for the tests (Figure 2). A K-type thermocouple (NiCr–Ni), welded to the side surface of the sample, was used to register and control changes in temperature. The tested material was heated with a heating rate 5 °C/s to the desired temperature, held for 10 s, then deformed and cooled freely.

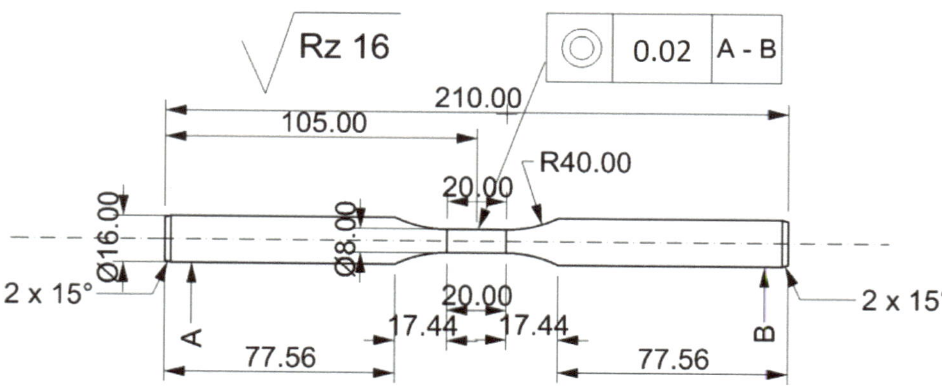

Figure 2. The 5019 aluminium alloy sample's technical specification.

Equation (5) was used to determine the true strain, the true strain rate was determined based on Equation (6), while the yield stress was calculated according to Equation (7) [2,22].

$$\varepsilon = \frac{2 \cdot \pi \cdot r \cdot N}{\sqrt{3} \cdot L} \quad (5)$$

$$\dot{\varepsilon} = \frac{2 \cdot \pi \cdot r \cdot \dot{N}}{\sqrt{3} \cdot 60 \cdot L} \quad (6)$$

$$\sigma_p = \frac{\sqrt{3} \cdot 3 \cdot M}{2 \cdot \pi \cdot r^3} \quad (7)$$

where r is the sample radius, L is the sample length, N is the number of sample twists (revolutions), $\dot{N}$ is the torsion speed (rpm), and M is the torque.

2.2.3. Metallographic Analysis and Microhardness Measurements

As part of the experimental work, metallographic tests and microhardness measurements of the tested 5019 aluminium alloy in the initial state were conducted after homogenisation before deformation and after the deformation process. The purpose of this part of the study was to determine the influence of the applied strain parameters on the distribution and size of the grain as well as the microhardness of the tested aluminium alloy for the selected temperature and strain rate. Observations of the microstructure were carried out using an Olympus GX51 light microscope equipped with digital image recording. Additionally, the grain size measurements on the cross-section were performed using the secant method.

Microhardness measurements were conducted using the Vickers method with the use of the Innovatest Nexus 4000 hardness tester, applying a 100 gramme load. Both the measurements of grain size and microhardness were carried out in two perpendicular directions on the cross-sections of samples made of the 5019 aluminium alloy (according to Figure 15). The test samples were mechanically polished according to the procedure of the Struers company. To reveal the microstructure, the samples were etched in Barker reagent with the composition of 2 mL HBF_4 + 100 mL of H_2O, and the observations were made under polarized light.

Metallographic analyses using the electron backscatter diffraction (EBSD) method were performed using a Hitachi S-3400N scanning electron microscope. An accelerating voltage of 15 kV was used for the tests. The sample was tilted at an angle of 70° at a distance of approximately 20 mm from the column. The detector and the software used for the EBSD analysis were from the HKL company. The electron backscatter diffraction analysis was performed to obtain maps of the crystal lattice orientation distribution on the surface of the test sample with a 1 µm step at a magnification of 200×. The colours obtained were assigned to the obtained orientations according to the basic IPF triangle. The results are presented both in the form of a map of the crystallographic orientation distribution on the sample surface and in the form of pole figures (PFs) along with inverse pole figures (IPFs). The texture results were also calculated for orientation intensity to both PF and IPF with 10 × 10° clustering. The disorientation map was also determined, considering the grain boundaries above 15° (marked in black) and the sub-grain boundaries above 5° (marked in blue). Measurements were made in three areas: (1) the centre of the sample, (2) at 0.67 r (2.68 mm from the centre), and (3) at 0.724 r (2.9 mm from the centre), where r is the sample's radius.

3. Results

In order to determine the actual temperature distribution along the sample's length, temperature measurements were performed using the contact method with the use of three K-type thermocouples (NiCr–NiAl) (Figure 3). Examples of the results obtained during the tests at the temperature of 480 °C are presented in Figure 4. It was stated that the highest temperature value (480 °C) appeared in the centre of the sample. At the end of the measurement base, the temperature was only slightly lower (below 3%) at 466 °C. On this basis, it was found that the heating and temperature control system of the STD 812 torsion plastometer ensured an even temperature distribution along the length of the measurement base of the torsion samples.

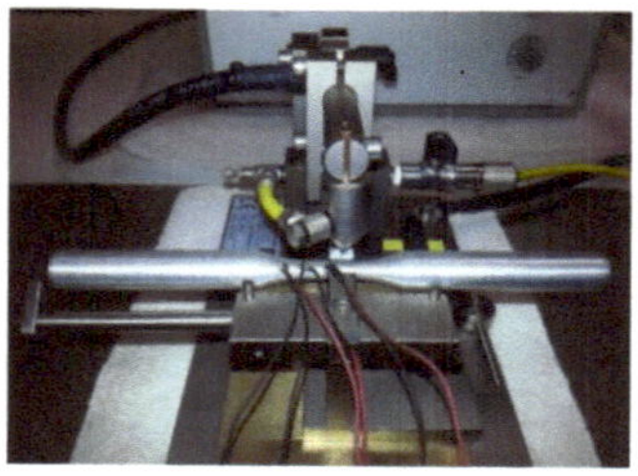

Figure 3. Example specimen during welding of the thermocouple with a lateral surface.

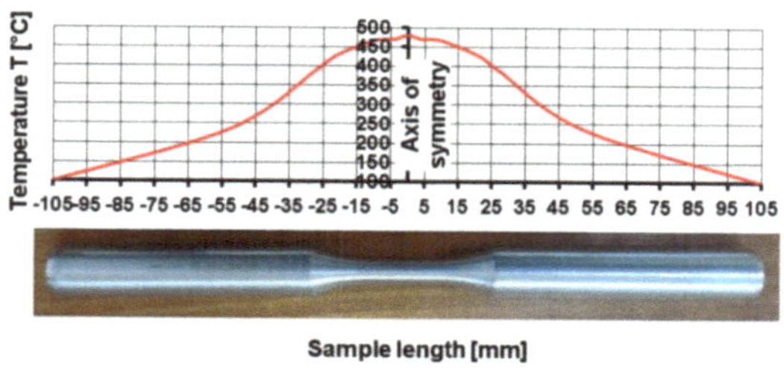

Figure 4. Actual temperature distribution along the sample's length determined by the contact method.

By analysing the temperature changes outside the working part of the samples, a much larger drop in temperature can be found. The main reason for the rapid decrease in temperature was the increasing cross-section of the samples and the change in heat exchange conditions. The test area of the samples was located in the centre of the induction coil heated to the set temperature, while the rest of the samples was outside the area of heat influence caused by the induction field. Additionally, heat was conducted towards the cold grips of the torsion plastometer.

3.1. Analysis of the Numerical Research Results: Problem with the Mathematical Model's Accuracy

Table 1 presents the values of the equation coefficients (4), approximating the test results of the 5019 aluminium alloy obtained in the tested range of the strain and temperature parameters.

Table 1. Coefficients of Equation (4) used to determine σ_p values of the 5019 aluminium alloy (Variant 1) [45].

A	m_1	m_2	m_3	m_4	m_5	m_7	m_8	m_9
0.271553	−0.00957753	−0.0823773	**−0.246413**	**−0.000573716**	**0.000622463**	**−0.164103**	0.00101501	1.65143

Sample diagrams of the course of yield stress during the torsion of samples at the temperatures of 440 °C, 480 °C, and 520 °C are presented in Figure 5.

By analysing the course of the actual and approximate hardening curves of the analysed aluminium alloy, it can be stated that the values of yield stress, approximated by Equation (4) using the coefficients in Table 1, were close to the values of yield stress determined experimentally.

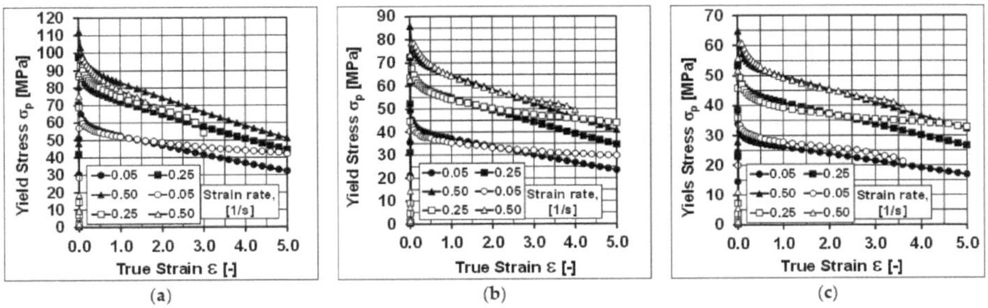

Figure 5. Flow curves of the 5019 aluminium alloy at temperatures of (**a**) 440 °C; (**b**) 480 °C; (**c**) 520 °C; empty symbols—plastometric test data; full symbols—results after approximation using coefficients from Table 1 (Variant 1).

Despite the small range of the strain rate (0.05–0.5), it can be noticed that during the torsion of the tested alloy with a strain rate of 0.05 s^{-1}, at the temperatures of 440 and 480 °C, the values of yield stress obtained as a result of the approximation, above the actual strain value of about three, are lower than the values determined experimentally. During the torsion of the tested alloy with a strain rate of 0.05 s^{-1}, at the temperature of 440 °C, the maximum differences between the values of yield stress determined experimentally and obtained as a result of the approximation occurred at the true strain of five and amounted to over 23%. When deforming samples with a strain rate of 0.05 s^{-1}, at a temperature of 480 °C, the maximum differences between the values of yield stress determined in torsion tests and the approximated values were about 21% (also for true strain of five). Similar differences between the actual and approximate values of yield stress were found on the basis of the analysis of the test results obtained during the torsion of the samples with a strain rate of 0.25 s^{-1} at temperatures of 480 and 520 °C. Moreover, in this case, the greatest differences between the actual and approximate values of the yield stress were noted for the true strain equal to five. During the deformation of the tested 5019 aluminium alloy at a temperature of 480 °C, they were approximately 22%, while during the torsion at a temperature of 520 °C, they were approximately 19%.

After implementing the determined coefficients of the mathematical model of changes in the rheological properties of the 5019 aluminium alloy to the material base of the FORGE 2011® program, numerical modelling of torsion tests (Variant 1) was performed in the next stage of the research. Examples of the results obtained during the torsion of samples with a strain rate of 0.25 s^{-1} at a temperature of 480 °C are presented in Figures 6–9. The analysis covers the temperature distribution, strain intensity, strain rate intensity and stress intensity.

Figure 6 shows the temperature distribution in the tested aluminium alloy, determined numerically.

By analysing the data presented in Figure 6, it was found that the initial temperature distribution in the material set at the beginning of the torsion test (Figure 6a) does not change during the entire hot torsion process (Figure 6b). The obtained distribution is consistent with the results obtained experimentally (Figure 4). By analysing the temperature distribution in the central part of the sample, on the cross-section (Figure 6c), it was found that the highest temperature values caused by the highest strain value occurred on the surface. In the tested case, the temperature difference on the cross-section of the tested alloy was 5 °C. Based on the analysis of the results of the temperature distribution, it can be stated that the conditions for conducting the tests at a constant temperature were met.

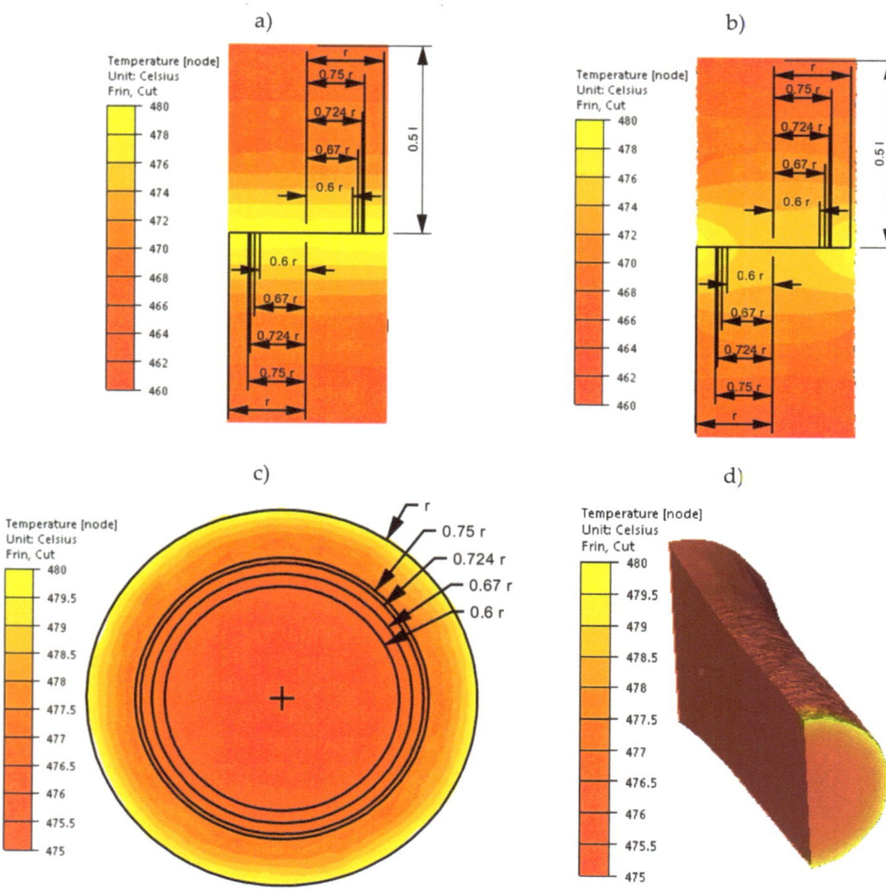

Figure 6. Temperature distribution of the 5019 aluminium alloy calculated numerically using coefficients from Table 1: (**a**) at the beginning of the torsion—longitudinal section, (**b**) at the end of the torsion—longitudinal section, (**c**) at the end of the torsion—cross-section in the centre of the working part of the sample and (**d**) at the end of the torsion—perspective.

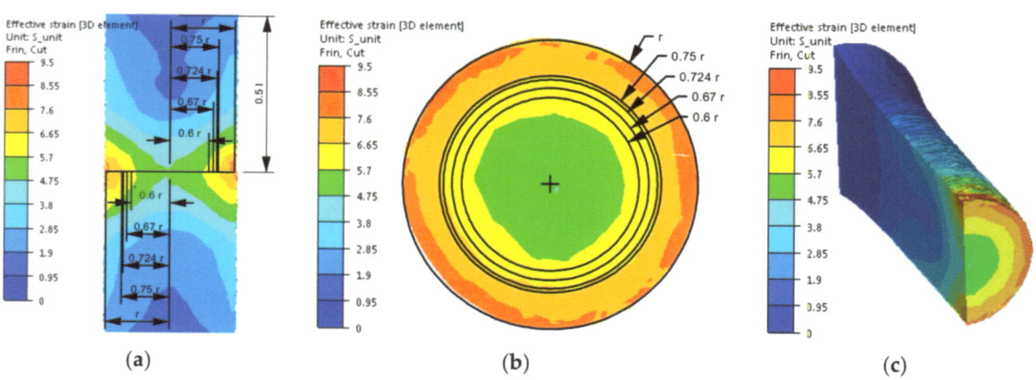

Figure 7. Distribution of the strain intensity of the 5019 aluminium alloy calculated numerically using coefficients from Table 1: (**a**) longitudinal section, (**b**) cross-section—centre of the working part and (**c**) perspective.

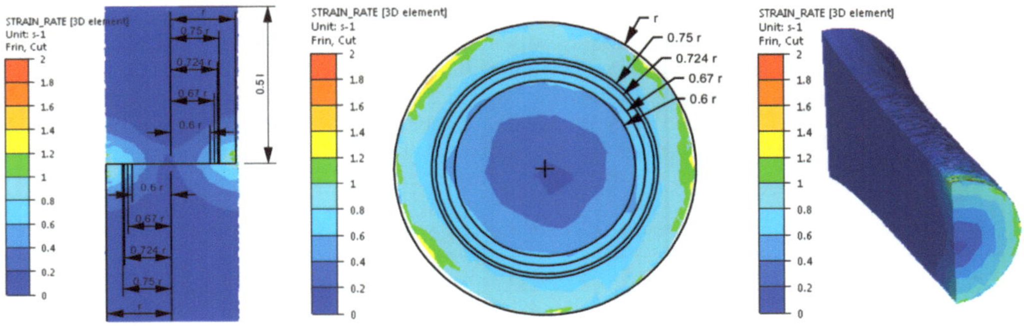

(a) (b) (c)

Figure 8. Distribution of the strain rate intensity of the 5019 aluminium alloy calculated numerically using coefficients from Table 1: (**a**) longitudinal section, (**b**) cross-section—centre of the working part and (**c**) perspective.

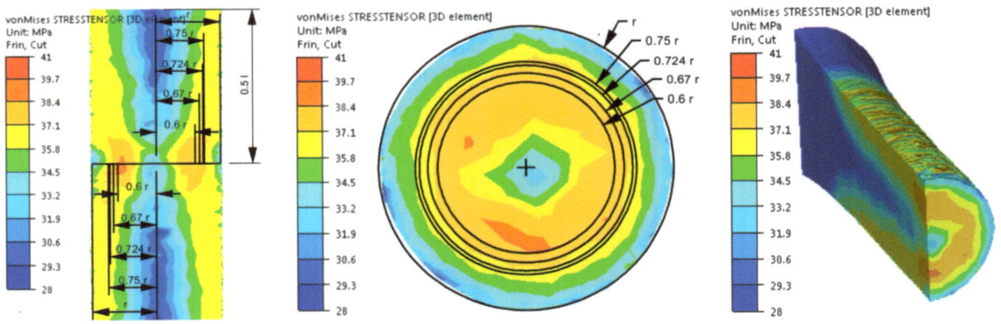

(a) (b) (c)

Figure 9. Distribution of the stress intensity of the 5019 aluminium alloy calculated numerically using coefficients from Table 1: (**a**) longitudinal section, (**b**) cross-section—centre of the working part and (**c**) perspective.

The distribution of strain intensity, strain rate intensity, and stress intensity in the entire volume of the working part of the tested material is presented in Figures 7–9.

On the basis of the data presented in Figures 7–9, inhomogeneity of the analysed parameters can be observed both on the cross-section and longitudinal sections of the deformed samples. The numerically determined values of the strain intensity and strain rate intensity were higher than the values obtained in the experimental tests for all the equivalent analysed radii. In the analysed case, the numerically calculated values of the strain intensity ranged from 6 (for the equivalent radius of the 0.6 r sample) to 7.2 (for the radius of 0.75 r), while in the experimental tests, the true strain was equal to five. The values of the strain rate intensity ranged from 0.6 s^{-1} (for the equivalent radius of the 0.6 r sample) to 0.78 s^{-1} (for the radius of 0.75 r), while in the experimental tests, the strain rate was 0.25 s^{-1}. By analysing the results of the stress intensity distribution, it was found that the nature of its distribution was incorrect. The highest values of this parameter occurred for the equivalent sample radius of 0.6 r and not on the surface, which is inconsistent with the data published in the literature. On the basis of the obtained test results, it was found that the direct cause of incorrectly calculated strain intensity distributions, strain rate intensity and stress intensity as well as higher values of these parameters compared to the experimental tests may be an inaccurate description of the rheological properties of the tested alloy with the use of the coefficients presented in Table 1. For the analysed case (strain rate of 0.25 s^{-1}, temperature of 480 °C), the approximated values of the yield stress used in the numerical tests were lower than the values obtained in the experimental tests. Consequently, the plastic deformation resistance was also lower. As a result, the numerically determined values of strain intensity and strain rate intensity were overestimated in relation to the values obtained in actual torsion tests.

Therefore, in the next stage of the work, the results of plastometric tests were re-approximated using the equation coefficients (4), obtained by approximation for a narrow temperature range (Table 2, Variant 2). The obtained results are presented graphically in Figure 10.

Table 2. Coefficients of Equation (4) used to determine σp values of the 5019 aluminium alloy (Variant 2) [45,46].

A	m1	m2	m3	m4	m5	m7	m8	m9
0.271553	−0.00957753	−0.0823773	**−0.2465**	**−0.002**	**0.0001**	**−0.032**	0.00101501	1.65143

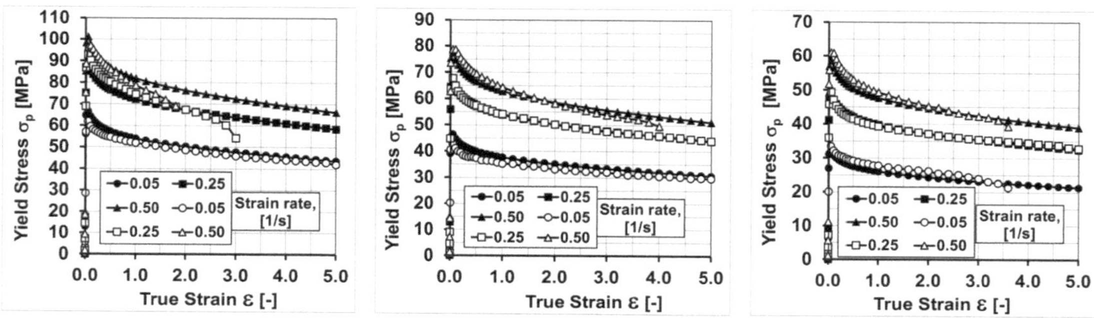

(a) (b) (c)

Figure 10. Flow curves of the 5019 aluminium alloy at temperatures of (**a**) 440 °C; (**b**) 480 °C; (**c**) 520 °C; empty symbols—plastometric test data; full symbols—results after approximation using coefficients from Table 2 (Variant 2).

Based on the analysis of the data presented in Figure 10, it can be found that by increasing the accuracy of the approximation (coefficients m3, m4, m5 and m7 of the approximating Equation (4)), the accuracy of the mathematical description (model) of the rheological properties of the tested aluminium alloy was increased. In the diagrams, no differences were observed between the approximated and true values of yield stress for a strain rate of 0.05 s^{-1} during the torsion at the temperatures of 440 and 480 °C and for a strain rate of 0.25 s^{-1} during the torsion at the temperatures of 480 and 520 °C, which were observed using the coefficients of the mathematical model of rheological properties presented in Table 1.

In the next stage of the research, numerical modelling of the torsion tests was conducted again using the coefficients of the mathematical model of the rheological properties presented in Table 2. Similar to the previous case, the results obtained during the torsion of the samples with a strain rate of 0.25 s^{-1} at the temperature of 480 °C are shown in Figures 11–13. As the results of the temperature distribution were similar to those presented in Figure 6, only the results concerning the strain intensity distribution, strain rate intensity and stress intensity in the entire volume of the working part of the tested alloy are presented below.

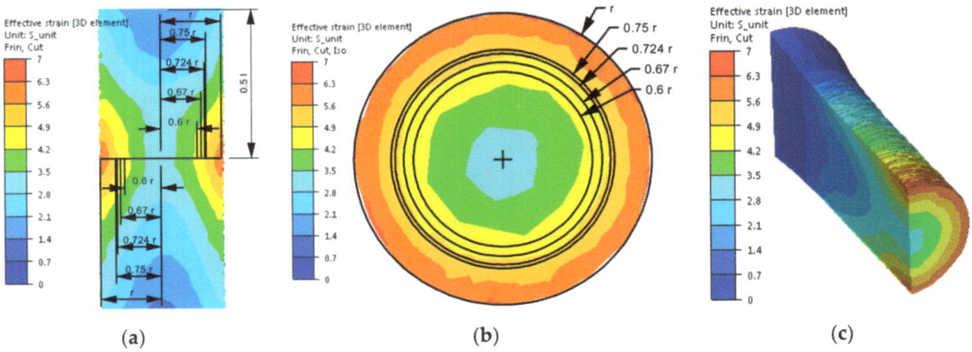

Figure 11. Distribution of the strain intensity of the 5019 aluminium alloy calculated numerically using coefficients from Table 2: (**a**) longitudinal section, (**b**) cross-section—centre of the working part and (**c**) perspective.

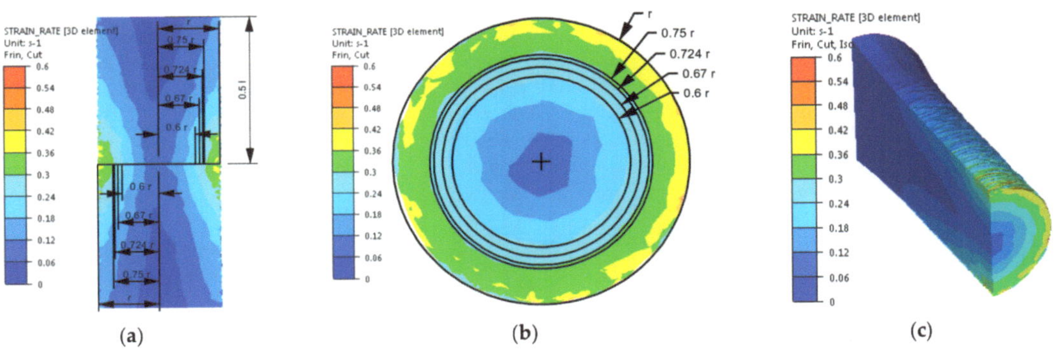

Figure 12. Distribution of the strain rate intensity of the 5019 aluminium alloy calculated numerically using coefficients from Table 2: (**a**) longitudinal section, (**b**) cross-section—centre of the working part and (**c**) perspective.

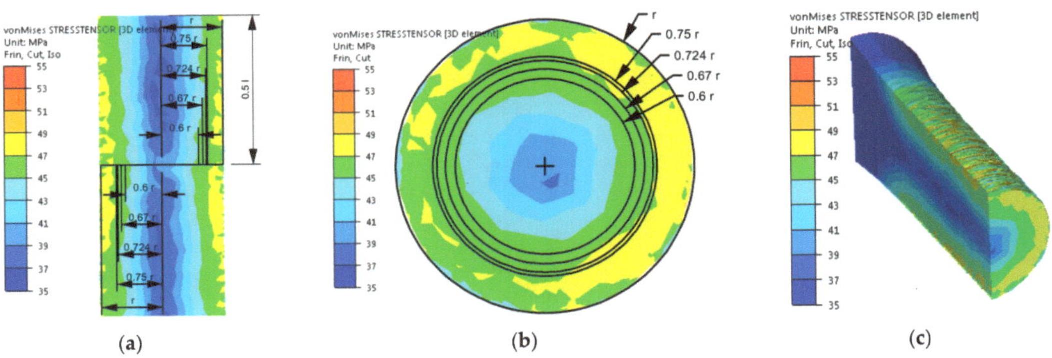

Figure 13. Distribution of the stress intensity of the 5019 aluminium alloy calculated numerically using coefficients from Table 2: (**a**) longitudinal section, (**b**) cross-section—centre of the working part and (**c**) perspective.

Based on the analysis of the data presented in Figure 11, it can be found that during the torsion of samples from the tested aluminium alloy with a strain rate of 0.25 s^{-1} at the temperature of 480 °C and up to the true strain value of five, the numerically determined values of the strain intensity changed from 4.3 (for the equivalent sample radius of 0.6 r) up to 5.1 (for the radius of 0.75 r). The highest accuracy between the values determined numerically and those obtained in experimental tests was obtained for the equivalent radii of the 0.67 r and 0.724 r samples.

Figure 12 presents that for the analysed case (strain rate of 0.25 s^{-1}, temperature of 480 °C, true strain of five), numerically determined values of strain rate intensity ranged from 0.25 s^{-1} (for the equivalent sample radius of 0.6 r) to 0.3 s^{-1} (for the radius of 0.75 r). The highest accuracy between the numerically determined values and those obtained in the experimental tests was obtained for the equivalent sample radius of 0.6 r. For the remaining equivalent radii, the numerically determined value of the strain rate intensity was slightly higher than the value obtained in the experimental tests. For the analysed case, during the experimental tests, the speed of the moving tool (grip), after reaching the value of 20.5 rpm, was slightly fluctuating, while during the numerical tests it was at a constant level. This could cause a slight overestimation of the numerically determined intensity of the strain rate.

By analysing the data on the stress intensity distribution (Figure 13), it can be found that during the torsion of the samples from the tested material with a strain rate of 0.25 s^{-1}, at the temperature of 480 °C, up to the true strain value equal to five, the numerically determined values of the stress intensity ranged from 44 MPa (for the equivalent sample radius of 0.6 r) to 47 (for the radius of 0.75 r). The highest accuracy between the numerically determined values and those obtained in the experimental tests was obtained for the equivalent sample radius of 0.6 r. For the remaining equivalent radii, the numerically determined stress intensity values only slightly differed from the yield stress obtained during the torsion tests. The reason for this, as before, may be the way of defining the rotational speed of the movable tool (grip) during numerical calculations. The numerically higher value of the strain rate intensity resulted in a slight increase in the stress intensity of the tested alloy.

By analysing the obtained results, it can be found that after applying the coefficients of the mathematical model of the rheological properties listed in Table 2, the numerically determined values of the analysed parameters corresponded with high accuracy to the values obtained in the true torsion tests. This confirms the significant influence of the accuracy of the mathematical model of the rheological properties on the distribution of the analysed parameters.

The correctness of the conclusions of the authors is also confirmed by the results of the experimentally and numerically calculated torsion moment as well as the results of the course of the yield stress (experimental and calculated on the basis of the numerically determined torsion moment values, using the dependency (7)), which are presented in Figure 14.

By analysing the data presented in Figure 14, it is possible to observe a characteristic peak in the value of the torsion moment and yield stress at the beginning of the torsion process, and then their decrease along with the increase in the deformation. Based on the analysis of the results of the torsion moment and yield stress, it was found that greater consistency between the values calculated and recorded during torsion tests was obtained by using the coefficients of the rheological properties model presented in Table 2 (Variant 2).

Based on the analysis of the results of the numerical tests, it was found that as representative radii, marking the area in which the values of strain and stress parameters corresponded to the average values on the cross-section of the torsion samples and, at the same time, to the values obtained in the experimental tests, the radii equal to 0.67 r and 0.724 r should be taken.

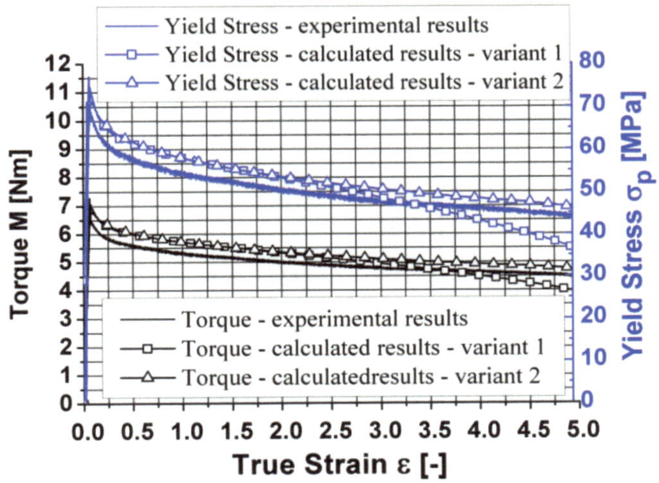

Figure 14. Changes in the torque and yield stress of the 5019 aluminium alloy—true and calculated values.

3.2. Analysis of the Experimental Research Results

Figure 15 shows the general diagram of the sample with the directions marked on the cross-section in which metallographic tests and microhardness measurements were carried out.

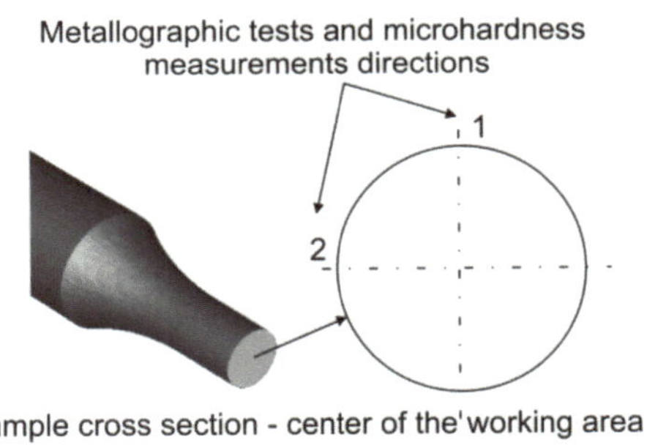

Figure 15. Cross-section of the sample with marked directions in which metallographic tests and microhardness measurements were carried out (general diagram).

Figure 16 presents photos showing the microstructure of the tested alloy in its initial state, before the deformation process. Figure 17 presents sample photos of the microstructure showing the centre (Figure 17a) and the edge (Figure 17b) of the sample.

Figure 18 presents a graph showing the changes in grain size on the cross-section of the tested samples in their initial state.

On the basis of the metallographic analysis, it was found that the tested alloy in its initial state (after the homogenisation process) had a homogeneous microstructure, both on the cross-section and longitudinal section. The revealed grains had a regular shape. The determined average grain diameter for the initial material before the deformation process was approximately 68 μm (Figure 18).

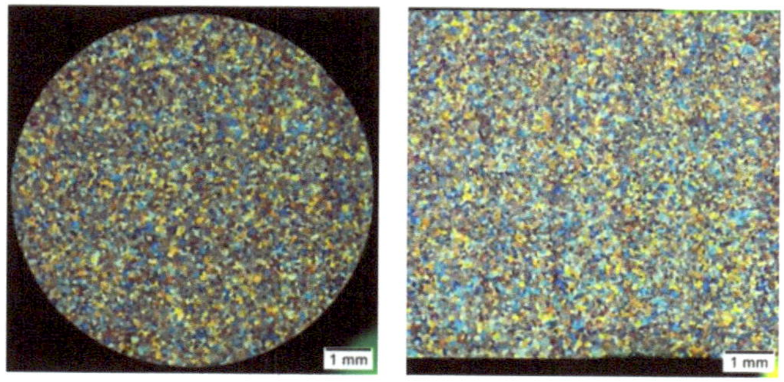

(a) (b)

Figure 16. Microstructure of the 5019 aluminium alloy in initial state, after the homogenisation process (before the deformation process): (**a**) cross-section; (**b**) longitudinal section.

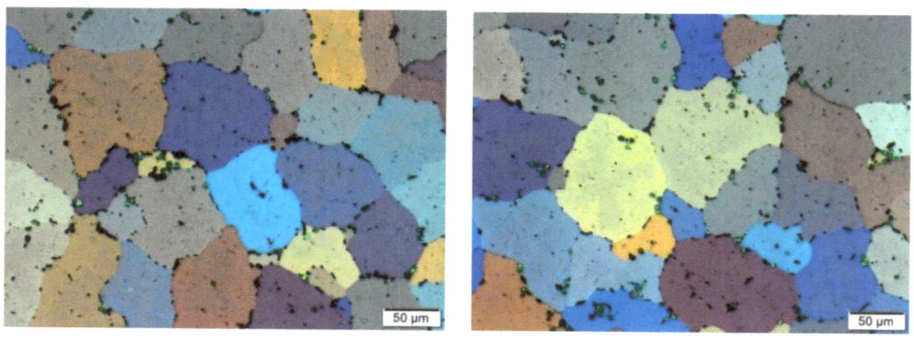

(a) (b)

Figure 17. Microstructure of the 5019 aluminium alloy in the initial state, after the homogenisation process (before the deformation process): (**a**) sample centre; (**b**) sample edge; cross-section, magnification 200×.

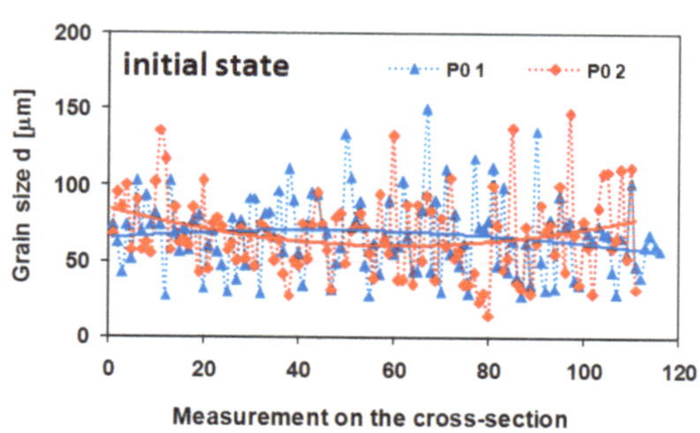

Figure 18. Changes in the grain size on the cross-section of the sample made of the 5019 aluminium alloy; the measurements were performed in two perpendicular directions (as shown in Figure 15); P0—the initial state of the material (after homogenisation).

Figures 19 and 20 present sample photos showing the microstructure of the tested aluminium alloy after plastic deformation at the temperature of 480 °C with a strain rate of 0.25 s^{-1}.

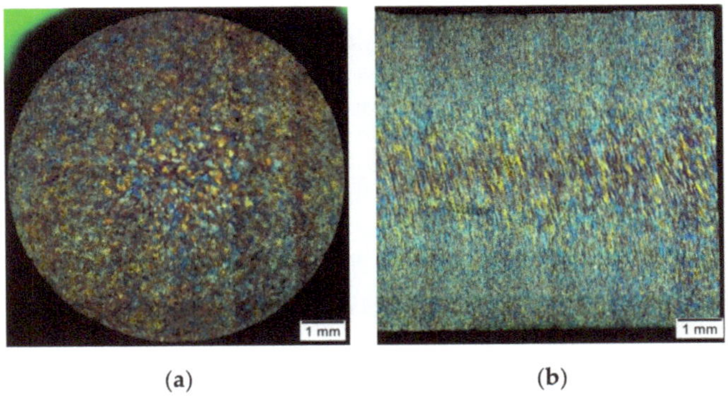

Figure 19. Microstructure of the 5019 aluminium alloy after plastic deformation: (a) cross-section; (b) longitudinal section.

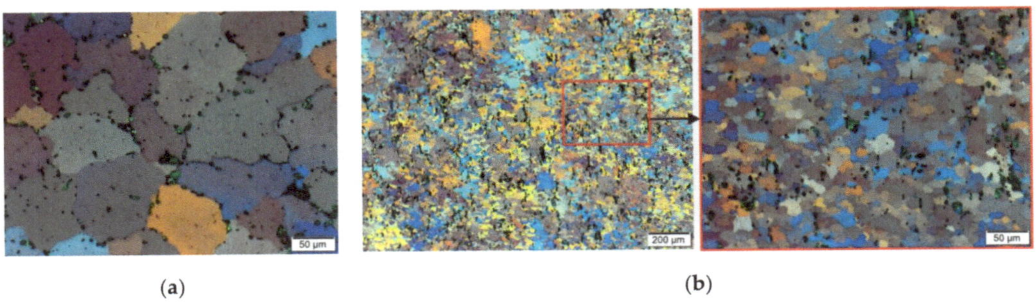

Figure 20. Microstructure of the 5019 aluminium alloy after plastic deformation: (a) centre of the sample; (b) edge of the sample; cross-section, magnification 200×.

Figure 21 presents a graph showing the changes in the grain size on the cross-section of the tested 5019 aluminium alloy samples after the deformation process.

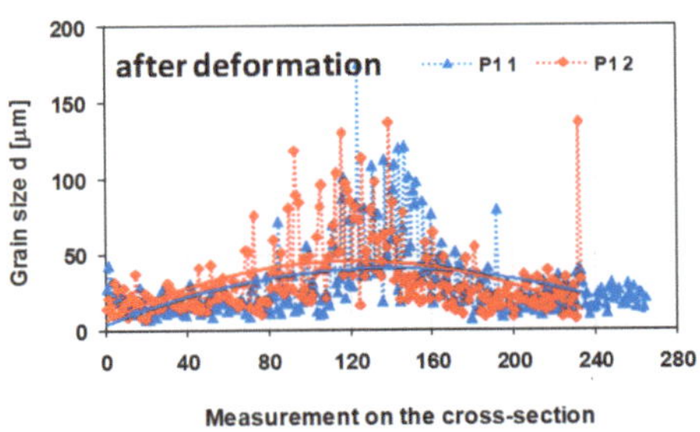

Figure 21. Changes in the grain size on the cross-section of a sample from the 5019 aluminium alloy after the deformation process; the measurements were made in two perpendicular directions (as shown in Figure 15); P1—material condition after hot torsion.

On the basis of the metallographic analysis, a large diversity of the microstructure on the cross-section and longitudinal section of the tested material was found. While analysing the results of the measurements of the grain size of the tested aluminium alloy on the cross-section after the hot torsion process at the temperature of 480 °C with a strain rate of 0.25 s^{-1}, its typical distribution for the torsion test was found. The largest grains were located in the central area (sample axis), where the strain intensity value was the smallest (3.20). The size of single grains in this area reached even 120–140 μm, while the average grain size was approximately 47 μm. For the radius of 0.67, the average grain size was about 27 μm, and for the radius of 0.724, the average grain size was approximately 26 μm. In the surface area, where the value of the strain intensity was the highest (6.65), small recrystallised grains with an average size of approximately 18 μm were visible. Additionally, in this area, coagulated separations arranged in stripes were observed. It was found that as a result of the deformation, the average grain size of the tested alloy decreased by approximately 21 μm (31%) in the sample axis, 41 μm (60%) for the 0.67 r radius and 42 μm (62%) for the 0.724 r radius. The average grain size of the 5019 aluminium alloy decreased by 50 μm (74%).

In turn, the determined average grain diameter of the tested alloy after the cross-sectional deformation was approx. approximately 32 μm.

On the basis of the conducted metallographic analyses, it was shown that the structure was highly diversified on the cross-section of the sample of the tested material after the hot torsion, mainly due to the high inhomogeneity of the deformation parameters (strain intensity, strain rate intensity). The obtained results of metallographic analyses confirm the necessity to define the so-called representative area for the assessment of the microstructure of the material after the torsion process. The presented results also indicate the need to analyse the microstructure of twisted samples in a strictly defined zone for which the value of local strain should be determined.

In the next stage of the research, microhardness measurements were performed on the cross-section of the samples. Exemplary results of the microhardness measurement on the cross-section of the samples of the tested aluminium alloy before and after the deformation process are presented in Figure 22.

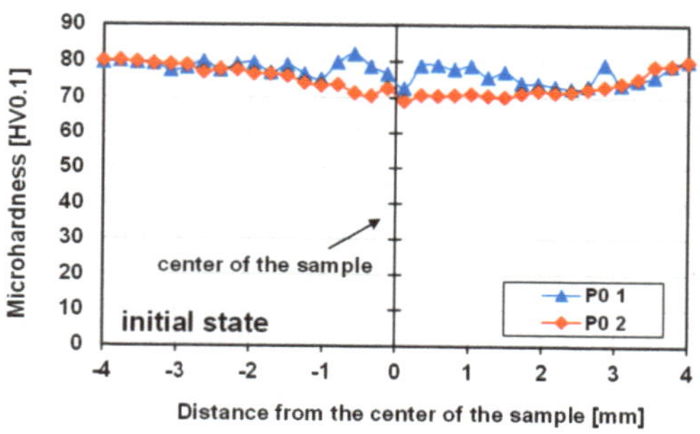

(a)

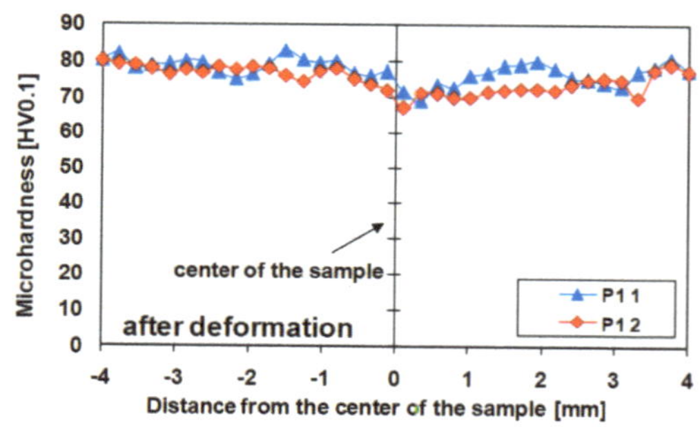

(b)

Figure 22. Changes in microhardness on the cross-section of the 5019 aluminium alloy samples (measurements made in two perpendicular directions—according to Figure 14): (**a**) undeformed material; (**b**) material after plastic deformation.

The obtained results did not show any differences in the level of microhardness for the material after homogenisation and after the hot torsion process, despite significant differences in the grain size on the cross-section. The average measured microhardness was approximately 76 HV0.1 (Figure 22).

However, diversity in microhardness on the cross-section of the tested samples were noted. The material after the plastic deformation process was marked by greater inhomogeneity of the microhardness distribution on the cross-section, which was mainly caused by a greater differentiation of the grain size on the cross-section. The nature of the microhardness distribution is consistent with the results of the grain size measurement. In the deformed axis of the sample, where the grain is the largest, the tested material had the lowest microhardness. On the other hand, in the surface areas, where the average grain size was the smallest, the average values of the microhardness of the tested alloy were the highest.

In the conditions of hot deformation of aluminium alloys, the strengthening effect resulting from deformation is counteracted by softening as a result of structural renewal processes. Therefore, despite the significant fragmentation of the microstructure, especially outside the centre, no increase in the level of microhardness was found.

During thermomechanical processing, mechanical properties of materials are affected by strain history, chemical composition and the microstructure. When aluminium alloys are strained at the elevated temperatures, they may experience work hardening and flow softening which result from dynamic recovery (DRV), dynamic recrystallisation (DRX) or dynamic precipitation transformations [47–49]. Due to the high stacking fault energy of aluminium alloys, dynamic recovery (DRV) was more likely to occur than dynamic recrystallisation (DRX) [50].

In order to characterize the microstructure of the samples more fully, studies were conducted using the EBSD. Since the radii assumed that the representative radii were equal to 0.67 r and 0.724 r, EBSD analyses were performed in these areas and additionally along the axis of the samples.

The EBSD analyses for the sample after homogenisation (starting material before the deformation process), as presented in Figure 23, were similar for each of the analysed areas. Basic triangles indicate the dominance of directions around the <101> pole, and the maximum intensity of the texture was 2.32–2.84. Pictures presenting the types of edge indicate the dominance of large angles, above 15°, while in the axis of the sample, large angles of disorientation constitute 85% of the studied population, and at a distance of 0.67 r and 0.724 r—over 90%. The size of grains/sub-grains in the area covered by the study determined on the basis of EBSD tests was 57 µm, and the average surface area was 3528 μm^2. The determined grain size at a distance of 0.67 r and 0.724 r was slightly smaller, and it was approximately 55 µm for both distances.

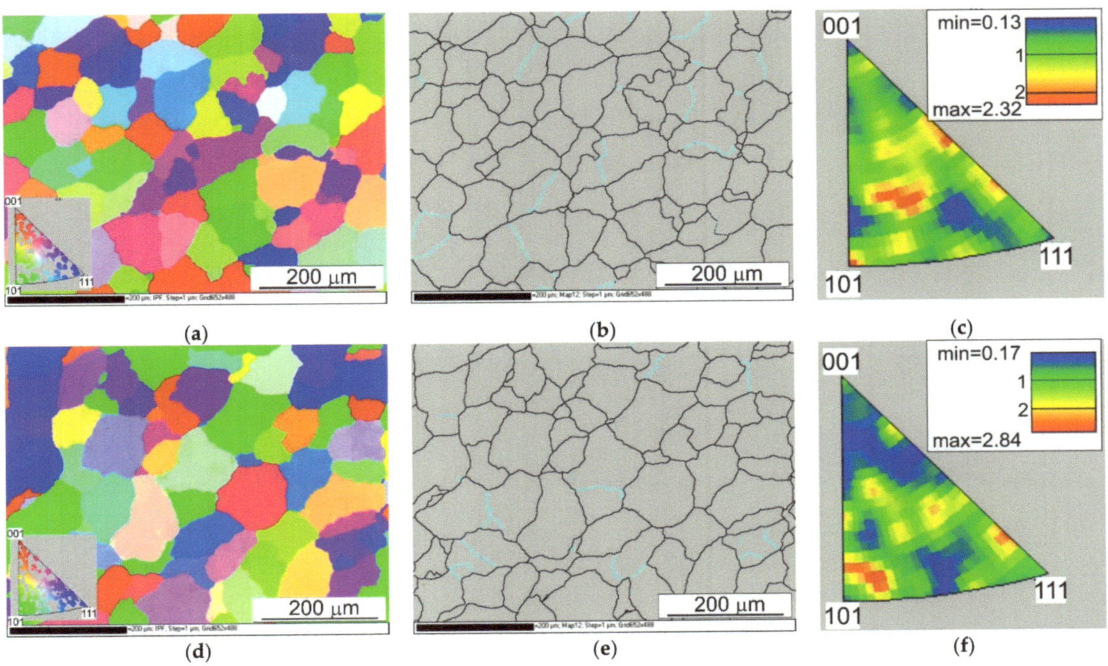

Figure 23. *Cont.*

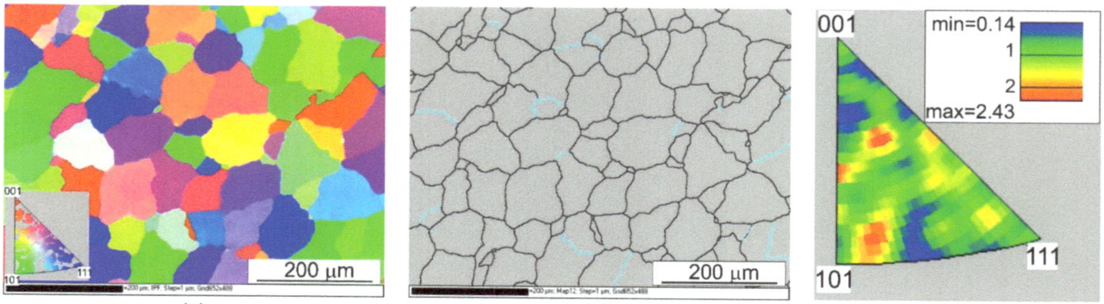

Figure 23. EBSD analysis results of the sample after homogenisation; (**a,b,c**) centre of the sample, (**d,e,f**) distance from the centre = 0.67 r, (**g,h,i**) distance of the centre = 0.72 r; (**a,d,g**) EBSD maps showing changes in orientation, (**b,e,h**) maps showing the types of edges (the edges of a large angle are marked in black, and the edges of a small angle are marked in blue) and a (**c,f,i**) basic triangle—orientation intensity.

The results of the EBSD analysis of the samples after hot torsion confirmed the earlier observations about the significant diversity of the microstructure on the cross-section (Figures 24 and 25).

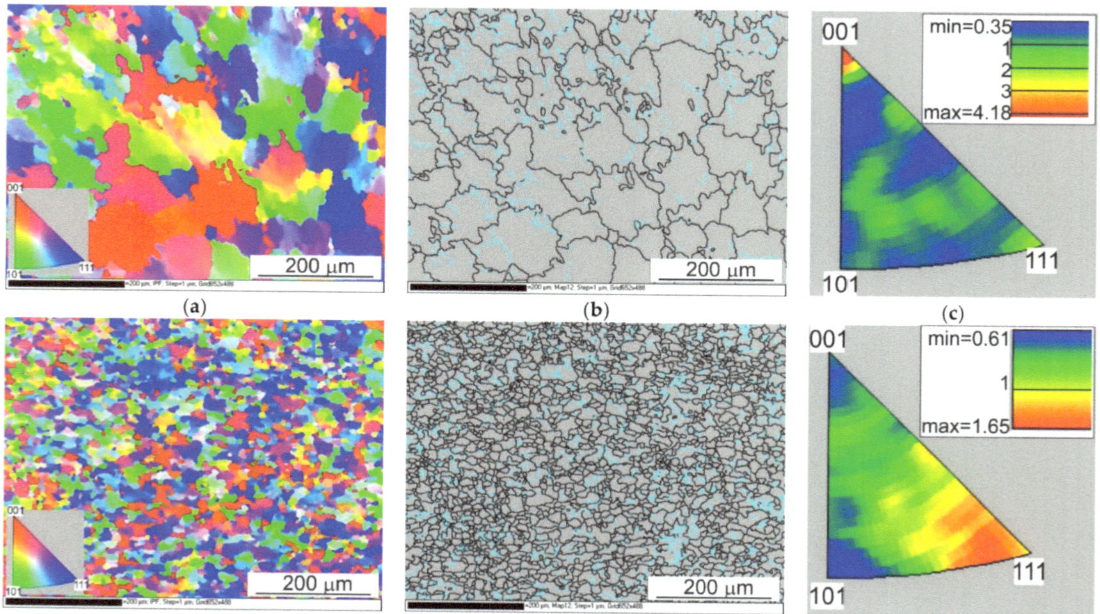

Figure 24. *Cont.*

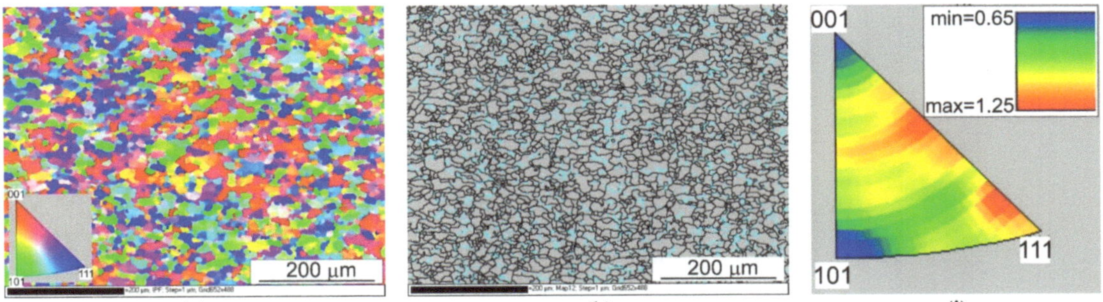

(g) (h) (i)

Figure 24. EBSD analysis results of the sample after deformation: (**a**,**b**,**c**) centre of the sample, (**d**,**e**,**f**) distance from the centre = 0.67 r, (**g**,**h**,**i**) distance of the centre = 0.72 r; (**a**,**d**,**g**) EBSD maps showing changes in orientation, (**b**,**e**,**h**) maps showing the types of edges (the edges of a large angle are marked in black, and the edges of a small angle are marked in blue) and a (**c**,**f**,**i**) basic triangle—orientation intensity.

(a)

	LABs, %	HABs, %
Center	39.86	60.14

(b)

	LABs, %	HABs, %
0.67 r	24.13	75.87

(c)

	LABs, %	HABs, %
0.72 r	23.09	76.91

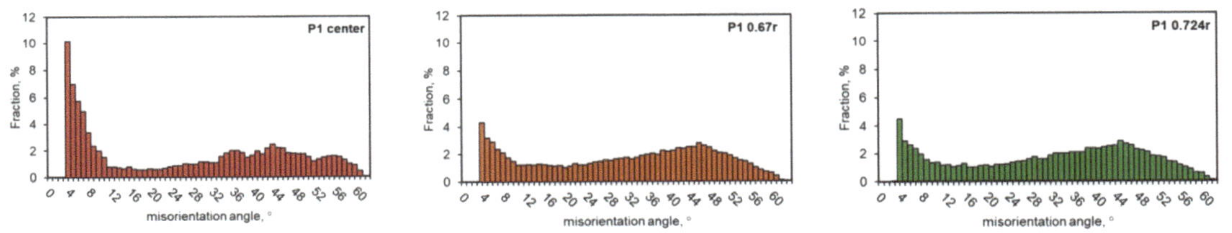

Figure 25. Changes in disorientation angles after hot torsion: (**a**) sample centre, (**b**) at a distance from the centre of 0.67 r and (**c**) at a distance from the centre of 0.72 r.

In the case of analyses performed on the axis of the sample, the directions around the <001> pole were found to be dominant. The maximum texture intensity was 4.18. Almost 40% of the measured grain edge population were small disorientation angles below 15°.

In the distance from the sample axis 0.67 and 0.724 r, fine, almost equiaxial grains were observed. Compared to the central part of the sample, the proportion of the large angle edges was above 15°; it was much larger and represented more than 75% of the measured grain edge population. At a distance of 2.68 mm (0.67 r) from the centre of the sample, the texture intensity distribution was centred around the <111> pole, the maximum intensity was 1.65. In turn, at a distance of 2.9 mm (0.724 r) from the centre, the intensity distributions were concentrated around the <112> and <111> poles, and the maximum intensity was 1.25.

The average size of grains and sub-grains measured along the axis of the sample using the EBSD method was approximately 20 μm (cross-sectional area, respectively: 840 μm^2). At a distance of 0.67 and 0.724 r from the axis of the sample, the size of grains/sub-grains was comparable and amounted to approximately 9 μm, and the cross-sectional area was approximately 90 μm^2.

The differences in the obtained results of the grain size measurements with the use of photos taken using light microscopy techniques and determined using the EBSD method were largely due to the fact that in the set of the results obtained with the EBSD technique, there were also sub-grains with limits below 15°. Low-angle edges were not distinguished while observing using the light microscopy technique.

4. Directions of Future Studies

Knowledge of the exact distribution of strain and stress parameters in the material is of particular importance in the case of the so-called complex deformation diagrams that are planned in the future for the physical modelling of selected combining processes of various materials. Figure 26 presents the preliminary results of the research on the applicability of complex deformation states for physical modelling of combining materials.

(**a**) (**b**)

Figure 26. 6XXX series aluminium alloy sample: (**a**) before the friction welding process; (**b**) after the friction welding process.

The test parameters were similar to those occurring during the friction welding process. The combining process was conducted in two stages at a temperature of 400 °C. In the first stage of the research, the material was twisted with compression for 3 s.

Sample diagrams of changes in parameters during the first stage of the friction welding process are presented in Figures 27–29.

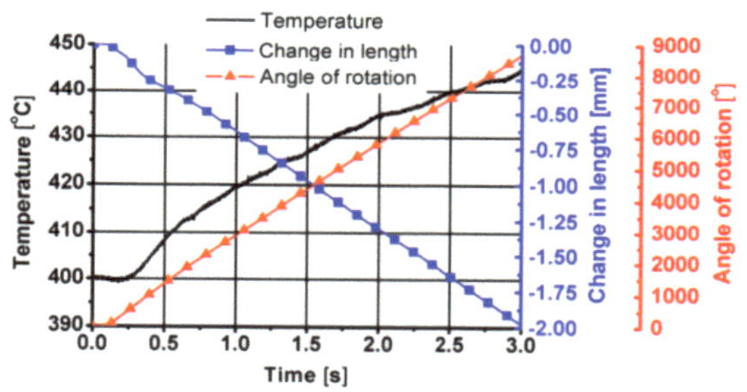

Figure 27. Temperature, length and angle of rotation changes over time in the first stage of friction welding of the aluminium 6XXX series—simultaneous torsion with compression.

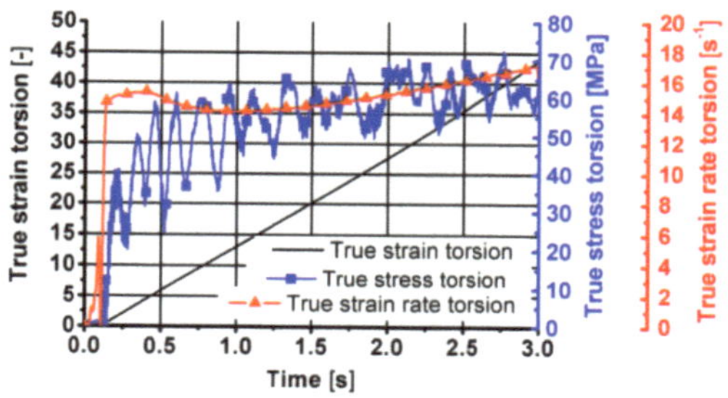

Figure 28. Changes in the true strain torsion, true stress torsion and true strain rate torsion over time in the first stage of friction welding of the 6XXX series aluminium—simultaneous torsion with compression components resulting from torsion.

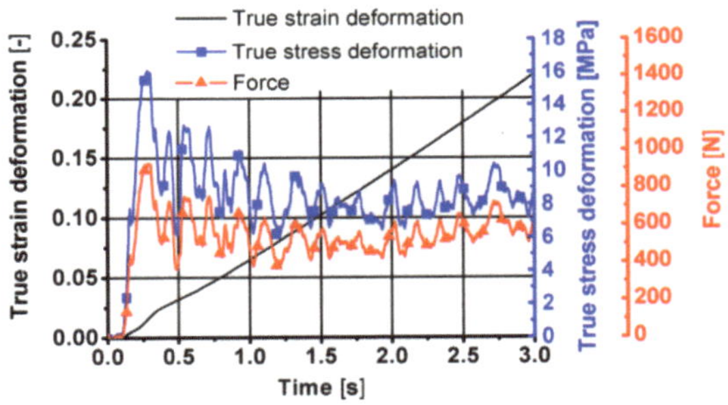

Figure 29. Changes in the true strain deformation, true stress deformation and force over time in the first stage of friction welding of the aluminium 6XXX series—simultaneous torsion with compression components resulting from compression.

As can be seen from the data presented in Figures 27–29, during this stage of the tests, the temperature of the tested material increased by approximately 40 °C, caused mainly by friction between the surfaces of the combined parts of the samples. The displacement along the sample axis was 2 mm, while the torsion angle was 8640° (24 turns). The strain caused by torsion was, in this case, approximately 45, while the strain rate was approximately 15 s^{-1}. By analysing the course of changes in stress caused by simultaneous torsion and compression, its rapid increase in the initial phase of this stage can be observed. In the further part of the simultaneous compression and torsion, the stress increase was small or remains constant. The reason for this may be additional slip mechanisms (shear lines) caused by a complex load pattern (surely, it requires additional metallographic tests). As shown in the research, the complex load condition also reduced the pressure force at this stage of the friction welding process.

In the second stage of the combining process, the tested aluminium alloy was subjected to compression (3 mm) as shown in Figures 30 and 31.

At this stage, a slight decrease in the temperature of the tested material was also observed. The reason for lowering the temperature at this stage of the friction welding process may be a low value of the true strain and the lack of friction. On the other hand, an increase in both the stress and the pressure force was observed. The increase in the pressure force at this stage results from the fact that, in this stage, the sample was only compressed; there were no additional slip lines lowering the energy and strength parameters.

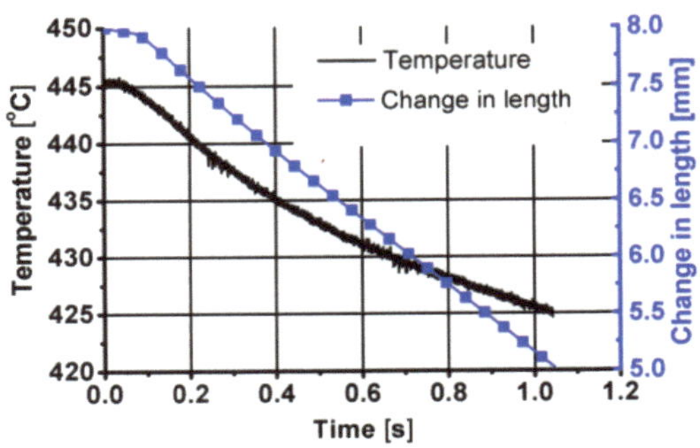

Figure 30. Temperature and length changes over time in the second stage of the friction welding of the 6XXX series aluminium—compression.

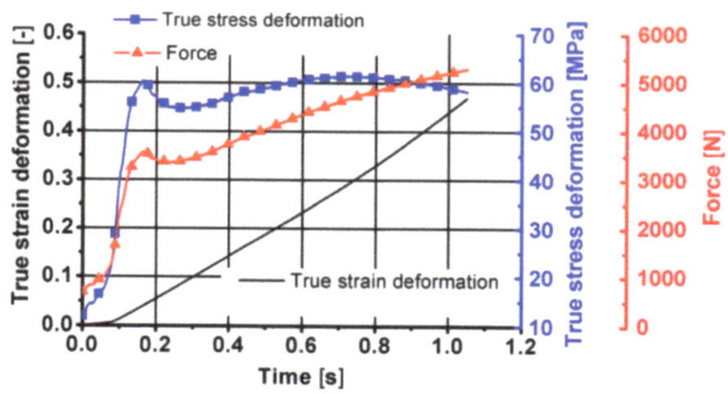

Figure 31. Changes in the true strain deformation, true stress deformation and force over time in the second stage of the friction welding of the 6XXX series aluminium—compression.

The parameters used enabled a permanent combination of the material. Therefore, it was found that it is possible to physically model the process of combining materials using complex deformation patterns using the STD 812 plastometer. However, in order to develop an accurate methodology for combining materials and physical modelling of actual technological processes (e.g., friction welding, rolling of bimetallic materials), numerical and physical modelling tests should be conducted in a wide range of strain parameters. It is necessary to study, for example, the distribution of strain components depending on the applied load condition (strain state) and the influence of the parameters used on the plastic flow of the material and, thus, on the microstructure and mechanical properties (deriving from the quality of the combination). Then, the possibility of adapting the obtained results to actual technological processes should be checked. It may also be necessary to plan a different geometry of the samples.

According to the knowledge of the author, this type of research has not been conducted in Poland so far. Developing the theoretical and experimental methodology of joining various materials with the use of complex load states will enable an accurate prediction of process parameters that guarantee a permanent combination of individual components, which will significantly reduce costs and accelerate the implementation of the technology of combining new materials in industrial conditions.

5. Discussion and Conclusions

The obtained test results showed a significant impact of the accuracy of the mathematical model of rheological properties on the distribution and values of strain parameters and on the stress intensity of the tested aluminium alloy. The correct determination of the strain parameters and the stress intensity in the material is particularly important during the numerical analysis of complex strain states and, in the case of tests, the parameters of which exceed the scope of the research possibilities of the used equipment.

After conducting the research and after analysing the obtained results, the following conclusions can be drawn:

1. The condition for the correct determination of the strain parameters' distribution and the stress intensity in the torsion test is the high accuracy of the mathematical model describing the rheological properties of the tested material and the correct determination of the initial and boundary conditions, consistent with the experimental tests;
2. As representative radii, defining the area in which the values of strain and stress parameters correspond to the average values on the cross-section of the torsion samples and, at the same time, to the values obtained in experimental tests, radii equal to 0.67 r and 0.724 r can be taken;
3. Using the coefficients of the rheological properties model presented in Table 2, a high agreement was obtained between the recorded and the calculated values of the torsion moment and the yield stress;

4. The heating and temperature control system installed in the STD 812 torsion plastometer ensures an even temperature distribution along the length and cross-section of the measurement base of the torsion samples;
5. As a result of deformation in the hot torsion process, a significant fragmentation of the microstructure of the tested alloy was obtained. It was proved that the grain size distribution on the sample cross-section was inhomogeneous and typical for a torsion test. The diversity of the microstructure on the cross-section after the hot torsion results from the inhomogeneity of the strain parameters such as strain intensity, strain rate intensity and stress inhomogeneity. The largest grains occurred in the sample axis, where the strain intensity value was the smallest, and the smallest in the surface area, where the strain intensity value reached its maximum value;
6. No significant differences were found in the level of the microhardness of the material after homogenisation and after the hot torsion process. However, it has been shown that the distribution of microhardness on the sample cross-section after the hot torsion was more inhomogeneous, which is caused by a greater differentiation of the grain size on the cross-section. The lack of diversity in the level of microhardness is related to the fact that under the conditions of hot deformation of aluminium alloys, the strengthening effect caused by plastic deformation was eliminated by softening as a consequence of the structural renewal processes;
7. The results of the EBSD analysis after the hot torsion present that the sample axis was dominated by the directions around the <001> pole. At a distance of 2.68 mm from the centre of the sample (0.67 r), the distribution of the texture intensity was focused around the <111> pole, while at a distance of 2.9 mm from the centre (0.724 r), the intensity distributions were centred around the <112> and <111> poles. In addition, it was proved that the fraction of large disorientation angles increased along with the distance from the sample axis, and the share of large disorientation angles above 15° was comparable for the distance from the sample axis of 2.68 mm (0.67 r) and 2.9 mm (0.724 r) and amounted to about 75% in the axis of the sample, large angles of disorientation constituted 60% of the studied population;
8. The obtained results of metallographic analyses confirm the necessity to define a representative area for the assessment of the microstructure of the material after the torsion process as well as indicate the need to analyse the microstructure of the torsion samples in a strictly defined zone, for which local values of the strain intensity, strain rate intensity and stress intensity should be determined.

Author Contributions: Conceptualisation, K.B.L.; methodology, K.B.L.; formal analysis, K.B.L. and B.L.-M.; investigation, K.B.L. and B.L.-M.; resources, K.B.L. and B.L.-M.; data curation, K.B.L. and B.L.-M.; writing—original draft preparation, K.B.L.; writing—review and editing, K.B.L. and B.L.-M.; visualisation, K.B.L. and B.L.-M.; project administration, K.B.L. All authors have read and agreed to the published version of the manuscript.

Funding: This research received no external funding.

Institutional Review Board Statement: Not applicable.

Informed Consent Statement: Not applicable.

Data Availability Statement: The data presented in this study are available on request from the corresponding author.

Acknowledgments: The authors would like to acknowledge PhD. Małgorzata Perek-Nowak (UST-AGH, Faculty of Non-Ferrous Metals) for support concerning EBSD investigations.

Conflicts of Interest: The authors declare no conflict of interest.

References

1. Dyja, H.; Gałkin, A.; Knapiński, M. *Reologia Metali Odkształcanych Plastycznie (Rheology of Plastically Deformed Metals)*; Czestochowa University of Technology: Częstochowa, Poland, 2010; ISBN 9788371934711.
2. Grosman, F.; Hadasik, E. *Technologiczna Plastyczność Metali, Badania Plastometryczne (The Technological Plasticity of Metals, Plastometric Testing)*; Silesian University of Technology: Gliwice, Poland, 2005; pp. 11–12, ISBN 837335204X.
3. Hadasik, E.; Schindler, I. *Plasticity of Metallic Materials—Deformation Behavior, Structure Development, Testing, Modeling*; Silesian University of Technology: Gliwice, Poland, 2004; pp. 39–64, 143–167, ISBN 8373351973.
4. Lenard, J.G. *Modelling Hot Deformations of Steel. An Approach to Understanding and Behavior*; Springer-Verlag: Berlin/Heidelberg, Germany, 1989; pp. 19–34, ISBN 9783642525179. [CrossRef]
5. Scheider, I.; Brocks, W.; Cornec, A. Procedure for the determination of true stress-strain curves from tensile tests with rectangular cross-section specimens. *J. Eng. Mater. Technol.* **2004**, *126*, 70–76. [CrossRef]
6. Sińczak, J. *Podstawy Procesów Przeróbki Plastycznej (Fundamentals of Plastic Forming Processes)*; Akapit Scientific Publisher: Krakow, Poland, 2010; pp. 473–524, ISBN 9788360958728.
7. Dróżdż, T.; Dyja, H.; Mrożek, J. Pomiar i sterowanie plastometrem skrętnym przy zastosowaniu techniki komputerowej (Measurement and control of a torsion plastometer using computer technology). *Metrol. Syst. Pomiar.* **1993**, *16*, 335–342.
8. Moreira Jorge, A.; Balancin, O. Prediction of steel flow stresses under hot working conditions. *Mater. Res.* **2005**, *8*, 309–315. [CrossRef]
9. López, B.; Urcola, J.J. Hot deformation characteristics of Inconel 625. *Mater. Sci. Technol.* **1996**, *12*, 673–678. [CrossRef]
10. Biswas, S.; Beausir, B.; Toth, L.S.; Suwas, S. Evolution of texture and microstructure during hot torsion of a magnesium alloy. *Acta Mater.* **2013**, *61*, 5263–5277. [CrossRef]
11. Eddahbi, M.; Borrego, A.; Mongea, M.A.; González-Doncel, G. Microstructure gradient after hot torsion deformation of powder metallurgical 6061 Al alloy. *Mater. Sci. Eng. A* **2012**, *555*, 154–164. [CrossRef]
12. Ducki, K.; Mendala, J.; Wojtynek, L. The characteristic of deformability of Fe-Ni superalloy during high-temperature deformation. In Proceedings of the 27th International Conference on Metallurgy and Materials—METAL 2018, Brno, Czech Republic, 23–55 May 2018; pp. 348–353.
13. Patil, R.D.; Gore, P.N. Review the effect of specimen geometry on torsion test results. *Int. J. Innov. Res. Sci. Eng. Technol.* **2013**, *2*, 7567–7574.
14. Laber, K.; Kawałek, A.; Sawicki, S.; Dyja, H.; Borowski, J.; Leśniak, D.; Jurczak, H. Investigations of plasticity of hard-deformed aluminium alloys of 5XXX series using torsion plastometer. *Arch. Metall. Mater.* **2016**, *61*, 1853–1860. [CrossRef]
15. Siciliano, F.; Allen, B.; Ferguson, D. Hot torsion tests—A reliable rolling simulation method for C-Mn steels. *Mater. Sci. Forum* **2017**, *879*, 1783–1787. [CrossRef]
16. Pernis, R.; Bidulská, J.; Kvačkaj, T.; Pokorný, I. Application of the torsion test in calculating the extrusion force. *Arch. Metall. Mater.* **2011**, *56*, 81–85. [CrossRef]
17. De Souza Nunes, L.G.; Pereira Machado, M.L. Simulation of the accumulative roll-bonding process through warm torsion test. *J. Mater. Res. Technol.* **2018**, *7*, 326–330. [CrossRef]
18. Mikloweit, A.; Bambach, M.; Pietryga, M.; Hirt, G. Development of a testing procedure to determine the bond strength in joining-by-forming processes. *Adv. Mater. Res.* **2014**, *966–967*, 481–488. [CrossRef]
19. Wang, S.; Dunlap, A.; Möhring, K.; Lohmar, J.; Schwedt, A.; Aretz, A.; Walther, F.; Hirt, G. Torsion plastometer trials to investigate the effect of non-proportional loading paths in caliber rolling on damage and performance of metal parts. *Prod. Eng.* **2020**, *14*, 17–32. [CrossRef]
20. Rule, J.R.; Lippold, J.C. Physical simulation of friction stir welding and processing of nickel-base alloys using hot torsion, *Metall. Mater. Trans. A* **2013**, *44*, 3649–3663. [CrossRef]
21. Laber, K.; Kułakowska, A.; Dyja, H. Physical modelling of the process of rolling AlZn5.5MgCu aluminium alloy bars on the RSP14/40 three-high reeling mill. In Proceedings of the 27th International Conference on Metallurgy and Materials—METAL Brno, Czech Republic, 23–25 May 2018; pp. 1599–1604, ISBN 9788087294840.
22. Horsinka, J.; Drozd, K.; Kliber, J.; Ostroushko, D.; Černý, M.; Mamuzic, I. Strain and strain rate in torsion test. In Proceedings of the 20th International Conference on Metallurgy and Materials—METAL, Brno, Czech Republic, 18–20 May 2011; pp. 383–389, ISBN 9788087294246.
23. Cepeda-Jiménez, C.; Hidalgo, P.; Carsí, M.; Ruano, O.; Carreño, F. Microstructural characterization by electron backscatter diffraction of a hot worked Al–Cu–Mg alloy. *Mater. Sci. Eng. A* **2011**, *528*, 3161–3168. [CrossRef]
24. Van Geertruyden, W.H.; Misiolek, W.Z.; Wang, P.T. Grain structure evolution in a 6061 aluminum alloy during hot torsion. *Mater. Sci. Eng. A* **2006**, *419*, 105–114. [CrossRef]
25. Gołębski, R.; Szarek, A. Diagnosis of the operational gear wheel Wear. *Teh. Vjesn. Tech. Gaz.* **2019**, *26*, 658–661. [CrossRef]
26. Blum, W.; Zhu, Q.; Merkel, R.; McQueen, H.J. Geometric dynamic recrystallization in hot torsion of A1-5Mg-0.6Mn (AA5083). *Mater. Sci. Eng. A* **1996**, *205*, 23–30. [CrossRef]
27. Shkatov, V.; Mazur, I. Modeling the dynamic recrystallization and flow curves using the kinetics of static recrystallization. *Materials* **2019**, *12*, 3024. [CrossRef]
28. Lv, J.; Zheng, J.H.; Yardley, V.A.; Shi, Z.; Lin, J. A review of microstructural evolution and modeling of aluminium alloys under hot forming conditions. *Metals* **2020**, *10*, 1516. [CrossRef]

29. Poletti, M.C.; Simonet-Fotso, T.; Halici, D.; Canelo-Yubero, D.; Montheillet, F.; Piot, D.; Kovács, Z.; Schell, N.; Tolnai, D. Continuous dynamic recrystallization during hot torsion of an aluminum alloy. *J. Phys. Conf. Ser.* **2019**, *1270*. [CrossRef]
30. Xia, X.; McQueen, H.J. Deformation behaviour and microstructure of a 20% Al2O3 reinforced 6061 Al composite. *Appl. Compos. Mater.* **1997**, *4*, 333–347. [CrossRef]
31. Maizza, G.; Pero, R.; Richetta, M.; Montanari, R. Continuous dynamic recrystallization (CDRX) model for aluminum alloys. *J. Mater. Sci.* **2017**, *53*, 4563–4573. [CrossRef]
32. Kawasaki, M.; Lee, H.-J.; Ahn, B.; Zhilyaev, A.P.; Langdon, T.G. Evolution of hardness in ultrafine-grained metals processed by high-pressure torsion. *J. Mater. Res. Technol.* **2014**, *3*, 311–318. [CrossRef]
33. Pereira, P.H.R.; Huang, Y.; Langdon, T.G. Examining the microhardness evolution and thermal stability of an Al–Mg–Sc alloy processed by high-pressure torsion at a high temperature. *J. Mater. Res. Technol.* **2017**, *6*, 348–354. [CrossRef]
34. Ahmadkhaniha, D.; Huang, Y.; Jaskari, M.; Järvenpää, A.; Sohi, M.H.; Zanella, C.; Karjalainen, L.P.; Langdon, T.G. Effect of high-pressure torsion on microstructure, mechanical properties and corrosion resistance of cast pure Mg. *J. Mater. Sci.* **2018**, *53*, 16585–16597. [CrossRef]
35. Laber, K.; Kawałek, A.; Sawicki, S.; Dyja, H.; Borowski, J.; Leśniak, D.; Jurczak, H. Application of torsion test for determination of rheological properties of 5019 Aluminium Alloy. *Key Eng. Mater.* **2016**, *682*, 356–361. [CrossRef]
36. Forge2011®. *Reference Guide, Sophia-Antipolis, Version 3.3 EN*; Transvalor: Sophia Antipolis, France, 2011.
37. Chenot, J.L.; Fourment, L.; Coupez, T.; Ducloux, R.; Wey, E. Forge3®: A general tool for practical optimisation of forging sequence of complex 3-D parts in industry. In *Proceedings of the International Conference on Forging and Related Technology (ICFT'98), Birmingham, UK, 27–28 April 1998*; Professional Engineering Publishing: Suffort, UK, 1998; pp. 113–122.
38. Gavrus, A.; EMassoni, E.; Chenot, J.L. An inverse analysis using a finite element model for identification of rheological parameters. *J. Mater. Process. Technol.* **1996**, *60*, 447–454. [CrossRef]
39. Hoff, N.J. Approximate analysis of structures in the presence of moderately large steps deformation. *Q. Appl. Mech.* **1954**, *2*, 49. [CrossRef]
40. Norton, F.H. *Creep of Steel at High Temperature*; McGraw Hill: New York, NY, USA, 1929.
41. Koczurkiewicz, B. *Podstawy technologii Wytwarzania Blach Grubych ze Stali Niskowęglowych z Mikrododatkami Przeznaczonych na Rury Przewodowe (Basics of the Technology for the Production of Low-Carbon Steel Plates with Micro-Additives for Line Pipes)*; Czestochowa University of Technology: Czestochowa, Poland, 2016; ISBN 9788363989361. ISSN 2391-632X.
42. Mróz, S. *Teoretyczno-Technologiczne Podstawy Walcowania Prętów Bimetalowych w Wykrojach (Theoretical and Technological Basics of the Bimetallic Bars Rolling Process in Grooves)*; Czestochowa University of Technology: Czestochowa, Poland, 2015; ISBN 9788363989231.
43. Suliga, M. *Analiza Wielostopniowego Ciągnienia Drutów Stalowych z Dużymi Prędkościami w Ciągadłach Konwencjonalnych i Hydrodynamicznych (Analysis of the Multi-Stage Dawing Process of Steel Wires at High Speeds in Conventional and Hydrodynamic Dies)*; Czestochowa University of Technology: Czestochowa, Poland, 2013; ISBN 9788363989046.
44. Hensel, A.; Spittel, T. *Strength and the Need for Labor*; Bildsomer Formgeburgs, Verfahren, VEB Deutscher Verlag für Grundstoffindustrie: Leipzig, Germany, 1979.
45. Laber, K.; Dyja, H.; Kawałek, A. Numerical modelling of the torsion testing on the STD 812 torsion plastometer. In Proceedings of the 26th International Conference on Metallurgy and Materials—METAL 2017, Brno, Czech Republic, 24–26 May 2017; pp. 465–471, ISBN 9788087294796.
46. Laber, K. The influence of the mathematical model accuracy of the rheological properties of a 5xxx series aluminium alloy on the strain parameters and stress intensity in hot torsion testing. *J. Chem. Technol. Metall.* **2021**, *56*, 417–423.
47. Jiang, F.; Zhang, H.; Su, J.; Sun, Y. Constitutive characteristics and microstructure evolution of 7150 aluminium alloy during isothermal and non-isothermal multistage hot compression. *Mater. Sci. Eng. A* **2015**, *636*, 459–469. [CrossRef]
48. McQueen, H.J. Development of dynamic recrystallization theory. *Mater. Sci. Eng. A* **2004**, *387–389*, 203–208. [CrossRef]
49. Huang, K.; Logé, R.E. A review of dynamic recrystallization phenomena in metallic materials. *Mater. Des.* **2016**, *111*, 548–574. [CrossRef]
50. McQueen, H.J.; Blum, W. Dynamic recovery: Sufficient mechanism in the hot deformation of Al (<99.99). *Mater. Sci. Eng. A* **2000**, *290*, 95–107.

materials

MDPI

Article

New Platforms Based on Frontal Cellular Automata and Lattice Boltzmann Method for Modeling the Forming and Additive Manufacturing

Łukasz Łach * and Dmytro Svyetlichnyy

AGH University of Science and Technology, Faculty of Metals Engineering and Industrial Computer Science, al. Mickiewicza 30, 30-059 Krakow, Poland
* Correspondence: lach@agh.edu.pl

Abstract: Materials science gives theoretical and practical tools, while new modeling methods and platforms provide rapid and efficient development, improvement, and optimization of old and new technologies. Recently, impressive progress has been made in the development of computer software and systems. The frontal cellular automata (FCA), lattice Boltzmann method (LBM), and modeling platforms based on them are considered in the paper. The paper presents basic information on these methods and their application for modeling phenomena and processes in materials science. Recrystallization, crystallization, phase transformation, processes such as flat and shape rolling, additive manufacturing technologies (Selective Laser Sintering (SLS)/ Selective Laser Melting (SLM)), and others are examples of comprehensive and effective modeling by the developed systems. Selected modeling results are also presented.

Keywords: recrystallization; phase transformation; rolling; additive manufacturing technologies; frontal cellular automata; lattice Boltzmann method

Citation: Łach, Ł.; Svyetlichnyy, D. New Platforms Based on Frontal Cellular Automata and Lattice Boltzmann Method for Modeling the Forming and Additive Manufacturing. *Materials* **2022**, *15*, 7844. https://doi.org/10.3390/ma15217844

Academic Editors: Jakub Zdarta, Agata Zdarta and Antonino Recca

Received: 31 July 2022
Accepted: 27 October 2022
Published: 7 November 2022

1. Introduction

Materials science and engineering are focused on creating new substances or changing the physical and chemical composition of existing materials to improve their properties. The important manufacturing techniques of product manufacturing are forming and additive manufacturing. Forming processes are important for the production of lightweight components made of metallic materials with defined geometric parameters and mechanical properties. Additive technologies have been introduced in many facilities in the preparation of short-term production (parts with complex geometry), and their use allows for saving materials, time, and production costs of various components. The preparation of forming processes and additive manufacturing using a trial-and-error method is a thing of the past. Today, methods that include computer-aided design systems (CAD) are more effective. The application of different numerical methods for the simulation of different phenomena in materials has become incredibly important in the last few years. The first method that is often used for such simulations is the phase field (PF) method. PF was recently used, for example, for the morphological and microstructural evolution of metallic materials under environmental attack [1]. DeWitt and Thornton presented a brief introduction to phase-field modeling and possible applications for simulations of precipitate evolution, grain growth, solidification, phase separation in battery electrodes, and deposition [2]. Recent applications of phase-field simulations of solid-state microstructure evolution and solidification that have been compared and/or validated with experiments were described by Tourret et al. [3]. The level-set (LS) method can be used, for example, to simulate grain growth with an evolving population of second phase particles [4] and for modeling multiphase thermo-fluid flow in additive manufacturing processes [5]. A review of level-set methods and some recent applications can be found elsewhere [6]. The

Monte Carlo (MC)–Potts model was used to model the primary recrystallization and grain growth in cold-rolled single-phase Al alloy [7] and grain growth simulation of single-phase systems [8]. The finite element method (FEM) was recently used for the analysis of microstructure evolution and mechanical properties during compression of open-cell Ni-foams with hollow struts [9], for simulations of microstructure evolution in single crystal and polycrystal shape memory alloys under uniaxial tension and compression [10], and for predicting phase transformations and microhardness for directed energy deposition of Ti6Al4V [11]. The cellular automata (CA) [12–14] method can be used, for example, for simulation of dynamic recrystallization behavior under hot isothermal compressions for as-extruded 3Cr20Ni10W2 heat-resistant alloy [12], modeling of solidification microstructure evolution in laser powder bed fusion-fabricated 316L stainless steel [13], simulation of coupled hydrogen porosity, and microstructure during solidification of ternary aluminum alloys [14]. Other examples of used numerical methods to model the evolution of microstructure in materials are analysis of metal extrusion by the finite volume method (FVM) [15], prediction of multidirectional forging microstructure evolution of GH4169 superalloy by the neural network [16], and prediction of microstructure evolution with convolutional recurrent neural networks [17]. Multiscale models, which are a combination of several methods (for example, finite element and cellular automata or cellular automata and finite volume), are also used. FE and CA methods were recently used for numerical prediction of microstructure for selective electron beam melting [18], while CA and FVM were used to predict the grain structure of an alloy, e.g., Inconel 718, fabricated by additive manufacturing [19].

Considering the use of a given method to model a given phenomenon or process, its advantages and disadvantages are always important. The phase-field method automatically takes into account changes in front topology and generalizes easily to 3D areas; however, there is a complicated theoretical side of the model, small model volumes, and the need for fine discretization in the front area. The level-set method allows for a direct representation of the interface and curvature of the grain boundary, and there is also no need to discretize explicitly the interface. LS has an inability to track the evolution of the texture, which can be solved by combining LS with the theory of crystal plasticity. The numerical implementation of the Monte Carlo algorithm is quite simple, and good computational performance can be achieved using parallel calculations; however, the MC solutions depend on the type of mesh used, a good representation of the curvature of the grain boundary, the proper determination of the simulation length, and the time scale. Taking into account the finite element method, it can be seen that complex shapes can be approximated with high accuracy, the sizes of the elements can be different, and nonlinear boundary conditions can be defined. FEM needs to control the numerical error because it depends on discretization parameters, material properties, and boundary conditions. CAs provide very high computing performance and a better accounting of many factors, including local, e.g., crystallographic orientation, or dislocation density. The simple rules and synchronous updating of cell states on the grid introduce additional simplifications in the modeled reality, and the calculation time is the main preference for the use of 2D solutions compared to 3D.

Recently, one of the most effective methods for simulations of different phenomena and processes in materials science is the frontal cellular automata (FCA) and the lattice Boltzmann method (LBM). FCA is a modified CA algorithm that allows for algorithmic acceleration of calculations with the same level of accuracy. The calculation time of the 3D FCA calculations is similar to the time of classical 2D CA. LBM is very feasible for modeling the flow of liquids, gases, mixtures, heat flow, and transfer as well as diffusion occurring simultaneously with the liquid (gas) flow. The wide application of the LBM method was limited by some of its features. The main disadvantages were the high memory volume requirements and a large number of iterations, i.e., the computation time even in two-dimensional and especially in three-dimensional calculations. The development of computer and computational technology related to the possibility of using parallel computing with the use of graphics cards has changed the perception of the LBM method from being a little useful to very effective. LBM was recently used, for example, for modeling the evolution of coherent vortices and periodic flow in a continuous casting mold [20], simulating fluid flow in multi-scale porous media [21], and simulating fluid flow in multi-scale porous media [22]. LBM comes from the CA, so a combination of these two methods in comprehensive solutions can be considered natural. Taking into account their algorithmic structure, these methods are almost ideal for a parallel programming environment involving CUDA (Compute Unified Device Architecture) graphics cards. Models based on LBM and FCA do not need a complicated interface. The solution can be implemented in a common domain, while the number of processes and phenomena modeled simultaneously depends on the number of variables associated with a point, cell, or node in the same model volume. Extending the model with a new process or phenomenon consists of adding variables and an algorithm without developing additional communication between different parts of the model. This is an undeniable advantage compared to multiscale models. FCA and LBM can be the base methods for the development of comprehensive platforms to model various technological processes.

A frontal cellular automata (FCA) algorithm was developed by Svyetlichnyy and for the first time was successfully applied to model microstructure evolution in 2010 [23]. Over the next years, the method has been successfully used to model different phenomena and processes in materials science. Crystallization, recrystallization, phase transformations, severe plastic deformation, and technological processes such as rolling or continuous casting were successfully modeled. Universal frontal cellular automata can be used for practically all possible processes consisting of nucleation and grain growth. Closed-transition circuits make the FCA suitable for modeling multiple nucleation and grain growth.

The introduction of the lattice Boltzmann equation to lattice gas automata (LGA) applied to viscosity calculations can be regarded as the first of the milestone steps leading to the LBM method. The next steps were the replacement of particles by the continuous Fermi–Dirac distribution for the equilibrium distribution, linearization of the collision operator, application of the Boltzmann distribution instead of the Fermi–Dirac distribution, and replacement of the collision operator with the BGK (Bhatnagar–Gross–Krook) approximation. After 1992, the LBM method can be considered as developed. Its further development is primarily related to the expansion of application areas and the development of new principles, methods, etc. Materials science and engineering are now important areas of FCA and LBM application, and further dynamic growth of interest in these methods is expected soon.

The paper presents the new modeling platforms that allow for the simulation of different phenomena and processes in materials science as well as forming processes (for example, rolling processes) and additive manufacturing technology. Basic information on the FCA and LBM methods, which are the basis of these platforms, is shown. The principles of calculations carried out with the use of these methods are presented. Finally, examples of modeling results of phenomena within the considered processes and the results of simulation of industrial processes in the general aspect obtained by using the developed platforms are shown. The presented computing platforms are based on their software developed based on FCA and LBM methods. Calculations can be performed on Windows 8, 10, and 11 and Linux operating systems. Some of the new algorithms developed include the possibility of parallel computing with the use of CUDA (Compute Unified Device Architecture), i.e., the universal architecture of multicore processors on Nvidia cards and the programming environment based on the C and C++ programming language, which is an integral part of this architecture. Some of the developed software is also a CPU-based sequential computing version, written in C++ or Fortran. For parallel calculations, graphics cards (GeForce RTX 2080 Ti, GeForce TITAN 6GB, TITAN Z 12 GB, GeForce 1080Ti, GeForce 1060, NVIDIA, Santa Clara, California, USA) are used. Processing and presentation of the results are carried out primarily with the use of OpenGL technology.

2. Frontal Cellular Automata

Frontal cellular automata (FCA) are a modernization of the classical cellular automata and their development began in the 2000s. FCA is very efficient The method is very effective for a certain class of tasks, in which changes can be presented and described as the movement of boundaries or the movement of the front of changes, while the rest, a significantly larger part of the space, remains unchanged. This principle can be applied, for example, to modeling the evolution of microstructure. The principle of considering only the front of changes gave the name to the modified method.

There are several differences between the classical CA and the FCA. Three of the most important modifications are increasing the number of states (multiple states), inverse information transfer, and using linked lists. The increasing number of states differentiates the processing procedures of cells in different states. The inverted transfer allows sending only critical information at an appropriate moment to the neighboring cell that influences its states rather than collecting mainly unused information from the whole neighborhood in each iteration. Figure 1 presents the direction of information transfer considering the Moore and von Neumann neighborhoods. The linked lists allow cells to proceed only in "active" states on the front of changes. These and other means exclude a huge number of cells from calculations in every iteration, reducing overall computation time by hundreds and even thousands of times.

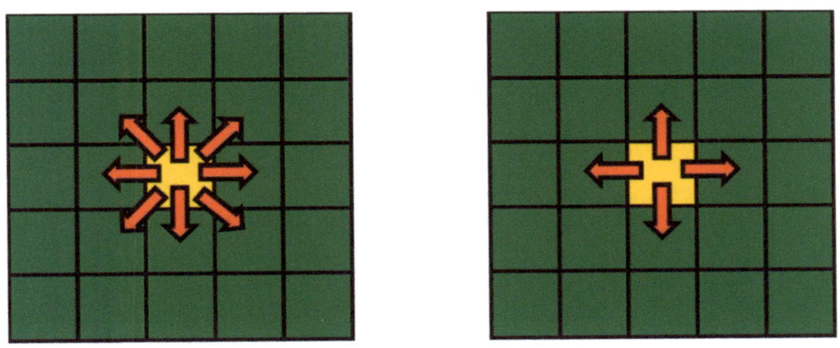

(a) **(b)**

Figure 1. FCA communication direction using the Moore (**a**) and von Neumann (**b**) neighborhoods.

Figure 2 shows a fragment of the space with cells in three states. The initial state of the cell does not change if there are no changes in its neighborhood and changes if a cell in the frontal state appears in its neighborhood. The cell in the frontal state changes its state in the final state immediately or with delay, depending on a modeling phenomenon or process. The final state never changes. As a result, only frontal cells are used for calculations. The introduction of the frontal state allowed us to resign from exploring the entire space and to concentrate activities on cells in which changes take place.

The described modifications reduce the computation time but increase the complexity of the algorithm. FCA creates, among other things, the possibility to take into account changes in shape and sizes of the cells, reorganize space, and introduce new structures that connect cells in the group of the same state, in the same grain, or according to other arbitrary features. In FCA, linked lists usually unite cells in the same state by adding and removing elements. It is a central point in the acceleration of calculation, especially for sequence programming. Reducing the calculation time of the frontal automata compared with the classical one is significant in the 2D models, and it is radical in the 3D models. More detailed information on FCA can be found elsewhere [23]. However, it should be noted that the advantage of FCA over classic CA is slightly smaller for parallel computations because there are no effective methods of linked-list maintenance.

Figure 2. The front of the growing grain with cells in the state: 1, initial; 2,frontal (transitional); 3, final.

3. Lattice Boltzmann Method

The lattice Boltzmann method (LBM) is very feasible for modeling the flow of liquids, gases, mixtures, heat flow, and transfer as well as diffusion occurring simultaneously with the liquid (gas) flow. LBM is derived from a variant of the CA method, which can be regarded as a simplified model of molecular dynamics that implements the discretization of space, time, particle velocity, and the statistical Boltzmann equation. The basis of the method is the solution of the transport Boltzmann equation:

$$\frac{\partial f}{\partial t} + \frac{\partial f}{\partial \boldsymbol{x}} \cdot \xi + \frac{\boldsymbol{F}}{m} \cdot \frac{\partial f}{\partial \boldsymbol{u}} = \Omega \quad (1)$$

where f—particle distribution function; $\boldsymbol{x}$ and ξ—phase space variables (coordinate and velocities); t—time; $\boldsymbol{F}$—external force (for example gravity); m—mass; $\boldsymbol{u}$—macroscopic velocity; Ω—the collision operator.

The particle distribution function along the directions and velocities is a Maxwell function, so along any direction of space, particles have velocities according to the Gaussian distribution:

$$f = \frac{\rho}{(2\pi RT)^{D/2}} exp\left(-\frac{(\xi - \boldsymbol{u})^2}{2RT}\right) \quad (2)$$

where ξ—microscopic velocities; R—gas constant; D—space dimension; ρ—gas density; T—temperature.

The particles velocities space ξ can be reduced to a finite (discrete) velocity system $\{e_i, i = 1, \ldots, b\}$, and a grid can be built on which calculations can be made. Therefore, the equation is approximated for the characteristic velocities of the determined grid and the selected directions:

$$f_i(\boldsymbol{x} + \boldsymbol{e_i}, t + 1) = f_i(\boldsymbol{x}, t) + \Omega_i \quad (3)$$

Equation (3) is called the lattice Boltzmann equation (LBE) and is the basis of the LBM method. An important element of this equation is the collision operator. One of the first and possibly the simplest and most frequently used is the operator proposed by Bhatnagar, Gross, and Krook (named after BGK), which has the following form:

$$\Omega_i(f) = \frac{\Delta t}{\tau}\left[f_i(\boldsymbol{x}, t) - f_i^{eq}(\boldsymbol{x}, t)\right] \quad (4)$$

where Δt = 1—time step; τ—relaxation time; f^{eq}—the distribution function in the appropriate direction in the equilibrium state.

The solution of Equation (3) with the collision operator (4) is divided into two stages: collision and advection (streaming). Calculations for the collision stage are preceded by the calculation of macroscopic variables: density ρ and velocity $\boldsymbol{u}$ (through momentum $\rho\boldsymbol{u}$):

$$\rho = \sum_{i=0}^{b} f_i \tag{5}$$

$$\rho\boldsymbol{u} = \sum_{i=0}^{b} f_i \boldsymbol{e_i} \tag{6}$$

and the equilibrium distribution f^{eq}:

$$f_i^{eq}(\boldsymbol{x}, t) = w_i\rho\left(1 + \frac{\boldsymbol{u}\cdot\boldsymbol{e_i}}{c_s^2} + \frac{(\boldsymbol{u}\cdot\boldsymbol{e_i})^2}{2c_s^4} - + \frac{\boldsymbol{u}\cdot\boldsymbol{u}}{2c_s^2}\right) \tag{7}$$

where c_s—sound speed in the modeled flow; w_i—directional weights.

The output distribution function after the collision has the following form:

$$f_i^{out}(\boldsymbol{x}, t) = f_i^{in}(\boldsymbol{x}, t) - \frac{1}{\tau}\left[f_i^{in}(\boldsymbol{x}, t) - f_i^{eq}(\boldsymbol{x}, t)\right] + F_i \tag{8}$$

On the other hand, at the advection stage, a simple transfer of the distribution function to neighboring nodes takes place:

$$f_i^{in}(\boldsymbol{x} + \boldsymbol{e_i}, t + 1) = f_i^{out}(\boldsymbol{x}, t) \tag{9}$$

An additional stage of the calculations is considering the boundary conditions.

The same LBM algorithm is used for all issues, including cyclically repeating steps (Figure 3):

1. Calculation of macroscopic values: density ρ, velocity u, temperature T, concentration C, etc.;
2. Determination of the equilibrium distribution function f^{eq} for the modeled variable;
3. Collision operation, determination of the output distribution function f^{out};
4. Streaming operation, transfer of the distribution function to the appropriate cells, site, or nodes;
5. Consideration of boundary conditions.

Some of these stages can be combined (for example, 1 + 2 + 3 or 4 + 5). The cycle can be started from anywhere although it starts with the determination of macroscopic values, the determination of equilibrium distributions functions, and the assignment of output functions not according to the collision operator but with the equal equilibrium distribution $f^{out} = f^{eq}$. In this way, the computation starts from the local equilibrium state in all nodes, and changes occur due to the global imbalance.

The basis of the method is the discretization of space and time. A regular square (two-dimensional) or cubic (three-dimensional) grid with a distance between adjacent nodes equal to one is superimposed on the space. The length of the time step is equal to one. An important element of the system is the selection of the velocity model. In one-dimensional space, the models D1Q2 and D1Q3 are used; in a two-dimensional space, D2Q4, D2Q5, and D2Q9 are used. For three-dimensional problems. D3Q6, D3Q7, D3Q15, D3Q19, and others are used. D indicates the dimensionality, while Q indicates the number of velocities.

More information on the LBM method can be found elsewhere [24].

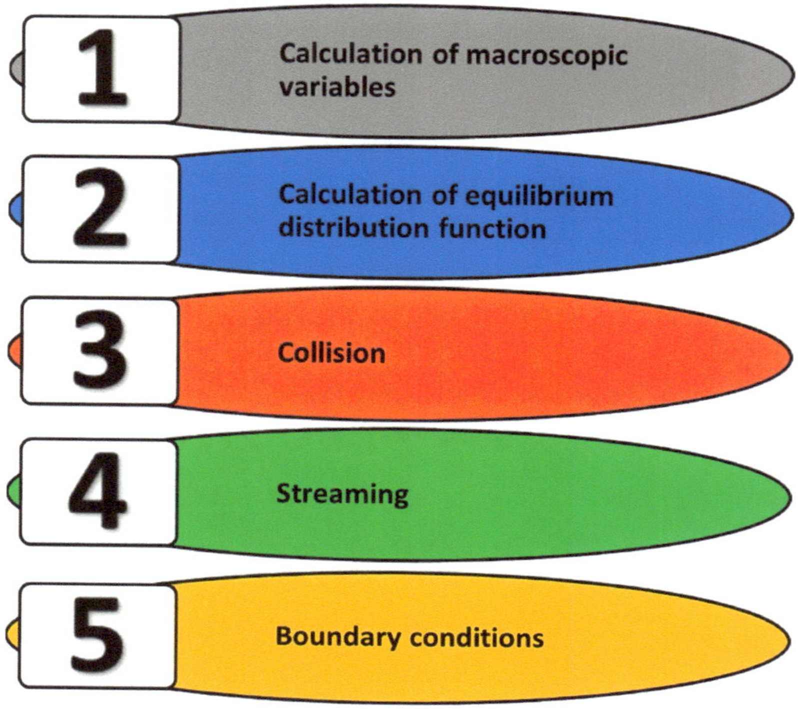

Figure 3. Structure of the LBM calculation algorithm.

4. Platform Based on Frontal Cellular Automata for Modeling the Microstructural Phenomena in Technological Processes

The use of any method, including cellular automata, to model a specific technological process requires formalization and solving several problems directly related to the limitations of its applicability. The accurate analysis allows for adequate and effective use of a given modeling method. By determining and accurately describing selected issues, microstructural phenomena, and processes, it is possible to develop an appropriate scheme of numerical modeling.

Choosing as a goal to create a universal tool for modeling microstructure evolution in various technological processes, e.g., casting, forming, heat treatment, etc., a platform based on frontal cellular automata was developed (Figure 4). This FCA platform was created as a hierarchical system and used for the study of microstructural phenomena to design, improve, and optimization of technological processes, including crystallization, recrystallization, grain growth, and phase transformation. The basis of the platform (first level) is frontal cellular automata (FCA) augmented with the transformation and reorganization of cells. The models creating the second level and responsible for the simulation of basic microstructural phenomena are based on FCA. The third level of the system includes models of technological processes and uses the models of the second level as building blocks. An additional element that complements the system is the material database.

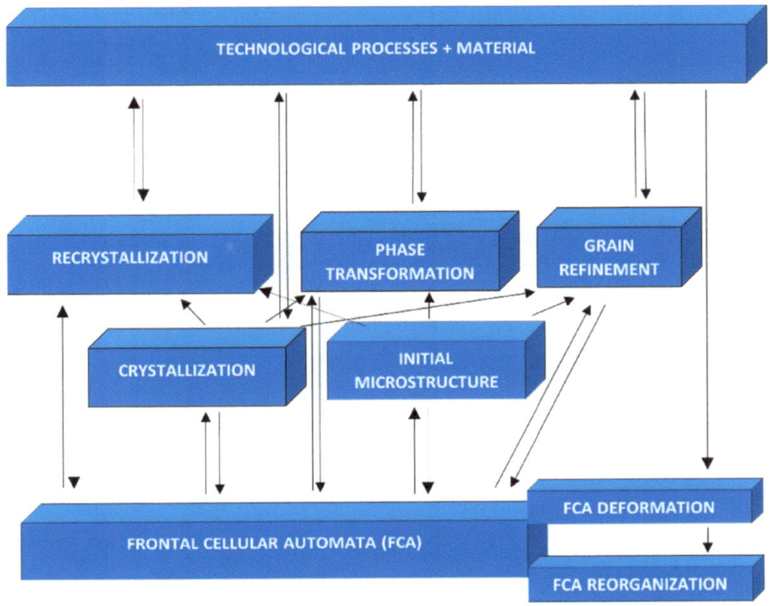

Figure 4. Structure of the platform for modeling microstructure evolution in technological processes.

The development of the platform included several stages according to its hierarchical structure, from the development of FCA algorithms and models of individual phenomena to the final comprehensive model of the technological process. In the beginning, the second-level models (Figure 4) were simplified, schematically reflecting real phenomena. Then, all elements of the platform were constantly expanded after a wide range of experimental and theoretical research. Models of the second level can be divided into two groups. The models of the first group create the original structure. They are the model of the «real» crystallization phenomenon (solidification) [25] and the «unreal» process of creating the initial microstructure. Both models do not require the initial microstructure, and the results of their simulation, i.e., the initial microstructure, can be used by the models of the second group. Models in the second group require initial microstructure, and the group includes models of recrystallization, phase transformations, and microstructure refinement. The phenomena modeled on the second level can be divided into two stages: nucleation and grain growth (crystals, crystallites, and a new phase), so they can be realized using the same FCA (considering the relevant details).

4.1. FCA Deformation and Reorganization

During the modeling by FCA, it is very important to define the space geometry and boundary conditions as one of the most important parameters. Under certain conditions of deformation, CA space may be distorted, and it may be necessary to reorganize it. It directly affects the boundary conditions, which should be changed. Calculations are optimal with cells of cubic shape, but then, the shape of the cells can be deformed. If the strain accumulated in the material is not large enough, the CA cells change shape and size. When the deformation reaches a predetermined level, the space of the CA should be reorganized to obtain the cubic shape of the cells. Reorganization in CA is similar to "remeshing", widely used in FEM codes. Different space reorganization methods have been developed (Figure 5). A detailed description of the solutions and algorithms developed in this area can be found elsewhere [26].

Reorganization Method	Microstructure	
	Before Reorganization	After Reorganization
Halving	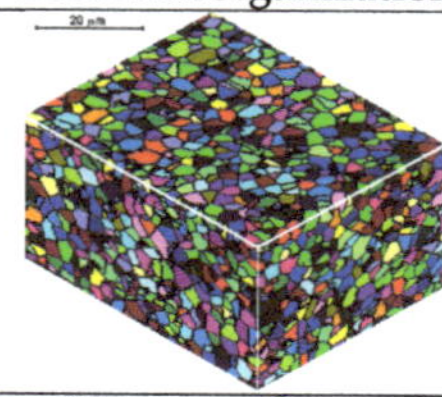	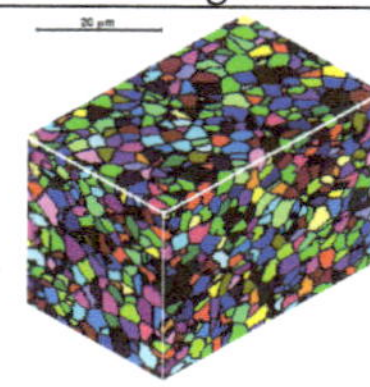
Cutting and Bonding	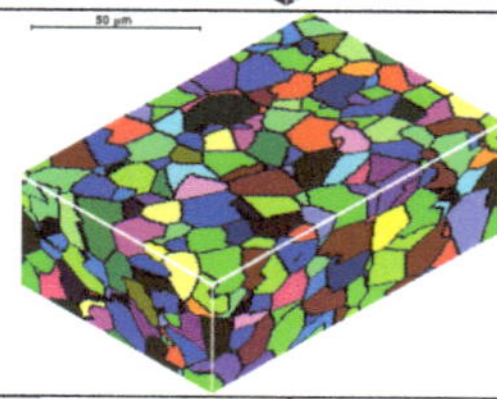	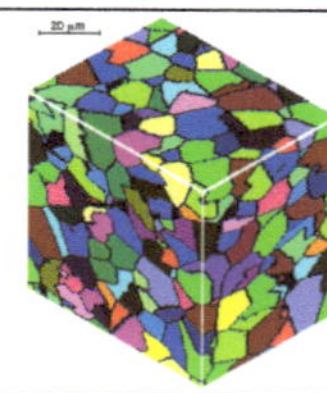
Doubling	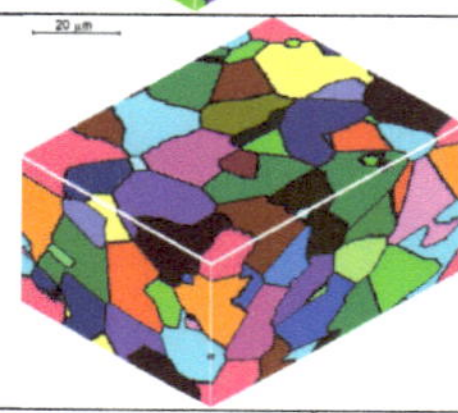	
Straightening	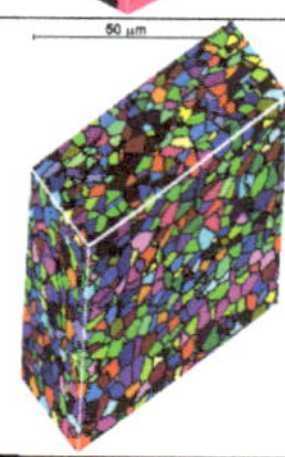	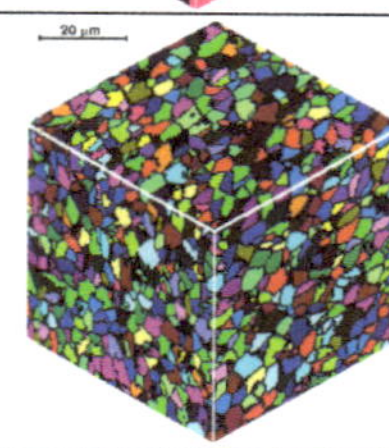

Figure 5. Space reorganization methods.

4.2. Initial Microstructure

Considering the modeling of microstructure evolution in any process, an initial microstructure is necessary, which under certain modeling conditions will be closest to the initial real structure. The developed model of universal frontal cellular automata makes it possible to obtain the initial microstructure, which is the result of two processes or stages: nucleation of grains and their growth. Figure 6 shows the general scheme for modeling the initial microstructure.

The first step in modeling the initial microstructure, based on the three-dimensional model of cellular automata, is the determination of the geometry of the cellular automata by defining the sizes of the three-dimensional space used. This parameter is directly related to the definition of the number of cells for each axis of the system and the actual sizes. An additional factor that is also considered is the shape of the FCA space. The developed model allows using the cuboid or cylinder space. The next step in the process of creating the initial microstructure is to apply the boundary conditions, which may be different in different directions. Information on nucleation is collected and stored by specifying the time and place of appearance of each nuclei and its spatial orientation. This information is ordered according to the time of nucleation. The orientation of the growing grains is also an element that should be considered during the process of creating the initial microstructure. The model considers two types of orientation, i.e., crystallographic orientation and spatial orientation of grain other than spherical. These orientations are combined when modeling the initial microstructure. The model allows for both random selections of the orientation for each grain and allocation of the same orientation for all grains, whose orientation can be chosen randomly or strictly defined. Taking into account all of the factors described above, the model allows modeling of the initial microstructure, which may be different by considering different modeling conditions.

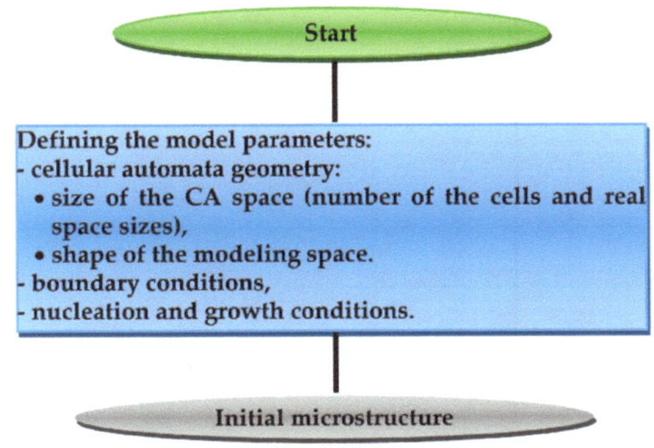

Figure 6. The modeling scheme of the initial microstructure.

Figure 7 presents the results of the initial microstructure simulation for the spherical shape of the growing grains and open and periodic boundary conditions, with the assumed space of 400 × 400 × 400 cells, the dimensions of the space 500 × 500 × 500 µm, the number of nuclei equal to 600; in each simulation, the nuclei were distributed in the same places in the cell space.

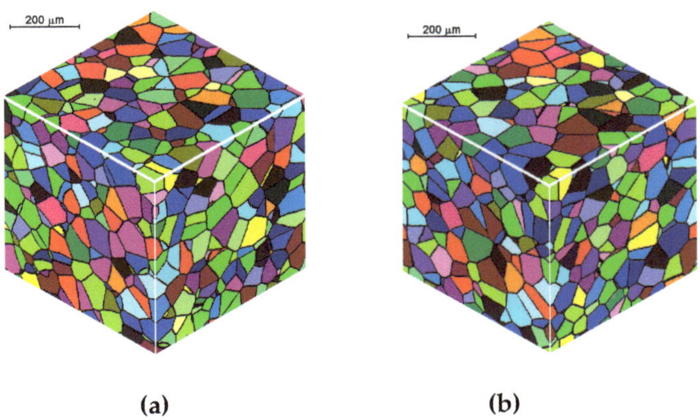

Figure 7. Initial microstructure under different boundary conditions: (**a**) open and (**b**) periodic.

The selection of the boundary conditions and the shape of the growing grains is a preliminary stage preparatory to obtaining the initial microstructure for the given parameters. Such a microstructure, apart from the grain shape, should also meet other requirements concerning, e.g., average grain size, grain size distribution, texture, and others. Obtaining the required grain size distribution in the microstructure can be done in two ways: either by correcting the existing microstructure or by modeling a new microstructure with a matching of the nucleation rate and grain growth. Figure 8 shows schematically the algorithm that allows to obtain a microstructure with a given distribution.

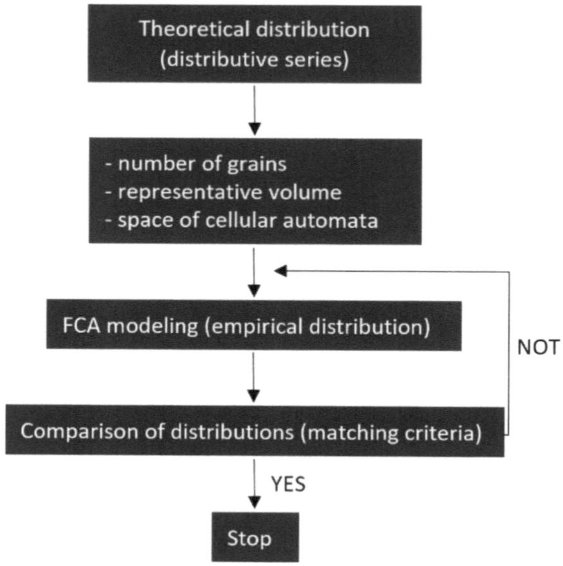

Figure 8. The algorithm for modeling the initial microstructure with a given grain size distribution.

The steps of the algorithm for creating a microstructure with a given grain size distribution are as follows:

1. Representation of the theoretical particle size distribution using a distributive series;
2. Determining the number of grains in a representative model volume and checking the size of the cell space;
3. Defining the nucleation conditions;
4. Modeling the microstructure and determining its empirical distribution;
5. Determining the error of fit by comparing the obtained empirical distribution with the theoretical one;
6. Return to point 3 and perform correction of nucleation conditions in case of failure to meet the matching criterion.

The fulfillment of the matching condition becomes tantamount to obtaining a digital representation of the microstructure with a given grain size distribution and determination of the nucleation conditions. In the case of correction of the existing microstructure, only the first three steps of the algorithm are carried out.

The description of the algorithm that allows obtaining a microstructure with given parameters and examples of simulation results can be found elsewhere [27,28].

4.3. Recrystallization

An important factor that influences the microstructure and properties of the material, without which it is impossible to comprehensively consider the processes that take place in many materials, is the phenomenon of recrystallization. In hot-deformed materials, one of the dominant microstructure-reconstruction processes is static recrystallization.

The developed FCA-based model for recrystallization contains several submodels: dislocation density and flow stress model, nucleation model, and grain growth model.

The main assumptions of the recrystallization model are the following:

1. The dislocation density is assigned to the grain, and the initial dislocation density is the same for all grains and equal to a minimal value for current thermal conditions.
2. Nucleation begins when the dislocation density reaches a critical value ρ_c.
3. The nucleation rate depends on the dislocation density, strain, strain rate, and temperature.

4. The migration rate of the grain boundaries is a function of temperature, dislocation density, crystallographic orientation, and other parameters.

Table 1 shows the basic relationships used in the submodels of the recrystallization model. A detailed description of the recrystallization model developed can be found elsewhere [29,30].

Table 1. Recrystallization model.

Submodels	Dependencies	
Dislocation Density and Flow Stress Model	$\sigma = \sigma_0 + \alpha\mu b\sqrt{\rho_{av}}$	(10)
	$\dot{\rho} = U(\varepsilon)\dot{\varepsilon} - \Omega(\rho)\dot{\varepsilon} - R(\rho)$	(11)
Nucleation Model	$\dot{n}_V = a_N\varepsilon^{l_N-1}D_0^{-k_N}\dot{\varepsilon}$	(12)
	$\Delta\dot{n}_V = a_N\varepsilon^{l_N-1}D_0^{-k_N}\dot{\varepsilon}(N_{Vmd} - N_V)\Delta t$	(13)
	$N_{Vmd} = a_{md}\dot{\varepsilon}^{-m}exp\left(\frac{Q}{RT}\right)^{-m}$	(14)
	$D_{srx} = A\varepsilon^a D_0^b Z^c$	(15)
	$Z = \dot{\varepsilon}exp\left(\frac{Q}{RT}\right)$	(16)
Grain Growth Model	$v = Mf(\theta)F$	(17)
	$\begin{cases} f = 0 & dla \quad \theta < 7^\circ \\ f = 3.72 * 10^{-4}exp\left\{8 * \left[1 - exp\left(-\frac{\theta}{2.2}\right)\right]^{10}\right\} & dla \quad 7^\circ < \theta < 15^\circ \\ f = 1 & dla \quad \theta > 15^\circ \end{cases}$	(18)

Figure 9 shows the initial microstructure and an example of the final microstructure after deformation at T = 1100 °C and complete static recrystallization. The microstructure was obtained for the following conditions: space sizes of 500 × 500 × 500 cells, the dimensions of the space 400 × 400 × 400 µm, and strain rate and strain equal to 1 s^{-1} and 0.18, respectively. The final average grain size was d_{av} = 75 µm (measured 78.5 µm). The calculations were made for C 45 steel.

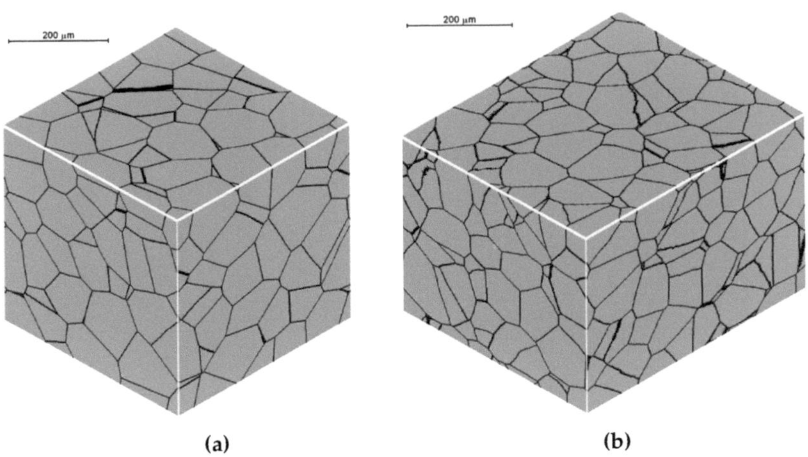

Figure 9. Initial microstructure with an average grain size d_{av} = 100 µm (**a**) and final microstructure d_{av} = 75 µm (**b**).

Figure 10 shows the results of the dynamic and metadynamic recrystallization modeling considering the geometry of the deformation. The initial microstructure before deformation (Figure 10a) and after deformation without recrystallization (Figure 10b) with partial dynamic recrystallization (Figure 10c) and after metadynamic recrystallization (Figure 10d) is shown. Cellular automata with dimensions $n_x \times n_y \times n_z$ = 200 × 300 × 400 cells and space dimensions $a_x \times a_y \times a_z$ = 400 × 600 × 800 µm were used for the simulation. The number of grains was 250. The deformation temperature T = 1000 °C, the deformation time t = 0.5 s, and the nonzero components of the strain rate tensor: $\dot{\varepsilon}_x = 1.0$, $\dot{\varepsilon}_z = -1.0$, and $\dot{\varepsilon}_{xz} = 0.07$. The shape of the space and each cell is not a cube in this case but a cuboid. Figure 10b shows the changes in the shape of the space and microstructure with uniform deformation. Figure 10c shows the microstructure with the dynamic recrystallization process that occurs after deformation in 0.5 s. Figure 10d shows the microstructure after full metadynamic recrystallization. The model assumes that nucleation occurs only during deformation, and Figure 10d shows the microstructure after metadynamic recrystallization as an effect of grain growth after dynamic recrystallization.

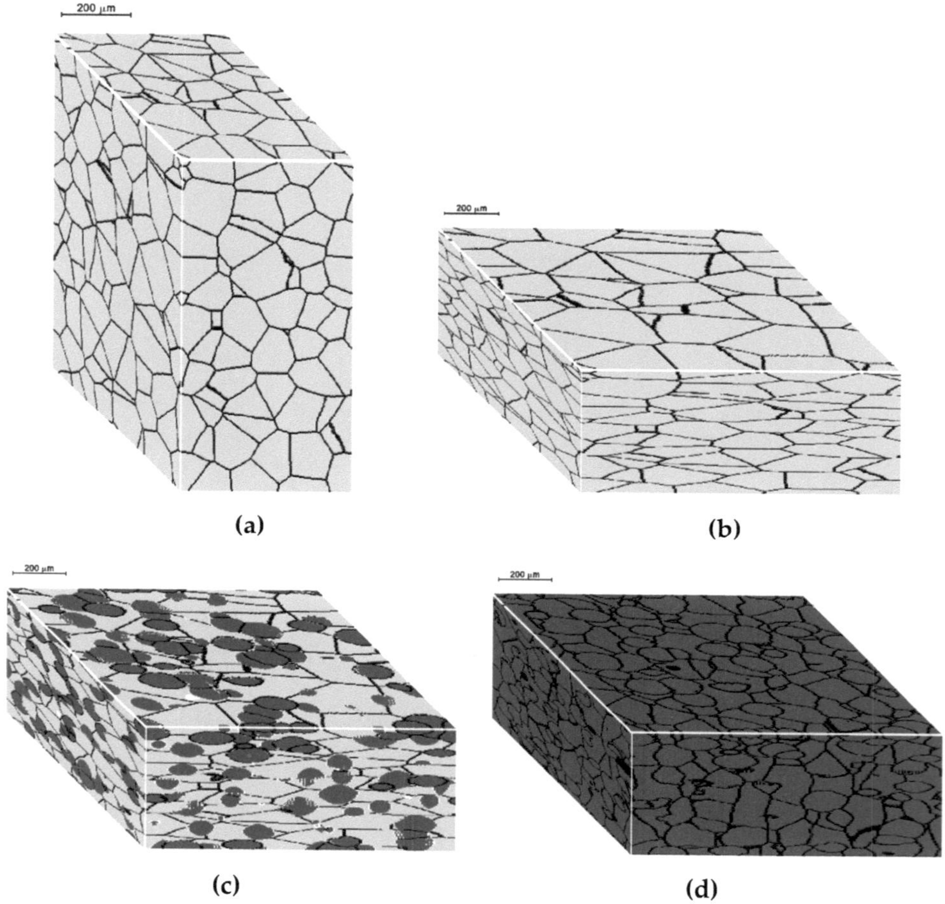

Figure 10. Microstructure before deformation (**a**) and after deformation: without recrystallization (**b**), with dynamic recrystallization (**c**), and after metadynamic recrystallization (**d**).

4.4. Phase Transformation

The main tools for the modeling of phase transformations are cellular automata, often combined with the finite difference method (CA+FDM), phase-field (PF), or multi-phase-field method (MPF). The structure of the hybrid model of diffusion phase transformations developed in carbon steels is shown in Figure 11. The model is based on two modeling methods: FCA and LBM. The model is developed to simulate the evolution of microstructures (FCA), carbon diffusion, and heat flow (LBM).

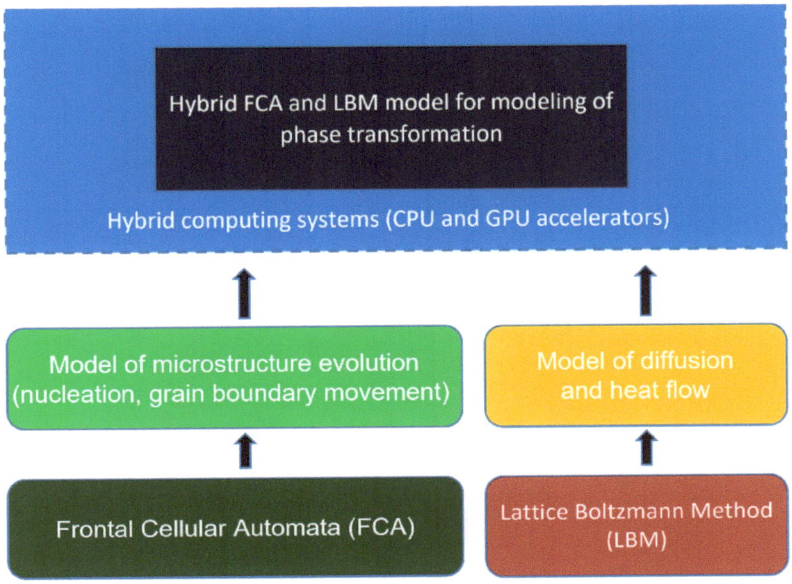

Figure 11. Structure of the hybrid model for modeling diffusional phase transformations.

The model was developed in several stages, from the 1D variant to the 3D variant. At the final stage of model development, the 3D Fourier equation in the following form was used in the calculations:

$$\frac{1}{\alpha}\frac{\partial T}{\partial t} = \left(\frac{\partial^2 T}{\partial x^2} + \frac{\partial^2 T}{\partial y^2} + \frac{\partial^2 T}{\partial z^2}\right) + Q(x,y,z,t); \frac{\partial c}{\partial t} = D\left(\frac{\partial^2 c}{\partial x^2} + \frac{\partial^2 c}{\partial y^2} + \frac{\partial^2 c}{\partial z^2}\right) \quad (19)$$

where T—temperature; c—carbon concentration; α—thermal diffusivity; $Q(x,y,z,t)$—the source of thermal energy; D—diffusion coefficient.

The BGK (Bhatnagar–Gross–Krook) model was used for simulations:

$$f_k(x+\Delta x, y+\Delta y, z+\Delta z, t+\Delta t) - f_k(x,y,z,t) = -\frac{\Delta t}{\tau}\left[f_k(x,y,z,t) - f_k^{eq}(x,y,z,t)\right] + \Delta t w_k S \quad (20)$$

$$f_k(x+\Delta x, y+\Delta y, z+\Delta z,\ t+\Delta t) - f_k(x,y,z,t) = -\frac{\Delta t}{\tau}\left[f_k(x,y,z,t) - f_k^{eq}(x,y,z,t)\right] \quad (21)$$

where $f_k(x,y,z,t)$, $f_k^{eq}(x,y,z,t)$—the particle and equilibrium distribution functions; τ—the single-relaxation-time parameter; S—the source term; w_k—weighting factor in the direction k.

The equilibrium distribution function f_k^{eq}:

$$f_k^{eq} = w_k \varphi(x,y,z,t) \quad (22)$$

where ϕ—the dependent variable (temperature T, concentration c).

Two steps can be considered in LBM calculations:

Collision:

$$f_i(x,y,z,t+\Delta t) = f_k(x,y,z,t)[1-\omega] + \omega f_k^{eq}(x,y,z,t) \quad (23)$$

where $\omega = \Delta t/\tau$.

Streaming:

$$f_i(x + \Delta x, y + \Delta y, z + \Delta z, t + \Delta t) = f_k(x, y, z, t + \Delta t) \tag{24}$$

Table 2 shows the developed numerical algorithms that are used in the 3D heat flow and diffusion models. (Steps 2–7 are repeated cyclically).

Table 2. Algorithms for 3D simulation of heat flow and carbon diffusion.

<table>
<tr><th>Heat Flow</th><th>Diffusion</th></tr>
<tr><td colspan="2">Initial parameters (1)</td></tr>
<tr><td colspan="2">Number of nodes: n_x, n_y, and n_z, node position: x, y, z, number of time steps: tsteps, grid step: Δx, Δy, Δz, time step: Δt, austenite and ferrite fractions in interface node: $\varphi_{IA} = 1$, $\varphi_{IF} = 0$ (1a)</td></tr>
<tr><td>Velocity and specific enthalpy coefficients: k_v and k_q, temperatures of phase transformation, in the interface, and in the node: T_P, T_I, and $T_{x,y,z}$ (1b)</td><td>Carbon concentration in nodes: c_i, diffusion coefficient: D_{cf} (1b)</td></tr>
<tr><td colspan="2">Calculations of a boundary velocity, the quantity, mass, or volume of the transformed material in the interface node; fraction of the new phase in the interface node; fraction checking (2)</td></tr>
<tr><td>$v = k_v(T_p - T_I)$ (2a)</td><td>$v = \Delta c$; (2a)</td></tr>
<tr><td colspan="2">$\Delta\varphi_I = v\Delta t$; $\varphi_{IF,t} = \varphi_{IF,t-1} + \Delta\varphi_I$ (2b)</td></tr>
<tr><td>If $\varphi_{IF,t} < 1 \Rightarrow$ go to step (2e) (heat source) (2c)</td><td>if $\varphi_{IF,t} < 1 \Rightarrow$ go to step (3) (c calculation) (2c)</td></tr>
<tr><td colspan="2">If $\varphi_{IF,t} \geq 1$ then: $\phi_{IF,t}' = 1$, change the state of node from interface to ferrite and the neighboring nodes from austenite to interface (according to neighborhood), set the value of the fraction for the new interface: $\Delta\varphi_{nI} = (\varphi_{IF,t} - \varphi_{IF,t}')/numA = \varphi_{IF,nI}$; $numA$—the number of nodes in phase austenite in the neighborhood of old interface, the $\Delta\varphi$ for the old interface (new ferrite) node: $\Delta\varphi_{oI} = 1 - \varphi_{IF,t-1}$; $\varphi_{IF,oI} = 1$; $\varphi_{IA,oI} = 0$ (2d)</td></tr>
<tr><td>Heat source calculation:
if $\varphi_t < 1 \Rightarrow Q_I = k_q\Delta\varphi_I$
if $\varphi_t \geq 1 \Rightarrow Q_{nI} = k_q\Delta\varphi_{nI}$; $Q_{oI} = k_q\Delta\varphi_{oI}$ (2e)</td><td></td></tr>
<tr><td>Determination of new temperature:
$T_{x,y,z} = \Sigma f_i + Q_{x,y,z}$ (3)</td><td>Determination of new concentration: $c = \Sigma f_i$ (3)</td></tr>
<tr><td colspan="2">f^{eq} calculations (4)</td></tr>
<tr><td colspan="2">Collision (5)</td></tr>
<tr><td colspan="2">Streaming (6)</td></tr>
<tr><td colspan="2">Boundary conditions (7)</td></tr>
</table>

Figure 12 presents the examples of modeling results for the selected values of the parameters defined in the heat flow algorithm developed: k_v and k_q values. As can be seen, these parameters have a direct impact on the rate of transformation. The k_v is used directly for boundary velocity calculations, while k_q determines the rate of temperature increase in the interface and has an impact on the boundary velocity. The growth of the grain placed in the center of the grid was modelled with the use of the Moore neighborhood and the vector normal to the surface. The calculations were performed on the NVIDIA GeForce GTX 1060 graphics card with the use of CUDA parallel programming. D3Q19 LBM scheme was used for the calculations. The main simulation parameters are as follows: $T_{x,y,z} = 750$ °C, $T_P = 800$ °C, $n_x \times n_y \times n_z = 128 \times 128 \times 128$, *steps* = 100, $\Delta x = 1$, $\Delta y = 1$, $\Delta z = 1$, $\Delta t = 1$, $\tau = 1$, $k_v = 0.003$, $k_q = 15$, and the bounce-back boundary conditions.

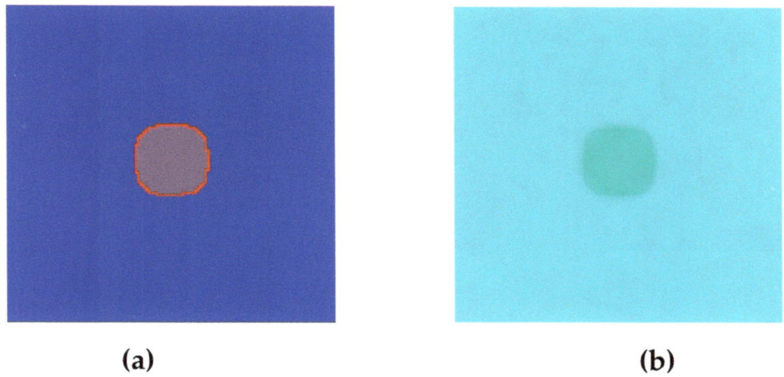

Figure 12. Growth of the new phase in the selected plane: (**a**) CA states and (**b**) temperature distribution.

The concept of the model and the first 1D modeling results of carbon diffusion and heat flow can be found elsewhere [31,32]. The results of 2D modeling can be found in subsequent work carried out in this area [33].

4.5. Modeling of Technological Processes—Rolling Processes

The search for modern technological solutions for the rolling process, which refers to the acquisition of higher-quality rolled products and reducing the cost of their production, becomes a necessity in the modern stage of development of this technological process. Often, during rolling, it is difficult to simultaneously ensure product dimensions and shape and the required microstructure, which is responsible for the final mechanical properties. The intensive development of computer technology makes it easier to use modern numerical methods for the simulation of complex forming processes. The developed modeling system can be applied effectively to model microstructure evolution in the rolling processes of various materials. General characteristics of flat rolling is required for the modeling of the process (Figure 13). Analytical solutions and advanced numerical models can be used to obtain the process parameters and their average values. FCA modeling is based on the information about the duration of plastic deformation, the time intervals between the deformations, temperature, strain rate, etc.

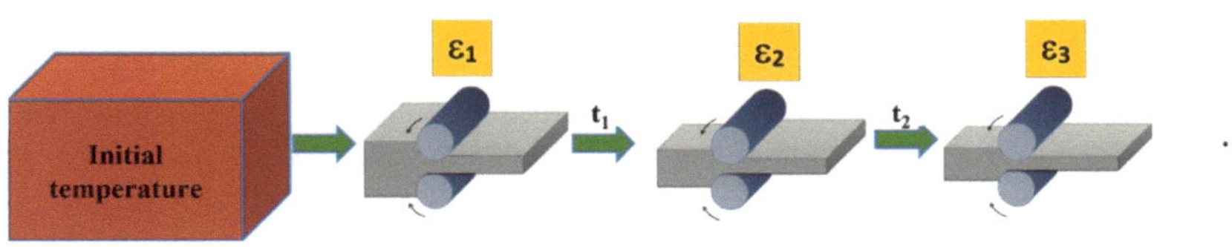

Figure 13. Flat rolling scheme.

The developed modeling system can be adapted to model various materials. The overall model of this process has been comprehensively verified, for example, for modeling the evolution of the microstructure in AISI 304L stainless steel [34].

Simulations of microstructure evolution can be a part of more complex modeling system or performed independently. It can be realized in three modeling stages (Figure 14): design of rolling schedule, finite element method simulation, and frontal cellular automata simulation of microstructure evolution. The second and third stages can be repeated several times.

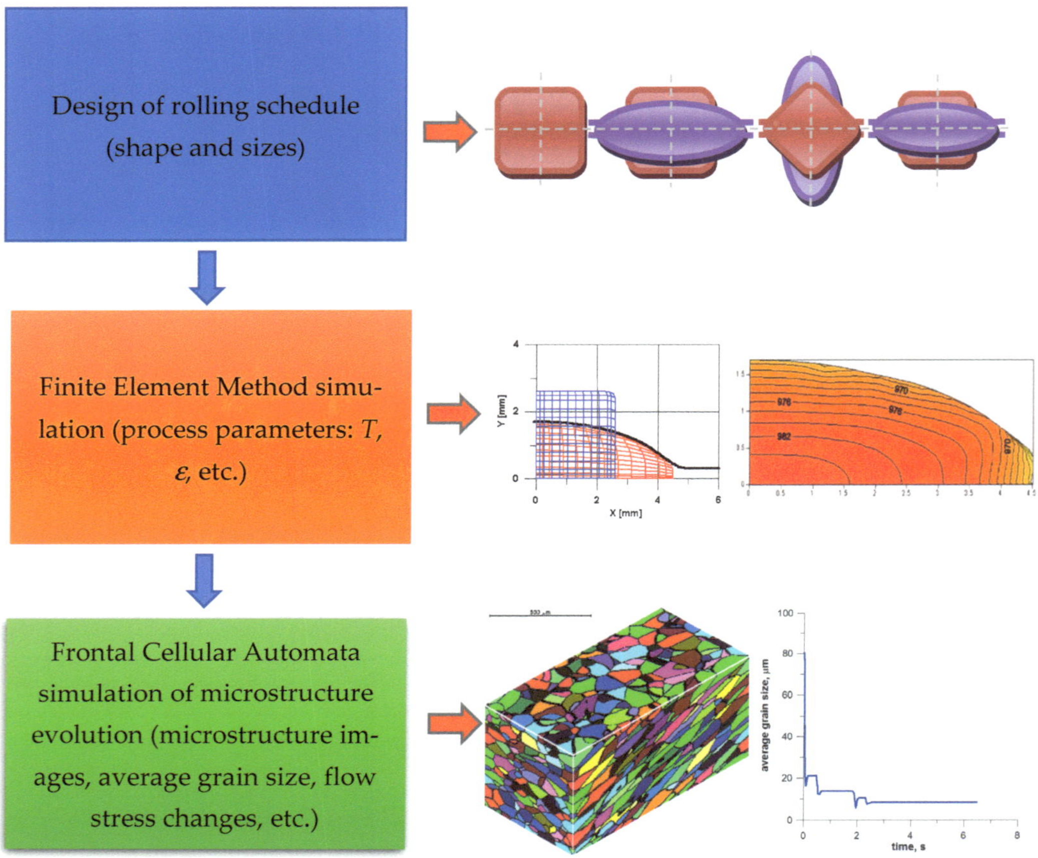

Figure 14. The general scheme of shape rolling modeling.

A detailed description of first stage and the designed schedule can be found elsewhere [35]. The examples of simulation results of the second stage can be found in detail elsewhere [36]. The third stage includes the FCA simulation of the microstructure evolution, which can be performed for any representative point (for which data can be obtained) using any FEM code. For the calculation, information on the time, temperature, and strain rate tensor are used. More information on this stage can be found elsewhere [37]. The developed shape rolling scheme can be applied, in principle, to any material. Simulation, prediction, and validation of the evolution of microstructures in AISI 304L stainless steel during the shape rolling process were presented in one of the unpublished works [38]. Other results can be found elsewhere [39].

4.6. Grain Refinement and Modeling of Severe Plastic Deformation

The main principles of severe plastic deformation (SPD) modeling based on grain refinement using the presented FCA platform are described in [40] and their references. Then, several SPDs were modeled: accumulative roll bonding process and MaxStrain technology [41,42].

Another innovative combined metal forming process, which can be treated as one of the SPD techniques, consists of three different modes of deformation: asymmetric drawing with bending, namely accumulated angular drawing, wire drawing, and wire flattening, which was also modeled on the FCA platform [43].

The evolution around highly reactive interfaces in the processing of nanocrystallized multilayered metallic materials has been investigated and discussed in [44].

5. LBM-Based Platform for Modeling of Advanced Additive Manufacturing Characterized by the Changes of State of Matter

Authors took part in a development of a new modeling platform presented firstly in [45]. Additive manufacturing (AM) technologies that contain phase transitions are a main subject for the three-dimensional simulation on the platform. It is based on LBM with CA elements and was primarily oriented on selective laser melting (SLM). An example of calculation of the model parameters for the real material (Ti-6Al-4V alloy) and the real process is presented in [45]. It contains also the first quantitative results. It can be used for analysis of new multipasses and multimaterials SLM processes and can be served for computer-aided design. The principles and evolution of the development platform can be found by tracking previous publications presented in the references of the cited publications [45]. Other aspects of this LBM-CA platform can be found elsewhere [46,47]. Some essential details of the platform are presented below.

Figure 15 represents the proposed scheme, where the process is divided into different stages according to the associated physical phenomena, which are related to the corresponding mathematical models.

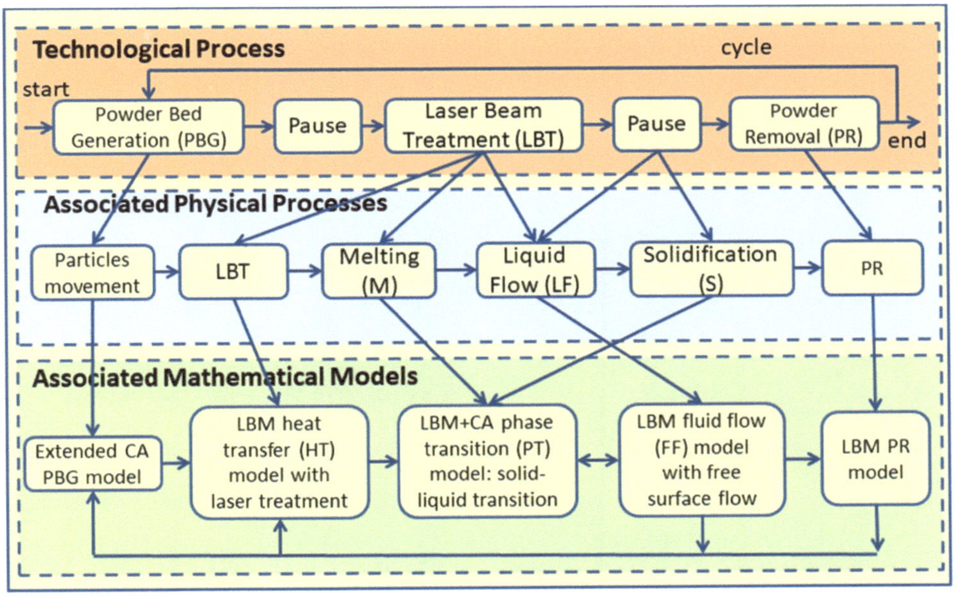

Figure 15. Schematic representation of interconnection between analyzed process physical phenomena and models [45].

The powder bed generation (PBG) model with some results was published in [46,48]. The first a simplified model of heat transfer was presented in [48,49]. The CA method was the basis of these two models. For the other models, the LBM was used [45,47,49].

In this paper, we present a model of powder particles removal that was developed on the general principles of the platform with the use of LBM, as in Section 3. It can be also used for optimization of micro vacuum material removal system. The main LBM blocks for gas flow in this model are the same (Figure 3). The same velocity model (D2Q9) is applied. The model contains equations that describe the particles motion in the gas. The new horizontal v_x and vertical v_y velocity components are calculated as follows:

$$v_x^{\mathrm{P}}(t+1) = mv_x^{\mathrm{P}}(t) + (1-m)v_x(t) \tag{25}$$

$$v_y^{\mathrm{P}}(t+1) = mv_y^{\mathrm{P}}(t) + (1-m)v_x(t) - g \tag{26}$$

where m—particle mass; g—gravity. The mass m and gravity g are dimensionless in the models. The mass m varies from zero (m = 0, very small, light particles) to unity (m = 1, a very heavy particle). The gravity g depends on the modeling parameters: lattice size, time step, and particle size. For very small particles, when the gas resistance is high, g = 0. For big, very heavy particles, the lattice size and time step define the gravity, $g = g_{\max}$.

The location of a particle is a continuous variable, while the velocity field is discretized. A new position is calculated according to simple Euler's integration scheme:

$$x^{\mathrm{P}}(t+1) = x^{\mathrm{P}}(t) + \frac{u_x^{\mathrm{P}}(t+1) + u_x^{\mathrm{P}}(t)}{2} \tag{27}$$

$$y^{\mathrm{P}}(t+1) = y^{\mathrm{P}}(t) + \frac{u_y^{\mathrm{P}}(t+1) + u_y^{\mathrm{P}}(t)}{2} \tag{28}$$

An analysis of simulation results shows that particle removal with the use of only a vacuum function is highly dependent on the gap. When the gap is wider, the effectiveness is lesser, and the control of the gap is difficult. To enforce movement in the opposite direction and turbulences, an additional inflow channel with much higher gas pressure was proposed. An example of a simulation result is presented in Figure 16. The modelling space was 128 × 128 cells or nodes. To make this scheme more effective, strong gas stream is forced in the horizontal direction. Other designs of removal systems can be analyzed by this model.

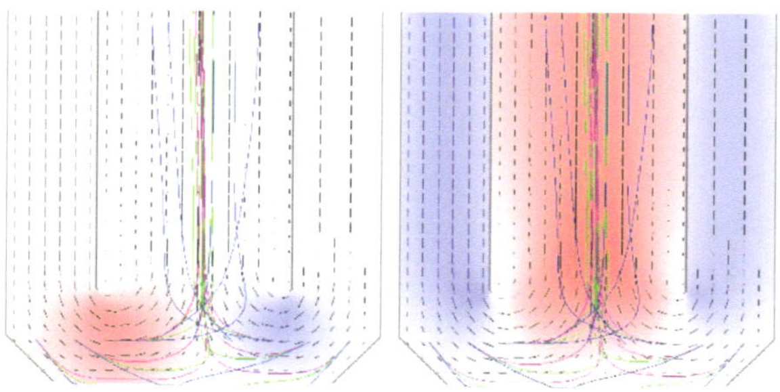

Figure 16. Modeling results showing inflows and outflows in the removal equipment.

In Figure 17, examples of two cases of SLM simulation are presented. The modeling space was 128 × 80 nodes. The presented results were obtained with the use of sequential FORTRAN code on PC with Intel Core i7-3930 K, and calculations lasted about 3 h. The similar simulations with parallel computations on GPU with CUDA technology using the graphic card GeForce 1060 with 1280 CUDA cores lasted about 1.5 min, that is, about 100 times faster. The solid material is presented in gray and black, while the liquid material is blue. The laser beam is shown by the green color. The intensity of heat transfer from the laser beam to the materials is correspondent to intensity of pink and red colors. The case of one material is presented in Figure 17a, whereas the case of two different materials is presented in Figure 17b. It can be seen that the processing of two materials with a significant difference in material properties is accompanied by the problem of melting the material with a higher melting temperature, and solving this problem is one of the main goals of here-presented development platform.

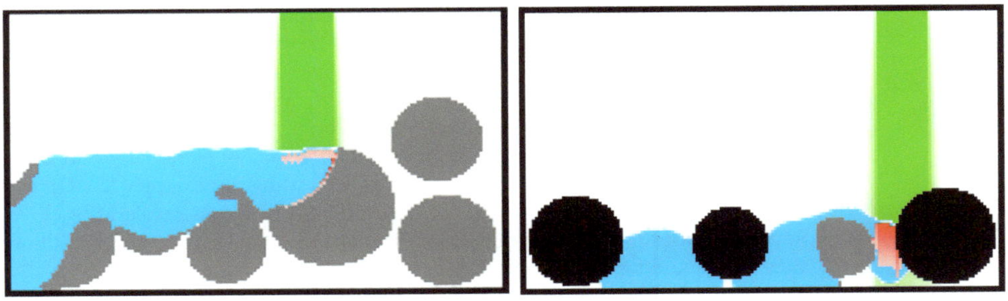

(a) (b)

Figure 17. Simulation results of the SLM process: (**a**) one material and (**b**) two different materials.

An example of modeling a 3D selective laser melting is presented in Figure 18. The following conditions were assumed for the calculations: the average particle size is about 35–40 μm, the model space is $x \times y \times z = 256 \times 128 \times 64$ cells, the laser travel velocity is 1 m/s, and the laser power was equal to 200 W. Three passes are presented. The gray color represents particles of powder and solidified material. The blue color corresponds to the liquid phase of the material. The laser beam is shown in green. Other examples can be found elsewhere [45].

(a) (b) (c)

Figure 18. 3D perspective views of the modeled SLM process with three passes: (**a**) the first, (**b**) the second, and (**c**) the third.

6. Conclusions

The paper presents basic information about modeling platforms based on relatively new and little-known methods for modeling phenomena and processes in materials science and engineering FCA and LBM.

The first platform for modeling microstructure evolution was first based on FCA, which is very suitable for modeling processes of multiple nucleations and grain growth in materials and allows for algorithmic acceleration of calculations with maintaining high accuracy. Then, this platform was supplemented by the LBM, which is widely used not only in hydrodynamics but also for modeling heat flow or diffusion processes. LBM can also become the basic method for modeling complex processes and phenomena containing not only flows but also different transformations of both the first and second type: both state and phase transformations, including flow, diffusion, heat transfer, latent heat, and many other phenomena that are difficult to consider with the use of other methods. The components of the systems and their role in the comprehensive modeling of various processes are shown. The first platform allows modeling of recrystallization, crystallization, and phase transformation, and it is mainly oriented to model forming processes (flat, shape rolling, drawing, SPD, etc.). The performance of the systems was verified in application to specific materials and process parameters, and the selected modeling results are presented.

The large variety of modeling phenomena creates opportunities for the expansion of complex models that are homogeneous and do not need multiscale modeling with a complex slow-acting interface between different methods. This aspect eliminates the possibility of efficient and fast modeling. An advantage of using FCA and LBM is the ease of parallelization of calculations on modern GPUs using, for example, CUDA software. The use of several or tens of thousands of GPU processors allows accelerating the calculations several hundred times. The developed systems also take into account solutions based on parallel calculations, and examples of the results are shown.

The second platform for the modeling, design, and optimization of additive manufacturing is in development and is based on LBM and CA. Elimination of complicated interfaces allows for modeling of the manufacturing process within the single integrated platform. The platform is ready to be used for computer-aided analysis, optimization, and design of multimaterial, multipass SLM cycles.

The second platform can be adapted for several other additive manufacturing (AM) technologies that contain phase transitions (melting-solidification). Mainly, they are included in the group of powder bed fusion techniques (PBF): DMLS (direct metal laser sintering), SLS (selective laser sintering), MJF (multi-jet fusion), and EBM (electron beam melting). It can be used for modeling other AM, for example, WAAM (wire arc additive manufacturing) and FFF (fused filament fabrication), also known as FDM (fused deposition modeling), etc. However, it requires additional efforts for the adaptation of existing models and the development of missed ones.

The second platform can be expanded with modeling by CA (FCA) of microstructure formation and evolution during the solidification.

Author Contributions: Conceptualization, Ł.Ł. and D.S.; methodology, Ł.Ł. and D.S.; software, Ł.Ł. and D.S.; validation, Ł.Ł. and D.S.; formal analysis, Ł.Ł. and D.S.; investigation, Ł.Ł. and D.S.; data curation, Ł.Ł. and D.S.; writing—original draft preparation, Ł.Ł.; writing—review and editing, D.S.; visualization, Ł.Ł. and D.S.; funding acquisition, Ł.Ł. and D.S. All authors have read and agreed to the published version of the manuscript.

Funding: Financial support of the Polish National Science Centre (grant no. 2018/31/B/ST8/00622 and 2016/21/D/ST8/01690) is greatly appreciated.

Institutional Review Board Statement: Not applicable.

Informed Consent Statement: Not applicable.

Data Availability Statement: Not applicable.

Conflicts of Interest: The authors declare no conflict of interest.

References

1. Ansari, T.Q.; Huang, H.; Shi, S.Q. Phase field modeling for the morphological and microstructural evolution of metallic materials under environmental attack. *npj Comput. Mater.* **2021**, *7*, 143. [CrossRef]
2. DeWitt, S.; Thornton, K. Phase field modeling of microstructural evolution. In *Computational Materials System Design*; Springer International Publishing: Berlin, Germany, 2017; pp. 67–87, ISBN 9783319682808.
3. Tourret, D.; Liu, H.; LLorca, J. Phase-field modeling of microstructure evolution: Recent applications, perspectives and challenges. *Prog. Mater. Sci.* **2022**, *123*, 100810. [CrossRef]
4. Alvarado, K.; Florez, S.; Flipon, B.; Bozzolo, N.; Bernacki, M. A level set approach to simulate grain growth with an evolving population of second phase particles. *Model. Simul. Mater. Sci. Eng.* **2021**, *29*, 035009. [CrossRef]
5. Lin, S.; Gan, Z.; Yan, J.; Wagner, G.J. A conservative level set method on unstructured meshes for modeling multiphase thermo-fluid flow in additive manufacturing processes. *Comput. Methods Appl. Mech. Eng.* **2020**, *372*, 113348. [CrossRef]
6. Gibou, F.; Fedkiw, R.; Osher, S. A review of level-set methods and some recent applications. *J. Comput. Phys.* **2018**, *353*, 82–109. [CrossRef]
7. Pauza, J.G.; Tayon, W.A.; Rollett, A.D.; Adam, K.; Zöllner, D.; Field, D.P. 3D microstructural evolution of primary recrystallization and grain growth in cold rolled single-phase aluminum alloys. *Model. Simul. Mater. Sci. Eng.* **2018**, *26*, 035011. [CrossRef]

8. Maazi, N.; Lezzar, B. An efficient Monte Carlo Potts method for the grain growth simulation of single-phase systems. *Comput. Methods Mater. Sci.* **2020**, *20*, 85–94. [CrossRef]
9. Lee, J.H.; Lee, G.Y.; Rha, J.J.; Kim, J.H.; Cho, J.H. Analysis of Microstructure Evolution and Mechanical Properties during Compression of Open-Cell Ni-Foams with Hollow Struts Using Micro-CT and FEM. *Materials* **2021**, *15*, 124. [CrossRef]
10. Ozsoy, I.B.; Babacan, N. Finite element simulations of microstructure evolution in stress-induced martensitic transformations. *Int. J. Solids Struct.* **2016**, *81*, 361–372. [CrossRef]
11. Baykasoğlu, C.; Akyildiz, O.; Tunay, M.; To, A.C. A process-microstructure finite element simulation framework for predicting phase transformations and microhardness for directed energy deposition of Ti6Al4V. *Addit. Manuf.* **2020**, *35*, 101252. [CrossRef]
12. Li, L.; Wang, L.Y. Simulation of Dynamic Recrystallization Behavior under Hot Isothermal Compressions for as-extruded 3Cr20Ni10W2 Heat-Resistant Alloy by Cellular Automaton Model. *High Temp. Mater. Process.* **2018**, *37*, 635–647. [CrossRef]
13. Zhang, Y.; Zhang, J. Modeling of solidification microstructure evolution in laser powder bed fusion fabricated 316L stainless steel using combined computational fluid dynamics and cellular automata. *Addit. Manuf.* **2019**, *28*, 750–765. [CrossRef]
14. Gu, C.; Lu, Y.; Ridgeway, C.D.; Cinkilic, E.; Luo, A.A. Three-dimensional cellular automaton simulation of coupled hydrogen porosity and microstructure during solidification of ternary aluminum alloys. *Sci. Rep. 2019 91* **2019**, *9*, 13099. [CrossRef] [PubMed]
15. Bressan, J.D.; Martins, M.M.; Button, S.T. Analysis of metal extrusion by the Finite Volume Method. *Procedia Eng.* **2017**, *207*, 425–430. [CrossRef]
16. Jin, Y.; Zhao, J.; Zhang, C.; Luo, J.; Wang, S. Research on Neural Network Prediction of Multidirectional Forging Microstructure Evolution of GH4169 Superalloy. *J. Mater. Eng. Perform.* **2021**, *30*, 2708–2719. [CrossRef]
17. Yang, K.; Cao, Y.; Zhang, Y.; Fan, S.; Tang, M.; Aberg, D.; Sadigh, B.; Zhou, F. Self-supervised learning and prediction of microstructure evolution with convolutional recurrent neural networks. *Patterns* **2021**, *2*, 100243. [CrossRef]
18. Koepf, J.A.; Soldner, D.; Ramsperger, M.; Mergheim, J.; Markl, M.; Körner, C. Numerical microstructure prediction by a coupled finite element cellular automaton model for selective electron beam melting. *Comput. Mater. Sci.* **2019**, *162*, 148–155. [CrossRef]
19. Lian, Y.; Gan, Z.; Yu, C.; Kats, D.; Liu, W.K.; Wagner, G.J. A cellular automaton finite volume method for microstructure evolution during additive manufacturing. *Mater. Des.* **2019**, *169*, 107672. [CrossRef]
20. Zhao, P.; Piao, R. Lattice Boltzmann Method Modeling of the Evolution of Coherent Vortices and Periodic Flow in a Continuous Casting Mold. *Metals* **2022**, *12*, 572. [CrossRef]
21. Zhu, J.; Ma, J. Extending a Gray Lattice Boltzmann Model for Simulating Fluid Flow in Multi-Scale Porous Media. *Sci. Rep. 2018 81* **2018**, *8*, 1–19. [CrossRef]
22. Rátkai, L.; Pusztai, T.; Gránásy, L. Phase-field lattice Boltzmann model for dendrites growing and moving in melt flow. *npj Comput. Mater.* **2019**, *5*, 113. [CrossRef]
23. Svyetlichnyy, D.S. Modelling of the microstructure: From classical cellular automata approach to the frontal one. *Comput. Mater. Sci.* **2010**, *50*, 92–97. [CrossRef]
24. Mohamad, A.A. *Lattice Boltzmann Method Fundamentals and Engineering Applications with Computer Codes*; Springer London: London, UK, 2011; ISBN 978-0-85729-454-8.
25. Svyetlichnyy, D.S. Modeling of Macrostructure Formation during the Solidification by using Frontal Cellular Automata. In *Cellular Automata—Innovative Modelling for Science and Engineering*; Salcido, A., Ed.; InTech: London, UK, 2011; pp. 179–196, ISBN 978-953-307-172-5.
26. Łach, Ł. Modeling of Microstructure Evolution during Deformation Processes by Cellular Automata—Boundary Conditions and Space Reorganization Aspects. *Materials* **2021**, *14*, 1377. [CrossRef] [PubMed]
27. Svyetlichnyy, D.S.; Łach, Ł. Digital material representation of given parameters. In Proceedings of the 14th international conference on Metal Forming, Krakow, Poland, 16–19 September 2012; pp. 1151–1154.
28. Svyetlichnyy, D.S. A three-dimensional frontal cellular automaton model for simulation of microstructure evolution—Initial microstructure module. *Model. Simul. Mater. Sci. Eng.* **2014**, *22*, 085001. [CrossRef]
29. Svyetlichnyy, D.S.; Nowak, J.; Łach, Ł. Modeling of Recrystallization with Recovery by Frontal Cellular Automata. *Lect. Notes Comput. Sci.* **2012**, *47*, 494–503. [CrossRef]
30. Svyetlichnyy, D.S. Simulation of Microstructure Evolution during Shape Rolling with the Use of Frontal Cellular Automata. *ISIJ Int.* **2012**, *52*, 559–568. [CrossRef]
31. Łach, Ł.; Svyetlichnyy, D. Development of hybrid model for modeling of diffusion phase transformation. *Eng. Comput.* **2020**, *37*, 2761–2783. [CrossRef]
32. Łach, Ł.; Straka, R.; Svyetlichnyy, D. Development of hybrid model based on Lattice Boltzmann Method and Cellular Automata devoted for phase transformation—Simulation of heat flow with consideration of enthalpy. *MATEC Web Conf.* **2018**, *240*, 01020. [CrossRef]
33. Łach, Ł.; Svyetlichnyy, D.; Straka, R. Heat flow model based on lattice Boltzmann method for modeling of heat transfer during phase transformation. *Int. J. Numer. Methods Heat Fluid Flow* **2020**, *30*, 2255–2271. [CrossRef]
34. Łach, Ł.; Nowak, J.; Svyetlichnyy, D. The evolution of the microstructure in AISI 304L stainless steel during the flat rolling—Modeling by frontal cellular automata and verification. *J. Mater. Process. Technol.* **2018**, *255*, 488–499. [CrossRef]
35. Łach, Ł.; Svyetlichnyy, D. Multiscale Model of Shape Rolling Taking into Account the Microstructure Evolution—Schedule Design by Finite Element Method. *Adv. Mater. Res.* **2014**, *871*, 263–268. [CrossRef]

36. Łach, Ł.; Svyetlichnyy, D. Multiscale Model of Shape Rolling Taking into Account the Microstructure Evolution—Finite Element Modeling. *Adv. Mater. Res.* **2014**, *1025–1026*, 379–384. [CrossRef]
37. Łach, Ł.; Svyetlichnyy, D. Multiscale Model of Shape Rolling Taking into Account the Microstructure Evolution—Frontal Cellular Automata. *Adv. Mater. Res.* **2014**, *998–999*, 545–548. [CrossRef]
38. Łach, Ł.; Nowak, J.; Svyetlichnyy, D. Experimental study and modeling of microstructure evolution in AISI 304L stainless steel during the shape rolling. *Metall. Mater. Trans. A* **2022**. *submitted for publication*.
39. Svyetlichnyy, D.S.; Kustra, P.; Milenin, A. Modeling with FCA-based model of microstructure evolution of MgCa08 alloy during drawing of thin wire in heated die. *Arch. Metall. Mater.* **2015**, *60*, 2721–2727. [CrossRef]
40. Svyetlichnyy, D.S. Modeling of grain refinement by cellular automata. *Comput. Mater. Sci.* **2013**, *77*, 408–416. [CrossRef]
41. Svyetlichnyy, D.S.; Majta, J.; Kuziak, R.; Muszka, K. Experimental and modelling study of the grain refinement of Fe-30wt%Ni-Nb austenite model alloy subjected to severe plastic deformation process. *Arch. Civ. Mech. Eng.* **2021**, *21*, 20. [CrossRef]
42. Svyetlichnyy, D.S.; Muszka, K.; Majta, J. Three-dimensional frontal cellular automata modeling of the grain refinement during severe plastic deformation of microalloyed steel. *Comput. Mater. Sci.* **2015**, *102*, 159–166. [CrossRef]
43. Majta, J.; Madej, Ł.; Svyetlichnyy, D.S.; Perzyński, K.; Kwiecień, M.; Muszka, K. Modeling of the inhomogeneity of grain refinement during combined metal forming process by finite element and cellular automata methods. *Mater. Sci. Eng. A* **2016**, *671*, 204–213. [CrossRef]
44. Bajda, S.; Svyetlichnyy, D.; Retraint, D.; Krzyzanowski, M. Numerical modelling of grain refinement around highly reactive interfaces in processing of nanocrystallised multilayered metallic materials by duplex technique. *Int. J. Adv. Manuf. Technol.* **2018**, *96*, 2893–2905. [CrossRef]
45. Svyetlichnyy, D.S. Development of the Platform for Three-Dimensional Simulation of Additive Layer Manufacturing Processes Characterized by Changes in State of Matter: Melting-Solidification. *Materials* **2022**, *15*, 1030. [CrossRef] [PubMed]
46. Svyetlichnyy, D.; Krzyzanowski, M.; Straka, R.; Lach, L.; Rainforth, W.M. Application of cellular automata and Lattice Boltzmann methods for modelling of additive layer manufacturing. *Int. J. Numer. Methods Heat Fluid Flow* **2018**, *28*, 31–46. [CrossRef]
47. Svyetlichnyy, D.; Krzyzanowski, M. Development of Holistic homogeneous model of selective laser melting based on Lattice Boltzmann method: Qualitative simulation. *AIP Conf. Proc.* **2020**, *2293*, 420100. [CrossRef]
48. Krzyzanowski, M.; Svyetlichnyy, D.; Stevenson, G.; Rainforth, W.M. Powder bed generation in integrated modelling of additive layer manufacturing of orthopaedic implants. *Int. J. Adv. Manuf. Technol.* **2016**, *87*, 519–530. [CrossRef]
49. Krzyzanowski, M.; Svyetlichnyy, D. A multiphysics simulation approach to selective laser melting modelling based on cellular automata and lattice Boltzmann methods. *Comput. Part. Mech.* **2022**, *9*, 117–133. [CrossRef]

materials

MDPI

Article

Equilibrium, Kinetic, and Thermodynamic Studies on Adsorption of Rhodamine B from Aqueous Solutions Using Oxidized Mesoporous Carbons

Michal Marciniak [1], Joanna Goscianska [1], Małgorzata Norman [2], Teofil Jesionowski [2], Aleksandra Bazan-Wozniak [1] and Robert Pietrzak [1,*]

[1] Faculty of Chemistry, Adam Mickiewicz University in Poznań, Uniwersytetu Poznańskiego 8, 61-614 Poznan, Poland
[2] Faculty of Chemical Technology, Poznan University of Technology, Berdychowo 4, 60-965 Poznan, Poland
* Correspondence: pietrob@amu.edu.pl; Tel.: +48-61-829-1560

Abstract: Oxidized mesoporous carbon $C_{SBA\text{-}15}$, obtained by the hard method, was applied to remove rhodamine B from the aqueous system. The process of carbon oxidation was performed using 0.5 and 5 M of nitric (V) acid solution at 70 and 100 °C. Functionalization of mesoporous carbon with HNO_3 solutions led to reduction in the surface area, pore volume, and micropore area, however, it also led to an increased number of oxygen functional groups of acidic character. The functional groups probably are located at the entrance of micropores, in this way, reducing the values of textural parameters. Isotherms of rhodamine B adsorption indicate that the oxidation of mesoporous carbons resulted in an increase in the effectiveness of the removal of this dye from aqueous solutions. The influence of temperature, pH, and contact time of mesoporous material/rhodamine B on the effectiveness of dye removal was tested. The process of dye adsorption on the surfaces of the materials studied was established to be most effective at pH 12 and at 60 °C. Kinetic studies of the process of adsorption proved that the equilibrium state between the dye molecules and mesoporous carbon materials is reached after about 1 h. The adsorption kinetics were well fitted using a pseudo-second-order model. The most effective in rhodamine B removal was the sample $C_{SBA\text{-}15}$-5-100, containing the greatest number of oxygen functional groups of acidic character. The Langmuir model best represented equilibrium data.

Keywords: ordered mesoporous carbons; hexagonal structure; hard template method; oxidation by nitric acid; cationic dye; adsorption

Citation: Marciniak, M.; Goscianska, J.; Norman, M.; Jesionowski, T.; Bazan-Wozniak, A.; Pietrzak, R. Equilibrium, Kinetic, and Thermodynamic Studies on Adsorption of Rhodamine B from Aqueous Solutions Using Oxidized Mesoporous Carbons. *Materials* **2022**, *15*, 5573. https://doi.org/10.3390/ma15165573

Academic Editor: Antonio Gil Bravo

Received: 1 July 2022
Accepted: 10 August 2022
Published: 13 August 2022

1. Introduction

Water, which is the element necessary to support all forms of life, covers most of Earth's surface. Its pollution is negative to human health and the entire hydrosphere [1–5]. There is a constant increase in the content of organic compounds in wastewater, especially in post-production waste produced by different types of industry [6,7], such as the textile industry, reaching several thousand tons a year, of which 10–20% is lost in the process of production [8–10]. Dyes are usually well soluble in water and resistant to degradation by various methods (biological, physical, and chemical) [9–12]. Therefore, they must be removed from the wastewater. At present, a few methods are used to remove organic compounds [12–14]. From among newly proposed methods, very attractive seems to be the method of membrane separation, offering simplicity of operation and low energy consumption [15]. On the other hand, membrane efficiency decreases with time because of the fouling of pollutant particles on the membrane surface and inside membrane pores. Moreover, molecules of many organic compounds are smaller than the membrane pores and the process of membrane separation is ineffective [16–18]. Currently, the most effective are the adsorption methods that can be used for the removal of organic and inorganic compounds from water solutions. The properties and quality of the adsorbent influence the adsorption efficiency. Many materials, for example, silica gels [19], alumina [20], and zeolite molecular sieves [21,22] are rather ineffective in the adsorption process. Activated carbons are used as organic pollutant adsorbents, although the presence of micropores in their structure limits the penetration of larger dye molecules into the pores. Recently, much attention was paid to ordered mesoporous carbon materials, as they show unique properties, such as thermal and mechanical stability, as well as good textural parameters [23]. These materials are easily modified, which allows for improved adsorption properties towards a range of pollutants [23–26]. Many methods for carbon material modification were proposed, but the most effective is that of wet oxidation, in which oxygen functional groups are introduced on the carbon surface [27]. A number of authors studied the adsorption of organic compounds on carbon materials, for instance, Liu et al. [28] proposed the removal of Acid Red 73 and Reactive Black 5 by CMK-3, CMK-5, and its carbon/silica composite Si-CMK-5, with different pore structures. The greatest sorption capacity towards these dyes was shown by CMK-5, which was related to double pore systems and large specific surface area. Asouhidou et al. [29] obtained mesoporous carbons, a highly ordered CMK-3 sample with hexagonal structure and a disordered mesoporous carbon (DMC), and tested them in the removal of Remazol Red 3BS, comparing their performance with that of commercial products (Takeda 5A, Calgon, and Norit SAE-2) and a HMS mesoporous silica with a wormhole pore structure. Their results show that the material structure and pore size have a significant impact on the effectiveness of adsorption. The resulting sorption capacities decreased in the order CMK-3 (0.531 mmol/g) > DMC (0.453 mmol/g) > SAE-2 (0.167 mmol/g) > Calgon (0.05 mmol/g) > Takeda 5A = HMS (0.007 mmol/g). The group of Peng [30] successfully applied the ordered mesoporous CMK-3 containing nitrogen functional groups as the adsorbent of Acid Black 1. According to their results, the modification of mesoporous carbon significantly improved its effectiveness in the removal of this dye. The maximum sorption capacities of the initial and functionalized materials were over 270 mg/g and nearly 500 mg/g, respectively.

One of the most popular textile dyes is rhodamine B, which is a cationic synthetic dye of green or red purple crystals [31]. With the release of rhodamine B into the water, several environmental and public health problems are caused [32,33]. Very popular and effective adsorbents of rhodamine B from water solutions are mesoporous carbons [34–36]. Therefore the aim of the study was to analyze the process of rhodamine B removal from water solutions by adsorption on oxidized mesoporous carbons of hexagonal structure. The effects of pH of the dye solution, adsorbent/adsorbate contact time, and temperature of the process were checked on the sorption capacity of obtained mesoporous carbons. The mesoporous carbon materials used as adsorbents were characterized in detail by a number of physicochemical methods.

2. Materials and Methods

2.1. Sample Preparation

2.1.1. Mesoporous Carbon Synthesis

Mesoporous carbon $C_{SBA\text{-}15}$ was obtained by the hard template method using the ordered silica SBA-15 as the solid template and sucrose as the carbon precursor [37,38]. The substrates for SBA-15 preparation were 50 mg of Pluronic P123 ($EO_{20}PO_{70}EO_{20}$; Aldrich, Saint Louis, MO, USA), 19 mL of 1.6 M HCl (Avantor Performance Materials Poland S.A., Gliwice, Poland), and 1.1 mL of TEOS (MERCK KGaA, Darmstadt, Germany). Pluronic P123 was dissolved in a water solution of HCl at 35 °C. To the mixture, TEOS was added dropwise upon stirring continued for 6 h. Then the mixture was subjected to hydrothermal treatment in poly-propylene bottles in a drier (1 day at 35 °C and the next 6 h at 100 °C). Then the material obtained was filtered off and dried (100 °C, 12 h). The template was removed by calcination (550 °C, 8 h). The procedure produced ordered mesoporous silica SBA-15 of hexagonal structure [37,38]. The mesoporous silica SBA-15 was subjected to twice repeated impregnation using sucrose solution. Next, 125 mg sucrose (MERCK KGaA,

Darmstadt, Germany) was dissolved in H_2SO_4 (0.14 mL, Avantor Performance Materials Poland S.A., Gliwice, Poland) and 5 mL of distilled water. The solution was added to the flask with the ordered silica. The contents were heated in the oven (6 h at 100 °C and 6 h at 160 °C). The obtained material was subjected to another impregnation with a solution of sucrose (800 mg), H_2SO_4 (0.09 mL), and distilled water (5 mL).

The composite was then subjected to carbonization by heating for 2 h at 900 °C. The remaining silica was washed out twice with 5% of HF solution (200 mL, Avantor Performance Materials Poland S.A., Gliwice, Poland). The material was filtered off, washed with C_2H_5OH, and dried (12 h at 100 °C) [39].

2.1.2. Sample Functionalization

The $C_{SBA\text{-}15}$ of the hexagonal structure was subjected to oxidation at 70 or 100 °C with the use of HNO_3 at the concentrations of 0.5 or 5 mol/L as the oxidizing agent. Next, 0.5 g of the carbon was placed in a round-bottomed flask and flooded with 30 mL of nitric (V) acid solution. Oxidation was performed under reflux (12 h). Next, the contents were filtered off and the carbon was washed with C_2H_5OH and distilled water. The oxidized carbon materials were labeled as $C_{SBA\text{-}15}$-0.5-70, $C_{SBA\text{-}15}$-5-70, $C_{SBA\text{-}15}$-0.5-100, and $C_{SBA\text{-}15}$-5-100, where 70 and 100 refer to the temperature of oxidation, while 0.5 and 5 to the concentration of HNO_3.

2.2. Analytical Procedures

The texture parameters of the samples obtained were characterized by low-temperature nitrogen adsorption/desorption isotherms measured on a sorptometer Quantachrome AutosorbiQ (Boynton Beach, FL, USA) [9].

X-ray diffraction patterns were obtained on a Bruker AXS DB Advance diffractometer (CuK_α radiation, λ = 0.154 nm, step size 0.02°).

The number of surface oxygen functional groups was determined by the Boehm method [40]. A portion of 0.25 g of adsorbent was placed in 25 mL of 0.1 mol/L solutions of either NaOH or HCl. The vials were sealed and shaken for 24 h and then 10 mL of each filtrate was pipetted and the excess of base or acid was titrated with 0.1 mol/L HCl or NaOH, as required. The numbers of acidic sites of various types were calculated assuming that NaOH neutralizes all acidic groups and HCl reacts with all basic groups.

Structural changes in the oxidized mesoporous carbon materials were determined by FT-IR spectroscopy. The preparation of samples is described in the paper [9]. The study was carried out on a Varian 640-IR spectrometer (Agilent, Santa Clara, CA, USA).

Zeta potential was determined using a Zetasizer Nano ZS instrument equipped with an autotitrator (Malvern Instruments Ltd., Malvern, United Kingdom) [41]. The electrophoretic mobility of the particles was measured and converted to the zeta potential according to the Henry Equation (1):

$$R_{ef} = \frac{2\varepsilon\zeta f(K_a)}{3\eta} \quad (1)$$

where R_{ef} is the electrophoretic mobility, ε the dielectric constant; ζ the electrokinetic (zeta) potential; η the viscosity, and f (K_a) the Henry function. The isoelectric point is a pH value at which zeta potential is zero, the surface has net electrical neutrality. When pH > i_{ep}, the surface charge is negative and pH < i_{ep}, it is positive.

2.3. Adsorption of Rhodamine B

The carbon material portions of 0.02 g were placed in flasks and flooded with 50 mL of a dye solution of a given concentration (25–250 mg/L), and the contents were shaken at 22 ± 1 °C for 24 h. Spectrophotometric measurements were carried out with a spectrometer Cary 100 Bio (Agilent, Santa Clara, CA, USA). Rhodamine B absorbs the irradiation of

λ_{max} = 553 nm. The amount of rhodamine B adsorbed on the oxidized mesoporous carbons was calculated from Equation (2):

$$q_e = \frac{C_0 - C_e}{m} \times V \tag{2}$$

where: C_0—initial rhodamine B concentration (mg/L); C_e—equilibrium rhodamine B concentration (mg/L); m—the mass of mesoporous carbon sample (g); V—volume of rhodamine B solution (L). The experimental adsorption studies were carried out twice and are shown with a standard deviation error.

The effects of pH (CP-401 pH-meter, ELMETRON, Zabrze, Poland) of the dye solutions, temperature, and contact time of the sample/rhodamine B on the sorption capacities of mesoporous carbons were studied.

2.4. Adsorption Modeling

Experimental data were fitted to pseudo-first-order (3) and pseudo-second-order (4) models [9]:

$$ln(q_e - q_t) = lnq_e - \frac{k_1 t}{2.303} \tag{3}$$

where: q_e—sorption capacity of the rhodamine B adsorbed at equilibrium state (mg/g); q_t —sorption capacity of the rhodamine B adsorbed in time (mg/g); and k_1—the rate constant for the pseudo-first order model (min^{-1}).

The pseudo-second-order model can be expressed by the equation:

$$\frac{t}{q_t} = \frac{1}{k_2 {q_e}^2} + \frac{t}{q_e} \tag{4}$$

where: k_2—the rate constant for pseudo-second order model (g/mg·min).

2.5. Thermodynamic Study

Thermodynamic parameters [42–44] were calculated by using the following Equation (5):

$$\Delta G^0 = -RTlnK_d \tag{5}$$

where: ΔG^0—Gibbs free energy ($J \cdot mol^{-1}$); R—universal constant (8.314 $J \cdot mol^{-1} \cdot K^{-1}$); T temperature (K); and K_d—thermodynamic equilibrium constant.

The Gibbs free energy of adsorption (ΔG^0) can be represented by the Equation (6):

$$\Delta G^0 = \Delta H^0 - T\Delta S^0 \tag{6}$$

where: ΔH^0—enthalpy change; ΔS^0—entropy change.

Thermodynamic parameters can be also calculated from Equation (7):

$$lnK_d = \frac{\Delta S^0}{R} + \frac{\Delta H^0}{RT} \tag{7}$$

ΔH^0 and ΔS^0 parameters were calculated (7) from the slope and intercept of the plot of lnK_d versus 1/T yields, respectively.

2.6. Adsorption Isotherms

In our work, we used the Langmuir and Freundlich models to explain the mechanism of rhodamine B adsorption on oxidized carbon materials [45,46].

The linear equation of Langmuir isotherm (8) is represented as follows:

$$\frac{C_e}{q_e} = \frac{1}{K_L \times q_{max}} + \frac{C_e}{q_{max}} \tag{8}$$

where: C_e—equilibrium concentration of rhodamine B (mg/L); q_e—sorption capacity of rhodamine B adsorbed onto the adsorbent at equilibrium (mg/g); q_m—maximum monolayer adsorption capacity of adsorbent (mg/g); and K_L—Langmuir adsorption constant (L/mg).

The Freundlich isotherm is expressed mathematically as [46] (9):

$$lnq_e = lnK_F + \frac{1}{n} lnC_e \tag{9}$$

where: q_e—sorption capacity of rhodamine B adsorbed at equilibrium (mg/g); C_e—equilibrium concentration of rhodamine B (mg/L); K_F—Freundlich adsorption constant $(mg/g \cdot (L/mg)^{1/n})$; and n—Freundlich constant indicates how favorable the adsorption process is.

Experimental data were also fitted to the non-linear Langmuir and Freundlich models.

3. Results and Discussion

3.1. Characterization of Adsorbents

The data on textural parameters are presented in Table 1. The sample of $C_{SBA\text{-}15}$ is characterized by a well-developed surface area (S_{BET} = 1203 m^2/g) and a total pore volume (V_t) of 1.32 cm^3/g. The carbon obtained in this way also has micropores, whose area (S_{micro}) is 682 m^2/g. The process of functionalization of sample $C_{SBA\text{-}15}$ leads to a decrease in these parameters. However, the changes are not uniform and depend on the conditions of the oxidation process. Although sample $C_{SBA\text{-}15}$-5-100 was obtained by treatment with a 5 mol/L solution of HNO_3 at 100 °C, it shows the largest surface area from among all oxidized materials (854 m^2/g). The micropore area and total pore volume of this material are 198 m^2/g and 1.10 cm^3/g, respectively. Most probably in the conditions applied (100 °C, 5 mol/L HNO_3), the micropores are unblocked again and the carbon compounds are removed, which increases the micropore area and total surface area [47]. The smallest S_{BET} of 685 m^2/g, S_{micro} of 106 m^2/g, and V_t of 0.89 cm^3/g were obtained for sample $C_{SBA\text{-}15}$-5-70. The reduced values of textural parameters are most probably caused by the localization of the newly generated oxygen functional groups in micropores, which leads to their blocking and decreases the surface area.

Table 1. Textural parameters of adsorbents obtained [1].

Adsorbent	S_{BET} (m^2/g)	V_t (cm^3/g)	S_{micro} (m^2/g)	Average Pore Diameter (nm)
$C_{SBA\text{-}15}$	1203	1.32	685	4.45
$C_{SBA\text{-}15}$-0.5-70	789	1.12	147	5.66
$C_{SBA\text{-}15}$-5-70	685	0.89	106	5.18
$C_{SBA\text{-}15}$-0.5-100	837	1.04	204	4.91
$C_{SBA\text{-}15}$-5-100	854	1.10	198	5.18

[1] Error range between 2% and 5%.

Recorded in the small angle range XRD, diffractograms of the adsorbents obtained are depicted in Figure 1. Diffractograms of all samples show one intensive peak characteristic of hexagonal pore arrangement [9]. The diffractograms of the initial material $C_{SBA\text{-}15}$ and samples $C_{SBA\text{-}15}$-0.5-100 and $C_{SBA\text{-}15}$-5-100 contain also the reflections in the range $2\Theta \approx 1.7$–2.5°, corresponding to the planes (100), (110), and (200) of P6 mm structure, evidencing good ordering of the materials. No analogous reflections are observed for $C_{SBA\text{-}15}$-0.5-70 and $C_{SBA\text{-}15}$-5-70, which indicates a partial disturbance in the mesoporous structure.

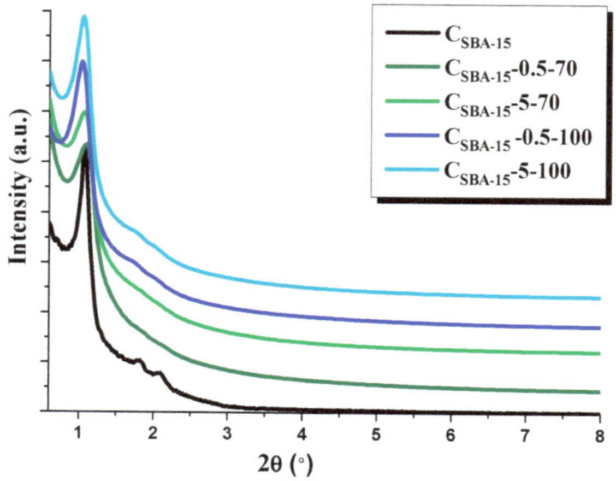

Figure 1. XRD patterns of adsorbents obtained.

Table 2 presents the results obtained from Boehm titration. The oxidation of mesoporous sample C_{SBA-15} with a solution of HNO_3 generates on its surface acidic functional groups whose content depends on the conditions of the process. When using 0.5 mol/L and 5 mol/L HNO_3 solution, the content of basic groups decreased to 0.13 mmol/g for samples C_{SBA-15}-0.5-70 and C_{SBA-15}-0.5-100, and to their total disappearance on samples C_{SBA-15}-5-70 and C_{SBA-15}-5-100. For the acidic functional groups, the results are different. The content of acidic groups in the initial carbon material C_{SBA-15} is 1.09 mmol/g. The content of such groups increases after oxidation and clearly depends on the functionalization conditions, which are the concentration of nitric (V) acid and process temperature. The highest content of the acidic functional groups of 4.88 mmol/g was observed for the sample oxidized with 5 mol/L HNO_3 solution at 100 °C (C_{SBA-15}-5-100). Reduction in the temperature of the process or in the concentration of the oxidizing agent led to lower content of acidic surface functional groups: C_{SBA-15}-0.5-70–2.14 mmol/g, C_{SBA-15}-5-70–3.24 mmol/g, and C_{SBA-15}-0.5-100–3.45 mmol/g.

Table 2. The results obtained from Boehm titration.

Adsorbent	Acidic Groups (mmol/g)	Basic Groups (mmol/g)	Total Content of Acidic and Basic Groups (mmol/g)
C_{SBA-15}	1.09 ± 0.01	0.74 ± 0.01	1.83
C_{SBA-15}-0.5-70	2.14 ± 0.02	0.13 ± 0.01	2.27
C_{SBA-15}-5-70	3.24 ± 0.02	0.00 ± 0.00	3.24
C_{SBA-15}-0.5-100	3.45 ± 0.02	0.13 ± 0.01	3.58
C_{SBA-15}-5-100	4.88 ± 0.03	0.00 ± 0.00	4.88

Figure 2 presents the transmission FT-IR spectra of the initial carbon C_{SBA-15} and oxidized samples of acidic surface nature.

The FT-IR spectra show a rather broad band at about 1200 cm^{-1} that can be assigned to the stretching vibrations of the C–O bond in ethers, acid anhydrides, or phenol. The bands about 1600 cm^{-1} correspond to the stretching vibrations of carbon–carbon bonds in the aromatic ring. There is also a clearly visible band at 3400 cm^{-1}, assigned to the stretching vibrations of O–H bonds in hydroxyl groups, whose presence can indicate the oxidation of mesoporous carbon surface and the presence of carboxyl or phenol oxygen groups [48,49]. For the oxidized materials, a new band appeared at about 1750 cm^{-1}, which was most pronounced for C_{SBA-15}-5-100. This band can be assigned to the carbonyl group of aldehyde, ester, or carboxyl acid. The most probable origin of this band is the presence of -COOH groups on the surface of oxidized carbons, which is supported by the simultaneous presence of a band at about 3400 cm^{-1} [48,49]. The bands of the highest intensity were obtained for sample $C_{SBA\text{-}15}$-5-100, which has the highest content of functional acidic groups. The bands recorded for the other samples were less intensive because of the lower content of functional groups.

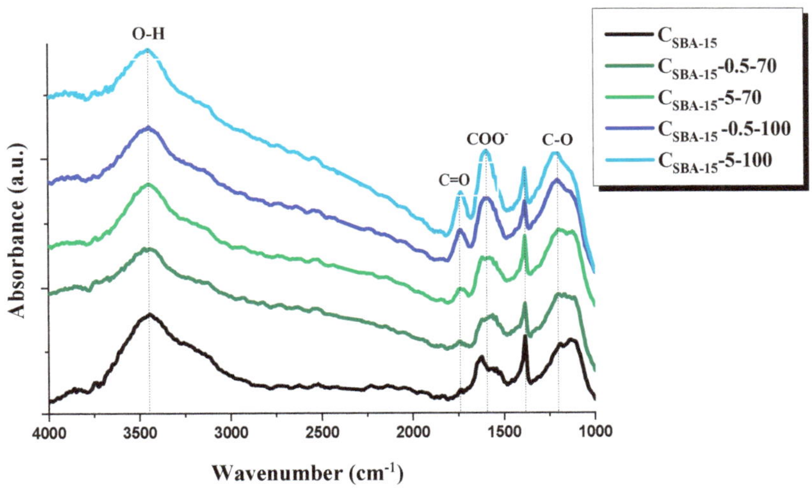

Figure 2. FT-IR spectra of carbon adsorbents obtained.

3.2. Adsorption Studies

In this work, we studied the kinetics of adsorption (Figure 3). The adsorption of rhodamine B on the surface of the carbon materials studied is very fast for the first 10 min. After this time, the majority of the active sites on the carbon surface is already occupied by the dye molecules and the rate of adsorption considerably decreases. After 1 h, no increase q_e was noted, which means that a state of equilibrium was reached and there are no more active sites on the carbon material studied. Next step, the experimental data were fitted to two kinetic models: pseudo-first-order (PFO) and pseudo-second-order models (PSO).

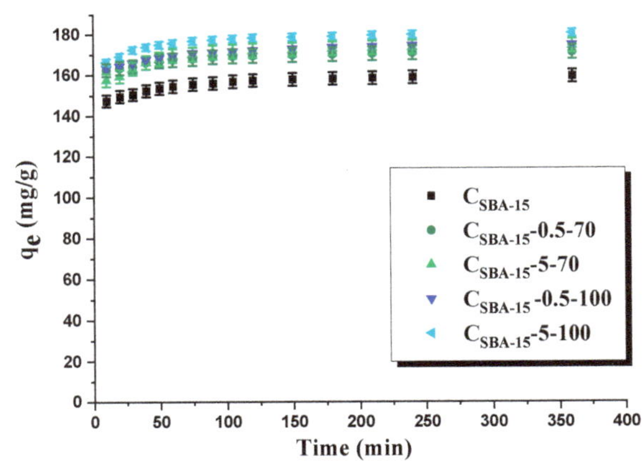

Figure 3. Amount of rhodamine B adsorbed on adsorbents obtained (C_0–75 mg/L).

According to data collected in Table 3, the R^2 for the PFO model takes values from the range 0.9265 to 0.9865. Much higher, and the same for all carbon material values of R^2 (0.9999), were obtained assuming the PSO model. Moreover, the sorption capacities calculated assuming this model are mostly in agreement with their experimental values. Therefore, we conclude that the kinetics of rhodamine B adsorption on the surface of the samples studied can be described by the PSO model.

Table 3. Kinetic models parameters.

Adsorbent	q_e (mg/g)	PFO Model			PSO Model		
		$q_{e\ [cal]}$ (mg/g)	k_1 (min^{-1})	R^2	$q_{e\ [cal]}$ (mg/g)	k_2 (g/mg min)	R^2
$C_{SBA\text{-}15}$	159.41 ± 3.19	9.94	0.019	0.9391	161.29	0.003	0.9999
C_{SBA15}-0.5-70	171.29 ± 3.43	7.35	0.018	0.9265	172.41	0.004	0.9999
$C_{SBA\text{-}15}$-5-70	174.46 ± 3.49	11.93	0.022	0.9253	175.44	0.003	0.9999
$C_{SBA\text{-}15}$-0.5-100	174.90 ± 3.49	11.44	0.026	0.9865	175.45	0.003	0.9999
$C_{SBA\text{-}15}$-5-100	180.71 ± 3.61	10.72	0.023	0.9687	181.81	0.003	0.9999

Figure 4 illustrates the effect of the adsorbate solution pH, changed in the range 2–12, on the amount of adsorbed rhodamine B, while Figure 5A presents the zeta potential curves vs. pH of mesoporous carbons before the dye adsorption, and Figure 5B, after its adsorption. As shown in Figure 5A, sample $C_{SBA\text{-}15}$ has the isoelectric point (i_{ep}) at 3.4. The functionalization changes the adsorbents' surface, for $C_{SBA\text{-}15}$-5-70 the isoelectric point is at pH 2.6 and for $C_{SBA\text{-}15}$-5-100 it does not exist; the zeta potential in the whole range of measurement is negative. The negative zeta potential indicates that oxidation treatment with HNO_3 introduces hydroxyl, carboxylic, and carbonyl groups on the surfaces of the samples, which dissociate generating the negatively charged surface [50,51]. The zeta potential value higher than 25 mV, positive or negative, is indicative of electrokinetic stability [52]. Pristine carbon sample $C_{SBA\text{-}15}$, and the samples after oxidation ($C_{SBA\text{-}15}$-5-70 and $C_{SBA\text{-}15}$-5-100), show good electrokinetic stability at pH values higher than 6 (Figure 5A). The data presented in Figure 4 suggest that the effectiveness of rhodamine B removal from water solutions depends on the pH of the adsorbate solution. The lowest sorption capacity was recorded at pH 2, which can be explained as a result of the protonation of carboxyl, amide, and hydroxyl groups, leading to the generation of a positive charge, which is engaged in repulsive interaction with the positive charge of the cationic dye (rhodamine B). In addition, at low pH, the H^+ cations compete with rhodamine B cations for the adsorption sites. The surface is more negatively charged when the pH increases as a result of deprotonation of the functional groups by the hydroxide anion OH-, which is more favorable for the adsorption of rhodamine B. Then, we have strong electrostatic attraction between the negatively charged surface of the adsorbent and the positively charged cationic dye. At pH 12, the sorption capacities of all adsorbents increased by about 20 mg/g in relation to that of the initial carbon sample. The zeta potential values of $C_{SBA\text{-}15}$-5-100 after adsorption of rhodamine B from the solutions of the concentration of 50 mg/L and 150 mg/L (Figure 5B) differ from those for $C_{SBA\text{-}15}$-5-100 before adsorption, and i_{ep} can be observed (changes in i_{ep} value suggest the chemical adsorption) [53].

The effect of temperature on the amount of rhodamine B adsorbed on the surface of the samples studied was also checked (Figure 6). The amount of rhodamine B adsorbed (q_e) on the surface of mesoporous carbon materials before and after their oxidation increases with temperature. It follows the consequence of increased mobility of dye molecules at higher temperatures. For instance, the amount of organic dye adsorbed on the surface of sample $C_{SBA\text{-}15}$-5-100 is 227 mg/g (25 °C), 254 mg/g (45 °C), and 261 mg/g (60 °C).

According to the data collected in Table 4, the results of $\Delta G°$ prove that the process of adsorption of dye is spontaneous. For all samples, the degree of spontaneity was the highest at 60 °C. The positive values of ΔH obtained reveal an endothermic adsorption process. In addition, the positive values of ΔSevidence increased the degree of randomness at the interface in the process of the rhodamine B adsorption.

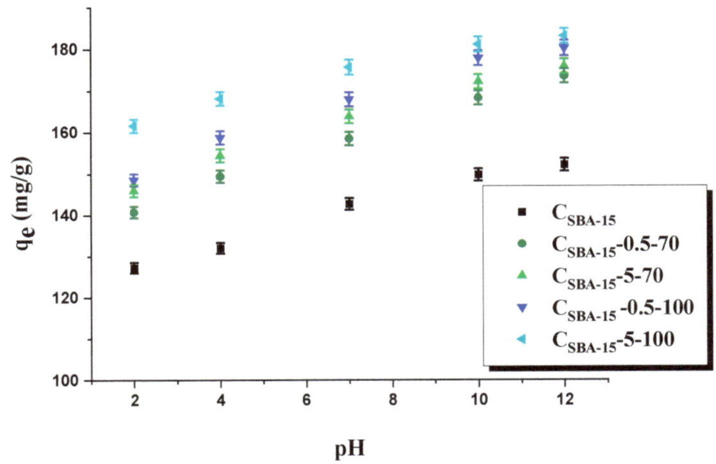

Figure 4. The pH influence on the sorption capacity of adsorbents obtained.

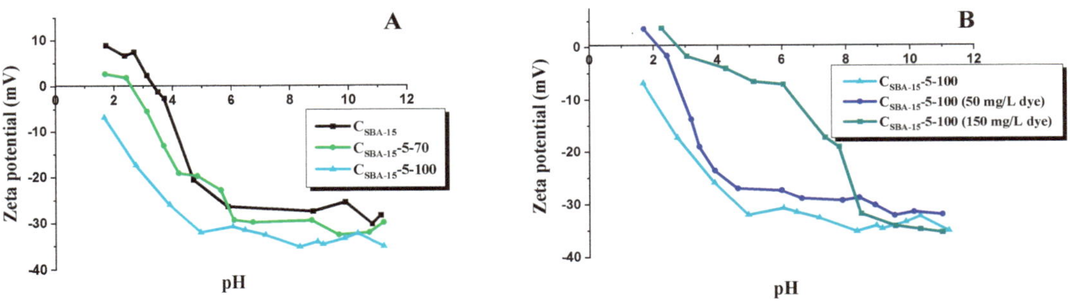

Figure 5. Zeta potential curves vs. pH for $C_{SBA\text{-}15}$, $C_{SBA\text{-}15}$-5-70, and $C_{SBA\text{-}15}$-5-100 mesoporous carbons before dye adsorption (**A**) and for $C_{SBA\text{-}15}$ and $C_{SBA\text{-}15}$-5-100 after adsorption of rhodamine B (**B**).

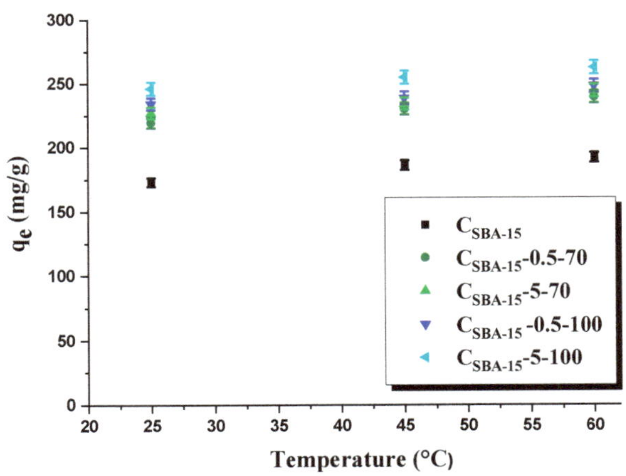

Figure 6. Effect of temperature on the adsorption of dye (100 mg/L) onto adsorbents obtained.

Table 4. Thermodynamic parameters obtained.

Adsorbent	Temperature (°C)	ΔG^0 (kJ·mol^{-1})	ΔH^0 (kJ·mol^{-1})	ΔS^0 (J·mol^{-1}·K^{-1})
$C_{SBA\text{-}15}$	25	−16.41	31.20	159.71
	45	−19.53		
	60	−22.01		
$C_{SBA\text{-}15}$-0.5-70	25	−19.51	23.96	145.66
	45	−22.18		
	60	−24.51		
$C_{SBA\text{-}15}$-5-70	25	−22.27	22.44	149.90
	45	−25.14		
	60	−27.53		
$C_{SBA\text{-}15}$-0.5-100	25	−19.12	43.69	210.76
	45	−23.24		
	60	−26.51		
$C_{SBA\text{-}15}$-5-100	25	−21.66	47.17	230.88
	45	−26.19		
	60	−29.73		

Figure 7 presents the isotherms of dye adsorption on the surface of carbon adsorbents. According to the results, the initial carbon $C_{SBA\text{-}15}$ proved to be the least effective in the removal of dye, despite having the largest surface area from among all samples studied (1203 m^2/g). However, this material also has the lowest number of surface functional groups of acidic character. Further data analysis allows us to conclude that the process of the adsorption of organic dye depends on surface functional groups of acidic character. Sample $C_{SBA\text{-}15}$-5-100, showing the largest sorption capacity towards rhodamine B (325 mg/g), has the greatest content of such functional groups (4.88 mmol/g). The process of its oxidation was performed with 5 mol/L nitric (V) acid solution at 100 °C. The amounts of rhodamine B adsorbed on the other oxidized carbon materials $C_{SBA\text{-}15}$-0.5-70, $C_{SBA\text{-}15}$-5-70, and $C_{SBA\text{-}15}$-0.5-100 were 248, 268, and 302 mg/g, respectively.

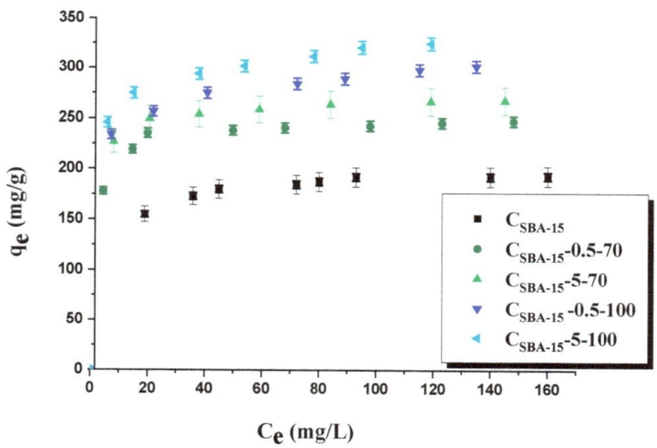

Figure 7. Adsorption of rhodamine B onto adsorbents obtained.

Analysis of the adsorption data was performed with the use of two theoretical models of Langmuir and Freundlich (Figures 8 and 9, Tables 5 and 6). Comparison of the experimental data with the predictions of a particular model provides information on the mechanism of adsorption and mesoporous carbon/rhodamine B interactions. The criterion of best fitting was the determination coefficient R^2.

The determination coefficient R^2 for the linear form of the Langmuir model takes values from 0.9987 to 0.9999 for all adsorbents obtained. The values of R^2 (Freundlich model) for particular carbon materials were: 0.8564 for C_{SBA-15}, 0.7993 for C_{SBA-15}-0.5-70, 0.8009 for C_{SBA-15}-5-70, 0.8611 for C_{SBA-15}-0.5-100, and 0.8598 for C_{SBA-15}-5-100. Therefore, the isotherms of rhodamine B adsorption on the surface of the carbon materials studied correspond to the Langmuir model. It should be noted that the experimental values of q_e are slightly lower from the theoretical q_m. For the oxidized mesoporous carbons, the coefficient K_L values are higher than for the initial material C_{SBA-15}, which means that the bonding between rhodamine B and the functionalized carbon material surfaces is stronger. The coefficient 1/n for the Freundlich model lies between (0 < 1/n < 1) which shows that this isotherm is favorable.

Table 5. The parameters of linear form of Langmuir and Freundlich models for rhodamine B adsorption onto carbon materials.

Material	Langmuir			Freundlich		
	q_m (mg/g)	K_L (L/mg)	R^2	K_F (mg/g (L/mg)$^{1/n}$)	1/n	R^2
C_{SBA-15}	196.07	0.369	0.9999	102.48	0.1373	0.8564
C_{SBA15}-0.5-70	250.00	0.727	0.9998	119.92	0.1701	0.7993
C_{SBA-15}-5-70	270.27	0.787	0.9998	127.98	0.1783	0.8009
C_{SBA-15}-0.5-100	303.03	0.493	0.9988	141.83	0.1717	0.8611
C_{SBA-15}-5-100	322.58	0.574	0.9987	153.34	0.1780	0.8598

Table 6. The parameters of non-linear Langmuir and Freundlich models for rhodamine B adsorption onto carbon materials.

Material	Langmuir			Freundlich		
	q_m (mg/g)	K_L (L/mg)	R^2	K_F (mg/g (L/mg)$^{1/n}$)	1/n	R^2
C_{SBA-15}	200.64	0.179	0.9999	122.13	0.0949	0.8983
C_{SBA15}-0.5-70	248.73	0.566	0.9986	171.94	0.0793	0.8053
C_{SBA-15}-5-70	268.28	0.660	0.9995	208.07	0.0530	0.9357
C_{SBA-15}-0.5-100	296.66	0.454	0.9942	197.63	0.0860	0.9953
C_{SBA-15}-5-100	319.83	0.505	0.9950	212.12	0.0897	0.9922

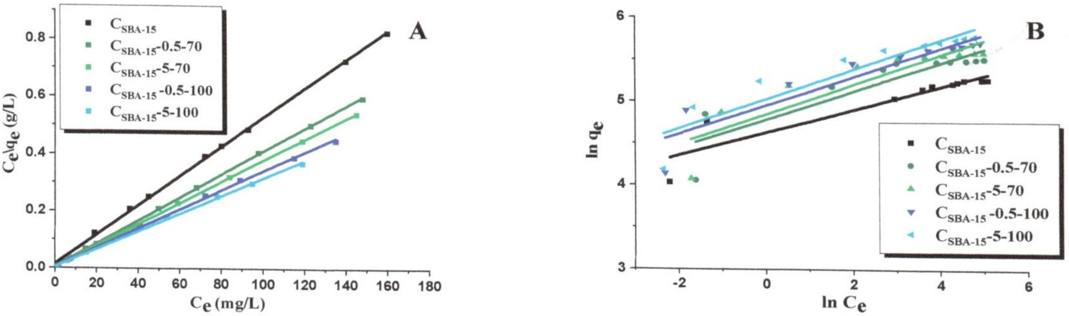

Figure 8. Linear fitting of rhodamine B adsorption isotherms onto carbon adsorbents to Langmuir (**A**) and Freundlich (**B**) models.

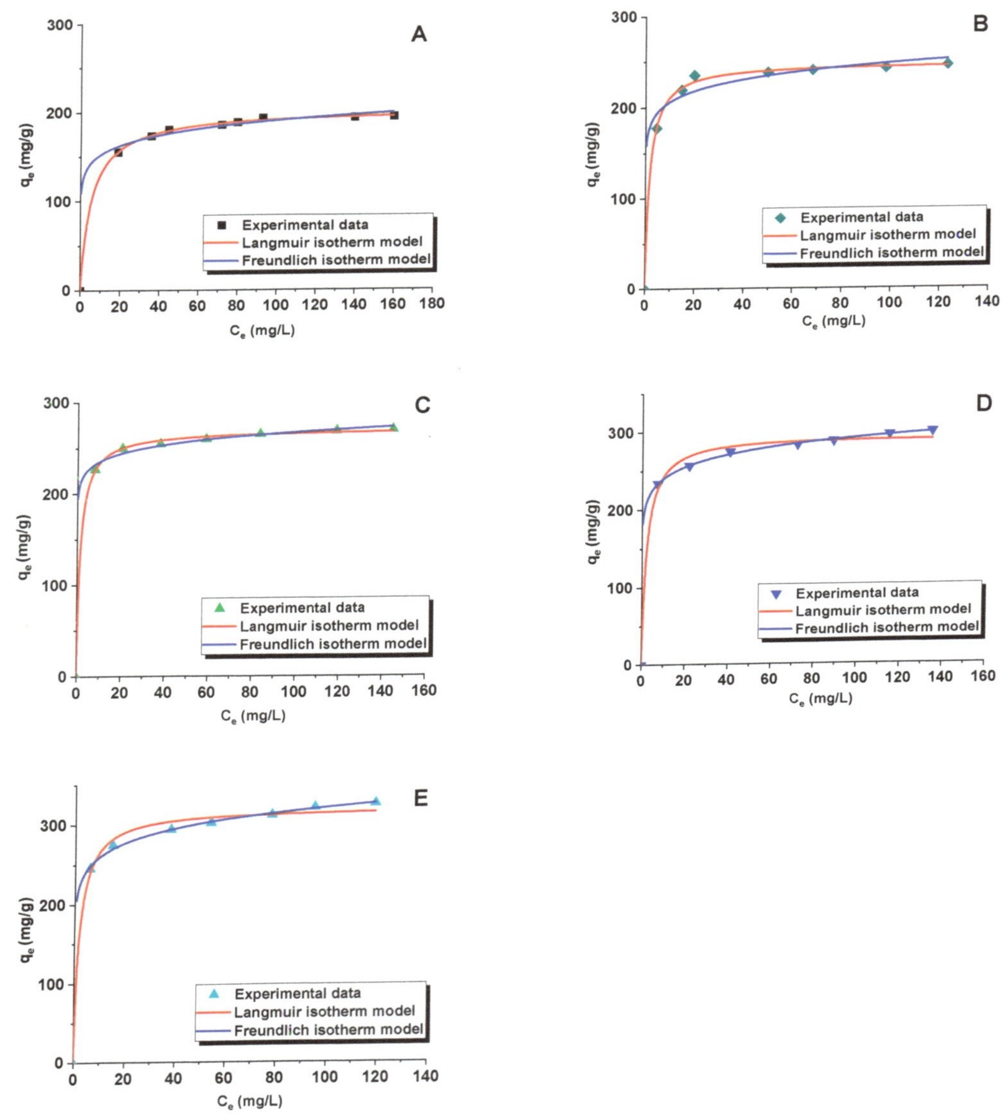

Figure 9. Non-linear fitting of rhodamine B adsorption isotherms to Langmuir and Freundlich models for samples: $C_{SBA\text{-}15}$ (**A**), C_{SBA15}-0.5-70 (**B**), $C_{SBA\text{-}15}$-5-70 (**C**), $C_{SBA\text{-}15}$-0.5-100 (**D**), and $C_{SBA\text{-}15}$-5-100 (**E**).

Additionally, the comparison of the non-linear form of the Langmuir and the Freundlich [54] isotherm models with experimental data for the adsorption of rhodamine B on the surface of carbon adsorbents is presented in Table 6 and Figure 9. It was established that the Langmuir adsorption model indicates a better fit to the experimental data than the Freundlich model. The R^2 values for the non-linear and linear Langmuir isotherms were similar.

Table 7 presents a comparison of the sorption capacities of carbon samples obtained towards tested dye with those of other adsorbents. The comparison implies that the ordered mesoporous carbon materials are very effective in the removal of rhodamine B from water solutions. The majority of literature reported adsorbents, including hierarchical SnS_2 nanostructure [5], TA-G [55], iron-pillared bentonite [56], sago waste activated carbon [57], kaolinite [58], $[Ni_{(bipy)2}]_2(HPW_{12}O_{40})$ [59], orange peel [60], whose sorption capacities are lower than for the oxidized carbon materials studied in this work. From among all materials mentioned in Table 6, the highest sorption capacity towards rhodamine B showed the magnetic mesoporous carbon materials (342–400 mg/g) [61], also high sorption capacities were noted for MoS_2/MIL-101-345 mg/g [62] and oxidized ordered mesoporous carbon material $C_{SBA\text{-}15}$-5-100 (325 mg/g). The hierarchical SnS_2 nanostructure [5] and TA-G [55] adsorption capacity was at a level of 200 mg/g, and the sorption capacities of the other samples listed in Table 7 were much lower: iron-pillared bentonite—99 mg/g [56], sago waste activated carbon—47 mg/g [57], kaolinite—46 mg/g [58], $[Ni_{(bipy)2}]_2(HPW_{12}O_{40})$—23 mg/g [59], and orange peel—14 mg/g [60].

Table 7. Comparison of sorption capacities of oxidized mesoporous carbons with other adsorbents presented in literature.

Adsorbent	Sorption Capacity (mg/g)	References
oxidized mesoporous carbon	248–325	This study
hierarchical SnS_2 nanostructure	200	[5]
TA-G	201	[55]
iron-pillared bentonite	99	[56]
sago waste activated carbon	47	[57]
kaolinite	46	[58]
$[Ni_{(bipy)2}]_2(HPW_{12}O_{40})$	23	[59]
orange peel	14	[60]
magnetic mesoporous carbon materials	342–400	[61]
MoS_2/MIL-101	345	[62]

4. Conclusions

According to the above-presented results, functionalization of the ordered mesoporous carbon of hexagonal structure ($C_{SBA\text{-}15}$) using a solution of nitric (V) acid, has brought about an increase in its effectiveness of rhodamine B removal from water solutions. Oxidization of mesoporous carbon reduced its textural parameters but increased the acidic character of the surface. Moreover, for the samples functionalized at 70 °C, the oxidation resulted in the disappearance of the reflections corresponding to the carbon structure ordering, as confirmed by XRD diffractograms in the small angle range. Sample $C_{SBA\text{-}15}$-5-100 was characterized by the highest content of surface functional groups of acidic nature and the largest surface area from among the functionalized carbon materials studied. This sample was the most effective in the removal of tested dye from water solutions (325 mg/g). The sorption capacity towards rhodamine B depended also on the pH of the solution and the process temperature. The adsorption of organic dye was more effective at higher pH because of deprotonation of the surface functional groups on the carbon samples by the hydroxide anion. The measurements of zeta potential, before and after adsorption, proved that with increasing pH, the surface charge on carbon samples changed. The amount of adsorbed rhodamine B was also found to increase with the increasing temperature of the process, which is related to increased mobility of the dye molecules at higher temperatures. The thermodynamic parameters showed that the adsorption process of dye was endothermic and proceeded spontaneously. The adsorption of rhodamine B onto oxidized mesoporous carbons was described by the Langmuir isotherm and pseudo-second-order kinetic model.

Author Contributions: Conceptualization, M.M. and J.G.; methodology, M.M.; validation, M.M., J.G., M.N. and A.B.-W.; formal analysis, M.M. and J.G.; investigation, M.M., M.N. and A.B.-W.; resources, J.G., T.J. and R.P.; data curation, M.M.; writing—original draft preparation, M.M. and J.G.; writing—review and editing, M.M., J.G., M.N., T.J., A.B.-W. and R.P.; visualization, M.M. and J.G.; supervision, R.P.; funding acquisition, R.P. All authors have read and agreed to the published version of the manuscript.

Funding: This research received no external funding.

Institutional Review Board Statement: Not applicable.

Informed Consent Statement: Not applicable.

Data Availability Statement: Data is contained within the article.

Conflicts of Interest: The authors declare no conflict of interest.

References

1. Liu, R.C.; Zhang, B.; Mei, D.D.; Zhang, H.Q.; Liu, J.D. Adsorption of methyl violet from aqueous solution by halloysite nanotubes. *Desalination* **2011**, *268*, 111–116. [CrossRef]
2. Liang, Z.J.; Zhao, Z.W.; Sun, T.Y.; Shi, W.X.; Cui, F.Y. Enhanced adsorption of the cationic dyes in the spherical CuO/meso-silica nano composite and impact of solution chemistry. *J. Colloid Interf. Sci.* **2017**, *485*, 192–200. [CrossRef] [PubMed]
3. Hao, O.J.; Kim, H.; Chiang, P.C. Decolorization of wastewater. *Crit. Rev. Environ. Sci. Technol.* **2000**, *30*, 449–505. [CrossRef]
4. Bulai, I.M.; Venturino, E. Biodegradation of organic pollutants in a water body. *J. Math. Chem.* **2016**, *54*, 1387–1403. [CrossRef]
5. Wang, S.; Yang, B.; Liu, Y. Synthesis of a hierarchical SnS_2 nanostructure for efficient adsorption of Rhodamine B dye. *J. Colloid Interf. Sci.* **2017**, *507*, 225–233. [CrossRef] [PubMed]
6. Huang, Z.; Li, Y.; Chen, W.; Shi, J.; Zhang, N.; Wang, X.; Li, Z.; Gao, L.; Zhang, Y. Modified bentonite adsorption of organic pollutants of dye wastewater. *Mater. Chem. Phys.* **2017**, *202*, 266–276. [CrossRef]
7. Raman, C.D.; Kanmani, S. Textile dye degradation using nano zero valent iron: A review. *J. Environ. Manag.* **2016**, *177*, 341–355. [CrossRef]
8. Patil, S.P.; Bethi, B.; Sonawane, G.H.; Shrivastava, V.S.; Sonawane, S. Efficient adsorption and photocatalytic degradation of Rhodamine B dye over Bi_2O_3-bentonite nanocomposites: A kinetic study. *J. Ind. Eng. Chem.* **2016**, *34*, 356–363. [CrossRef]
9. Ptaszkowska-Koniarz, M.; Goscianska, J.; Pietrzak, R. Adsorption of dyes on the surface of polymer nanocomposites modified with methylamine and copper(II) chloride. *J. Colloid Interf. Sci.* **2017**, *504*, 549–560. [CrossRef]
10. Li, S.; Jia, Z.; Li, Z.; Li, Y.; Zhu, R. Synthesis and characterization of mesoporous carbon nanofibers and its adsorption for dye in wastewater. *Adv. Powder Technol.* **2016**, *27*, 591–598. [CrossRef]
11. Firmino, P.I.M.; Da Silva, M.E.R.; Cervantes, F.J.; Dos Santos, A.B. Colour removal of dyes from synthetic and real textile wastewaters in one- and two-stage anaerobic systems. *Bioresour. Technol.* **2010**, *101*, 7773–7779. [CrossRef] [PubMed]
12. Cooper, C.; Burch, R. Mesoporous materials for water treatment processes. *Water Res.* **1999**, *33*, 3689–3694. [CrossRef]
13. Arshadi, M.; Mousavinia, F.; Amiri, M.J.; Faraji, A.R. Adsorption of methyl orange and salicylic acid on a nano-transition metal composite: Kinetics, thermodynamic and electrochemical studies. *J. Colloid Interf. Sci.* **2016**, *483*, 118–131. [CrossRef] [PubMed]
14. Errais, E.; Duplay, J.; Darragi, F. Textile dye removal by natural clay-case study of Fouchana Tunisian Clay. *Environ. Technol.* **2010**, *31*, 373–380. [CrossRef]
15. Marcucci, M.; Ciardelli, G.; Matteucci, A.; Ranieri, L.; Russo, M. Experimental campaigns on textile wastewater for reuse by means of different membrane processes. *Desalination* **2002**, *149*, 137–143. [CrossRef]
16. Alzahrani, S.; Mohammad, A.W. Challenges and trends in membrane technology implementation for produced water treatment: A review. *J. Water Process Eng.* **2014**, *4*, 107–133. [CrossRef]
17. Damodar, R.A.; You, S.; Ou, S. Coupling of membrane separation with photocatalytic slurry reactor for advanced dye wastewater treatment. *Sep. Purif. Technol.* **2010**, *76*, 64–71. [CrossRef]
18. Ou, W.; Zhang, G.; Yuan, X.; Su, P. Experimental study on coupling photocatalytic oxidation process and membrane separation for the reuse of dye wastewater. *J. Water Process Eng.* **2015**, *6*, 120–128. [CrossRef]
19. Kanagaraj, J.; Senthilvelan, T.; Panda, R.C. Degradation of azo dyes by laccase: Biological method to reduce pollution load in dye wastewater. *Clean. Technol. Environ. Policy* **2015**, *17*, 1443–1456. [CrossRef]
20. Kannan, C.; Sundaram, T.; Palvannan, T. Environmentally stable adsorbent of tetrahedral silica and non-tetrahedral alumina for removal and recovery of malachite green dye from aqueous solution. *J. Hazard. Mater.* **2008**, *157*, 137–145. [CrossRef]
21. Li, C.Z.; Zhong, H.; Wang, S.; Xue, J.R.; Zhang, Z.Y. Removal of basic dye (methylene blue) from aqueous solution using zeolite synthesized from electrolytic manganese residue. *J. Ind. Eng. Chem.* **2015**, *23*, 344–352. [CrossRef]
22. Dong, K.; Qiu, F.X.; Guo, X.R.; Xu, J.C.; Yang, D.Y.; He, K.C. Polyurethane-attapulgite porous material: Preparation, characterization, and application for dye adsorption. *J. Appl. Polym. Sci.* **2013**, *129*, 1697–1706. [CrossRef]
23. Goscianska, J.; Olejnik, A.; Pietrzak, R. In vitro release of L-phenylalanine from ordered mesoporous materials. *Micropor. Mesopor. Mater.* **2013**, *177*, 32–36. [CrossRef]

24. Derylo-Marczewska, A.; Marczewski, A.W.; Winter, S.; Sternik, D. Studies of adsorption equilibria and kinetics in the systems: Aqueous solution of dyes–mesoporous carbons. *Appl. Surf. Sci.* **2010**, *256*, 5164–5170. [CrossRef]
25. Goscianska, J.; Ptaszkowska, M.; Pietrzak, R. Equilibrium and kinetic studies of chromotrope 2R adsorption onto ordered mesoporous carbons modified with lanthanum. *Chem. Eng. J.* **2015**, *270*, 140–149. [CrossRef]
26. Goscianska, J.; Marciniak, M.; Pietrzak, R. Ordered mesoporous carbons modified with cerium as effective adsorbents for azo dyes removal. *Sep. Purif. Technol.* **2015**, *154*, 236–245. [CrossRef]
27. Bazula, P.A.; Lu, A.H.; Nitz, J.T.; Schüth, F. Surface and pore structure modification of ordered mesoporous carbons via a chemical oxidation approach. *Micropor. Mesopor. Mater.* **2008**, *108*, 266–275. [CrossRef]
28. Liu, F.; Guo, Z.; Ling, H.; Huang, Z.; Tang, D. Effect of pore structure on the adsorption of aqueous dyes to ordered mesoporous carbons. *Micropor. Mesopor. Mater.* **2016**, *227*, 104–111. [CrossRef]
29. Asouhidou, D.D.; Triantafyllidis, K.S.; Lazaridis, N.K.; Matis, K.A.; Kim, S.S.; Pinnavaia, T.J. Sorption of reactive dyes from aqueous solutions by ordered hexagonal and disordered mesoporous carbons. *Micropor. Mesopor. Mater.* **2009**, *117*, 257–267. [CrossRef]
30. Peng, X.; Hu, X.; Fu, D.; Lam, F.L.Y. Adsorption removal of acid black 1 from aqueous solution using ordered mesoporous carbon. *Appl. Surf. Sci.* **2016**, *294*, 71–80. [CrossRef]
31. Cai, T.; Liu, Y.; Wang, L.; Dong, W.; Chen, H.; Zeng, W.; Xia, X.; Zeng, G. Activation of persulfate by photoexcited dye for antibiotic degradation: Radical and nonradical reactions. *Chem. Eng. J.* **2019**, *375*, 122070. [CrossRef]
32. Sulistina, D.R.; Martini, S. The effect of Rhodamine B on the cerebellum and brainstem tissue of *Rattus norvegicus*. *J. Public Health Res.* **2020**, *9*, 1812. [CrossRef]
33. Chen, L.; Liu, H.; Ren, X. Synthesis and characterization of magnetic metal–organic framework for the adsorptive removal of rhodamine B from aqueous solution. *J. Ind. Eng. Chem.* **2016**, *34*, 278–285. [CrossRef]
34. Deng, J.; Chen, Y.J.; Lu, Y.A.; Ma, X.Y.; Feng, S.F.; Gao, N.; Li, N. Synthesis of magnetic $CoFe_2O_4$/ordered mesoporous carbon nanocomposites and application in Fenton-like oxidation of rhodamine B. *Environ. Sci. Pollut. Res.* **2017**, *24*, 14396–14408. [CrossRef] [PubMed]
35. Moradi, S.E.; Nasrollahpour, A. Competitive adsorption and photodegradation of Methyl orange and Rhodamine B by TiO_2modified mesoporous carbon photo-catalyst on UV irradiation. *Mater. Technol.* **2017**, *32*, 716–723. [CrossRef]
36. Tripathi, N.K. Porous carbon spheres: Recent developments and applications. *AIMS Mater. Sci.* **2018**, *5*, 1016–1052. [CrossRef]
37. Zhao, D.; Huo, Q.; Feng, J.; Chmelka, B.F.; Stucky, G.D. Nonionic triblock and star diblock copolymer and oligomeric surfactant syntheses of highly ordered hydrothermally stable mesoporous silica structures. *J. Am. Chem. Soc.* **1998**, *120*, 6024–6036. [CrossRef]
38. Zhao, D.; Feng, J.; Huo, Q.; Melosh, N.; Fredrickson, G.H.; Chmelka, B.F.; Stucky, G.D. Triblock copolymer syntheses of mesoporous silica with periodic 50 to 300 angstrom pores. *Science* **1998**, *279*, 548–552. [CrossRef]
39. Joo, H.; Choi, J.; Oh, I.; Kwak, J.; Liu, Z.; Terasaki, O.; Ryoo, R. Ordered nanoporous arrays of carbon supporting high dispersions of platinum nanoparticles. *Nature* **2001**, *412*, 169–172. [CrossRef]
40. Boehm, H.P. Some aspects of the surface chemistry of carbon blacks and other carbons. *Carbon* **1994**, *32*, 759–769. [CrossRef]
41. Olejnik, A.; Gosicianska, J. On the importance of physicochemical parameters of copper and aminosilane functionalized mesoporous silica for hydroxychloroquine release. *Mater. Sci. Eng. C* **2021**, *130*, 112438. [CrossRef] [PubMed]
42. Nashine, A.L.; Tembhurkar, A.R. Equilibrium, kinetic and thermodynamic studies for adsorption of As(III) on coconut (*Cocos nucifera* L.) fiber. *J. Environ. Chem. Eng.* **2016**, *4*, 3267–3273. [CrossRef]
43. Konggidinata, M.I.; Chao, B.; Lian, Q.; Subramaniam, R.; Zappi, M.; Gang, D.D. Equilibrium, kinetic and thermodynamic studies for adsorption of BTEX onto Ordered Mesoporous Carbon (OMC). *J. Hazard. Mater.* **2017**, *336*, 249–259. [CrossRef] [PubMed]
44. Dursun, A.Y.; Kalayci, C.S. Equilibrium, kinetic and thermodynamic studies on the adsorption of phenol onto chitin. *J. Hazard. Mater.* **2005**, *123*, 151–157. [CrossRef] [PubMed]
45. Langmuir, I. The constitutional and fundamental properties of solids and liquids. *J. Am. Chem. Soc.* **1916**, *38*, 2221–2295. [CrossRef]
46. Freundlich, H. Überdie Adsorption in Lösungen. *Z. Phys. Chem.* **1907**, *57*, 385–470. [CrossRef]
47. Wu, Z.; Webley, P.A.; Zhao, D. Comprehensive Study of Pore Evolution, Mesostructural Stability, and Simultaneous Surface Functionalization of Ordered Mesoporous Carbon (FDU-15) by Wet Oxidation as a Promising Adsorbent. *Langmuir* **2010**, *26*, 10277–10286. [CrossRef]
48. Cohn, A.P.; Erwin, W.R.; Share, K.; Oakes, L.; Westover, A.S.; Carter, R.E.; Bardhan, R.; Pint, C.L. All Silicon Electrode Photocapacitor for Integrated Energy Storage and Conversion. *Nano Lett.* **2015**, *15*, 2727–2731. [CrossRef]
49. Prahas, D.; Kartika, Y.; Indraswati, N.; Ismadji, S. Activated carbon from jackfruit peel waste by H_3PO_4 chemical activation: Pore structure and surface chemistry characterization. *Chem. Eng. J.* **2008**, *140*, 32–42. [CrossRef]
50. Li, Y.; Dua, Q.; Liua, T.; Peng, X.; Wang, J.; Sun, J.; Wang, Y.; Wu, S.; Wang, Z.; Xia, Y.; et al. Comparative study of methylene blue dye adsorption onto activated carbon, graphene oxide, and carbon nanotubes. *Chem. Eng. Res. Des.* **2013**, *91*, 361–368. [CrossRef]
51. Galán, J.; Rodríguez, A.; Gómez, J.M.; Allen, S.J.; Walker, G.M. Reactive dye adsorption onto a novel mesoporous carbon. *Chem. Eng. J.* **2013**, *219*, 62–68. [CrossRef]
52. Goscianska, J.; Olejnik, A.; Nowak, I. APTES-functionalized mesoporous silica as a vehicle for antipyrine–adsorption and release studies. *Colloid Surf. A* **2017**, *533*, 187–196. [CrossRef]

53. Dawood, S.; Sen, T.K.; Phan, C. Synthesis and Characterisation of Novel-Activated Carbon from Waste Biomass Pine Cone and Its Applicationin the Removal of Congo Red Dye from Aqueous Solution by Adsorption. *Water Air Soil Pollut.* **2014**, *225*, 1818. [CrossRef]
54. Alswieleh, A.M. Efficient Removal of Dyes from Aqueous Solution by Adsorption on L-Arginine-Modified Mesoporous Silica Nanoparticles. *Processes* **2022**, *10*, 1079. [CrossRef]
55. Liu, K.; Li, H.; Wang, Y.; Gou, X.; Duan, Y. Adsorption and removal of rhodamine B from aqueous solution by tannic acid functionalized graphene. *Colloid Surf. A Physicochem. Eng. Asp.* **2015**, *477*, 35–41. [CrossRef]
56. Hou, M.F.; Ma, C.X.; Zhang, W.D.; Tang, X.Y.; Fan, Y.N.; Wan, H.F. Removal of rhodamine B using iron-pillared bentonite. *J. Hazard. Mater.* **2011**, *186*, 1118–1123. [CrossRef]
57. Kadirvelu, K.; Karthika, C.; Vennilamani, N.; Pattabhi, S. Activated carbon from industrial solid waste as an adsorbent for the removal of rhodamine-B from aqueous solution: Kinetic and equilibrium studies. *Chemosphere* **2005**, *60*, 1009–1017. [CrossRef]
58. Khan, T.A.; Dahiya, S.; Ali, I. Use of kaolinite as adsorbent: Equilibrium, dynamics and thermodynamic studies on the adsorption of Rhodamine B from aqueous solution. *Appl. Clay Sci.* **2012**, *69*, 58–66. [CrossRef]
59. Li, F.; Chen, Y.; Huang, H.; Cao, W.; Li, T. Removal of rhodamine B and Cr(VI) from aqueous solutions by a polyoxometalate adsorbent. *Chem. Eng. Res. Des.* **2015**, *100*, 192–202. [CrossRef]
60. Annadurai, G.; Juang, R.S.; Lee, D.J. Use of cellulose-based wastes for adsorption of dyes from aqueous solutions. *J. Hazard. Mater.* **2002**, *92*, 263–274. [CrossRef]
61. Dong, Y.; Lin, H.; Qu, F. Synthesis of ferromagnetic ordered mesoporous carbons for bulky dye molecules adsorption. *Chem. Eng. J.* **2012**, *193–194*, 169–177. [CrossRef]
62. Yang, C.; Cheng, J.; Chen, Y.; Hu, Y. Enhanced adsorption performance of MoS_2 nanosheet-coated MIL-101 hybrids for the removal of aqueous rhodamine B. *J. Colloid Interface Sci.* **2017**, *504*, 39–47. [CrossRef] [PubMed]

materials

Article

Increasing the Mechanical Strength and Corrosion Resistance of Aluminum Alloy 7075 via Hydrostatic Extrusion and Aging

Marta Orłowska [1,*], Ewa Ura-Bińczyk [2], Lucjan Śnieżek [1], Paweł Skudniewski [2], Mariusz Kulczyk [3], Bogusława Adamczyk-Cieślak [2] and Jarosław Mizera [2]

1 Faculty of Mechanical Engineering, Military University of Technology, gen. S. Kaliskiego 2, 00-908 Warsaw, Poland; lucjan.sniezek@wat.edu.pl
2 Faculty of Materials Science and Engineering, Warsaw University of Technology, Wołoska 141, 02-507 Warsaw, Poland; ewa.ura@pw.edu.pl (E.U.-B.); pskudniewski@gmail.com (P.S.); boguslawa.cieslak@pw.edu.pl (B.A.-C.); jaroslaw.mizera@pw.edu.pl (J.M.)
3 Institute of High Pressure Physics, Polish Academy of Sciences, Sokolowska 29/37 St., 01-142 Warsaw, Poland; mariusz.kulczyk@unipress.waw.pl
* Correspondence: marta.orlowska@wat.edu.pl; Tel.: +48-261-839-245

Abstract: The present study investigates the correlation between mechanical properties and resistance to corrosion of hydrostatically extruded aluminum alloy 7075. Supersaturated solid solutionized samples undergo a plastic deformation process, followed by both natural and artificial aging. Furthermore, two types of hydrostatic extrusion are applied to the samples: single-stepped and double-stepped. This process is shown to influence grain refinement and the precipitation process, resulting in changes in the electrochemical properties of the samples. Hydrostatic extrusion combined with aging is shown to cause an increase in mechanical strength ranging from 50 MPa to 135 MPa in comparison to coarse-grained sample subjected to T6 heat treatment. The highest value of tensile strength is obtained for a sample subjected to single-step hydrostatic extrusion followed by natural aging. This strength increase is caused by refinement of the microstructure, in addition to the small size and number of precipitates at the grain boundaries, which are coarsened by artificial aging. Hydrostatic extrusion is also shown to increase resistance to corrosion, with the T6-treated coarse-grained sample being most susceptible to corrosion attack.

Keywords: hydrostatic extrusion; AA 7075; microstructure; mechanical properties; corrosion

Citation: Orłowska, M.; Ura-Bińczyk, E.; Śnieżek, L.; Skudniewski, P.; Kulczyk, M.; Adamczyk-Cieślak, B.; Mizera, J. Increasing the Mechanical Strength and Corrosion Resistance of Aluminum Alloy 7075 via Hydrostatic Extrusion and Aging. *Materials* **2022**, *15*, 4577. https://doi.org/10.3390/ma15134577

Academic Editors: Jakub Zdarta and Agata Zdarta

Received: 7 June 2022
Accepted: 26 June 2022
Published: 29 June 2022

1. Introduction

Composed of Al-Zn-Mg(-Cu), 7xxx aluminum alloys (AAs) are of high interest to the aerospace industry due to their unique combination of light weight, high strength, and high resistance to corrosion [1]. These properties are superior to those of other AAs [2]. The complex microstructure of such materials reveals substantial potential for further improvement of its properties and manufacturing applications [3,4]. These AAs are age-hardenable, and therefore the precipitates are the primary strengthening factor. Additionally, the complexity of the microstructure is seen in numerous intermetallic particles with varying chemical compositions [5]. The application of an additional strengthening mechanism, such as grain refinement, has been shown to improve the properties of the material [6]. A group of methods known as severe plastic deformation (SPD) processes [7] can be used to refine the microstructure. SPD techniques are characterized by the application of a large strain, which introduces structural defects to the microstructure. It has been shown that SPD processing, applied to AA 7075 [8] via high-pressure torsion and equal channel angular pressing (ECAP), significantly reduced grain size to 310 nm. Correspondingly, a considerable increase in microhardness was obtained: from approximately 100 HV to 140 HV following ECAP alone, and to 250 HV after the combination of the two processes. Further research used ECAP to refine the microstructure of AA 7075 [9]. Four passes of the process (ε = 4), resulting in an average grain size of approximately 500 nm, increased the microhardness of a coarse-grained (CG) sample from 68 HV to 143 HV, and the tensile strength from 230 MPa to 440 MPa. The proportion of deformation determines the final grain size, fraction of high-angle grain boundaries (HAGBs), and dislocation density. This influences mechanical properties, as such structural defects being effective barriers for moving dislocations. Grain refinement also influences other properties, such as electrical conductivity [10], magnetic properties, or corrosion resistance [11]. In the case of the latter, the influence of microstructure refinement is ambiguous, and depends on parameters such as environmental conditions and the examined material. In the case of pure aluminum, as grain size decreases, the corrosion rate also decreases [12]. This correlation arises due to the following phenomenon—the refined microstructure has more reactive surfaces with respect to oxide formation and film ion conduction. From the other side, a higher number of grain boundaries can accelerate corrosion, as these are favorable paths of corrosion propagation [13]. Therefore, the deformation process influences the corrosion resistance, as it changes the characteristics of grain boundaries. The important aspect in this case is a value of the equivalent strain—ε. In SPD processing, it directly influences the grain refinement. In, e.g., the ECAP process, the value of ε can be estimated based on the relationship between the intersection angles of the channel in a die. This is described in [14]. In simplified terms, it can be assumed that in the case of a die with two channels with equal cross-section and with the intersecting channel angles equal to 90°, the outer arc of curvature equals 0°, and during one pass the $\varepsilon \approx 1$.

Hydrostatic extrusion (HE) [15] is an SPD method based on pressing a billet in a chamber using a fluid. A scheme of the process is presented in Figure 1. Due to the hydrostatic pressure, the size of the billet (e.g., the diameter in the case of rods) is reduced. The equivalent strain of hydrostatic extrusion is given by Equation (1) [16]:

$$\varepsilon = 2\ln\left(\frac{d'}{d''}\right) \tag{1}$$

where d′ is the initial diameter of the rod and d″ is the exit diameter.

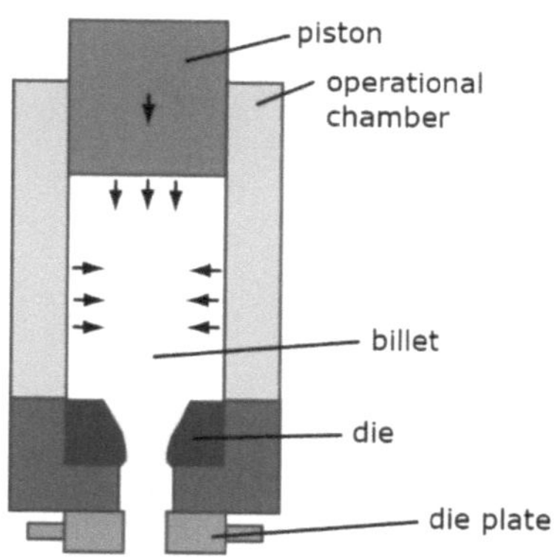

Figure 1. The scheme of hydrostatic extrusion process.

Unlike other SPD methods, such an approach can both obtain products of substantial length and also reduce strain, which is needed to refine the microstructure. However, due to the nature of the process, anisotropy is introduced to the microstructure, as grains are significantly elongated along the extrusion direction. Hence, mechanical anisotropy is observed [17]. Previous work has investigated the application of HE to 7xxx AAs. Lewandowska et al. [18] showed that applying HE to AA 7475 influenced the precipitation process, as an elevated number of nucleation sites caused a larger number of precipitates along boundaries than within grain interiors. Furthermore, the sizes of such precipitates were reduced. As a result, precipitate strengthening was reduced, as precipitates within grains interiors more strongly inhibit the movement of dislocations compared to those at grain boundaries. The results further revealed that appropriate heat treatment following HE can improve mechanical strength [6] due to the combination of grain boundaries and precipitation strengthening. Moreover, the resistance to pitting corrosion of AA 7475 samples that underwent HE was improved following aging [19]. Such an improvement is caused by the coarsening of stable intergranular $MgZn_2$ precipitates and the formation of metastable $MgZn_2$ in grain interiors during aging. As a result, corrosion resistance improved as the open circuit potential (E_{OCP}) was shifted to more noble values and the corrosion current density (i_{corr}) was reduced.

The literature study suggests there is high potential for 7xxx AAs to be endowed with substantial mechanical strength and superior corrosion resistance via appropriate thermo-mechanical treatment. Previous works have only investigated these two properties individually. Therefore, this paper adopts a more comprehensive approach, and attempts to correlate appropriate heat treatment of 7xxx AAs that have undergone HE with the development of high mechanical strength and high resistance to corrosion in an aggressive environment. This work is a continuation of our previous work, where the influence of heat treatment on the single-stepped hydrostatically extruded AA7075 was investigated [20].

2. Materials and Methods

AA 7075 was selected as the material under investigation. The exact chemical composition given by a producer is shown in Table 1. The material was delivered in the form of hot extruded bars with a diameter of 20 mm. The bars were supersaturated solid solutionized by annealing at 520 °C for 2 h, and then water-cooled. The billets were then plastically deformed via HE. Two HE approaches were chosen. In the single-stepped approach, the material was deformed in a single step with a reduction in diameter to 10 mm (ϕ10, $\varepsilon = 1.41$). In the multi-stepped approach, deformation consisted of two steps—an initial reduction to 12 mm, followed by a reduction to 8 mm (ϕ8). This approach produced an equivalent strain of $\varepsilon = 1.85$.

Table 1. The chemical composition of AA 7075.

Element	Zn	Mg	Cu	Cr	Ti	Si	Fe	Mn	Al
Content (wt. %)	5.70	2.40	1.50	0.19	0.04	0.09	0.23	0.06	balanced

The aging process was performed with the parameters shown in Table 2. The CG T6 sample was initially subjected to single-step HE, and then, to coarsen the grains, heat treated by supersaturated solid solutionizing and aged to a T6 state. In the case of the ultra-fine-grained (UFG) samples, two temperatures were examined, 100 °C and 120 °C, for 24 h. Additionally, samples after natural aging were investigated. Furthermore, retrogression and reaging (RRA) processes were applied to the ϕ10 UFG sample.

Table 2. Sample designations with respect to the applied heat treatments.

	Naturally Aged/180 Days	100 °C/24 h	120 °C/24 h (T6)	RRA 120 °C/24 h + 200 °C/40 min + 120 °C/24 h	RRA 120 °C/24 h + 240 °C/40 min + 120 °C/24 h
ϕ10 CG	-	-	CG T6	-	-
ϕ10 UFG	HE10	HE10 HT1	HE10 HT2	HE10 RRA1	HE10 RRA2
ϕ8 UFG	HE8	HE8 HT1	HE8 HT2	-	-

The sample microstructure was characterized using a Hitachi Su70 scanning electron microscope (SEM) with electron backscatter diffraction (EBSD) detector, and hence produced orientation maps (OIMs) and investigated average grain size, the fraction of HAGBs and low-angle grain boundaries (LAGBs), and texture. Pole figures (PFs) were generated during the analysis of the latter. Energy-dispersive X-ray spectroscopy (EDS) was used to determine the chemical composition of the intermetallic particles. A more detailed characterization of the microstructure was performed using a JEOL JEM 1200 transmission electron microscope (TEM). Samples that underwent microstructure characterization were electropolished using a Struers Tenupol-5 device using standard electrolyte A2. Moreover, X-ray diffraction (XRD) measurements were conducted with a use of Bruker D8 Advance diffractometer with Cu Kα radiation. The 2ϕ range of the XRD measurements was 25° to 90°.

The mechanical properties of the samples were investigated via static tensile tests and microhardness measurements. In the first case, flat mini-samples with a cross section of 0.6 × 0.8 mm and a gauge length of 5 mm were used [21,22]. Tests were performed at a strain rate of 10^{-3} s^{-1} with the use of the digital image correlation for the measurement of the strain. The tests produced quantitative results for average values of ultimate tensile strength (UTS), yield strength (YS), and elongation at break (E_b). Samples were cut longitudinally to the extrusion direction. For each state, three samples were tested. Microhardness measurements were conducted using a Vicker's method with a load of 200 g (HV0.2). Five measurements per sample were conducted.

The corrosion resistance of the samples was investigated via electrochemical experiments in quiescent 0.1M NaCl solution. The samples underwent cyclic polarization in potentiodynamic (PP) mode. Before each test, each specimen was immersed within a prepared electrolyte for 10 min to stabilize the rest potential. The tests were performed using a NOVA AutoLab PGSTAT302N potentiostat/galvanostat in three electrode configurations, with the sample acting as the working electrode, a silver chloride reference electrode, and a platinum sheet counter electrode. A 20 mm^2 area on the surface of each sample was examined. Corrosion tests were conducted on the surfaces perpendicular to the HE direction. PP was performed with a scan rate of 1 mV/s starting from −0.05 mV relative to E_{OCP}, and reversed when the current reached 0.25 mA. PP measurements were repeated three times to ensure reproducibility of the results. Average values of corrosion potential (E_{corr}), corrosion current density (i_{corr}), and repassivation potential (E_{rep}) were obtained. All tests were performed at room temperature.

3. Results

3.1. Microstructure Evolution

3.1.1. SEM

Figure 2 shows OIMs of CG T6, HE10 and HE8 samples on two planes: transverse and longitudinal to the extrusion direction. The distribution of grain size and the grain boundary misorientation angles are presented in Figure 3. The average grain size for the CG T6 sample is approximately 17 μm on both planes. The grain shape differs, with equiaxial grains on the transverse plane and elongated grains on the longitudinal plane. Both planes contain more than 90% of HAGBs. Following HE, substantial grain refinement can be observed, primarily indicated by a considerable increase in the number of LAGBs. For both deformed samples HE10 and HE8, the microstructure along the transverse plane displays equiaxial grains with a considerable fraction of LAGBs, which are randomly distributed to create a network of subgrains within the grains, which are surrounded by HAGBs. The average grain/subgrain size is decreased to 1.5 μm and 1.0 μm for HE10 and HE8, respectively. Those samples also have HAGB fractions of 24% and 15%, respectively. As seen in Figure 3, the most frequent grain sizes for the HE8 sample are below 3 μm. In addition, the majority of grain boundary misorientation angles are below 10°. These findings indicate that further grain refinement was primarily due to the formation of subgrains separated by LAGBs. A substantial difference in the microstructure can be observed on the longitudinal plane. Fibrous grains are formed, manifesting as highly elongated shapes. Straight HAGBs are aligned along the extrusion direction. Within the fibrous grains, a network of LAGBs can be observed. The average grain size is larger than that on the transverse plane, although the distance between HAGBs is smaller.

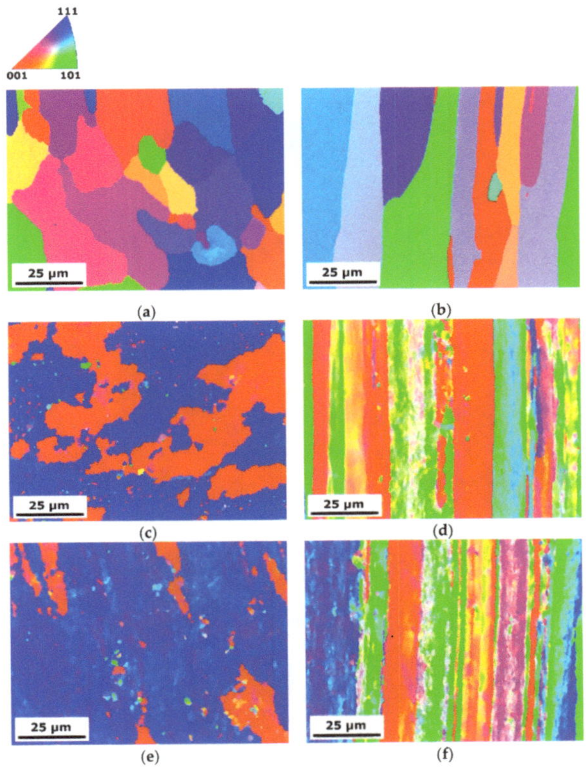

Figure 2. OIMs of the: (**a**) CG T6 transverse plane, (**b**) CG T6 longitudinal plane, (**c**) HE10 transverse plane, (**d**) HE10 longitudinal plane, (**e**) HE8 transverse plane and (**f**) HE8 longitudinal plane.

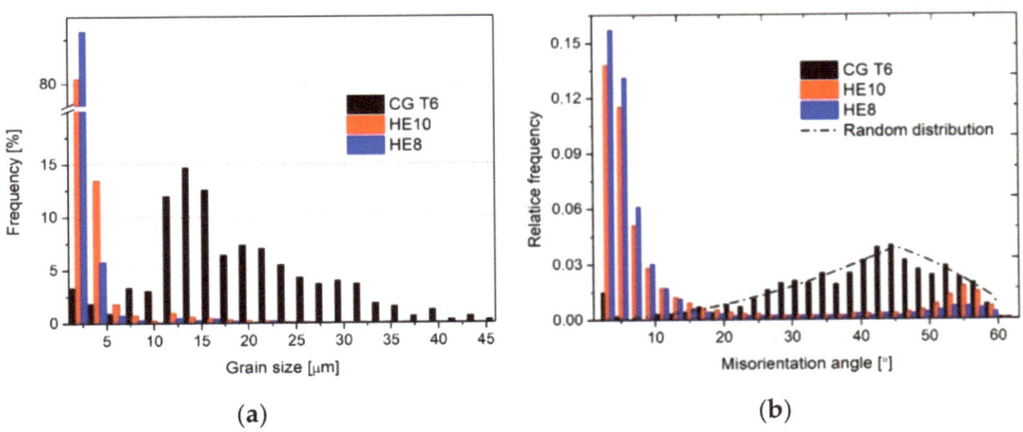

Figure 3. The distribution of: (**a**) grain size and (**b**) misorientation angle on the transverse plane of the CG T6, HE10, and HE8 samples.

Figure 4 shows PFs for the transverse plane of the samples. The CG T6 sample exhibits random grain orientations without a distinct texture. Following HE, the HE10 and HE8 samples display substantially changed grain orientations, with <111> and <001> dominant for HE10 and <111> dominant for HE8. Moreover, a higher intensity can be observed for the HE8 sample, corresponding with stronger deformation.

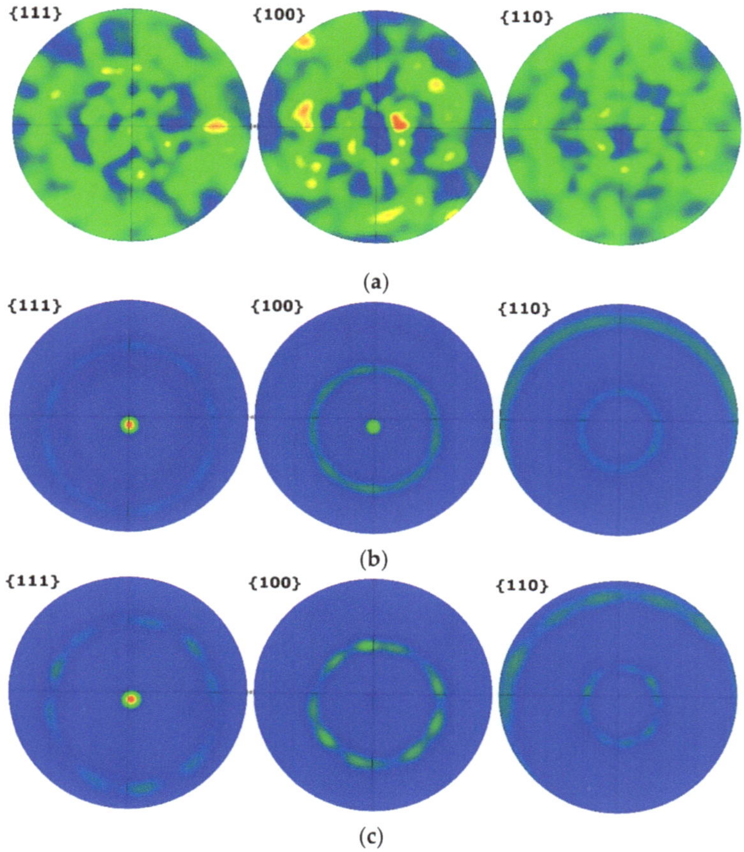

Figure 4. PFs of the: (**a**) CG T6, (**b**) HE10, and (**c**) HE8 samples.

Figure 5 presents the results of EDS analysis, used to characterize the intermetallic particles within the microstructure. A diverse range of particles can be observed, with the majority rich in Cu and Fe, in addition to Mn and Si. In terms of electrochemical potential, both Fe and Cu are nobler than the Al matrix [23]. This influences resistance to pitting corrosion, as pit nucleation is initiated in the vicinity of intermetallic particles. In terms of mechanical properties, such particles do not contribute to the increase in strength of the material.

Moreover, for the characterization of second-phase precipitates, the EDS analysis was also performed. The results are shown in Figure 6, where a line scan of the exemplary precipitate is presented. It can be seen that the precipitate is rich Zn and Mg. Both η and η' precipitate are rich in these elements. For AA7075, these two precipitates are the major precipitates in the microstructure, as can be seen in numerous papers, e.g., [24–27].

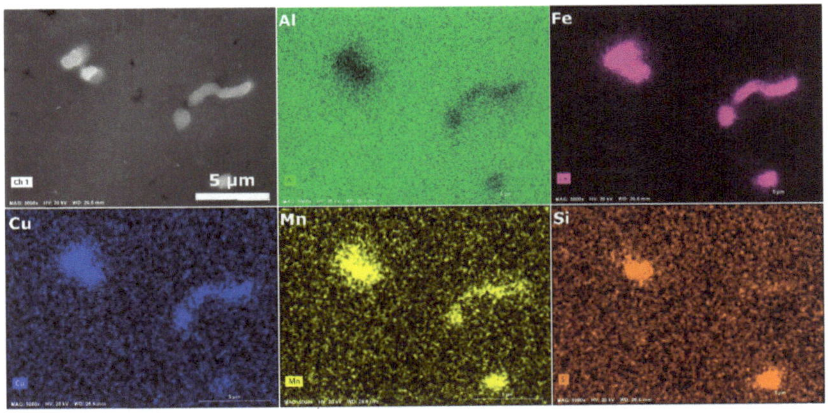

Figure 5. SEM/EDS analysis of the intermetallic particles.

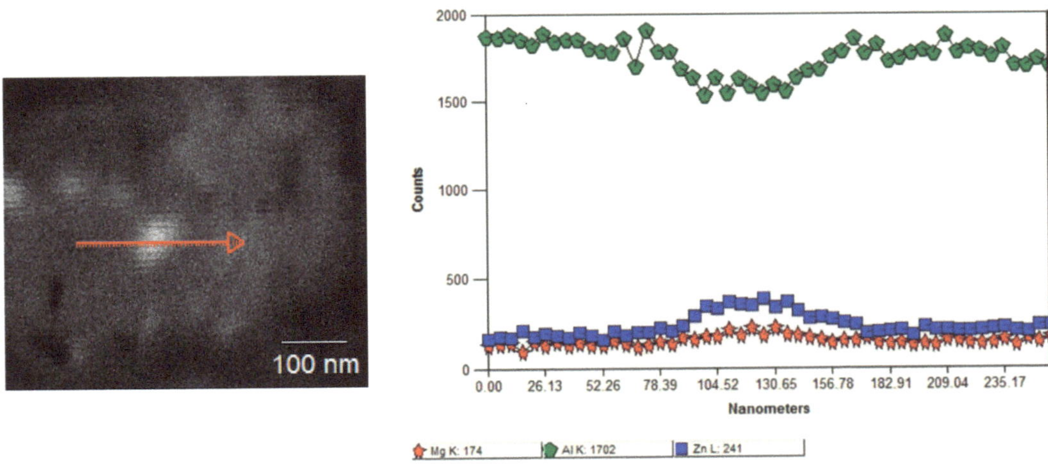

Figure 6. SEM/EDS analysis of the second-phase precipitate.

3.1.2. TEM

A detailed characterization of the microstructure was performed using TEM. Figure 7 presents example micrographs of the transverse plane of the undeformed and deformed samples further aged at 120 °C/24 h. The micrographs reveal differences between the samples in terms of plastic deformation and heat treatment. Samples display different grain size and density of dislocations. The CG T6 sample is characterized by a coarse-grained microstructure with a high density of η precipitates within the grain interiors, with additional η′ precipitates visible as small dots within the matrix. Similar observation of the microstructure components were found in, e.g., [24,28,29] The microstructures of the HE10 HT2 and HE8 HT2 samples are substantially deformed, which can be seen by the grain refinement, but also feature significant increases in dislocation density. Moreover, it is seen in selected area diffraction images, which are inside the TEM micrographs. For samples after hydrostatic extrusion, the diffractions are close to rings, which indicates a higher number of orientations in comparison to CG T6 sample. In addition, the size of η precipitates is reduced in comparison to the CG T6 sample.

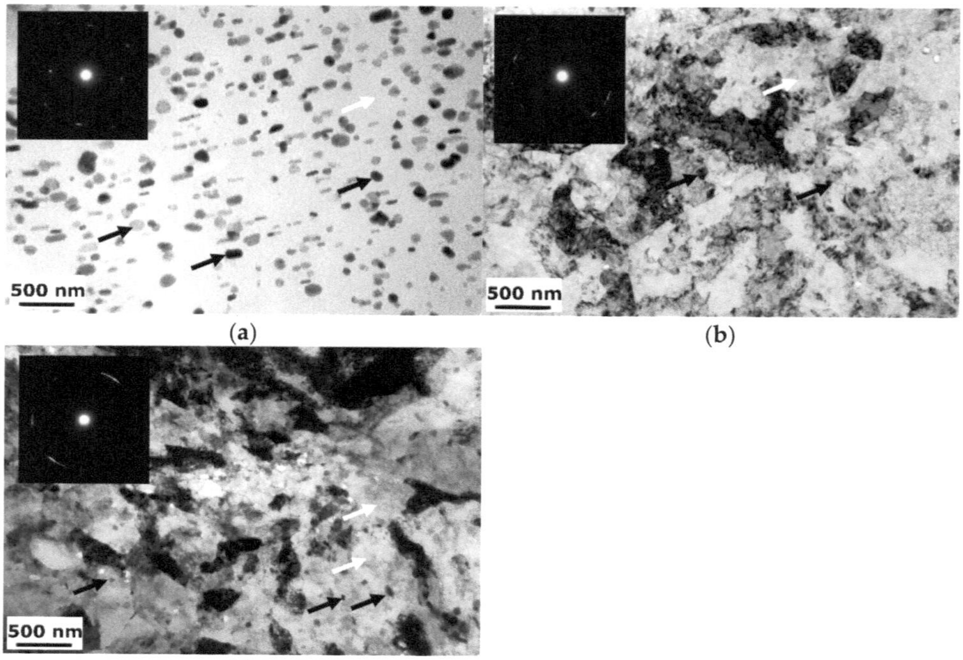

Figure 7. TEM micrographs of the: (**a**) CG T6, (**b**) HE10 HT2, and (**c**) HE8 HT2 samples, black arrows indicate η precipitates, white arrows—η′ precipitates.

The TEM micrographs were further analyzed at higher magnification. Figure 8 shows micrographs of the ϕ10 samples. The size and number of the η precipitates are substantially reduced within the naturally aged sample in comparison to the CG T6 sample (Figure 7). Artificial aging leads to the intensification of precipitation process, and the number and size of precipitates increase. Precipitates are observed at both grain boundaries but also grain interiors. With increasing the aging temperature, the size of the η precipitates increases. What can be noticed is that RRA heat treatment led to over-aging of the material. This can be seen by the significant coarsening of η precipitates, most likely as a cost of η′ precipitates. A higher number of the precipitates can be observed at the grain boundaries. Moreover, precipitate-free zones (PFZ) next to the grain boundaries can be observed with a thickness up to 30 nm.

Figure 9 shows TEM micrographs of the ϕ8 samples. A similar phenomenon can be observed as for the ϕ10 samples. Again, as the aging temperature increases, the size of the precipitates increases. This effect is stronger for the ϕ8 samples than for the ϕ10 samples, with more pronounced precipitate growth at the same aging temperature. Furthermore, PFZs are already observed at an aging temperature of 120 °C, indicating over-aging. The HE8 sample displays more pronounced grain refinement in comparison to the HE10 sample. Grains of nanometer size (below 100 nm) can already be observed, which was not the case for the HE10 sample. Artificial aging resulted in a precipitation process. In addition, the number of grain boundary precipitates is higher in comparison to samples with lower deformation rate.

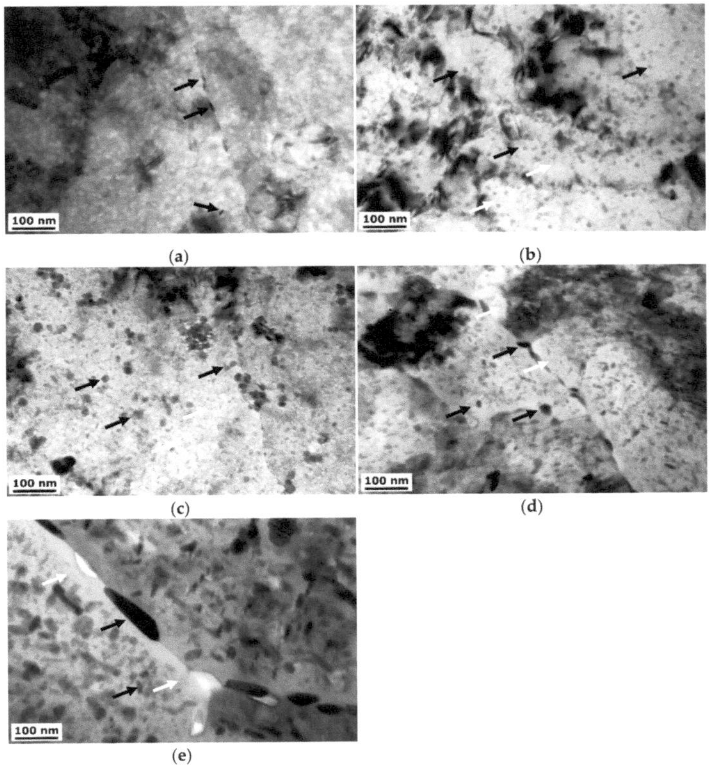

Figure 8. TEM micrographs of the: (**a**) HE10, (**b**) HE10 HT1, (**c**) HE10 HT2, (**d**) HE10 RRA1, and (**e**) HE10 RRA2 samples, black arrows indicate η precipitates, white arrows—η′ precipitates.

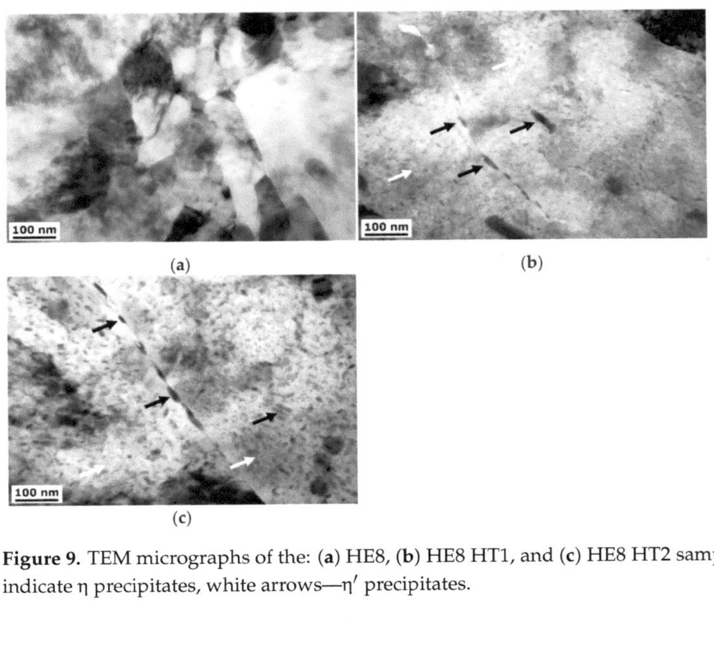

Figure 9. TEM micrographs of the: (**a**) HE8, (**b**) HE8 HT1, and (**c**) HE8 HT2 samples, black arrows indicate η precipitates, white arrows—η′ precipitates.

3.1.3. XRD

XRD analysis was performed to confirm TEM observations. Figure 10 shows results of the analysis for samples CG T6, HE10, HE10 HT2, HE8, and HE8 HT2. For all the samples, the same phases have been identified. For each sample, the most prominent phases are Al solid solution, η precipitates (labeled $MgZn_2$), $MgCuAl_2$, and Al_3Cu_2. In addition, η′ precipitates were detected, but their very small size resulted in diminished XRD peaks. Similar observations were found in work [30], where XRD measurements of AA7075 revealed peaks for $MgZn_2$ and Al_2Cu; however, the microstructure observations show the presence of $MgZn_2$. As in the present study, this indicates that these precipitates constitute the majority of the microstructure.

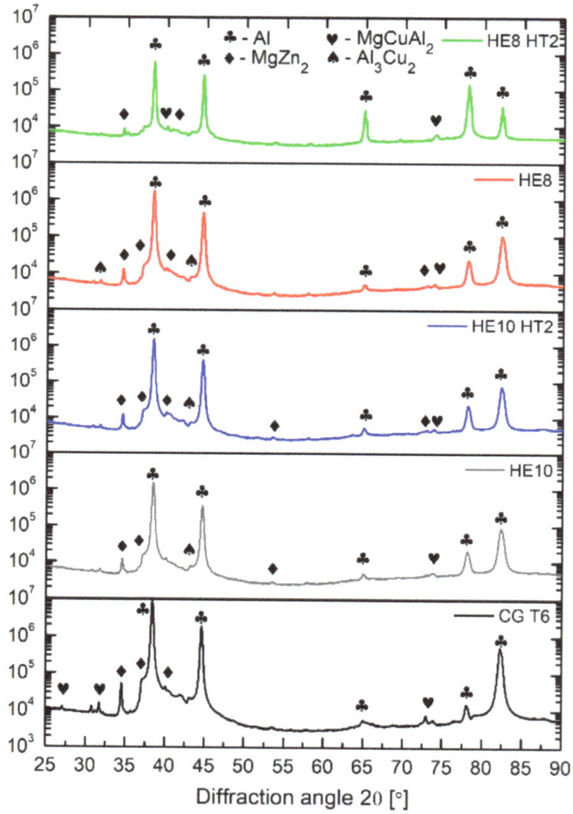

Figure 10. XRD analysis of the CG T6, HE10, HE10 HT2, HE8 and HE8 HT2 samples (PDF no. of the phases identified: PDF 00-004-0787—Al, PDF 01-073-5874—$MgCuAl_2$, PDF 00-034-0457—$MgZn_2$, PDF 01-071-5716—Al_3Cu_2).

3.2. Mechanical Properties

3.2.1. Microhardness Measurements

Figure 11 presents a graph of the microhardness measurement results. The CG T6 sample has an average hardness of 181 HV0.2. Following HE and natural aging, the microhardness changes are insignificant. The HT1 and HT2 treatments improved the microhardness of both samples. However, in the case of the ϕ8 samples, microhardness saturation was already observed for HT1, and a further increase in aging temperature caused a slight decrease in microhardness. In the case of the ϕ10 samples, the highest microhardness value of 202 HV0.2 was obtained for the HE10 HT2 sample. RRA treatment substantially decreased microhardness below that of the CG T6 sample. Based on these results, samples RRA1 and RRA2 were not investigated further.

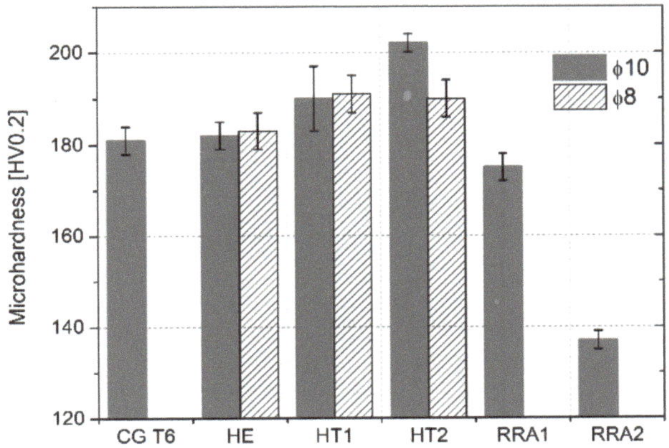

Figure 11. Microhardness results of the samples.

3.2.2. Tensile Tests

The representative stress–strain curves obtained by the tensile tests are shown in Figure 12. Table 3 presents the quantitative results. The samples were tested in the direction longitudinal to the extrusion direction. The CG T6 sample displays the lowest UTS value of about 540 MPa. All the samples that underwent HE show an improvement in UTS and YS, but a reduction in E_b. The increase in UTS, depending on a sample, equals from 50 MPa to 135 MPa, indicating a significant improvement. The tensile strength of the plastically deformed samples is dependent on the aging temperature. For the ϕ10 samples, the highest results are observed for the naturally aged sample, with a UTS and E_b of 674 MPa and 17%, respectively. Both HT1 and HT2 show a slight decrease in YS, UTS, and E_b. Among the ϕ8 samples, the highest strength of 625 MPa was obtained for the HE8 HT1 samples. However, it has to be emphasized that for the ϕ8 samples, the variations in the results is considerable, indicating the inhomogeneity of the microstructure.

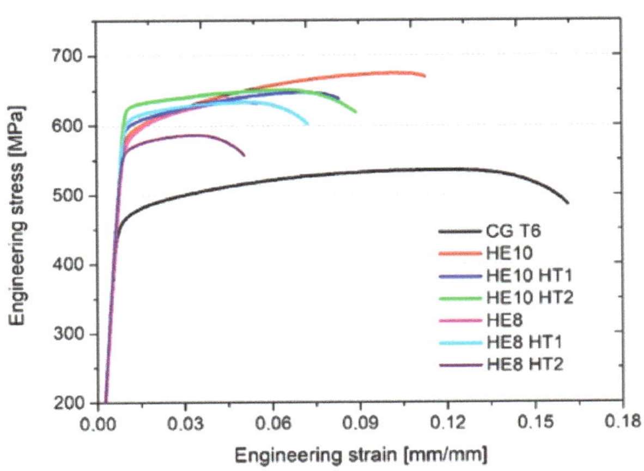

Figure 12. Representative engineering stress–strain curves for the reference CG T6 sample and samples after hydrostatic extrusion with subsequent natural aging (HE10 and HE8) and artificial aging (HT1 and HT2).

Table 3. Results obtained from the tensile tests.

Sample	YS [MPa]	UTS [MPa]	E_b [%]
CG T6	463 ± 4	539 ± 3	16.9 ± 1.4
HE10	582 ± 2	674 ± 2	11.0 ± 0.2
HE10 HT1	603 ± 11	652 ± 8	8.8 ± 0.5
HE10 HT2	608 ± 22	639 ± 19	9.3 ± 0.2
HE8	552 ± 33	608 ± 53	4.7 ± 2.8
HE8 HT1	588 ± 24	625 ± 18	7.5 ± 1.9
HE8 HT2	563 ± 16	590 ± 11	5.4 ± 1.1

3.3. Electrochemical Properties

Figure 13 shows the curves generated from the PP tests, and Figure 14 shows the average electrochemical parameters for each sample. The polarization curves for the HE and CG T6 samples are generally similar in shape, with the principal difference being the size of hysteresis loop (Figure 13a). As deformation increases, the hysteresis loop becomes larger. This indicates that the repassivation process is slower for the refined samples, and consequently suggests that the morphology of corrosion attacks may differ. The cathodic current density during the forward scan substantially decreases with increasing plastic deformation of the microstructure (Figure 13a). The corrosion potentials (E_{corr}) of the HE samples are lower than those of the CP T6 sample (Figure 14). The anodic behavior is dominated by active dissolution, as characterized by the lack of a passive region and an abrupt increase in current density above the corrosion potential.

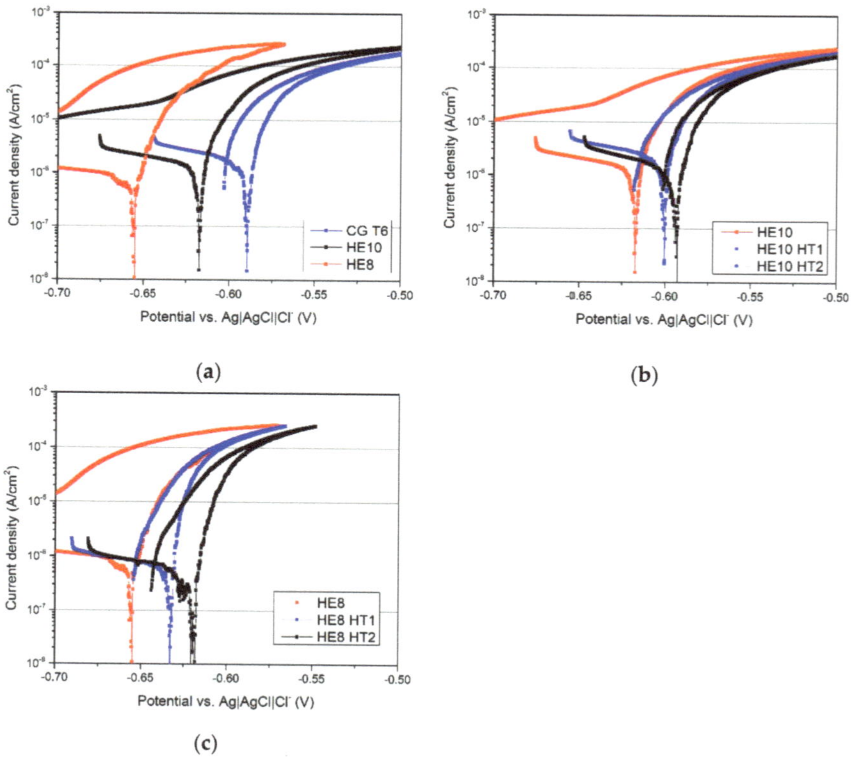

Figure 13. Results of PP tests in 0.1M NaCl, for the (**a**) CG T6, HE10, and HE8 samples; (**b**) the HE10 sample before and after aging; and (**c**) HE8 sample after aging.

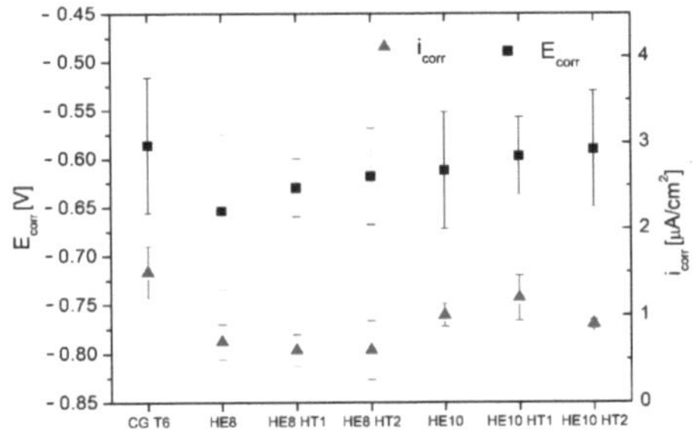

Figure 14. The electrochemical parameters E_{corr}, E_{rep} (left Y axis), and i_{corr} (right Y axis) for each sample.

Following artificial aging, the repassivation rates of both refined samples increase, as the hysteresis loop defined by the reverse scan curve becomes smaller (Figure 13b,c). This suggests a change in the morphology of the corrosion attack [31]. In the cathodic branch, the current density of the HE8 sample remains unaffected by aging, while that of the HE10 sample increases slightly. As aging temperature increases, E_{corr} shifts to more noble values. Most notably, the corrosion currents (i_{corr}) of all samples which are either refined or both aged and refined are lower than that of the CG T6 sample, indicating a slight improvement in corrosion resistance. Note that the differences in E_{corr} and i_{corr} are small, and the major dissimilarity in corrosion resistance lies in the size of the hysteresis loop, and hence the morphology of the corrosion attack.

Following PP testing, the microstructure was investigated further. Figure 15 shows SEM micrographs of the surfaces and cross-sections of the CG T6, HE10, and HE8 samples. Numerous pits can be observed on the CG T6 sample, where the whole examined surface was covered with them. For the samples that underwent HE, the corrosion attack is limited and more localized. For the HE8 sample, the corrosion attack on the surface looks like an intergranular, where the grain boundaries are more susceptible to the attack. Moreover, corrosion products can be observed on the surfaces. The cross-sections of all three samples display corrosion attack, which can be considered as combined intergranulars with pitting. Pits initiate in the vicinity of the intermetallic particles and further propagate along the extrusion direction, and correlate with highly elongated grains. Dissolution occurs along grain boundaries. As such, the dissolute networks of the HE10 and HE8 samples are more developed, as those samples possess a higher density of grain boundaries. In the case of the length of the corrosion attack, it varied between the samples. It was the most significant in the case of the CG T6 sample and equaled up to 100 μm. For the refined samples, this value was reduced and did not exceed 70 μm. Differences between HE10 and HE8 samples are negligible.

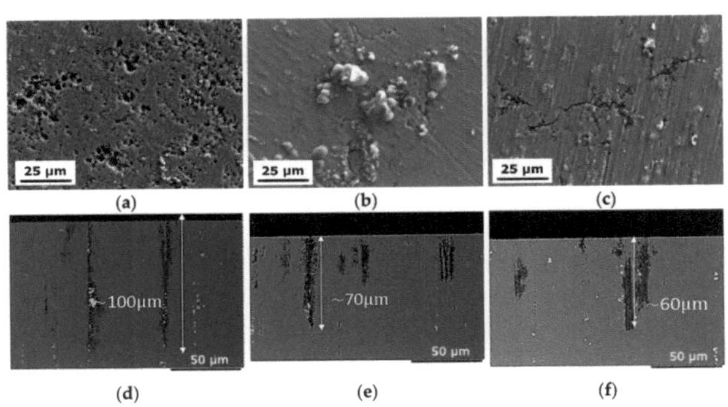

Figure 15. SEM micrographs of the surfaces of: (**a**) CG T6, (**b**) HE10, (**c**) HE8 and cross-sections of (**d**) CG T6, (**e**) HE10, (**f**) HE8 samples. 50 μm.

4. Discussion

4.1. The Influence of Microstructure Evolution on Mechanical Properties

Plastic deformation via HE in the present study with the application of low values of strain [32] led to the formation of substructures, within which the majority of grain boundaries were LAGBs. Higher strain values would be required to generate HAGBs by rotating the subgrains over 15°. The evolution of grain boundary angles during HE has been analyzed previously [33]. The misorientation angles are greatest when between grains oriented in the <111> and <001> directions. When the boundary lies between grains of similar orientations, then the misorientation angle in substantially lower. However, as the strain increases during deformation, the misorientation angles of such boundaries are subject to the largest increase. The grain refinement of the ϕ8 samples was slightly greater than that of the ϕ10 samples. This was reflected in an increased number of LAGBs, which resulted in more pronounced substructure formation. The grain orientation was somewhat invariant, with the majority of grains oriented along the <111> direction.

When compared to other SPD processes, such as ECAP or high-pressure torsion, HE produces grains of a distinctive shape. The grains are strongly elongated along the extrusion direction (longitudinal plane), while on the transverse plane the grains are equiaxial (see Figure 2). On the longitudinal plane the fibrous grains are defined by HAGBs, while their interiors feature networks of LAGBs. In terms of mechanical strength, the Hall–Petch equation describes an increase in yield strength as grain size decreases [34,35]. However, strength is also influenced by structure—namely, the grain boundary misorientation angle [36]. Therefore, the ϕ8 sample displayed no further increase in mechanical strength when compared to the ϕ10 sample, as the majority of grains were LAGBs. Such grains provide a weaker barrier to moving dislocations in comparison to HAGBs, and hence impart a lower contribution to strengthening the material [37]. Grain boundaries act as the primary strengthening mechanism in the case of plastic deformation; as the applied strain increases—creating HAGBs with smaller grain sizes—the mechanical strength increases. Accordingly, the results in this paper show that YS and UTS were significantly higher for the samples that underwent HE than the CG T6 sample. Note also that the increase in applied strain (from 1.41 to 1.85), which resulted in more substantial grain refinement, did not cause further improvement in mechanical strength. In addition to the misorientation angle between grains, precipitates are also a strengthening factor, which needs to be discussed.

In work [38], AA 7075 samples with both CG and UFG microstructure were aged. The results showed that, in equivalent conditions, the UFG sample obtained higher mechanical strength. However, precipitation strengthening was more impactful for the CG sample. Following T6 aging, the yield strength increment was 160% for the CG sample and only 26% for the UFG sample. The difference was caused by the size of the η and η' phase precipitates within the grain interiors, as these precipitates pin dislocations. This is known as the Orowan mechanism. The size of the strengthening precipitates was less than 5 nm within the UFG sample; such precipitates within the CG sample were much larger, at approximately 60 nm. Further work obtained similar findings when examining AA 7475 following HE [18]. Grain boundaries and dislocations are favorable locations for precipitation nucleation; however, such precipitates do not contribute to the strengthening of the material. Therefore, to obtain a high mechanical strength, the number of boundary precipitates should be reduced. This phenomenon explains the smaller values of YS and UTS obtained for the ϕ8 samples in comparison with the ϕ10 samples. In the case of the latter one, due to reduced grain refinement resulting from lower applied strain, the volume fraction of both grain boundaries and boundary precipitates was smaller. The HE10 sample developed the highest UTS value due to an optimal correlation between the number of grain boundaries and natural aging, engendered by a slower precipitation process. Precipitation was accelerated by the artificial aging process, resulting in a higher number of boundary precipitates and therefore a decrease in mechanical strength. In terms of mechanical strength, the elevated number of grain boundaries was unfavorable for the ϕ8 samples, as they caused a higher number of boundary precipitates at the expense of interior precipitates, which act as strengthening factors.

The ϕ10 samples were additionally subjected to RRA. This approach was taken to obtain high mechanical strength [39] together with excellent corrosion resistance. Following heat treatment to a T6 state, the CG T6 sample was heated to retrogression temperature and aged for a short time. Following RRA treatment, the resultant microstructure was more thermodynamically stable, and consisted of a greater number of bigger precipitates when compared to those formed due to the T6 heat treatment only [40]. Previous work showed that RRA reduced the corrosion depth of AA 7075, while mechanical strength remained high if parameters were optimized [41]. Due to adiabatic heating, which momentarily causes the temperature to increase to approximately 100–150 °C, the HE process influences precipitation. Further work, using AA 6082, showed that the precipitation sequence changed during HE, creating an additional phase capable of nucleation, which was not present within a CG sample [42]. The precipitation process differs also because of the increased number of structural defects, such as grain boundaries and dislocations, on which nucleation can occur [18]. Hence, RRA treatment of the HE samples caused a substantial decrease in microhardness. Following HE, the microstructure is much more susceptible to the precipitation process. Precipitation both along the grain boundaries and within the grain interiors underwent significant coarsening. Due to the increased number of grain boundaries, the number of boundary precipitates also increased. Following RRA treatment, the growth of such precipitates caused a substantial decrease in microhardness, and the obtained results showed that in the case of UFG sample, the RRA treatment is not effective and leads to extensive over-aging.

4.2. The Evolution of Resistance to Corrosion

The microstructure evolution during HE with the following aging influences the electrochemical properties of the AA 7075, as presented in Figure 14. The CG T6 sample displayed the highest E_{corr} values. For the samples that underwent HE, these parameters were reduced. However, the value of i_{corr}, which is directly correlated with corrosion rate, was reduced after HE process. The lowest values of i_{corr} were obtained for the ϕ8 samples, where the artificial aging resulted in further slight improvement. In the case of the ϕ10 samples, HE10 and HE HT12 displayed similar values of i_{corr}, with the latter having increased values of E_{corr} and E_{rep}. The quantitative results obtained from the electrochemical tests together with data from the literature are gathered in Table 4. The results for AA7075 varies in dependance of the corrosion environment—for higher concentration of NaCl, lower values of E_{corr} and higher values of i_{corr} are achieved. Nevertheless, the obtained results in a present study are within the results obtained in the literature. Upon investigation of the sample surfaces following PP testing, the CG T6 sample had the most pronounced susceptibility to corrosion attack. The corrosion attack was much more localized for the samples that underwent HE, with a limited number of pits observed on their surfaces. The entire examined area of the CG T6 sample was damaged, indicating that the corrosion was more uniform, rather than producing individual pits. However, upon examination of the cross-sections, all samples displayed pitting/intergranular corrosion. The pits were spread along the HE direction. A greater number of surface pits means a larger number of potential locations for further propagation of the corrosion into the material. The smaller number of pits observed for the samples that underwent HE may be caused by a denser passive layer. A more stable and integral passive layer has been observed for UFG materials in relation to the corrosion resistance of materials following SPD processing [13,43]. This phenomenon is caused by an elevated number of structural defects such as grain boundaries and dislocations, which are favorable locations for oxidation.

Table 4. Results of electrochemical properties with comparison to the literature data.

Sample	E_{corr} [mV]	i_{corr} [μA/cm^2]	Environment	Reference
CG T6	−585	1.5		
HE10	−612	1.0		
HE10 HT1	−597	1.2		
HE10 HT2	−590	0.9	0.1 M NaCl	Present study
HE8	−654	0.7		
HE8 HT1	−630	0.6		
HE8 HT2	−618	0.6		
AA7075 with different tempers	−800 to −770	-	3.5% NaCl	[44]
AA7075	−815	4.01	3.5% NaCl	
	−853	4.3	1 M NaCl	[45]
	−945	6.76	2 M NaCl	
	−1038	9.58	3 M NaCl	
AA7075 CG	−1165	76.2	3.5% NaCl	[26]
AA7075 after rolling	−1162 to −1088	5.4 to 55.9		
AA7075	−704	71.5	3.5% NaCl	[46]
AA7075	−553 to −497	0.1 to o.5	0.001 M NaCl	[47]
AA7075 welds	−1350 to −750	-	3.5% NaCl	[48]

Although corrosion attacks were observed in the vicinity of intermetallic particles, such particles themselves remained untouched. This is due to the higher electrochemical potential of these particles, which primarily contain Fe and Cu, in comparison to the Al matrix [23,49]. Further dissolution of the material takes place via grain boundaries. Previous work showed that HAGBs provide preferable locations for the propagation of corrosion due to the presence of enhanced excess energy [13]. Further work found that AA 5182 was susceptible to intergranular corrosion due to β-phase precipitates at grain boundaries, and that LAGBs provide resistance to attack [50]. HAGBs provided a variable resistance to attack, dependent upon the grain boundary plane. This could be caused by the crystallography of β-phase precipitation, which determines whether precipitation along boundaries is continuous or discontinuous. For 7xxx AAs, intergranular corrosion was shown to be caused by the presence of PFZs [51], and susceptibility to intergranular corrosion was reduced by decreasing the sizes of the PFZs. Higher susceptibility to pitting corrosion is unfavorable as it can initiate cracking, as shown in [52], where authors investigated the transition of corrosion pits to cracks under fatigue for AA 7075.

Naturally aged, fine-grained AA 7075 was less susceptible to the growth of stable pits compared to CG AA 7075 in [47]. This finding was attributed to the size of the intermetallic particles, which were larger for the CG sample. The potential difference between the cathodic particles and the matrix was higher in the CG sample than in the refined sample. This was caused by a higher content of alloying elements within particles, resulting in a greater driving force for pitting initiation and propagation in the CG sample. However, in this paper, the size of the intermetallic particles was constant across all samples. Therefore, the primary cause of the enhanced corrosion resistance of the deformed samples is grain refinement, which causes a better protective passive layer on the surface and therefore the number of corrosion nucleation sites is smaller in comparison to the CG T6 sample. Further propagation of the corrosion occurs along HAGB, and the observed corrosion damage is similar for the samples.

5. Conclusions

In this paper, AA 7075 was subjected to HE and aging. The microstructure evolution, mechanical properties, and resistance to corrosion were examined, and the results were compared to a CG T6 sample. From this, the following conclusions can be drawn:

- The HE process caused substantial grain refinement; however, the applied strain values of $\varepsilon = 1.41$ and $\varepsilon = 1.85$ resulted in the majority of grain boundaries being LAGBs;

- The samples that underwent HE displayed different precipitation phenomena—the increased number of grain boundaries resulted in a higher number of boundary precipitates, which do not contribute to an increase in mechanical strength;
- HE with an aging process caused an increase in tensile strength in a range of 50–135 MPa in comparison to CG T6 sample. The highest value of 675 MPa was obtained for singe-stepped HE with natural aging;
- Due to the increase in applied strain during HE, a lower aging temperature was required to develop high mechanical strength;
- The samples which underwent HE showed lower susceptibility to localized corrosion in comparison to the CG T6 sample as a result of the grain refinement and higher fraction of LAGBs, which are resistant to corrosion attack.

Author Contributions: Conceptualization, M.O. and E.U.-B.; methodology, M.O., E.U.-B. and M.K.; validation, E.U.-B., L.Ś. and M.K.; formal analysis, M.O.; investigation, M.O., E.U.-B., P.S., M.K. and B.A.-C.; resources, E.U.-B.; data curation, M.O., P.S. and B.A.-C.; writing—original draft preparation, M.O., E.U.-B. and L.Ś.; writing—review and editing, M.O. and E.U.-B.; visualization, M.O.; supervision, E.U.-B., L.Ś. and J.M.; project administration, E.U.-B.; funding acquisition, E.U.-B. and L.Ś. All authors have read and agreed to the published version of the manuscript.

Funding: This research was funded by POB "Technologie Materiałowe" of Warsaw University of Technology (Poland) within the Excellence Initiative: Research University (IDUB) program under the project entitled: Optimization of the corrosion resistance and mechanical properties of precipitation strengthened 7075 aluminum alloy subjected to hydrostatic extrusion.

Institutional Review Board Statement: Not applicable.

Informed Consent Statement: Not applicable.

Data Availability Statement: The data presented in this study are available on request from the corresponding author.

Acknowledgments: The authors would like to thank Kamil Majchrowicz from WUT for conducting tensile tests.

Conflicts of Interest: The authors declare no conflict of interest.

References

1. Polmear, I.J. *Solute Partitioning to Enhance Mechanical Properties of Aged Aluminium Alloys*; Woodhead Publishing Limited: Cambridge, UK, 2011. [CrossRef]
2. Majchrowicz, K.; Pakiela, Z.; Giżyński, M.; Karny, M.; Kulczyk, M. High-cycle fatigue strength of ultrafine-grained 5483 Al-Mg alloy at low and elevated temperature in comparison to conventional coarse-grained Al alloys. *Int. J. Fatigue* **2018**, *106*, 81–91. [CrossRef]
3. Klein, T.; Reiter, L.; Schnall, M. Wire-arc additive manufacturing of Al-Zn5.5-Mg-Cu (ML7075): Shifting paradigms of additive manufacture-ability. *Mater. Lett.* **2022**, *313*, 131841. [CrossRef]
4. Klein, T.; Schnall, M.; Gomes, B.; Warczok, P.; Fleischhacker, D.; Morais, P.J. Wire-arc additive manufacturing of a novel high-performance Al-Zn-Mg-Cu alloy: Processing, characterization and feasibility demonstration. *Addit. Manuf.* **2021**, *37*, 101663. [CrossRef]
5. Kairy, S.K.; Turk, S.; Birbilis, N.; Shekhter, A. The role of microstructure and microchemistry on intergranular corrosion of aluminium alloy AA7085-T7452. *Corros. Sci.* **2018**, *143*, 414–427. [CrossRef]
6. Wawer, K.; Lewandowska, M.; Kurzydłowski, K.J. Improvement of mechanical properties of a nanoaluminium alloy by precipitate strengthening. *Arch. Met. Mater.* **2012**, *57*, 877–881. [CrossRef]
7. Valiev, R.Z.; Islamgaliev, R.K.; Alexandrov, I. Bulk nanostructured materials from severe plastic deformation. *Prog. Mater. Sci.* **2000**, *45*, 103–189. [CrossRef]
8. Sabbaghianrad, S.; Langdon, T.G. A critical evaluation of the processing of an aluminum 7075 alloy using a combination of ECAP and HPT. *Mater. Sci. Eng. A* **2014**, *596*, 52–58. [CrossRef]
9. Darban, H.; Mohammadi, B.; Djavanroodi, F. Effect of equal channel angular pressing on fracture toughness of Al-7075. *Eng. Fail. Anal.* **2016**, *65*, 1–10. [CrossRef]
10. Murashkin, M.Y.; Sabirov, I.; Sauvage, X.; Valiev, R.Z. Nanostructured Al and Cu alloys with superior strength and electrical conductivity. *J. Mater. Sci.* **2016**, *51*, 33–49. [CrossRef]
11. Ralston, K.D.; Birbilis, N. Effect of Grain Size on Corrosion: A Review. *Corrosion* **2010**, *66*, 075005-13. [CrossRef]

12. Ralston, K.D.; Fabijanic, D.; Birbilis, N. Effect of grain size on corrosion of high purity aluminium. *Electrochim. Acta* **2011**, *56*, 1729–1736. [CrossRef]
13. Orłowska, M.; Ura-Bińczyk, E.; Olejnik, L.; Lewandowska, M. The effect of grain size and grain boundary misorientation on the corrosion resistance of commercially pure aluminium. *Corros. Sci.* **2019**, *148*, 57–70. [CrossRef]
14. Aida, T.; Matsuki, K.; Horita, Z.; Langdon, T.G. Estimating the equivalent strain in equal-channel angular pressing. *Scr. Mater.* **2001**, *44*, 575–579. [CrossRef]
15. Lewandowska, M.; Kurzydlowski, K.J. Recent development in grain refinement by hydrostatic extrusion. *J. Mater. Sci.* **2008**, *43*, 7299–7306. [CrossRef]
16. Topolski, K.; Pachla, W.; Garbacz, H. Progress in hydrostatic extrusion of titanium. *J. Mater. Sci.* **2013**, *48*, 4543–4548. [CrossRef]
17. Przybysz, S.; Kulczyk, M.; Pachla, W.; Skiba, J.; Wróblewska, M.; Mizera, J.; Moszczynska, D. Anisotropy of mechanical and structural properties in aa 6060 aluminum alloy following hydrostatic extrusion process. *Bull. Pol. Acad. Sci. Tech. Sci.* **2019**, *67*, 709–717. [CrossRef]
18. Lewandowska, M.; Wawer, K.; Kozikowski, P.; Ohnuma, M.; Kurzydlowski, K.J. Precipitation in a Nanograined 7475 Aluminium Alloy-Processing, Properties and Nanoanalysis. *Adv. Eng. Mater.* **2014**, *16*, 482–485. [CrossRef]
19. Ura-Bińczyk, E. Improvement of Pitting-Corrosion Resistance of Ultrafine-Grained 7475 Al Alloy by Aging. *Materials* **2022**, *15*, 360. [CrossRef]
20. Orłowska, M.; Ura-Bińczyk, E.; Śnieżek, L.; Skudniewski, P.; Kulczyk, M.; Adamczyk-Cieślak, B.; Majchrowicz, K. The Influence of Heat Treatment on the Mechanical Properties and Corrosion Resistance of the Ultrafine-Grained AA7075 Obtained by Hydrostatic Extrusion. *Materials* **2022**, *15*, 4343. [CrossRef]
21. Brynk, T.; Molak, R.M.; Janiszewska, M.; Pakiela, Z. Digital Image Correlation measurements as a tool of composites deformation description. *Comput. Mater. Sci.* **2012**, *64*, 157–161. [CrossRef]
22. Molak, R.M.; Paradowski, K.; Brynk, T.; Ciupinski, L.; Pakiela, Z.; Kurzydlowski, K.J. Measurement of mechanical properties in a 316L stainless steel welded joint. *Int. J. Press. Vessel. Pip.* **2009**, *86*, 43–47. [CrossRef]
23. Birbilis, N.; Buchheit, R.G. Electrochemical Characteristics of Intermetallic Phases in Aluminum Alloys. *J. Electrochem. Soc.* **2005**, *152*, B140–B151. [CrossRef]
24. Li, G.; Liang, Y.; Chen, F.; Han, Y.; Sun, L. Evolution of Microstructure and Mechanical Properties in a Single-Stage Aged 7075 Aluminum Alloy Pulse Variable Polarity Plasma Arc Welded Joint. *Metals* **2019**, *9*, 428. [CrossRef]
25. Liu, W.; Man, Q.; Li, J.; Liu, L.; Zhang, W.; Wang, Z.; Pan, H. Microstructural evolution and vibration fatigue properties of 7075-T651 aluminum alloy treated by nitrogen ion implantation. *Vacuum* **2022**, *199*, 110931. [CrossRef]
26. Sun, Q.; Yang, M.; Jiang, Y.; Lei, L.; Zhang, Y. Achieving excellent corrosion resistance properties of 7075 Al alloy via ultrasonic surface rolling treatment. *J. Alloys Compd.* **2022**, *911*, 165009. [CrossRef]
27. Huo, W.; Sun, T.; Hou, L.; Zhang, W.; Zhang, Y.; Zhang, J. Effect of heating rate during solution treatment on microstructure, mechanical property and corrosion resistance of high-strength AA 7075 alloy. *Mater. Charact.* **2020**, *167*, 110535. [CrossRef]
28. Xu, Y.; Zhuang, X.; Zhang, W.; Li, Q.; Zhao, Z. Mechanical behaviors and microstructure characteristics of W-tempered and peak-aged 7075 alloy sheets under low frequency vibration–assisted tension. *Mater. Sci. Eng. A* **2022**, *833*, 142338. [CrossRef]
29. Du, Z.; Deng, Z.; Xiao, A.; Cui, X.; Yu, H.; Feng, Z. Effect of the aging process on the micro-structure & properties of 7075 aluminum alloy using electromagnetic bulging. *J. Manuf. Process.* **2021**, *70*, 15–23. [CrossRef]
30. Ma, J.; Wang, Q.; Zhang, T.; Cao, H.; Yang, Y.; Zhang, Z. Effect of natural aging time on tensile and fatigue anisotropy of extruded 7075 Al alloy. *J. Mater. Res. Technol.* **2022**, *18*, 4683–4697. [CrossRef]
31. Brunner, J.G.; May, J.; Höppel, H.W.; Göken, M.; Virtanen, S. Localized corrosion of ultrafine-grained Al–Mg model alloys. *Electrochim. Acta* **2010**, *55*, 1966–1970. [CrossRef]
32. Prangnell, P.B.; Bowen, J.R.; Apps, P.J. Ultra-fine grain structures in aluminium alloys by severe deformation processing. *Mater. Sci. Eng. A* **2004**, *375–377*, 178–185. [CrossRef]
33. Lewandowska, M. Mechanism of Grain Refinement in Aluminium in the Process of Hydrostatic Extrusion. *Solid State Phenom.* **2006**, *114*, 109–116. [CrossRef]
34. Hall, E.O. The Deformation and Ageing of Mild Steel: III Discussion of Results. *Proc. Phys. Soc. Sect. B* **1951**, *64*, 747–753. [CrossRef]
35. Petch, N.J. The Cleavage Strength of Polycrystals. *J. Iron Steel Inst.* **1953**, *174*, 25–28.
36. Hansen, N. Hall–Petch relation and boundary strengthening. *Scr. Mater.* **2004**, *51*, 801–806. [CrossRef]
37. Bazarnik, P.; Huang, Y.; Lewandowska, M.; Langdon, T.G. Structural impact on the Hall–Petch relationship in an Al–5Mg alloy processed by high-pressure torsion. *Mater. Sci. Eng. A* **2015**, *626*, 9–15. [CrossRef]
38. Ma, K.; Wen, H.; Hu, T.; Topping, T.D.; Isheim, D.; Seidman, D.N.; Lavernia, E.J.; Schoenung, J.M. Mechanical behavior and strengthening mechanisms in ultrafine grain precipitation-strengthened aluminum alloy. *Acta Mater.* **2014**, *62*, 141–155. [CrossRef]
39. Karaaslan, A.; Kaya, I.; Atapek, H. Effect of aging temperature and of retrogression treatment time on the microstructure and mechanical properties of alloy AA 7075. *Met. Sci. Heat Treat.* **2007**, *49*, 443–447. [CrossRef]
40. Luiggi, N.J.; Valera, V. Kinetic study of an AA7075 alloy under RRA heat treatment. *J. Therm. Anal. Calorim.* **2017**, *130*, 1885–1902. [CrossRef]
41. Özer, G.; Karaaslan, A. Effect of RRA heat treatment on corrosion and mechanical properties of AA7075. *Mater. Corros.* **2019**, *70*, 2064–2072. [CrossRef]

42. Chrominski, W.; Lewandowska, M. Precipitation phenomena in ultrafine grained Al–Mg–Si alloy with heterogeneous microstructure. *Acta Mater.* **2016**, *103*, 547–557. [CrossRef]
43. Song, D.; Ma, A.B.; Jiang, J.H.; Lin, P.H.; Shi, J. Improving corrosion resistance of pure Al through ECAP. *Corros. Eng. Sci. Technol.* **2011**, *46*, 505–512. [CrossRef]
44. Andreatta, F.; Terryn, H.; de Wit, J.H.W. Corrosion behaviour of different tempers of AA7075 aluminium alloy. *Electrochim. Acta* **2004**, *49*, 2851–2862. [CrossRef]
45. Kumar, S.; Kumar, A.; Vanitha, C. Corrosion behaviour of Al 7075/TiC composites processed through friction stir processing. *Mater. Today Proc.* **2019**, *15*, 21–29. [CrossRef]
46. Zhang, C.; Huang, G.; Liu, Q. Research on local corrosion behavior of thermo-mechanically affected zone in dissimilar AA2024/7075 friction stir welds. *Intermetallics* **2021**, *130*, 107081. [CrossRef]
47. Tian, W.; Li, S.; Wang, B.; Liu, J.; Yu, M. Pitting corrosion of naturally aged AA 7075 aluminum alloys with bimodal grain size. *Corros. Sci.* **2016**, *113*, 1–16. [CrossRef]
48. Kumar, P.V.; Reddy, G.M.; Rao, K.S. Microstructure, mechanical and corrosion behavior of high strength AA7075 aluminium alloy friction stir welds–Effect of post weld heat treatment. *Def. Technol.* **2015**, *11*, 362–369. [CrossRef]
49. Buchheit, R.G. A Compilation of Corrosion Potentials Reported for Intermetallic Phases in Aluminum Alloys. *J. Electrochem. Soc.* **1995**, *142*, 3994–3996. [CrossRef]
50. Davenport, A.J.; Yuan, Y.; Ambat, R.; Connolly, B.J.; Strangwood, M.; Afseth, A.; Scamans, G.M. Intergranular Corrosion and Stress Corrosion Cracking of Sensitised AA5182. *Mater. Sci. Forum* **2006**, *519–521*, 641–646. [CrossRef]
51. Cao, C.; Zhang, D.; Wang, X.; Ma, Q.; Zhuang, L.; Zhang, J. Effects of Cu addition on the precipitation hardening response and intergranular corrosion of Al-5.2Mg-2.0Zn (wt.%) alloy. *Mater. Charact.* **2016**, *122*, 177–182. [CrossRef]
52. Sabelkin, V.; Perel, V.Y.; Misak, H.E.; Hunt, E.M.; Mall, S. Investigation into crack initiation from corrosion pit in 7075-T6 under ambient laboratory and saltwater environments. *Eng. Fract. Mech.* **2015**, *134*, 111–123. [CrossRef]

materials

Article

The Volumetric Wear Assessment of a Mining Conical Pick Using the Photogrammetric Approach

Jan Pawlik , Aleksandra Wróblewska-Pawlik and Michał Bembenek *

Faculty of Mechanical Engineering and Robotics, AGH University of Science and Technology, A. Mickiewicza 30, 30-059 Kraków, Poland
* Correspondence: bembenek@agh.edu.pl

Abstract: The rapid wear of conical picks used in rock cutting heads in the mining industry has a significant economic impact in cost effectiveness for a given mineral extraction business. Any mining facility could benefit from decreasing the cost along with a substantial durability increase of a conical pick; thus, the hardfacing method of production and regeneration should be taken into account. In order to automatize the regeneration, the wear rate assessment is necessary. This paper presents a methodology used to create a 3D photogrammetric model of most of the commercially available tangential-rotary cutters in their before and after abrasive exploitation state. An experiment of three factors on two levels is carried out to indicate the proper setup of the scanning rig to obtain plausible results. Those factors are: light level, presence of polarizing filter and the distance from the scanned object. The 3D scan of the worn out specimen is compared to the master model via algorithm developed by the authors. This approach provides more detailed information about the wear mechanism and can help either in roadheader cutting head diagnostics or to develop a strategy and optimize the toolpath for the numerically controlled hardfacing machine.

Keywords: conical pick; photogrammetry; mining; hardfacing; photogrammetric method

Citation: Pawlik, J.; Wróblewska-Pawlik, A.; Bembenek, M. The Volumetric Wear Assessment of a Mining Conical Pick Using the Photogrammetric Approach. *Materials* **2022**, *15*, 5783. https://doi.org/10.3390/ma15165783

Academic Editor: Liaoliang Ke

Received: 21 June 2022
Accepted: 18 August 2022
Published: 22 August 2022

1. Introduction

Wear of any kind of industrial tool is an inevitable phenomenon. The tools are subjected to many divergent wear mechanisms [1], depending mainly on the material selection and working environment [2]. The mining industry can be considered to be one of the extreme cases, considering the high loads and abrasive conditions to which the "end tools" are subjected. Considering tool wear measurement, one can stumble upon many approaches. Some of them are based on measurement of the response for a vibratory excitation [3,4]; others utilize mass measurement or harness a vision system [5].

A conical pick (sometimes named "tangential-rotary cutter") is a good example of a mining tool, designed to withstand harsh working conditions and crush hard rocks, yet it still tends to lose its primal shape and durability after a certain operational period. Replacement of the tool is problematic and expensive, but is sometimes needed even after a few hours of work, which leads to the need for remanufacturing of the tool [6,7]. The main challenge of the regeneration process of worn out tools at the first stage is the difficulty of classifying its wear rate and deciding whether the tool can still be in use or should be discarded. Whenever the cavities are excessive, it is crucial to determine the possibility of regeneration. If the particular pick is suitable for reconstruction, the next step is to develop a strategy for fabrication of a hardfaced coating [8]. This process is usually performed by skilled welders manually; however, today one can incorporate a numerically controlled hardfacing machine.

Previous research such as [9–11] proposes a method to define the C2 parameter which determines the wear rate that relies on the conical pick's loss in weight after the machine tool process. The method consists of disassembly of picks and measurement of their mass and the volume of output cut material obtained during the work of picks subjected to testing.

Using only a weighing sensor and measurements of the cut mineral makes it a suitable method of obtaining the parameter for rough analysis in difficult environment conditions such as mining facilities. This, however, does not bring about any further information on spatial deformation but only on the overall wear when compared to the master model, which can be misleading, especially during analysis of tools that have local cavities correlated to tools that have uniform abrasions (Figure 1). The products of plastic deformations of the working part can still be attached to the pick's body; therefore, theoretically, they can also add distortions to the mass measurement. Additionally, the tungsten carbide tip can have up to double the density of the steel, which can make mass measurements even less informative [12].

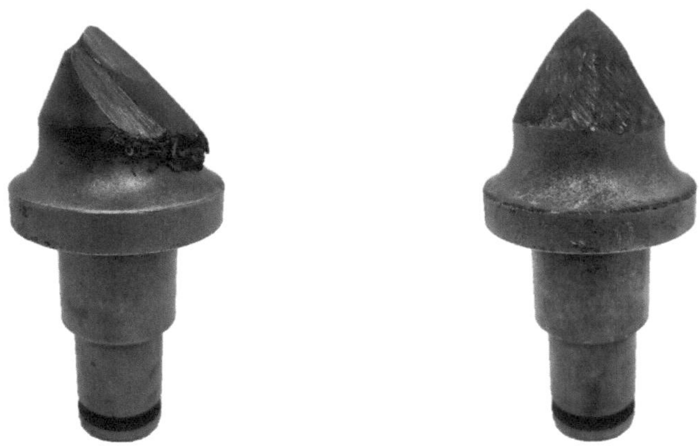

(a) (b)

Figure 1. Tools with local cavities—(**a**); uniform abrasions—(**b**).

In [13], the quality assessment procedure for picks was presented. Apart from calculating the C2 parameter, classification takes into account other criteria. Some of the measurements described in the paper require laboratory conditions; therefore, this process cannot be conducted in a difficult environment.

The next method of determining wear level is to analyze the mining power consumption signal [14]. The classification is carried out with the use of a wavelet transform for noise reduction of measurement data of the mining power signal and artificial intelligence. Despite the results being satisfactory, attention should be paid to selection of the base wavelet, which influences the obtained final results. This means that every sample should be analyzed by a person who should decide on the mathematical operator of the calculations, which can complicate the process.

Another method of obtaining the digital parameters of tool wear is image processing [15,16]. This gives some or all of the information that can be gathered by the human eye, creating a digital model that can be processed further on. For an industrial company that has the possibility to conduct LIDAR scanning, such a model can be obtained using the methodology described in [17–19]. However, this type of measurement uses the method of physically hitting a model's feature with light and measuring the reflection so that the texture of the object is not included in the results.

The solution to this problem that was chosen by the authors is the usage of a photogrammetry process that relies on images captured by a camera to reconstruct the 3D model coordinates using image overlap [20]. The photogrammetric model method brings forth enough information for conical picks regeneration, as the output data include information about local cavities of the operating part, its texture and its ability to be automated. This method does not require any expensive equipment or personnel training.

There are many photogrammetric methods, but one can distinguish three main ones considering the dynamic of an object that is projected and the cameras that are needed for image acquisition [21]:

- Static object, many cameras are triggered at the same moment, placed around the object;
- Static object, one camera moving around the object while taking pictures;
- Rotating object, static camera.

The authors have chosen the last method because of feasibility of the automation and low cost of the module capture setup. Data acquisition for photogrammetry should be conducted in a controlled environment [22–24] to ensure that the image quality is not influenced by factors such as extensive light, transparent surfaces, light reflections and surrounding objects. To do so, the choice of light source should be adapted to the object surface as well. It is best when the object diffuses light which is common as the outcome of the object's roughness. For metallic surfaces that can reflect light easily, the source light should be uniform and diffused [25]; additionally, cross-polarization and chromatic filters may enhance the results [26–28]. The baseline/distance ratio, i.e., the distance between two camera positions to the distance between the camera to the object ratio should be between 1/15 and 1/20 [29]. Crucial parameters that have a great impact on the suitability for photogrammetric synthesis are:

- Exposure time (shutter speed);
- Aperture;
- Depth of field;
- Sensitivity of light to camera (ISO).

Those parameters should be selected as a compromise, to result in capturing sharp images that are the appropriate input for the processing algorithm. The use of a tripod usually facilitates the process, especially for underexposed objects or long exposure time camera settings. The aperture parameter and shutter speed should be set according to ISO, the value of which should be the lowest possible to avoid digital noise in the images.

In the current study, the authors attempt to build a low-cost photogrammetric scanning setup for conical picks and estimate the process capabilities and general robustness, where the main goal of this paper is to develop and study a potential pick wear assessment algorithm scheme. The acquired data can serve many purposes, both as a practical tool for mining consumable evaluation and regeneration and as a mathematical model verification benchmark. The results of the scanning can be compared with the numerical models, such as EDEM approach by Liu et al. [30] or the peak cutting force model built by Kuidong et al. [31] or other tangential-rotary cutter theoretical approaches [32–39]. All of the cited research could benefit from introducing the method of confirmation of calculated mechanical properties with the output geometry of an exploited tool, existing in reality.

In the following sections, the authors describe the studied conical picks at various stages of wear and the scanning setup, both from the hardware and software point of view. Afterwards, the aligning and scaling algorithm is presented and tested on the divergent geometry. Lastly, the authors test robustness of their method with the help of the Taguchi Orthogonal Arrays design, in which they estimate which factor (light level, presence of polarizing film and the distance from camera to scanned object) has the biggest impact on the process stability.

2. Materials and Methods

2.1. Studied Specimen

In this study, the authors examine road cutting conical picks, manufactured from 34CrMo4 steel with geometry that is compliant with Figure 2. The tool in Figure 3 noted as CP0 is a brand-new pick, which will serve as a reference specimen for the degraded tools which had contact with abrasive rock.

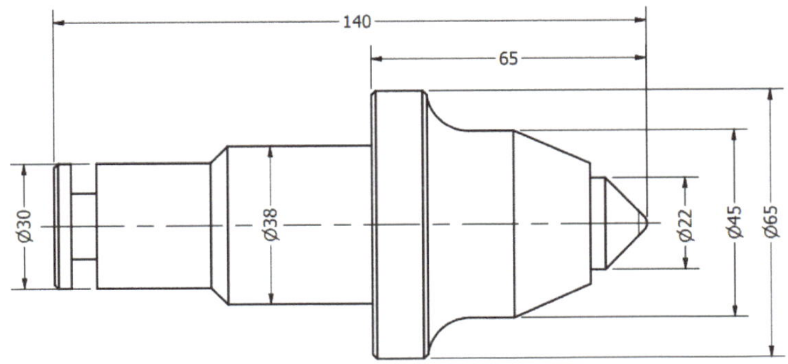

Figure 2. Technical drawing of the inspected conical pick. This geometry corresponds with the master model (CP0).

(**a**) (**b**) (**c**) (**d**) (**e**)

Figure 3. Image of all inspected conical pick specimens. The authors annotated them in the following manner: (**a**)—CP0 (master model), (**b**)—CP1, (**c**)—CP2, (**d**)—CP3, (**e**)—CP4.

2.2. Measurement Setup

The setup consisted of the phone camera, motorized turntable and the lightbox (Figure 4) and a computer with the required software installed.

(**a**) (**b**) (**c**)

Figure 4. Measurement setup: (**a**)—lightbox, (**b**)—turntable, (**c**)—phone stand with 3D printed holder. Please note that the above images present a pre-hardfaced pick, which is not included in this study.

In order to ensure the stable movement with constant velocity, the authors designed and manufactured a special rotary mount. This turntable was built from a stator base, a bearing and a belt-driven rotor pick holder. The motion was created by a bipolar stepper motor, controlled by a TMC2208 stepper driver (Trinamic GmBH, Hamburg, Germany), connected to an Arduino Nano microcontroller (Arduino, Sommerville, MA, USA). The pick holder was rotating at n = 0.6 rot/min. The golden and silver symbols on each of the 12 walls of the rotor holder serve as the reference pattern for the 3D reconstruction algorithm—if the reader was to rerun the experiment, he can apply any kind of non-repeating pattern. That pattern brings forth an especially valuable contribution when one scans a reflective and undeformed revolving solid.

2.3. Data Acquisition

Instead of image acquisitions, the authors decided to capture a video and split it into 60 frames. The video clip was captured with a 12.2 MP Sony IMX333 sensor (Sony, Tokyo, Japan), integrated in a Samsung Galaxy S8 smartphone (Samsung, Suwon, South Korea). The authors did not use the Samsung proprietary video-capturing software due to the possible presence of unknown filters and video enhancements; instead, they utilized IVCam software (e2esoft, Shanghai, China) and the capture was executed via a client installed on the computer. Measurement series were conducted using groups of parameters of values from the settings limit in Table 1. Selected parameters groups are described in Section 2.4.

Table 1. Camera and setup limits.

Light Level [lux]	Shutter [s]	ISO	Distance from Camera to Object [mm]	Polarizing Filters	Aperture
102–288	1/1000–1/20	200	155–250	yes/no	f/1.7

2.4. Image Processing

The authors used open-source software Meshroom, version 2021.1.0 (AliceVision Association, Paris, France) to obtain 3D models. It is a program based on the AliceVision framework with a specific pipeline for a project. Single steps of the pipeline consist of 3D model calculating algorithm steps:

- Natural feature extraction [40–44];
- Image matching [45];
- Features matching [46];
- Structure from motion [47];
- Depth maps estimation [48];
- Meshing [49];
- Texturing [50].

The first step was to extract 60 frames from captured videos. To do so, authors prepared the MATLAB (MathWorks, Natick, MA, USA) script that enabled saving images in a certain directory. Afterwards, a new project in Meshroom was created and the images saved were determined as the input for the algorithm.

The results of consecutive steps are presented in Section 3. The final step was to save the 3D model of the working part of the scanned picks in .stl format and analyze its geometry in the MATLAB environment (Section 2.5).

2.5. Statistical Analysis

The optimization process was planned as presented in Table 2. The first step was to choose appropriate parameter values for parametric analysis. The ISO value of 200 was set a priori and the light levels were set with the use of LED lights switch. The rest of the parameters were selected accordingly, for two of the light levels as in Table 3. The light level was measured at the position of the conical pick's tip with the GH59-14759A light sensor.

Table 2. Experimental scenarios.

Run	Light Level	Distance from Camera to Object	Polarizing Filters
1	−	−	−
2	−	−	+
3	−	+	−
4	−	+	+
5	+	−	−
6	+	−	+
7	+	+	−
8	+	+	+

Table 3. Camera and setup settings.

	Light Level [lux]	Distance from Camera to Object [mm]	Polarizing Filters Included
Lower limit (−)	102	155	No
Upper limit (+)	288	250	Yes

The impact of light level, distance to object and polarizing filter was analyzed by performing eight tests on the master model to choose optimal settings for the process. Optimization consisted of analysis of the influence of three factors with two levels (L8) on the resulting output function values that determined the model quality.

The process was optimized for best output model accuracy. For this purpose:

The algorithm for cavities classification was made and objective Function (1) was constructed for maximization target.

$$f(n_i, n_p, a) = \frac{n_i}{N} + \frac{n_p}{P_{max}} + a_m, \tag{1}$$

where:

n_i—number of images classified as proper, $N \geq n_i \geq 0$;
N—number of all input images, $N = 60$;
n_p—number of characteristic points matched, $P_{max} \geq n_i \geq 0$;
P_{max}—number of maximal amount of characteristic features matched points achieved;
a_m—accuracy of the 3D model, $a_m \subset \{0; 0.5; 1\}$.

The a_m parameter was determined by the authors following the rules: if the 3D model properly projects the geometry of the conical pick and is the suitable input for the wear classification algorithm, the a_m value is equal to 1. If it is conditionally suitable, the a_m value is equal to 0.5. If the model is improper, the a_m value is equal to 0.

2.6. Wear Classification

Five samples of conical picks after exploitation were evaluated. Their 3D models were examined in the geometry analysis process. Their symmetrical wear was stated (2) as below.

$$S = \begin{cases} 1; \sigma_A \leq 0.3A_M \\ 0; \sigma_A > 0.3A_M \end{cases}, \tag{2}$$

where:

S—symmetry determinant;
σ_A—standard deviation of cross-sections of 3D model;
A_M—area of the cross-section of the master model.

3. Results

3.1. Parametric Optimization

The parametric optimization process for obtaining the model of the conical pick resulted in the eight 3D models presented in Table 4. In Table 5, the properly and improperly projected picks are presented. Observations of output are described below:

- Run 3, 7—geometry projected on a plane, cameras detected improperly. In both runs, image is far from camera.
- Run 1, 5—geometry is generally proper, the carbide part has geometry artifacts from the reflected line as it is smooth material.
- Run 4, 8—no valid initial pair found automatically.
- Run 2, 6—geometry is proper.

Table 4. Parametric optimization results.

Run	Light Level (A)	Distance (B)	Polarizing Filter (C)	Randomized Trial [-]	Images Classified n_i	Points Matched n_p	Accuracy a_m	Objective Function f_i	SN_i
1	−	−	−	1	60	5806	0.5	2.5	7.9588
2	−	−	+	6	60	3050	1	2.525	8.0452
3	−	+	−	4	59	886	0	1.136	1.1076
4	−	+	+	8	0	0	0	0	−100
5	+	−	−	2	60	5529	0.5	2.452	7.7904
6	+	−	+	5	60	5262	1	2.906	9.2659
7	+	+	−	3	60	734	0	1.126	1.0308
8	+	+	+	7	0	0	0	0	−100

Table 5. Properly projected geometry, f = 2.525; improperly projected geometry, f = 1.126.

Run	Mesh	Texture	Structure from Motion
2	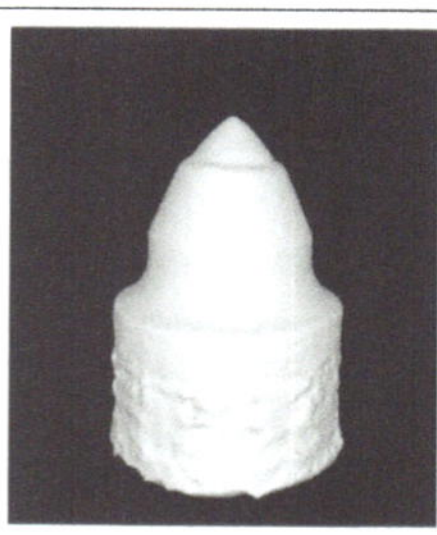		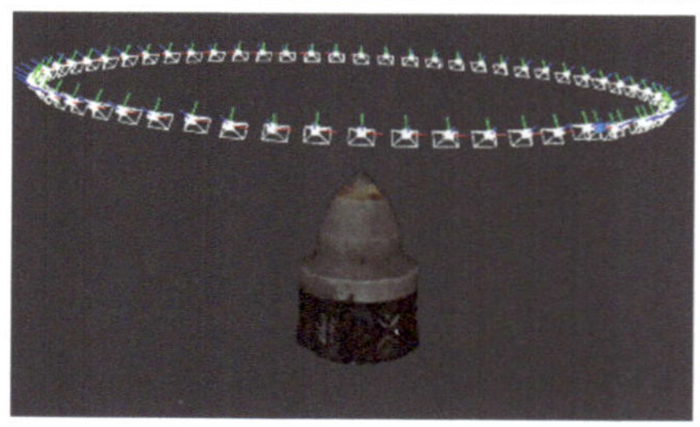
7			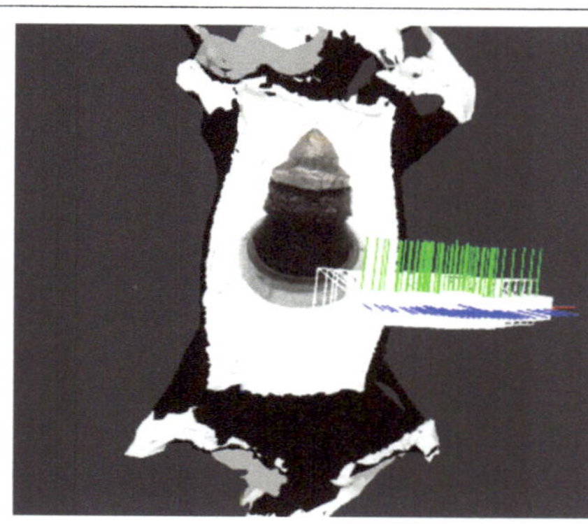

- Calculation of signal to noise ratio according to the rule "the-larger-is-better":

$$\mathrm{SN_i} = -10 * \log_{10} \frac{\sum_1^n \left(\frac{1}{Y^2}\right)}{n} \tag{3}$$

- Calculation of impact of each factor on the subjective function value:

$$\mathrm{F(A)} = \frac{f5 + f6 + f7 + f8}{4} - \frac{f5 + f6 + f7 + f8}{4} = 0.0807$$

$$\mathrm{F(B)} = \frac{f3 + f4 + f7 + f8}{4} - \frac{f5 + f6 + f1 + f2}{4} = -2.0303$$

$$\mathrm{F(C)} = \frac{f2 + f4 + f6 + f8}{4} - \frac{f1 + f3 + f5 + f7}{4} = -0.4458$$

- Calculation of the resistance of each factor to noise:

$$\mathrm{SN(A)} = \frac{\mathrm{SN5 + SN6 + SN7 + SN8}}{4} - \frac{\mathrm{SN5 + SN6 + SN7 + SN8}}{4} = 0.2439$$

$$\mathrm{SN(B)} = \frac{\mathrm{SN3 + SN4 + SN7 + SN8}}{4} - \frac{\mathrm{SN5 + SN6 + SN1 + SN2}}{4} = -57.7305$$

$$\mathrm{SN(C)} = \frac{\mathrm{SN2 + SN4 + SN6 + SN8}}{4} - \frac{\mathrm{SN1 + SN3 + SN5 + SN7}}{4} = -50.1441$$

Parametric optimization results are shown in Figure 5 and the parameters' influence on the objective function value is shown in Figure 6.

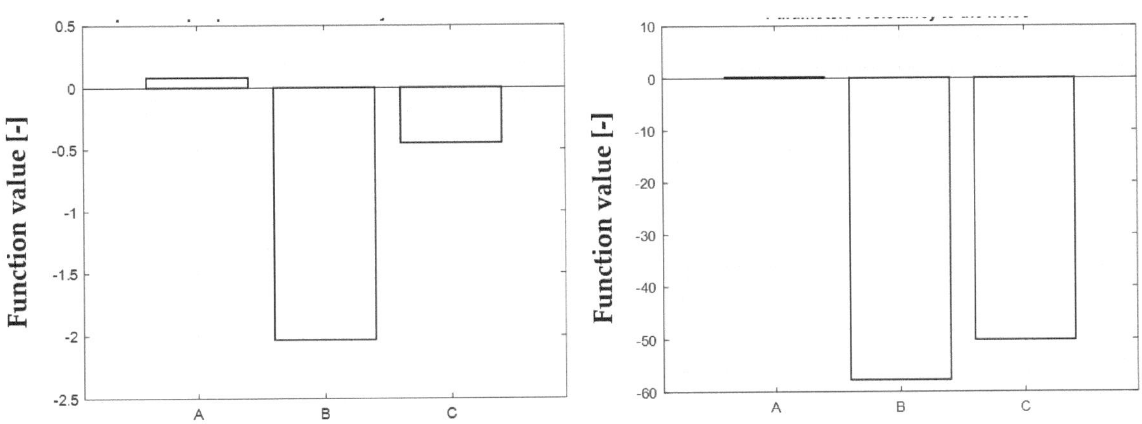

(a) (b)

Figure 5. Parametric optimization results; (**a**)—impact of input parameters with impact increasing: B, C, A; (**b**)—input parameters' resistance to noise with resistance increasing: A, C, B.

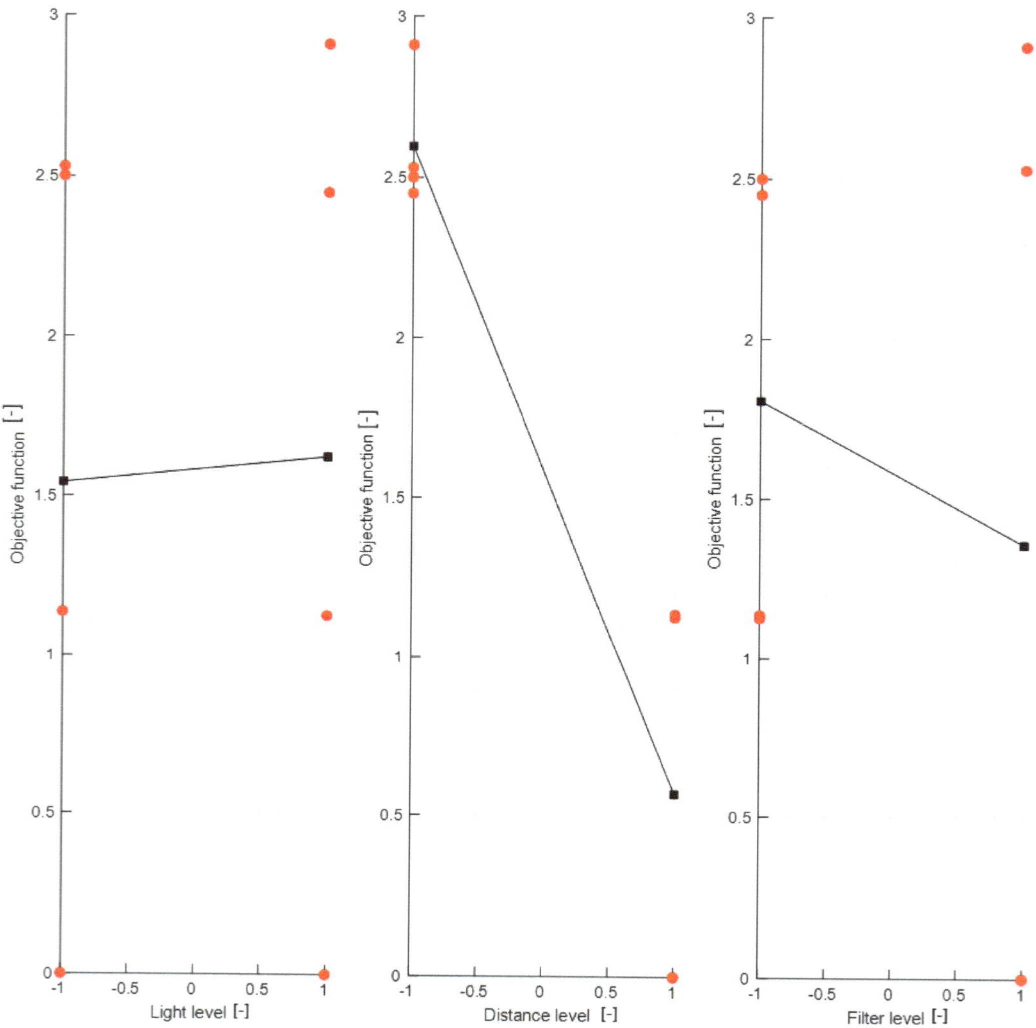

Figure 6. Parameters' influence on the objective function value. The red points correspond to the values on the boundary conditions before averaging.

3.2. Features Extraction

Individual model features were determined using MATLAB scripts to perform calculations on the .stl models. The first step was to center the data (3D points cloud) at zero. Next, the direction of most variance and rotation of the data was found to align it to the Z axis and translate it afterwards so that all of the data points are aligned so that z values are greater than 0 (Figure 7).

As can be seen in Figure 7, the resulting geometry may have a different orientation along the X axis. To change the pick's position so that the carbide has the X coordinate equal to 0 and the rest of its geometry lies on the right side of the axis, linear regression of density in the domain of the X coordinate was calculated and the slope value was checked. If a model is represented by regression with a slope value bigger than zero, all of the points were rotated along the Y axis (Figure 8).

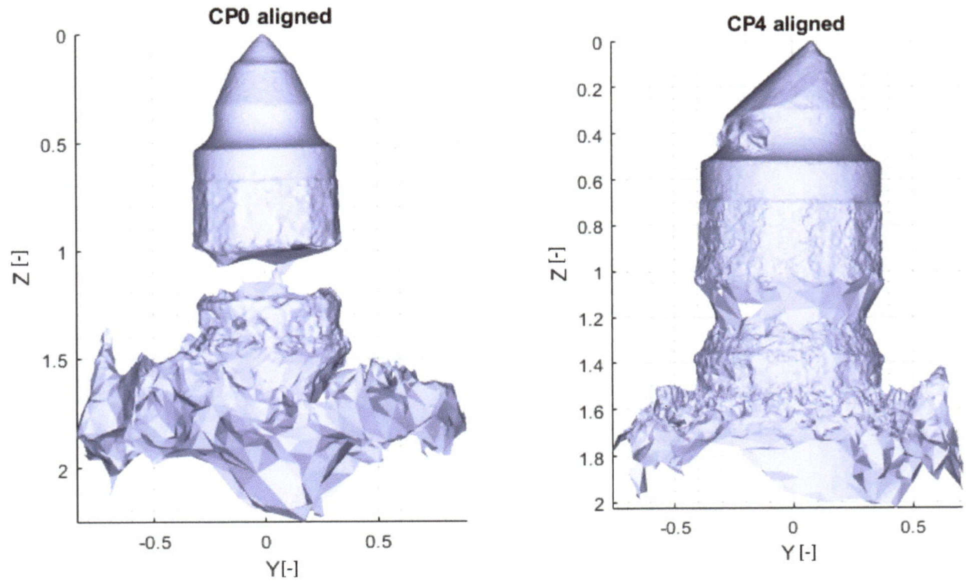

(**a**) (**b**)

Figure 7. Aligned .stl model of the pick; (**a**)—master model, (**b**)—pick after exploitation.

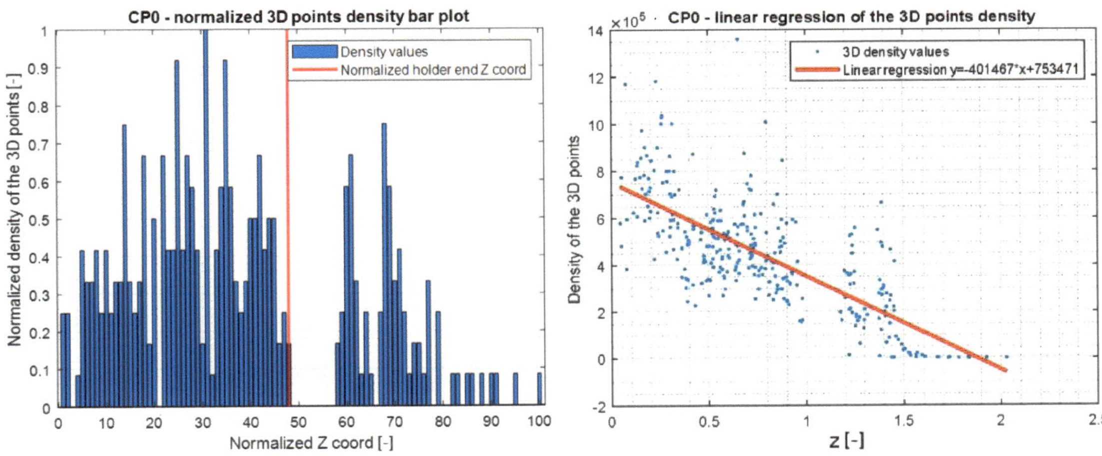

(**a**) (**b**)

Figure 8. Three-dimensional space density; (**a**)—bar plot in Z coordinate domain, (**b**)—regression.

The next step was to find the X axis coordinate of the pick's holder end and then to scale the model. It was made using calculation of data point density and the scale factor comparing maximal diameter values of the holder and its projected geometry (Figure 9).

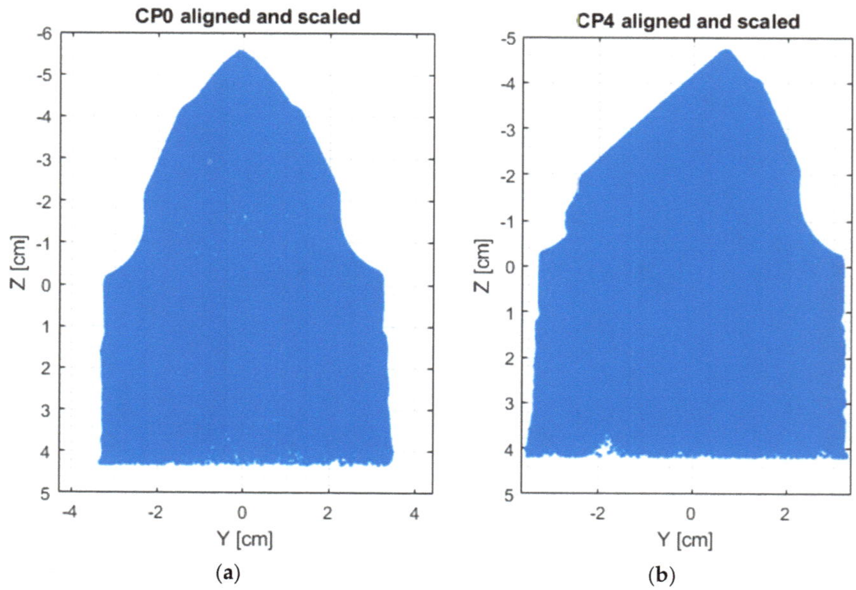

Figure 9. Scaled and cropped data points; (**a**)—master model, (**b**)—pick after exploitation.

The symmetry was calculated for 37 areas of cross-sections of the model with 10° of difference between: ϕ = [0°, 10°, . . . , 350°, 360°]. The 2D boundary of the model was determined and its area was calculated. Additionally, the plastic deformation area was defined, determining each sum of cross-section boundaries of the master and the rest of the picks (Figure 10). Afterwards, the difference between the summed boundary and the pick's original boundary was calculated (Table 6). Figure 11 presents the aggregated results for the examined picks.

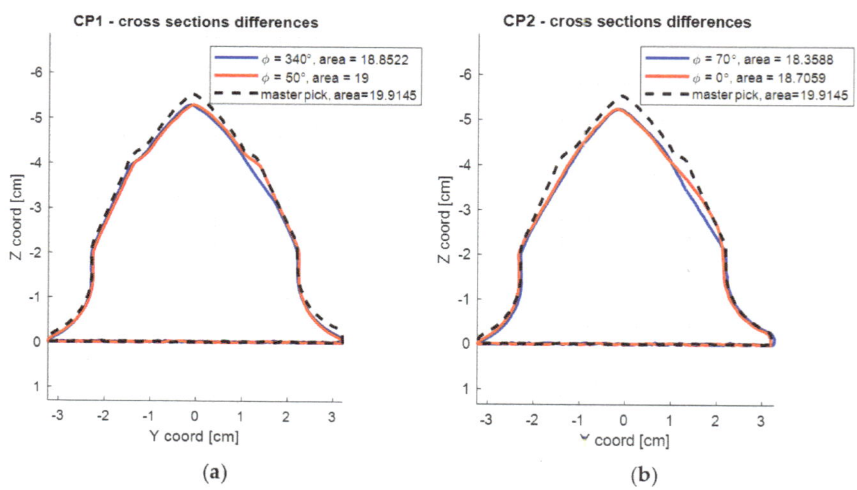

Figure 10. *Cont.*

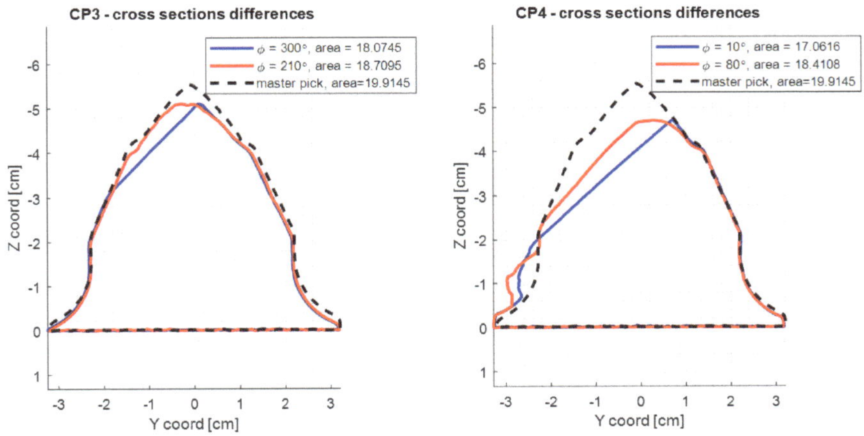

(c) (d)

Figure 10. The difference between the minimal and maximal cross-section area of the scanned picks. (**a**) shows the min/max cross-sectional area of CP01 pick, (**b**) of CP02, (**c**) of CP03 and (**d**) of CP04. Their physical representation is shown in Figure 3.

Table 6. Table containing the calculated data regarding the examined picks. The S parameter is a Boolean value assigned according to set threshold—in this case, the authors set the critical value at 3%. The threshold value may vary in the case of different tool geometry. The maximum plastic deformation coefficient calculates the area of surplus material, exceeding the contour of the master model cross-section.

Pick	Mean [cm^2]	Std Dev. [cm]	Max. Plastic Def. Area [cm^2]	Max. Area Diff. [cm^2]	Area Diff. as a Part of Mean Area [%]	S (Symmetrical Parameter)
CP0	19.869	0.03048	0	0.1080	0.5	1
CP1	18.911	0.038211	0	0.1480	0.7	1
CP2	18.496	0.11077	0	0.3470	1.9	1
CP3	18.416	0.22452	0	0.6360	3.5	0
CP4	17.791	0.48366	0.4910	1.3490	7.5	0

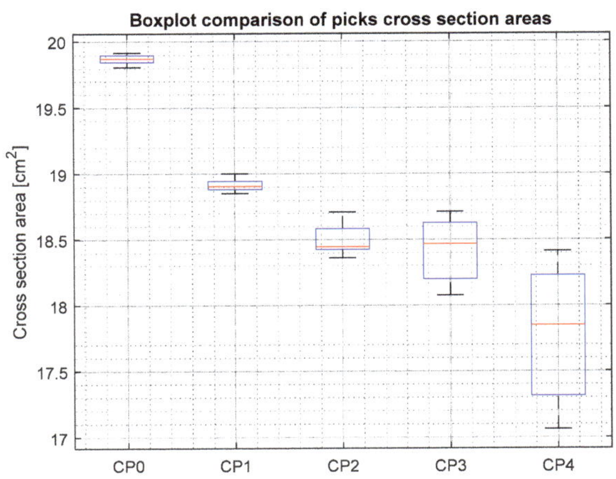

Figure 11. Cross-section areas of picks.

Cross-section boundary points of the pick nr 4 are considered to be worn out and are not suitable for regeneration.

As expected, the CP0, being the reference specimen, has low deviation and the highest average cross-section value. The eccentrically placed mean value (red line) in the CP2, CP3 and CP4 provides information about the asymmetrical wear of the picks. The bigger the box, the bigger the value of asymmetry.

4. Discussion

The scanned conical picks are typical examples of a set of tools removed during maintenance of the cutting head. If they were to be rated for their further usability, one could divide them among subsequent categories:

- Eligible for use: CP0, CP1;
- Eligible for hardfacing: CP0, CP1, CP2;
- Catastrophic wear: CP3, CP4.

The photogrammetric measurement of volume followed by the symmetry assessment script provides plausible results, especially considering its low cost and high reliability. The authors believe that implementing this method in the mining repair plants could be beneficial indeed. This approach could also serve as a basic quality control unit in a mining tool factory or another facility which manufactures parts with an axis of symmetry. Yet, it is noteworthy to consider the surface roughness (or reflectiveness, to be more precise) of the scanned part. The very reflective, polished objects with the R_a parameter below 2.5–5 μm cause some issues, namely, unexpected bumps or cavities in the place of a very bright spot. The solution for those issues is either to have those surfaces dulled with talc or another powder or to incorporate a different method for obtaining the 3D geometry.

Overall, this approach to tool wear characterization is relatively easy to use and provides much information about the wear mechanism. The current algorithm used in this paper will be applicable only to objects which are solids of revolution, since one of the steps is to find the axis of symmetry and align scanned objects according to the found axis. Nevertheless, after some modifications, a similar approach can be utilized to assess the wear rate (or even metrological compliance of the physical object with the designed virtual model) of other tools or parts. The setup that the authors used was supposed to be affordable for most populations, thus some efficiency-related concerns were a trade-off. Yet, increasing the efficiency is theoretically simple, since one has only to add more cameras to capture the images "at once", with little to no scanned object rotation.

One of the strengths of the current setup is that it separates the scanned object from the environment, being a method of scanning which is—quite literally—as robust as the walls of the lightbox. On the other hand, the necessity of the lightbox limits the maximum size of the scanned part.

5. Conclusions

Described methods of classification of the picks' wear are presented in the table below (Table 7).

Table 7. Limits of the methods of classification of the picks' wear and its features.

Type of Scanning Method	Implementation in a Difficult Environment	Automation Possibilities	Enough Output Data for Regeneration
C2 parameter	+	+	−
Parametric factors	−	−	+
Fuzzy neural network	+	+/−	−
LIDAR measurements	+	+/−	−
Photogrammetric model	+	+	+

By analyzing the presented methods from previous research, it can be stated that the parametric factors method is not suitable to be implemented in harsh conditions, i.e., in mining factories. The parametric factors method consists of laboratory measurements, including microscopic image analysis, which cannot be conducted in such conditions. Parametric factors, LIDAR and fuzzy neural network methods are not feasible to automate because of the large number of various processes included when it comes to the parametric factors method and the need for selection of the mathematical operator for a sample when it comes to the neural network method. The LIDAR method does not bring forth information about texture that can be the input for the algorithm of classification. The crucial disadvantage of this method is the poor availability of equipment compared to photogrammetry. The chosen method tends to be universal, as it can be performed using a phone camera. The C2 parameter assessment is still the quickest approach to conical pick wear rate evaluation; nevertheless, it does provide the user only with basic information about the wear characteristics.

The most influential of the studied parameters appeared to be the distance between the camera and the object. Nevertheless, the distances in this study were adjusted to the quality and focal point of the particular lens. Having an image-capturing device with a lens able to zoom in without any image distortion, the distances could vary significantly. The presence of the polarizing filter appeared to decrease the number of bright reflections, which resulted in 3D reconstructions with greater fidelity.

The goal of further work is to reduce the scanning and data processing time. Additionally, the authors plan to build a portable scanning device, which will enable the mining company maintenance team to gather the data on-site. This task, however, would require some extra steps in order to meet the requirements of underground heavy industry, e.g., a dustproof casing, spark-proof design of the drive and electronics and perhaps a conical pick initial cleaning device.

Another goal for future study might be connected with the advantage of photogrammetry over LIDAR or other laser-based scanning techniques, namely, the texture analysis. Since photogrammetry provides some otherwise lost information on the color of the surface of the scanned object; the textured 3D file could be subjected to more sophisticated analysis, such as hardfaced material overheat detection. Another great use of the method studied by the authors is the possibility to scan and instantly send a 3D textured file to a locally unavailable wear expert for analysis.

The algorithm, after development, could also serve as a low-cost linear and angular measurement system for the manufactured tools. In the case of conical picks, the geometry of the working part makes it difficult to utilize conventional means of measurement.

Author Contributions: Conceptualization, J.P., A.W.-P. and M.B.; methodology, J.P. and A.W.-P.; software, A.W.-P.; validation, J.P. and A.W.-P.; formal analysis, J.P.; investigation, J.P, A.W.-P. and M.B.; resources, J.P. and M.B.; data curation, J.P. and A.W.-P.; writing—original draft preparation, J.P., and A.W.-P.; writing—review and editing, M.B.; visualization, A.W.-P.; supervision, M.B.; project administration, J.P and M.B.; funding acquisition, M.B. and J.P. All authors have read and agreed to the published version of the manuscript.

Funding: This work was not supported by any funding. The APC was covered from the personal awards of Michał Bembenek and Jan Pawlik, funded by the Dean of the Faculty of Mechanical Engineering and Robotics of AGH UST for their earlier publication achievements.

Institutional Review Board Statement: Not applicable.

Informed Consent Statement: Not applicable.

Data Availability Statement: The data presented in this study are available upon request from the corresponding author.

Acknowledgments: We would like to express our gratitude towards Krzysztof Krauze, Tomasz Wydro and Kamil Mucha for supplying the materials to scan.

Conflicts of Interest: The authors declare no conflict of interest.

References

1. Feng, K.; Ji, J.; Ni, Q.; Beer, M. A review of vibration-based gear wear monitoring and prediction techniques. *Mech. Syst. Signal Process.* **2023**, *182*, 109605. [CrossRef]
2. Krawczyk, J.; Bembenek, M.; Pawlik, J. The Role of Chemical Composition of High-Manganese Cast Steels on Wear of Excavating Chain in Railway Shoulder Bed Ballast Cleaning Machine. *Materials* **2021**, *14*, 7794. [CrossRef] [PubMed]
3. Wang, D.; Zhang, S.; Wang, L.; Liu, Y. Developing a Ball Screw Drive System of High-Speed Machine Tool Considering Dynamics. *IEEE Trans. Ind. Electron.* **2021**, *69*, 4966–4976. [CrossRef]
4. Zhu, K.; Liu, T. Online Tool Wear Monitoring Via Hidden Semi-Markov Model with Dependent Durations. *IEEE Trans. Ind. Inform.* **2018**, *14*, 69–78. [CrossRef]
5. Fong, K.M.; Wang, X.; Kamaruddin, S.; Ismadi, M.-Z. Investigation on universal tool wear measurement technique using image-based cross-correlation analysis. *Measurement* **2021**, *169*, 108489. [CrossRef]
6. Bołoz, Ł. Directions for increasing conical picks' durability. *New Trends Prod. Eng.* **2019**, *2*, 277–286. [CrossRef]
7. Bołoz, Ł. Results of a Study on the Quality of Conical Picks for Public Procurement Purposes. *New Trends Prod. Eng.* **2018**, *1*, 687–693. [CrossRef]
8. Bembenek, M.; Prysyazhnyuk, P.; Shihab, T.; Machnik, R.; Ivanov, O.; Ropyak, L. Microstructure and Wear Characterization of the Fe-Mo-B-C—Based Hardfacing Alloys Deposited by Flux-Cored Arc Welding. *Materials* **2022**, *15*, 5074. [CrossRef]
9. Krauze, K.; Skowronek, T.; Mucha, K. Influence of the Hard-Faced Layer Welded on Tangential-Rotary Pick Operational Part on to Its Wear Rate. *Arch. Min. Sci.* **2016**, *61*, 779–792. [CrossRef]
10. Krauze, K.; Bołoz, Ł.; Wydro, T.; Mucha, K. Investigations into the wear rate of conical picks with abrasion-resistant coatings in laboratory conditions. In Proceedings of the IOP Conference Series: Materials Science and Engineering, Szczyrk, Poland, 14–16 October 2019; Volume 679, p. 012012. [CrossRef]
11. Krauze, K.; Mucha, K.; Wydro, T. Evaluation of the Quality of Conical Picks and the Possibility of Predicting the Costs of Their Use. *Multidiscip. Asp. Prod. Eng.* **2020**, *3*, 491–504. [CrossRef]
12. Choi, S.-W. Durability Evaluation Depending on the Insert Size of Conical Picks by the Field Test. *J. Korean Tunn. Undergr. Space Assoc.* **2019**, *21*, 49–59. [CrossRef]
13. Krauze, K.; Bołoz, Ł.; Wydro, T. Parametric Factors for the Tangential-Rotary Picks Quality Assessment/Wskaźniki Parametryczne Oceny Jakości Noży Styczno-Obrotowych. *Arch. Min. Sci.* **2015**, *60*, 265–281. [CrossRef]
14. Gajewski, J.; Jedliński, Ł.; Jonak, J. Classification of wear level of mining tools with the use of fuzzy neural network. *Tunn. Undergr. Space Technol.* **2013**, *35*, 30–36. [CrossRef]
15. Guerra, M.; Lavecchia, F.; Maggipinto, G.; Galantucci, L.; Longo, G. Measuring techniques suitable for verification and repairing of industrial components: A comparison among optical systems. *CIRP J. Manuf. Sci. Technol.* **2019**, *27*, 114–123. [CrossRef]
16. Bembenek, M.; Buczak, M. The Fine-Grained Material Flow Visualization of the Saddle-Shape Briquetting in the Roller Press Using Computer Image Analysis. *J. Flow Vis. Image Process.* **2021**, *28*, 69–78. [CrossRef]
17. Braun Neto, J.A.; Lima, J.L.; Pereira, A.I.; Costa, P. Low-Cost 3D LIDAR-Based Scanning System for Small Objects. In Proceedings of the 2021 22nd IEEE International Conference on Industrial Technology (ICIT), Valencia, Spain, 10–12 March 2021; pp. 907–912.
18. Hormann, K. Geometry Processing. In *Encyclopedia of Applied and Computational Mathematics*; Engquist, B., Ed.; Springer: Berlin/Heidelberg, Germany, 2015; pp. 593–606, ISBN 978-3-540-70528-4.
19. Valerga, A.; Batista, M.; Bienvenido, R.; Fernández-Vidal, S.; Wendt, C.; Marcos, M. Reverse Engineering Based Methodology for Modelling Cutting Tools. *Procedia Eng.* **2015**, *132*, 1144–1151. [CrossRef]
20. Rodríguez-Martín, M.; Rodríguez-Gonzálvez, P. Suitability of Automatic Photogrammetric Reconstruction Configurations for Small Archaeological Remains. *Sensors* **2020**, *20*, 2936. [CrossRef]
21. Qin, R.; Gruen, A. The role of machine intelligence in photogrammetric 3D modeling—An overview and perspectives. *Int. J. Digit. Earth* **2020**, *14*, 15–31. [CrossRef]
22. Pilar Valerga Puerta, A.; Aletheia Jimenez-Rodriguez, R.; Fernandez-Vidal, S.; Raul Fernandez-Vidal, S. Photogrammetry as an Engineering Design Tool. In *Product Design*; Alexandru, C., Jaliu, C., Comşit, M., Eds.; IntechOpen: London, UK, 2020; ISBN 978-1-83968-212-4.
23. Schenk, T. *Introduction to Photogrammetry*; Ohio State University: Columbus, OH, USA, 2005; p. 106.
24. Ackermann, F. Digital Image Correlation: Performance and Potential Application in Photogrammetry. *Photogramm. Rec.* **2006**, *11*, 429–439. [CrossRef]
25. James, M.R.; Robson, S. Straightforward reconstruction of 3D surfaces and topography with a camera: Accuracy and geoscience application. *J. Geophys. Res. Earth Surf.* **2012**, *117*, 03017. [CrossRef]
26. Wells, J.; Jones, T.; Danehy, P. Polarization and Color Filtering Applied to Enhance Photogrammetric Measurements of Reflective Surfaces. In Proceedings of the 46th AIAA/ASME/ASCE/AHS/ASC Structures, Structural Dynamics and Materials Conference, Austin, TX, USA 18–21 April 2005; American Institute of Aeronautics and Astronautics: Reston, VA, USA, 2005.
27. Conen, N.; Hastedt, H.; Kahmen, O.; Luhmann, T. Improving Image Matching by Reducing Surface Reflections Using Polarising Filter Techniques. *ISPRS—Int. Arch. Photogramm. Remote Sens. Spat. Inf. Sci.* **2018**, *XLII-2*, 267–274. [CrossRef]
28. Guidi, G.; Gonizzi, S.; Micoli, L.L. Image pre-processing for optimizing automated photogrammetry performances. *ISPRS Ann. Photogramm. Remote Sens. Spat. Inf. Sci.* **2014**, *II-5*, 145–152. [CrossRef]
29. Shashi, M.; Jain, K. Use of Photogrammetry in 3d Modeling and Visualization of Buildings. *ARPN J. Eng. Appl. Sci.* **2007**, *2*, 5.

30. Liu, J.; Ma, C.; Zeng, Q.; Gao, K. Discrete Element Simulation of Conical Pick's Coal Cutting Process under Different Cutting Parameters. *Shock Vib.* **2018**, *2018*, 7975141. [CrossRef]
31. Kuidong, G.; Changlong, D.; Hongxiang, J.; Songyong, L. A theoretical model for predicting the Peak Cutting Force of conical picks. *Frat. Ed Integrità Strutt.* **2013**, *8*, 43–52. [CrossRef]
32. Averin, E.A.; Zhabin, A.B.; Polyakov, A.V.; Linnik, Y.N.; Linnik, V.Y. Transition between relieved and unrelieved modes when cutting rocks with conical picks. *J. Min. Inst.* **2021**, *249*, 329–333. [CrossRef]
33. Wang, S.; Sun, L.; Li, X.; Zhou, J.; Du, K.; Wang, S.; Khandelwal, M. Experimental investigation and theoretical analysis of indentations on cuboid hard rock using a conical pick under uniaxial lateral stress. *Geomech. Geophys. Geo-Energy Geo-Resour.* **2022**, *8*, 34. [CrossRef]
34. Cheluszka, P.; Mikuła, S.; Mikuła, J. Theoretical consideration of fatigue strengthening of conical picks for rock cutting. *Tunn. Undergr. Space Technol.* **2022**, *125*, 104481. [CrossRef]
35. Li, X.; Wang, S.; Ge, S.; Malekian, R.; Li, Z. A Theoretical Model for Estimating the Peak Cutting Force of Conical Picks. *Exp. Mech.* **2018**, *58*, 709–720. [CrossRef]
36. Yasar, S. A General Semi-Theoretical Model for Conical Picks. *Rock Mech. Rock Eng.* **2020**, *53*, 2557–2579. [CrossRef]
37. Li, H.S.; Liu, S.Y.; Xu, P.P. Numerical simulation on interaction stress analysis of rock with conical picks. *Tunn. Undergr. Space Technol.* **2019**, *85*, 231–242. [CrossRef]
38. Liu, S.; Ji, H.; Liu, X.; Jiang, H. Experimental research on wear of conical pick interacting with coal-rock. *Eng. Fail. Anal.* **2017**, *74*, 172–187. [CrossRef]
39. Wang, X.; Wang, Q.-F.; Liang, Y.-P.; Su, O.; Yang, L. Dominant Cutting Parameters Affecting the Specific Energy of Selected Sandstones when Using Conical Picks and the Development of Empirical Prediction Models. *Rock Mech. Rock Eng.* **2018**, *51*, 3111–3128. [CrossRef]
40. Lowe, D.G. Distinctive Image Features from Scale-Invariant Keypoints. *Int. J. Comput. Vis.* **2004**, *60*, 91–110. [CrossRef]
41. Otero, I.R.; Delbracio, M. Anatomy of the SIFT Method. *Image Process. Line* **2014**, *4*, 370–396. [CrossRef]
42. Yu, G.; Morel, J.-M. ASIFT: An Algorithm for Fully Affine Invariant Comparison. *Image Process. Line* **2011**, *1*, 11–38. [CrossRef]
43. Li, Y.; Wang, S.; Tian, Q.; Ding, X. A survey of recent advances in visual feature detection. *Neurocomputing* **2015**, *149*, 736–751. [CrossRef]
44. Alcantarilla, P.; Nuevo, J.; Bartoli, A. Fast Explicit Diffusion for Accelerated Features in Nonlinear Scale Spaces. In Proceedings of the Proceedings of the British Machine Vision Conference 2013, Bristol, UK, 9–13 September 2013; British Machine Vision Association: Bristol, UK, 2013; pp. 13.1–13.11.
45. Nister, D.; Stewenius, H. Scalable Recognition with a Vocabulary Tree. In Proceedings of the 2006 IEEE Computer Society Conference on Computer Vision and Pattern Recognition—Volume 2 (CVPR'06), New York, NY, USA, 17–22 June 2006; Volume 2, pp. 2161–2168.
46. Lowe, D.G. Object Recognition from Local Scale-Invariant Features. In Proceedings of the Seventh IEEE International Conference on Computer Vision, Kerkyra, Greece, 20–27 September 1999; Volume 2, pp. 1150–1157.
47. Moulon, P.; Monasse, P.; Marlet, R. Global Fusion of Relative Motions for Robust, Accurate and Scalable Structure from Motion. In Proceedings of the 2013 IEEE International Conference on Computer Vision, Sydney, Australia, 1–8 December 2013; pp. 3248–3255.
48. Schönberger, J.L.; Zheng, E.; Frahm, J.-M.; Pollefeys, M. Pixelwise View Selection for Unstructured Multi-View Stereo. In *Computer Vision—ECCV 2016*; Leibe, B., Matas, J., Sebe, N., Welling, M., Eds.; Springer International Publishing: Cham, Switzerland, 2016; Volume 9907, pp. 501–518, ISBN 978-3-319-46486-2.
49. Labatut, P.; Pons, J.-P.; Keriven, R. Robust and Efficient Surface Reconstruction from Range Data. *Comput. Graph. Forum* **2009**, *28*, 2275–2290. [CrossRef]
50. Lévy, B.; Petitjean, S.; Ray, N.; Maillot, J. Least squares conformal maps for automatic texture atlas generation. In Proceedings of the 29th Annual Conference on Computer Graphics and Interactive Techniques, San Antonio, TX, USA, 23–26 July 2002; pp. 362–371. [CrossRef]

materials

MDPI

Article

Amine-Modified Carbon Xerogels as Effective Carbon-Based Adsorbents of Anionic Dye from Aqueous Solutions

Magdalena Ptaszkowska-Koniarz, Joanna Goscianska , Aleksandra Bazan-Wozniak and Robert Pietrzak *

Faculty of Chemistry, Adam Mickiewicz University in Poznań, Uniwersytetu Poznańskiego 8, 61-614 Poznań, Poland

* Correspondence: pietrob@amu.edu.pl; Tel.: +48-61-829-1560

Abstract: Carbon xerogels were obtained by polycondensation of resorcinol and formaldehyde in a water medium. Their surface was oxidized by ammonium persulfate and then modified with amine groups. Four amines were used: methylamine, ethylamine, propylamine, and ethylenediamine, differing in carbon chain length and number of amine groups. The materials were characterized by low-temperature nitrogen sorption, elemental analysis, thermal analysis, X-ray diffraction, infrared spectroscopy, and determination of the surface oxygen group content with the use of the Boehm method. The final carbon adsorbents had surface areas ranging from 172–663 m^2/g and acid–base nature. They were applied for adsorption of thymol blue from water solution. The sorption capacities of the studied adsorbents ranged from 83 to 140 mg/g. The presence of amine groups on the xerogel surface was found to increase its sorption capacity towards the dye studied. The dye adsorption process is endothermic and spontaneous, as indicated by the positive values of ΔH and the negative values of ΔG, respectively. The kinetics of adsorption of thymol blue was established to be described by the pseudo-second-order model. The equilibrium data were analyzed by the Langmuir and Freundlich models. The character of thymol blue adsorption is much better described by the Langmuir isotherm.

Keywords: carbon materials; amine functionalization; thymol blue; Langmuir isotherm; pseudo-second-order kinetic model

Citation: Ptaszkowska-Koniarz, M.; Goscianska, J.; Bazan-Wozniak, A.; Pietrzak, R. Amine-Modified Carbon Xerogels as Effective Carbon-Based Adsorbents of Anionic Dye from Aqueous Solutions. *Materials* **2022**, *15*, 5736. https://doi.org/10.3390/ma15165736

Academic Editor: Gerald Anthony Murphy

Received: 5 July 2022
Accepted: 17 August 2022
Published: 19 August 2022

1. Introduction

Water, in addition to oxygen and nutrients, is necessary to sustain life. Its appreciation increases with the development of urbanization, industry, and agriculture. Water pollution has become a problem of top priority on a global scale. The most troublesome pollutants include: dyes, nitrates (V), phosphates (V), chlorates, sulfates (V), heavy metal ions, aromatic amines, pesticides, detergents, and bacteria. The largest amounts of toxic pollutants are introduced to the water systems with wastewater. Other important sources are water and land transport, pesticides and artificial fertilizers used in agriculture, and municipal and industrial waste [1–3]. All water pollutants are hazardous to human health, so maintenance of water purity is a priority among environmental protection issues. Appropriate measures should be taken to limit the amount of pollutants entering the natural water system, as well as to effectively remove already-introduced contamination. The latter is realized through wastewater purification. Depending on the type of pollutants, different methods are applied: mechanical, biological, physical, and chemical. As follows from the literature, many methods of water treatment have been proposed to deal with the problem, including: ion-exchange, coagulation, reversed osmosis, ozonation, electrochemical methods, adsorption, and chemical precipitation [4–9]. In order to remove pollutants from post-industrial wastewater by sorption methods, it is necessary to apply effective adsorbents [10–12], represented by, e.g., carbon xerogels, characterized by well-developed surface area and offering a possibility of easy modulation of pore size distribution [13]. According to the literature, they have been successfully used for sorption of orange II [14], chromotrope 2R [14], methylene blue [15], and rhodamine B [16]. Satisfactory results, obtained using carbon xerogels as adsorbents towards cationic dyes, have been reported by Girgis et al. [17]. The high sorption capacities of carbon xerogels, which increased with extended internal porosity, were comparable with those of conventional activated carbons. Figueiredo's group [18] analyzed the sorption capacities of carbon xerogels oxidized by nitric (V) acid towards anionic dyes. According to the data, the sorption capacities of these adsorbents were comparable to those of commercial activated carbons, in the same conditions of the process, and ranged from approximately 90 to 160 mg/g.

The aim of our research was to synthesize carbon xerogels modified with different amines, perform comprehensive characterization of them, and test their performance as adsorbents of thymol blue from water solutions. Amine groups, as strong electron donors, enhance the basic character of the carbon xerogels surface, so their presence was expected to facilitate the adsorption of thymol blue, which is an anionic dye [16]. An important task was to establish the effect of pH of the adsorbate solutions and temperature of the process on the efficiency of adsorption of carbon xerogels. Thymol blue is a chemical pH indicator, as it changes colors from red to yellow in the pH range 1.2–2.8 and from yellow to blue in the pH range 8.0–9.6 [19–21]. The two ranges of the color changes of this indicator are related to the two stages of its dissociation (Figure 1).

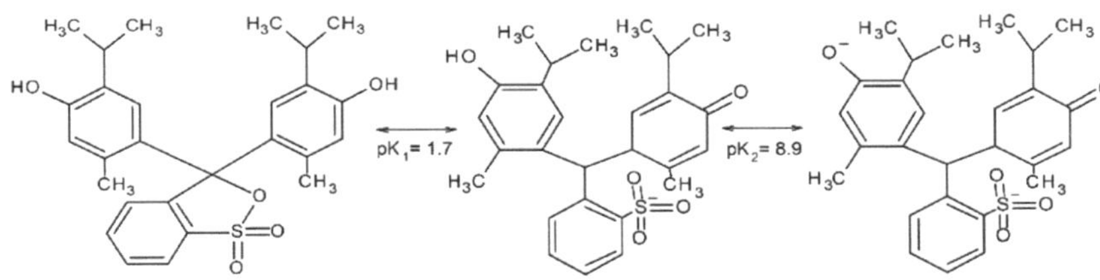

Figure 1. Structures of thymol blue at different pH.

2. Materials and Methods

2.1. Samples

Carbon xerogel (CX) was obtained as follows: 0.040 L of distilled water was added to 25 g of resorcinol (Merck SA, Darmstadt, Germany) while stirring [22]. When the resorcinol was dissolved, a solution of NaOH (2 mol/L, Chempur, Piekary Śląskie, Poland) was added in order to bring the pH to 5.3. Then, 34 mL of formaldehyde (Chempur, Piekary Śląskie, Poland) was added while stirring, and the pH was adjusted to 5.3 by adding a solution of HCl (0.1 mol/L, Avantor Performance Materials Poland S.A., Gliwice, Poland). The gelation was accomplished in a water bath over three days (85 °C). The gel was crushed and dried in an oven for 4 days. The dried gel was carbonized at 700 °C in a tubular furnace (N_2—150 mL × min^{-1}, heating rate—2 °C × min^{-1}). Pyrolysis was realized in the following stages: (1) up to 200 °C, held for 2 h; (2) up to 300 °C, held for 1 h; (3) up to 700 °C, held for 2 h. The carbon xerogel obtained was oxidized by 1 mol/L solution of $(NH_4)_2S_2O_8$ (Merck SA, Darmstadt, Germany) in a 2 mol/L solution of H_2SO_4 for 6 h (30 °C)—the obtained adsorbent was denoted by the symbol CX-APS. At subsequent stages, the sample CX-APS was modified by methylamine (MA), ethylamine (EA), propylamine (PA), and ethylenediamine (EDA) (Merck SA, Darmstadt, Germany). For the modification, 2 g of each amine was mixed with 30 g of methanol (Avantor Performance Materials Poland S.A., 99.8%, Gliwice, Poland), and, after 30 min, 2 g of the adsorbent CX-APS was added to each mixture; the mixtures were stirred at 40 °C for 6 h. Then, the mixtures were filtered off, and the samples were washed with methanol and dried at 70 °C. The carbon xerogels functionalized with amines were labelled as CX-APS-MA, CX-APS-EA, CX-APS-PA, and CX-APS-EDA.

2.2. Samples Characterization

2.2.1. XRD

The samples CX, CX-APS, and CX-APS-EDA were characterized by powder X-ray diffraction in the wide-angle range using a D8 Advance Diffractometer (Bruker, Billerica, MA, USA, CuK_{α} radiation, $\lambda = 0.154$ nm, step size 0.05°).

2.2.2. Nitrogen Sorption, Elemental Analysis and Acid–Base Properties

Elemental analyses of the samples were performed with the use of a Vario ELIII elemental analyzer (Elementar Analysen systeme GmbH, Langenselbold, Germany). The textural parameters of the obtained adsorbents were characterized on the basis of low-temperature nitrogen adsorption/desorption, using an Autosorb iQ instrument, provided by Quantachrome Instruments (Boynton Beach, FL, USA). Before adsorption measurements, the samples were degassed under vacuum for 8 h, at 200 °C. The surface areas of the samples (S_{BET}) were evaluated in the range of relative pressure p/p_0 from 0.05 to 0.30 (according to the Brunauer–Emmett–Teller method). The total pore volume (V) of each sample was estimated based on the amount of liquid nitrogen adsorbed at a relative pressure $p/p_0 = 0.99$. The mean pore diameter (D) was calculated from the equation $D = 4\ V/S_{BET}$. Moreover, the commonly known t-plot method was applied to determine the micropore volume and area. The content of the surface oxygen functional groups, both acidic and basic, was determined by standard neutralization titration, with HCl and NaOH, according to the Boehm method [16,23,24].

2.2.3. Thermogravimetric Analysis and Infrared Spectroscopy

Thermal stabilities of the samples were measured on a Setsys 1200 Setaram (Setaram, Lyon, France). The measurements were performed under flowing nitrogen at a heating rate of 10 °C/min, over a temperature range of 25−1000 °C, with an initial material weight of approximately 10 mg. Structural changes in the carbon xerogel, after modification with amine groups, were detected by FT-IR spectroscopy. Prior to measurement, the samples were mixed with dried potassium bromide at a rate of 0.5 mg of the adsorbents to 200 mg of KBr. The spectra, in the range of 400−4000 cm^{-1}, were recorded on an FT-IR spectrometer Bruker IFS 66v/S (Billerica, MA, USA) [24].

2.3. Adsorption of Dyes

Portions of 20 mg of each sample were added to 50 mL of thymol blue solutions (concentrations range of dye: 12.5–150 mg/L). The flooded samples were stirred by a magnetic stirrer for 1 day (temperature 22 ± 1 °C). Then, the samples were filtered off and the absorbance of the filtrate was measured by a UV-Vis spectrophotometer, model Carry 100 Bio (Agilent, Santa Clara, CA, USA). The measurement was made twice at $\lambda_{max} = 435$ nm. On the basis of the standard curve, the final concentrations of the dye were determined and the sorption capacity of each sample was calculated. The amount (q_e) of thymol blue adsorbed on a given carbon adsorbent, expressed in mg/g, was calculated from formula (1):

$$q_e = \frac{C_0 - C_e}{m} \cdot V \tag{1}$$

where C_0 is the initial concentration of thymol blue (mg/L), C_e is the residual concentration of thymol blue (mg/L), V is the volume of thymol blue (L), and m is the mass of carbon xerogel adsorbent (g).

The effect of the pH of the thymol blue solution on the sorption capacities of the carbon xerogel adsorbents was determined, as the pH values varied in the range of 2–12 (pH-meter ELEMETRON, Zabrze, Poland). The pH of each solution was adjusted with 0.1 M HCl and 0.1 M NaOH solution. Measurements were performed for 20 mg of the carbon xerogel adsorbents and 50 mL of the dye solution, with 80 mg/L concentration.

The adsorption isotherms were fitted to the two models proposed by Langmuir and Freundlich. The Langmuir adsorption isotherm [25] is described by the following linear Equation (2):

$$\frac{C_e}{q_e} = \frac{1}{K_L \times q_{max}} + \frac{C_e}{q_{max}} \quad (2)$$

where q_e is the experimentally determined amount of the thymol blue adsorbed on the carbon samples (mg/g), q_m is the theoretically predicted amount of the adsorbed dye (mg/g), C_e is the equilibrium concentration of the dye solution (mg/L), and K_L is the Langmuir constant.

The linear form of Freundlich isotherm equation [16] is expressed as (3):

$$lnq_e = lnK_F + \frac{1}{n} lnC_e \quad (3)$$

where q_e is the experimentally determined amount of the thymol blue adsorbed on the carbon samples (mg/g), C_e is the equilibrium concentration of the dye solution (mg/L), and K_F and $1/n$ are the Freundlich constants, characteristic of a given system.

The value of $1/n$ (varying from 0 to 1) is a measure of the intensity of adsorption or heterogeneity of the surface. The closer it is to zero, the more heterogeneous the surface [16].

In order to check the effect of contact time on the effectiveness of the adsorption of the dye studied, 20 mg portions of each sample were added to 50 mL of thymol blue solutions (150 mg/L). The samples were stirred by a magnetic stirrer. At selected time intervals (10, 20, 30, 40, 50, 60, 120, 150, 180, 240, 300, 360, and 1440 min), the absorbance of the dye solution was measured. Kinetics of the dye adsorption on carbon xerogels was characterized on the basis of comparison to two kinetic models: pseudo-first-order (4) and pseudo-second-order (5):

$$ln(q_e - q_t) = lnq_e - \frac{k_1 t}{2.303} \quad (4)$$

$$\frac{t}{q_t} = \frac{1}{k_2 {q_e}^2} + \frac{t}{q_e} \quad (5)$$

where q_t is the amount of the dye adsorbed on the surface of carbon xerogels in a given time (mg/g), k_1 is the adsorption constant in the pseudo-first-order equation (1/min), and k_2 is the adsorption constant in the pseudo-second-order equation (g/mg × min) [6].

Sorption studies were also carried out at different temperatures (25, 45, and 60 °C). The procedure of sample preparation was the same as was used when studying the effect of the pH of the dye solution on the adsorption capacities of the carbon adsorbents. The thermodynamics of the thymol blue adsorption on the carbon xerogels adsorbents was characterized based on the enthalpy, entropy, and Gibbs free energy, calculated from the formulas described in our work [24].

3. Results and Discussion

Results of the elemental analysis of the carbon xerogels, modified with different amine groups, are given in Table 1.

Table 1. Elemental analysis of the samples studied (wt. %).

Material	C	H	N	S	O *
CX	78.77	2.60	0.00	0.00	18.63
CX-APS	31.18	3.35	1.86	19.32	44.29
CX-APS-MA	70.97	2.06	2.20	2.79	21.98
CX-APS-EA	77.23	1.96	1.16	1.21	18.44
CX-APS-PA	74.55	2.36	1.26	1.33	20.50
CX-APS-EDA	50.32	3.61	9.23	10.43	26.41

* by difference; method error ≤ 0.3%.

Pure carbon xerogel (CX) and carbon xerogel modified with methylamine (CX-APS-MA), ethylamine (CX-APS-EA), and propylamine (CX-APS-PA) showed elemental carbon content above 70 wt. %. The lowest content of elemental carbon was found in CX-APS. This sample had the highest content of sulfur (19.32 wt. %) and oxygen (44.29 wt. %) among all xerogel samples studied. Pure carbon xerogel did not contain nitrogen and sulfur. The sample modified with ethylenediamine (having two amine groups) was distinguished from the others by higher nitrogen (9.23 wt. %) and hydrogen (3.61 wt. %) contents.

Figure 2 displays the wide-angle range of XRD profiles of the three samples: CX, CX-APS, and CX-APS-EDA. The diffractogram of CX sample showed no reflections. The XRD profile of the sample CX-APS exhibited peaks assigned to ammonium hydrosulfate [26]. After the modification of CX-APS with ethylenediamine, the XRD profile contained additional peaks, assigned to ethylenediamine sulfate ($C_2H_{10}N_2O_4S$) [27]. In the case of the samples functionalized with the other amines, no reflections were observed.

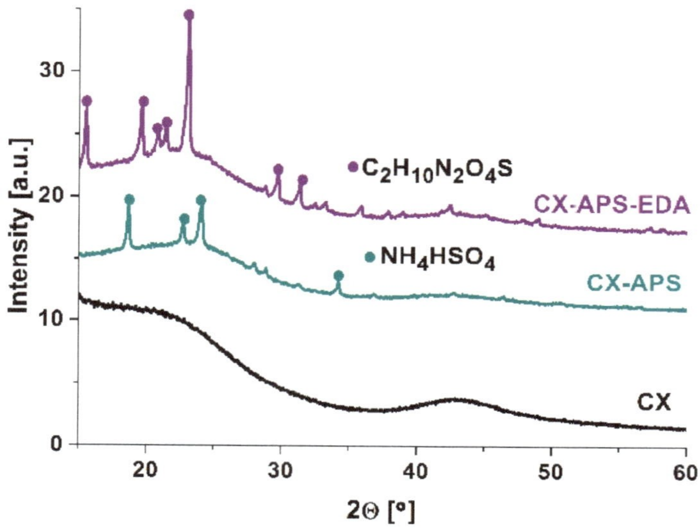

Figure 2. Wide-angle XRD patterns of carbon xerogels: CX, CX-APS, and CX-APS-EDA, recorded at room temperature with a step size of 0.05°.

Table 2 presents the textural parameters of all adsorbents studied. Pure carbon xerogel (CX) was found to have the largest specific surface area and the greatest pore volume. CX oxidation, by ammonium persulfate, led to the reduction of textural parameters: surface area, pore volume, and micropores area. It is supposed that the process of modification takes place mainly inside small mesopores and micropores. As a consequence of this process, oxygen-containing groups may partly block the pores of carbon xerogel, leading to a decrease in the specific surface area of CX-APS [16,27]. Functionalization with amines results in an increase in these parameters, with respect to those of sample CX-APS. This may lead to deblocking of the pores, and thus an increase in the specific surface area and pore volume [16,27]. After the modification of CX-APS with ethylenediamine (CX-APS-EDA), only the pore diameter was reduced (20.46 nm) compared to the pore diameter of CX-APS sample (21.19 nm). Most probably, upon functionalization with ethylenediamine (with two amine groups in a molecule), the xerogel pores are blocked with amine groups.

Table 2. Textural properties of the samples studied.

Material	Total Surface Area [1] (BET) [m^2/g]	Micropore Area [m^2/g]	Total Pore Volume [cm^3/g]	Micropore Volume [cm^3/g]	Average Pore Diameter [nm]
CX	663	401	1.49	0.171	20.30
CX-APS	155	8	0.82	0.002	21.19
CX-APS-MA	450	229	1.28	0.123	33.65
CX-APS-EA	368	142	1.21	0.075	33.14
CX-APS-PA	340	113	1.14	0.058	26.65
CX-APS-EDA	172	13	0.88	0.005	20.46

[1] Error range between 2–5%.

The TG (thermogravimetric) curves obtained upon heating the xerogel samples in a nitrogen atmosphere are depicted in Figure 3A, while Figure 3B presents the DTG (derivative thermogravimetric) curves that illustrate the rate of mass loss with increasing temperature. For pure carbon xerogel, the greatest mass loss was observed in the range of 500–600 °C, which is interpreted as corresponding to the removal of O_2, H_2, and CO_2 (Figure 3) [24]. For sample CX-APS, the mass loss observed was the most pronounced. The mass loss of the CX-APS sample, visible in Figure 3, up to about 100 °C, most probably corresponds to the elimination of the physically adsorbed water. The mass loss noted in the range of 180–300 °C is related to decomposition of carboxylic groups [28–30]. The TG and DTG curves of the samples modified with methylamine, ethylamine, and propylamine (CX-APS-MA, CX-APS-EA, CX-APS-PA) have similar smooth courses. The mass loss is visible for them in the range of 50–250 °C; the first takes place up to 100 °C and is ascribed to the elimination of physisorbed water. The mass loss at higher temperatures (180–250 °C) was assigned to the degradation of carboxylic groups. For the sample modified with ethylenediamine (CX-APS-EDA), a significant mass loss was detected at about 330 °C, which most probably corresponds to the decomposition of amine groups (Figure 3).

The FT-IR spectrum (Figure 4) of CX sample shows bands at ~2925, 2858, 1441, and 1384 cm^{-1}, assigned to the stretching and bending vibrations of C-H. The bands appearing at ~1715 and 1615 cm^{-1}, on the basis of the literature, are attributed to the vibrations of C=O [28]. The bands in the range of 1217–1094 cm^{-1} come from the stretching vibrations of C-O. The spectrum of the sample oxidized with ammonium persulfate (CX-APS) contains an additional band with a maximum at 3148 cm^{-1}, assigned to the stretching vibrations of O-H. It should be added that the FT-IR spectrum of sample CX-APS, besides the bands at ~1715 and 1615 cm^{-1} assigned to the vibrations of C=O, also shows bands in the ranges of 1450–1381 cm^{-1} and 950–910 cm^{-1}, corresponding to the deformational vibrations of O-H. The FT-IR spectra of carbon xerogels modified with amine groups show bands assigned to the deformation vibrations of N-H at 1624 cm^{-1}. Subsequent bands, characteristic of amines, appear in the range of 910–665 cm^{-1}. Bands corresponding to the stretching vibrations of C–N in aliphatic amines are of medium intensity, and appear in the wavenumber range of 1250–1044 cm^{-1} (Figure 4) [31–33].

According to data presented in Table 3, the carbon xerogels differ in acid-base properties. The pure CX sample has both basic and acidic functional oxygen surface groups. Its oxidation leads to a considerable increase in the content of acidic groups (6.16 mmol/g), at the complete disappearance of basic groups. Functionalization of carbon xerogels with amines causes the generation of surface oxygen functional groups of basic nature. Their amount is in the range of 0.50–1.49 mmol/g, depending on the type of amine applied for modification (Table 3).

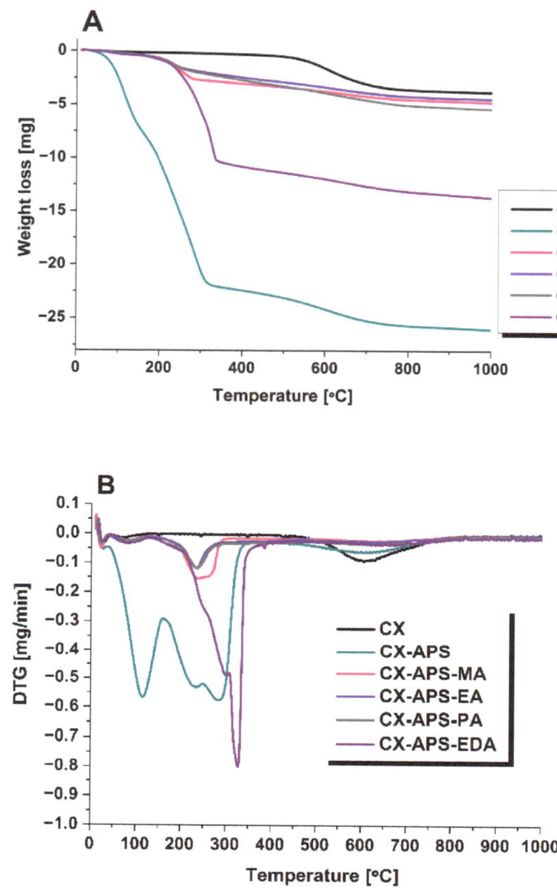

Figure 3. TG (**A**) and DTG (**B**) curves from 25 to 1000 °C of pure and amine-modified carbon xerogels registered under N_2 atmosphere.

Table 3. Acid–base properties of carbon xerogels.

Material	Acidic Oxygen Functional Groups [mmol/g]	Basic Oxygen Functional Groups [mmol/g]
CX	1.27 ± 0.02	0.49 ± 0.01
CX-APS	6.16 ± 0.05	0.00 ± 0.00
CX-APS-MA	0.91 ± 0.01	0.50 ± 0.01
CX-APS-EA	1.23 ± 0.02	0.74 ± 0.01
CX-APS-PA	0.91 ± 0.01	0.74 ± 0.01
CX-APS-EDA	0.73 ± 0.01	1.49 ± 0.02

Figure 5 illustrates the effect of the contact time on the sorption capacity of the xerogel samples. In the first 40–50 min, the adsorption process is fast, which can be explained by a large number of free adsorption centers available for the colorant [34]. In the time from 60 to about 240 min, adsorption is much slower, which could be explained by the saturation of available adsorption sites. According to the results presented in Figure 5, after about 240–260 min the state of equilibrium is achieved.

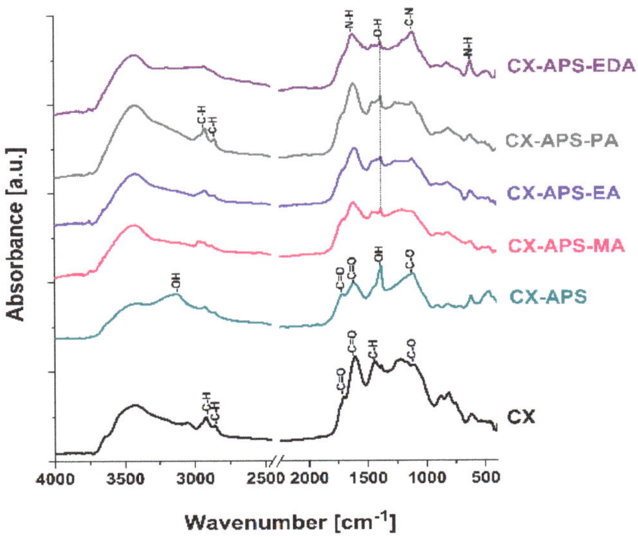

Figure 4. FT-IR spectra of samples studied.

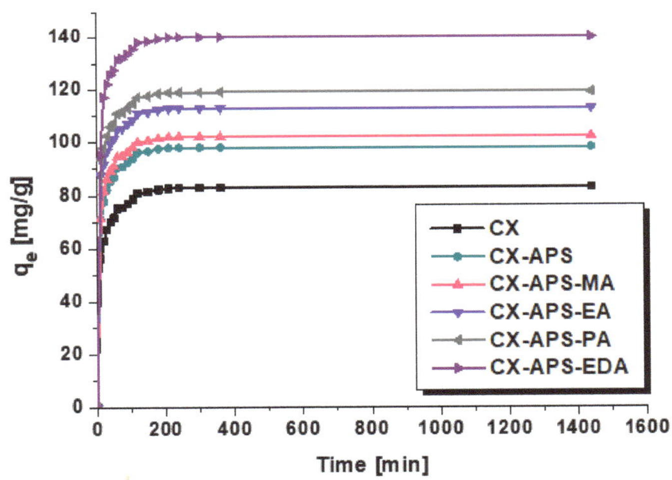

Figure 5. Effect of contact time on the sorption capacity of carbon xerogels (adsorbent mass, 20 mg; initial dye solution concentration, 150 mg/L; volume of dye solution, 50 mL, pH = 4; temperature, 22 ± 1 °C).

The data fitted to the two kinetic models are given in Table 4. Much higher values of R^2 were obtained for the pseudo-second-order kinetic model, so this model better describes the thymol blue adsorption. Additionally, for all carbon xerogels, the values of $q_{e(exp)}$ are identical to the theoretical values of $q_{e(cal)}$ calculated assuming pseudo-second-order kinetics.

Table 4. Kinetic parameters for the adsorption of thymol blue on the surface of carbon xerogels.

Material	$q_{e(exp)}$ [mg/g]	Pseudo-First-Order Model			Pseudo-Second-Order Model		
		$q_{e(cal)}$ [mg/g]	k_1 [1/min]	R^2	$q_{e(cal)}$ [mg/g]	k_2 [g mg^{-1} × min^{-1}]	R^2
CX	83 ± 1.6	50.90	0.0524	0.9811	83.40	0.0025	0.9999
CX-APS	98 ± 1.9	52.21	0.0533	0.9784	98.43	0.0025	0.9999
CX-APS-MA	102 ± 2.0	54.60	0.0532	0.9707	102.46	0.0025	0.9999
CX-APS-EA	113 ± 2.3	95.64	0.0700	0.9392	113.51	0.0024	0.9999
CX-APS-PA	119 ± 2.4	82.57	0.0817	0.9197	119.47	0.0025	0.9999
CX-APS-EDA	140 ± 2.8	76.30	0.0734	0.9426	140.65	0.0022	0.9999

Figure 6 presents the isotherms of thymol blue adsorption on the surface of all samples. The amount of thymol blue adsorbed on the surface of the xerogel samples increases with increasing initial concentrations of the colorant solutions. The amount of thymol blue adsorbed by sample CX (83 mg/g) increases after its oxidation with ammonium sulfate (CX-APS) to 98 mg/g. Moreover, functionalization of CX-APS with amine groups also leads to increased sorption capacity towards thymol blue. Analysis of the results shows that, for CX-APS-MA, CX-APS-EA, and CX-APS-PA, with increasing length of the carbon chain the sorption capacity towards thymol blue increases. However, as the differences in the sorption capacity values are small, it seems that further modification with amines that have longer carbon chains is pointless.

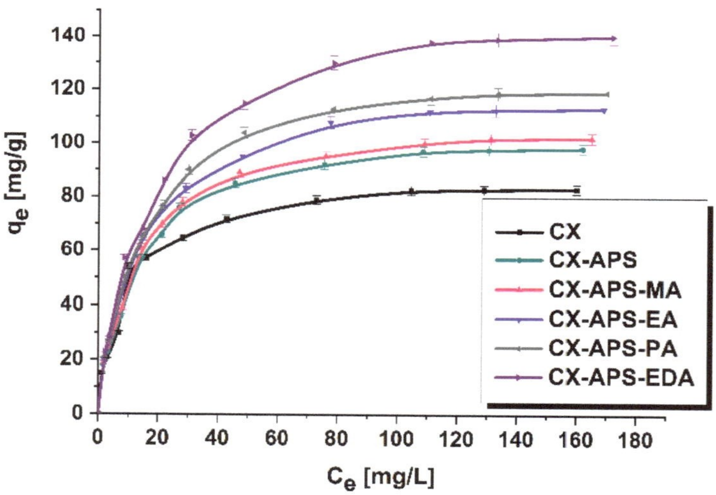

Figure 6. Isotherms of thymol blue adsorption on carbon xerogels (adsorbent mass, 20 mg; initial dye solution concentration, 12.5–150 mg/L; volume of dye solution, 50 mL; pH = 4; temperature, 22 ± 1 °C).

Analysis of the data depicted in Figure 6 leads to the conclusion that the most effective adsorbent is sample CX-APS-EDA, modified with ethylenediamine, which has two amine groups in the molecule, while the other modifying compounds have only one such group. The greatest sorption capacity of CX-APS-EDA, at a level of 140 mg/g, is confirmed by the highest nitrogen content and the basic character of the surface of this adsorbent. The lowest sorption capacity towards thymol blue was found for pure carbon xerogel, containing no nitrogen, which can explain its worse affinity to the dye. Electrostatic attraction and hydrogen bond formation between the colorant molecules and xerogel surface of basic nature are supposed to be the dominant mechanism of adsorption. During the adsorption of thymol blue in a water medium, the hydroxide groups from the dye molecules dissociate. Thus, the sorption capacity of samples towards the dye studied is substantially determined by the interaction of deprotonated hydroxide groups (anions –O-) coming from the dye molecules with the amine groups present on the surface of modified adsorbents [35–38]. The adsorption of the anionic dye for CX-APS-EDA was enhanced by the presence of a greater number of amine groups on the surface of this sample.

In order to determine the mechanism of thymol blue on the surface of the adsorbents, we used the equations of Langmuir and Freundlich isotherms (Table 5). The value of R^2 was higher for the fit to the Langmuir isotherm, and ranged from 0.9991 to 0.9994. The data presented in Table 5 are much poorer fitted to the Freundlich isotherm. The values of $1/n$, for all carbon xerogels, were in the range 0.348–0.409, i.e., lower than 1.

Table 5. The parameters calculated from Langmuir and Freundlich models.

Material	Langmuir			Freundlich		
	q_m [mg/g]	K_L [L/mg]	R^2	K_F [mg/g (L/mg)$^{1/n}$]	$1/n$	R^2
CX	88.2	0.11	0.9991	16.90	0.348	0.8926
CX-APS	105.3	0.09	0.9993	19.16	0.351	0.9280
CX-APS-MA	109.8	0.09	0.9993	18.47	0.367	0.9306
CX-APS-EA	121.9	0.08	0.9994	19.54	0.377	0.9241
CX-APS-PA	128.4	0.08	0.9993	20.07	0.383	0.9188
CX-APS-EDA	153.4	0.07	0.9991	20.51	0.409	0.9332

Another problem considered was the influence of the pH of the thymol blue solutions on the sorption capacities of the samples obtained (Figure 7). With increasing pH of these solutions, the amount of the dye adsorbed on the carbon xerogels surfaces increased. This effect was more pronounced for the xerogels functionalized with amines, for which q_e increased by about 7–8 mg/g. Thymol blue is negatively charged at pH > 1.7; therefore, the interaction takes place between the sulfonic group or deprotonated hydroxyl group of the dye (Figure 1) and the amine groups, that are usually Lewis bases, present on the xerogels surfaces. This interaction can be stronger in basic environments than in an acidic ones (pH < 7) [39–42].

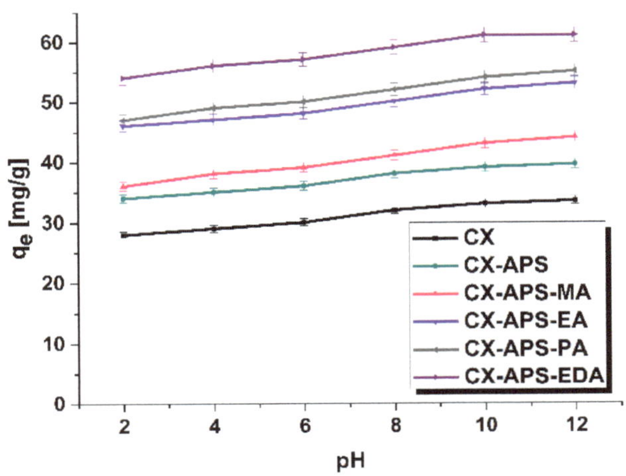

Figure 7. The influence of pH on the adsorption of thymol blue (adsorbent mass, 20 mg; initial dye solution concentration, 80 mg/L; volume of dye solution, 80 mL; temperature, 22 ± 1 °C).

The impact of temperature on the sorption capacity of adsorbents towards thymol blue was also studied (Figure 8). According to the results, with increasing temperature, the sorption capacity increased (q_e).

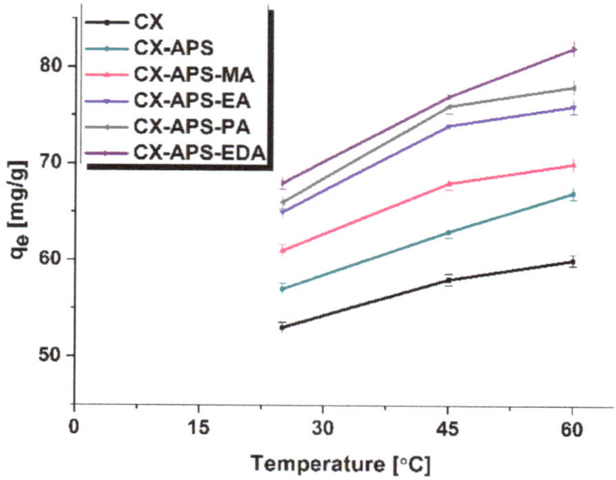

Figure 8. The influence of temperature on the adsorption of thymol blue (adsorbent mass, 20 mg; initial dye solution concentration, 80 mg/L; volume of dye solution, 50 mL; pH = 4; temperature, 25, 45, and 60 ± 1 °C).

This can be explained by the greater mobility of the dye molecules in the solution, which leads to increased binding of thymol blue molecules to the adsorption sites on carbon xerogels. This effect is the most pronounced for sample CX-APS-EDA, whose sorption capacity increases by 14 mg/g (Figure 8). For pristine xerogel, CX, the increase in q_e is the smallest, only by 7 mg/g. Table 6 presents the calculated values of ΔG, ΔH, and ΔS for the adsorption of thymol blue. The positive values of ΔH indicate the endothermic nature of the process, while the negative values of ΔG imply that it is a spontaneous process whose spontaneity increases with increasing temperature [43,44]. The values of entropy vary in the range of 34.60−44.28 J/K mol (Table 6).

Table 7 presents a comparison of the sorption capacities of the CX-APS-EDA sample towards thymol blue, tested in this study, with results obtained for other adsorbents. The comparison implies that CX-APS-EDA is very effective in the removal of thymol blue from water solutions. From the analysis of Table 7 data, it can be observed that the adsorption capacity of the prepared CX-APS-EDA sample is comparable, or superior, to the values obtained in previous studies. Only activated carbon, prepared by chemical activation of garcinia cola nutshell impregnated with H_3PO_4, showed higher sorption capacity towards thymol blue.

Adsorption of organic dyes may be influenced by the structure of the dye, the properties of the surface of carbon adsorbents, the formation of hydrogen bonds, and electrostatic interactions. The carbon xerogels synthesized in this study have amine groups on their surfaces. Electrostatic interactions are possible between the amine functional groups on the surface of the adsorbents and the sulfonic groups of thymol blue (Figure 9) [40]. The mechanism of adsorption may also involve the formation of hydrogen bonds and π-π interactions.

Table 6. Thermodynamical parameters calculated for the adsorption of thymol blue into carbon xerogels.

Material	Temperature [°C]	ΔG [kJ/mol]	ΔH [kJ/mol]	ΔS [J/mol K]
CX	25	−4.13	6.57	35.99
	45	−4.93		
	60	−5.38		
CX-APS	25	−3.57	6.75	34.60
	45	−4.23		
	60	−4.78		
CX-APS-MA	25	−3.65	6.64	34.62
	45	−4.46		
	60	−4.84		
CX-APS-EA	25	−3.71	7.78	38.73
	45	−4.66		
	60	−5.04		
CX-APS-PA	25	−3.72	8.38	40.80
	45	−4.73		
	60	−5.11		
CX-APS-EDA	25	−3.78	9.40	44.28
	45	−4.41		
	60	−4.76		

Table 7. Comparison of sorption capacities of the carbon xerogel modified with ethylenediamine to other adsorbents, presented in the literature, towards thymol blue.

Adsorbent	q_{max} [mg/g]	References
CX-APS-EDA	153.4	This study
activated carbon from *Trachycarpusfortunei* seeds	130.38	[41]
pomegranate peel	5.28	[45]
bentonite	117.6471	[46]
activated carbon from garcinia cola nutshells	396.04	[47]

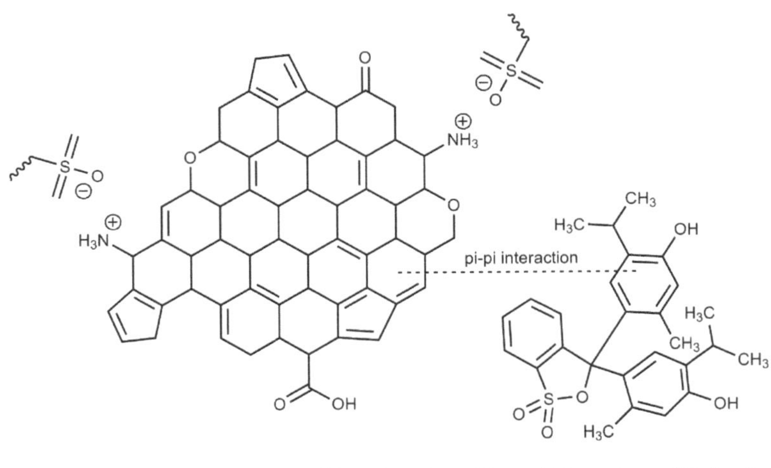

Figure 9. The proposed mechanism of the adsorption of thymol blue onto modified carbon xerogels.

4. Conclusions

Pristine carbon xerogel was subjected to surface oxidation followed by modification with four different amines. The xerogels functionalized with amine groups showed smaller surface areas and pore volumes, but greater average pore diameters, than the unmodified sample. In addition, the sample functionalized with ethylenediamine indicated the highest content of nitrogen and oxygen functional groups compared to the other samples.

The most effective adsorbent of thymol blue from water solutions was the sample modified with ethylenediamine, whose surface exhibited the most basic character of all adsorbents. The value of q_e obtained for sample CX-APS-EDA was 140 mg/g. The high sorption capacity of this carbon xerogel was due to its surface properties. The basic surface functional groups enhanced the interaction between the carbon xerogel and the anionic thymol blue. Amine groups (Lewis bases) derived from ethylenediamine sulfate on the surface of carbon adsorbent undergo protonation in an acidic solution of the dye, and therefore may interact electrostatically with the sulfonic groups of thymol blue. In addition, hydrogen bonds and π-π interactions may form during adsorption process.

It was found that the sorption capacities of the samples increased with increasing initial concentration of thymol blue in water solutions. The Langmuir isotherm model and pseudo-second-order kinetic model fitted well to the data of thymol blue adsorption. With increasing temperature, the sorption capacities of the carbon xerogel adsorbents towards the dye studied increased. The thermodynamic analysis suggested that the thymol blue adsorption process was spontaneous and endothermic in nature. Moreover, thymol blue sorption was more effective in a basic environment.

Author Contributions: Conceptualization, M.P.-K. and J.G.; methodology, M.P.-K.; validation, M.P.-K., J.G. and A.B.-W.; formal analysis, M.P.-K. and J.G.; investigation, M.P.-K. and A.B.-W.; resources, J.G. and R.P.; data curation, M.P.-K.; writing—original draft preparation, M.P.-K. and J.G.; writing—review and editing, M.P.-K., J.G., A.B.-W. and R.P.; visualization, M.P.-K. and J.G.; supervision, R.P. and J.G.; funding acquisition, R.P. All authors have read and agreed to the published version of the manuscript.

Funding: This research received no external funding.

Institutional Review Board Statement: Not Applicable.

Informed Consent Statement: Not Applicable.

Data Availability Statement: Data is contained within the article.

Conflicts of Interest: The authors declare no conflict of interest.

References

1. Taka, A.L.; Pillay, K.; Mbianda, X.Y. Nanospongecyclodextrin polyurethanes and their modification with nanomaterials for the removal of pollutants from wastewater: A review. *Carbohydr. Polym.* **2017**, *159*, 94–104. [CrossRef]
2. Goscianska, J.; Ptaszkowska-Koniarz, M.; Frankowski, M.; Franus, M.; Panek, R.; Franus, W. Removal of phosphate from water by lanthanum-modified zeolites obtained from fly ash. *J. Colloid Interface Sci.* **2018**, *513*, 72–81. [CrossRef]
3. Lan, T.; Cao, F.; Cao, L.; Wang, T.; Yu, C.; Wang, F. A comparative study on the adsorption behavior and mechanism of pesticides on agricultural film microplastics and straw degradation products. *Chemosphere* **2022**, *303*, 135058. [CrossRef]
4. Goscianska, J.; Fathy, N.A.; Aboelenin, R.M.M. Adsorption of solophenyl red 3BL polyazo dye onto amine-functionalized mesoporous carbons. *J. Colloid Interface Sci.* **2017**, *505*, 593–604. [CrossRef]
5. Afshari, M.; Dinari, M. Synthesis of new imine-linked covalent organic framework as high efficient absorbent and monitoring the removal of direct fast scarlet 4BS textile dye based on mobile phone colorimetric platform. *J. Hazard. Mater.* **2020**, *385*, 121514. [CrossRef]
6. Goscianska, J.; Ptaszkowska, M.; Pietrzak, R. Equilibrium and kinetic studies of chromotrope 2R adsorption onto ordered mesoporous carbons modified with lanthanum. *Chem. Eng. J.* **2015**, *270*, 140–149. [CrossRef]
7. Afshari, M.; Dinari, M.; Zargoosh, K.; Moradi, H. Novel triazine-based covalent organic framework as a superadsorbent for the removal of mercury(II) from aqueous solutions. *Ind. Eng. Chem. Res.* **2020**, *59*, 9116–9126. [CrossRef]
8. Goscianska, J.; Marciniak, M.; Pietrzak, R. Ordered mesoporous carbons modified with cerium as effective adsorbents for azo dyes removal. *Sep. Purif. Technol.* **2015**, *154*, 236–245. [CrossRef]
9. Afshari, M.; Dinari, M.; Moradi, H.; Noori, Z. Polyaniline/sulfonated-covalent organic polymer nanocomposite: Structural and dye adsorption properties. *Polym. Adv. Technol.* **2020**, *31*, 2433–2442. [CrossRef]
10. Yagub, M.T.; Sen, T.K.; Afroze, S.; Ang, H.M. Dye and its removal from aqueous solution by adsorption: A review. *Adv. Colloid Interface Sci.* **2014**, *209*, 172–184. [CrossRef] [PubMed]

11. Mashkoor, F.; Nasar, A. Preparation, characterization and adsorption studies of the chemically modified luffa aegypticapeel as a potential adsorbent for the removal of malachite green from aqueous solution. *J. Mol. Liq.* **2019**, *274*, 315–327. [CrossRef]
12. Zhang, H.; Pan, Y.; Wang, Z.; Wu, A.; Zhang, Y. Synthesis of hollow mesoporous manganese dioxide nanoadsorbents with strong negative charge and their ultra-efficient adsorption for cationic dyes. *Sep. Purif. Technol.* **2022**, *29*, 121241. [CrossRef]
13. Zhou, G.; Tian, H.; Sun, H.; Wang, S.; Buckley, C.E. Synthesis of carbon xerogels at varying sol–gel pHs, dye adsorption and chemical regeneration. *Chem. Eng. J.* **2011**, *171*, 1399–1405. [CrossRef]
14. Ribeiro, R.S.; Fathy, N.A.; Attia, A.A.; Silva, A.M.T.; Faria, J.L.; Gomes, H.T. Activated carbon xerogels for the removal of the anionic azo dyes Orange II and Chromotrope 2R by adsorption and catalytic wet peroxide oxidation. *Chem. Eng. J.* **2012**, *195–196*, 112–121. [CrossRef]
15. Paéz, C.A.; Contreras, M.S.; Léonard, A.; Blacher, S.; Olivera-Fuentes, C.-G.; Pirard, J.P.; Job, N. Effect of CO_2 activation of carbon xerogels on the adsorption of methylene blue. *Adsorption* **2012**, *18*, 199–211. [CrossRef]
16. Ptaszkowska-Koniarz, M.; Goscianska, J.; Pietrzak, R. Removal of rhodamine B from water by modified carbon xerogels. *Colloids Surf. A Physicochem. Eng. Asp.* **2018**, *543*, 109–117. [CrossRef]
17. Girgis, B.S.; Attia, A.A.; Fathy, N.A. Potential of nano-carbon xerogels in the remediation of dye-contaminated water discharges. *Desalination* **2011**, *265*, 169–176. [CrossRef]
18. Figueiredo, J.L.; Sousa, J.P.S.; Orge, C.A.; Pereira, M.F.R.; Órfão, J.J.M. Adsorption of dyes on carbon xerogels and templated carbons: Influence of surface chemistry. *Adsorption* **2011**, *17*, 431–441. [CrossRef]
19. Zaggout, F.R.; El-Nahhal, I.M.; Qaraman, A.E.A.; Al-Dahoudi, N. Behavior of *thymol blue* analytical pH-indicator entrapped into sol–gel matrix. *Mater. Lett.* **2006**, *60*, 3463–3467. [CrossRef]
20. Hopkins, A.E.; Sell, K.S.; Soli, A.L.; Byrne, R.H. In-situ spectrophotometric pH measurements: The effect of pressure on *thymolblue* protonation and absorbance characteristics. *Mar. Chem.* **2000**, *71*, 103–109. [CrossRef]
21. Hankare, P.P.; Patil, R.P.; Jadhav, A.V.; Garadkar, K.M.; Sasikala, R. Enhanced photocatalytic degradation of methyl red and *thymolblue* using titania–alumina–zinc ferrite nanocomposite. *Appl. Catal. B Environ.* **2011**, *107*, 333–339. [CrossRef]
22. Sousa, J.P.S.; Pereira, M.F.R.; Figueiredo, J.L. Carbon Xerogel Catalyst for NO Oxidation. *Catalysts* **2012**, *2*, 447–465. [CrossRef]
23. Boehm, H.P. Some aspects of the surface chemistry of carbon blacks and other carbons. *Carbon* **1994**, *32*, 759–769. [CrossRef]
24. Ptaszkowska-Koniarz, M.; Goscianska, J.; Pietrzak, R. Adsorption of dyes on the surface of polymer nanocomposites modified with methylamine and copper(II) chloride. *J. Colloid Interface Sci.* **2017**, *504*, 549–560. [CrossRef]
25. Langmuir, I. The adsorption of gases on plane surfaces of glass, mica and platinum. *J. Am. Chem. Soc.* **1918**, *40*, 1361–1403. [CrossRef]
26. Huang, J.; Zhang, W.; Huang, H.; Liu, Y.; Yang, Q.; Li, L. Facile synthesis of N,S-Codoped hierarchically porous carbon with high volumetric pseudocapacitance. *ACS Sustain. Chem. Eng.* **2019**, *7*, 16710–16719. [CrossRef]
27. Ptaszkowska-Koniarz, M.; Goscianska, J.; Pietrzak, R. Synthesis of carbon xerogels modified with amine groups and copper for efficient adsorption of caffeine. *Chem. Eng. J.* **2018**, *345*, 13–21. [CrossRef]
28. Goscianska, J.; Olejnik, A.; Nowak, I.; Marciniak, M.; Pietrzak, R. Stability analysis of functionalized mesoporous carbon materials in aqueous solution. *Chem. Eng. J.* **2016**, *290*, 209–219. [CrossRef]
29. Moreno, A.H.; Arenillas, A.; Calvo, E.G.; Bermúdez, J.M.; Menéndez, J.A. Carbonisation of resorcinol–formaldehyde organic xerogels: Effect of temperature, particle size and heating rate on the porosity of carbon xerogels. *J. Anal. Appl. Pyrol.* **2013**, *100*, 111–116. [CrossRef]
30. Álvarez, S.; Ribeiro, R.S.; Gomes, H.T.; Sotelo, J.L.; García, J. Synthesis of carbon xerogels and their application in adsorption studies of caffeine and diclofenac as emerging contaminants. *Chem. Eng. Res. Des.* **2015**, *95*, 229–238. [CrossRef]
31. Laginhas, C.; Nabais, J.M.V.; Titirici, M.M. Activated carbons with high nitrogen content by a combination of hydrothermal carbonization with activation. *Micropor. Mesopor. Mater.* **2016**, *226*, 125–132. [CrossRef]
32. Allahbakhsh, A.; Khodabadi, F.N.; Hosseini, F.S.; Haghighi, A.H. 3-Aminopropyl-triethoxysilane-functionalized rice husk and rice husk ash reinforced polyamide 6/graphene oxide sustainable nanocomposites. *Eur. Polym. J.* **2017**, *94*, 417–430. [CrossRef]
33. Goscianska, J.; Olejnik, A. Dispersion stability of the aminosilane-grafted mesoporous carbons in different solvents. *Microporous Mesopor. Mater.* **2018**, *265*, 149–161. [CrossRef]
34. Goscianska, J.; Marciniak, M.; Pietrzak, R. Mesoporous carbons modified with lanthanum(III) chloride for methyl orange adsorption. *Chem. Eng. J.* **2014**, *247*, 258–264. [CrossRef]
35. Hernández-Morales, V.; Nava, R.; Acosta-Silva, Y.J.; Macías-Sánchez, S.A.; Pérez-Bueno, J.J.; Pawelec, B. Adsorption of lead (II) on SBA-15 mesoporous molecular sieve functionalized with $-NH_2$ groups. *Micropor. Mesopor. Mater.* **2012**, *160*, 133–142. [CrossRef]
36. Qiu, J.; Feng, Y.; Zhang, X.; Jia, M.; Yao, J. Acid-promoted synthesis of UiO-66 for highly selective *adsorption* of *anionicdyes*: *Adsorption* performance and mechanisms. *J. Colloid Interface Sci.* **2017**, *499*, 151–158. [CrossRef]
37. Wang, W.; Huang, G.; An, C.; Zhao, S.; Zhang, P. *Adsorption* of *anionic* azo *dyes* from aqueous solution on cationic gemini surfactant-modified flax shives: Synchrotron infrared, optimization and modeling studies. *J. Clean. Prod.* **2018**, *172*, 1986–1997. [CrossRef]
38. Jung, K.W.; Choi, B.H.; Hwang, M.J.; Jeong, T.U.; Ahn, K.H. Fabrication of granular activated carbons derived from spent coffee grounds by entrapment in calcium alginate beads for adsorption of acid orange 7 and methylene blue. *Bioresour. Technol.* **2016**, *219*, 185–195. [CrossRef] [PubMed]

39. Vucurović, V.M.; Razmovski, R.N.; Miljić, U.D.; Puskas, V.S. Removal of cationic and anionic azo dyes from aqueous solutions by adsorption on maize stem tissue. *J. Taiwan Inst. Chem. Eng.* **2014**, *45*, 1700–1708. [CrossRef]
40. Cheng, Y.; Feng, Q.; Ren, X.; Yin, M.; Xue, Z. *Adsorption* and removal of sulfonic dyes from aqueous solution onto a coordination polymeric xerogel with amino groups. *Colloids Surf. A Physicochem. Eng. Asp.* **2015**, *485*, 125–135. [CrossRef]
41. Bakhsh, E.M.; Bilal, M.; Ali, M.; Ali, J.; Wahab, A.; Akhtar, K.; Fagieh, T.M.; Danish, E.Y.; Asiri, A.M.; Khan, S.B. Synthesis of Activated Carbon from Trachycarpusfortunei Seeds for the Removal of Cationic and Anionic Dyes. *Materials* **2022**, *15*, 1986. [CrossRef]
42. Shehata, H.M.; Ali, D.A.; Al-Akraa, I.M.; Elazab, H.A.; Elsawy, H.A. Development of novel adsorbent for industrial waste water treatment. *Int. J. Adv. Trends Comput. Sci. Eng.* **2020**, *9*, 704–712. [CrossRef]
43. Song, W.; Gao, B.; Xu, X.; Xing, L.; Han, S.; Duan, P.; Song, W.; Jia, R. Adsorption-desorption behavior of magnetic amine/Fe_3O_4 functionalized biopolymer resin towards anionic dyes from wastewater. *Bioresour. Technol.* **2016**, *210*, 123–130. [CrossRef]
44. Haddad, M.E.; Mamouni, R.; Saffaj, N.; Lazar, S. Evaluation of Performance of Animal Bone Meal as a new low cost adsorbent for the removal of a cationic dye RhodamineB from aqueous solutions. *J. Saudi Chem. Soc.* **2016**, *20*, S53–S59. [CrossRef]
45. Hasan, H.M.I.; Alfutisi, H.M.M. Removing of thymol blue from aqueous solutions by pomegranate peel. *Int. J. Appl. Sci.* **2018**, *1*, 111–119.
46. Koyuncu, H.; Aldemir, A.; Kul, A.R.; Canayaz, M. Removal of thymol blue from aqueous solution by natural and modified bentonite: Comparative analysis of ANN and ANFIS models for the prediction of removal percentage. *Pollution* **2021**, *7*, 959–980. [CrossRef]
47. Kuete, I.-H.T.; Tchuifon, D.R.T.; Ndifor-Angwafor, G.N.; Kamdem, A.T.; Anagho, S.G. Kinetic, isotherm and thermodynamic studies of the adsorption of thymol blue onto powdered activated carbons from garcinia cola nut shells impregnated with H_3PO_4 and KOH: Non-linear regression analysis. *J. Encapsul. Adsorpt. Sci.* **2020**, *10*, 1–27. [CrossRef]

materials

Article

Determination of Relative Density and Degree of Saturation in Mineral Soils Based on In Situ Tests

Simon Rabarijoely *, Mariusz Lech and Marek Bajda

Department of Geotechnics, Institute of Civil Engineering, Warsaw University of Life Sciences, Nowoursynowska 159 St., 02-776 Warsaw, Poland; mariusz_lech@sggw.edu.pl (M.L.); marek_bajda@sggw.edu.pl (M.B.)
* Correspondence: simon_rabarijoely@sggw.edu.pl; Tel.: +48-22-5935229

Abstract: Based on the results of dynamic probing (DP), time-domain reflectometry (TDR/MUX/MPTS), resistivity cone penetration tests (RCPT), Marchetti dilatometer tests (DMT), and seismic dilatometer tests (SDMT), it is possible to develop a relationship to calculate the relative density (D_r) and degree of saturation (S_r) of selected sandy soils. Compiled databases from documented research points for selected sandy soils were used to construct and develop direct correlations between the measured pressures p_0 and p_1 from DMT and shear wave velocity (V_s) from SDMT, along with pore water pressures (u_0) and atmospheric pressure (Pa). The results allowed us to make a preliminary prediction when evaluating the parameters. Further, they allowed limiting the use of an additional device, especially in the case of multilayer heavy preconsolidated subsoils. Moreover, soil physical and mechanical characteristics (temperature, humidity, pressure, swelling, salinity) measured from TDR/MUX/MPTS (laboratory/field-operated meter for simultaneous measurements of soil moisture, matric potential, temperature, and salinity—bulk electrical conductivity) were assessed. The main achievement of this paper is the original proposal of using a new nomogram chart to determine the relative density and degree of saturation based on DMT and SDMT tests.

Keywords: geotechnical engineering; new nomogram chart

Citation: Rabarijoely, S.; Lech, M.; Bajda, M. Determination of Relative Density and Degree of Saturation in Mineral Soils Based on In Situ Tests. *Materials* **2021**, *14*, 6963. https://doi.org/10.3390/ma14226963

Academic Editors: Jakub Zdarta and Agata Zdarta

Received: 29 September 2021
Accepted: 15 November 2021
Published: 17 November 2021

1. Introduction

In accordance with the applicable construction law (Eurocode 7), each project applying for a building permit should include, depending on the needs, the results of geological and engineering research. This documentation consists of the developed results of field and laboratory tests. The advantage of field tests is the fact that they take place in the natural environment, which is often difficult to recreate in laboratory conditions. Probing is an example of such research work carried out to determine the ground condition. Among numerous available methods, dynamic probing tests (DP), cone penetration tests (CPT), and dilatometer Marchetti tests (DMT) are commonly applied in leading research centers. In order to reduce the necessity to use various types of equipment, methods of field research are being sought for to enable the interpretation of the obtained results in a wide range [1–5]. One of the field tests that meets this requirement is the Marchetti dilatometer test [6], whose use in the world is significantly increasing.

The greatest advantage of dilatometer testing is its quick and relatively simple measurement, on the basis of which it is possible to determine the soil parameters. Interpretation of geotechnical parameters is based on the use of empirical relationships related to pressure values measured directly in the field. This article is based on this type of in situ research. Based on the results of DP, DMT, SDMT, and RCPT tests carried out in the Antoniny, Koszyce, and Nielisz sites with organic subsoils under embankment, and the Stegny and WULS-SGGW campus sites with a sandy soil layer (Figure 1), new relationships were determined to assess the parameters describing the state of selected mineral soils. Since cohesive and organic soils were not taken into account to develop the formulas for determining the parameters of D_r and S_r, these parameters were not included in their physical properties in this article.

Figure 1. Location of the test sites in Poland.

This paper presents the test results of mineral and organic subsoils obtained from the following sites (WGS-84): Antoniny (WGS-84: 53.06742, 17.07214), Koszyce (WGS-84: 53.17045, 16.74924), Nielisz (WGS-84: 50.80396, 23.03075), Stegny (WGS-84: 52.18215, 21.04864), and the WULS-SGGW campus (WGS-84: 52.16245, 21.03838).

The Antoniny test embankment was designed and tested in the frame of cooperation between the Department of Geotechnical Engineering SGGW and the Swedish Geotechnical Institute (SGI). The physical properties of the soils at the Koszyce, Nielisz, Stegny, and WULS-SGGW campus test sites were determined as part of a research program conducted at these sites in previous years.

The Antoniny embankment and Koszyce test dam are located in the Ruda river valley. A layer of soft organic soils was discovered in the sublayer of both objects. The organic soils are Quaternary deposits of an oxbow lake. The thickness of organic soils in this region generally exceeds 10 m and locally even 20 m. Dense sand occurs under the organic soils. The Nielisz site is located in eastern Poland in the Wieprz river valley in Lublin Province. The layer of soft subsoil has a thickness of 3 to 5 m; the soils are slightly preconsolidated. Two layers of organic subsoil were distinguished at the Nielisz site. The Stegny and SGGW campus sites are located in the southern part of Warsaw, where a few sedimentation cycles, from sands to clays, were observed in vertical succession. The entire complex of Pliocene clays comprises clays, silty clays (60–70%), silts (10–25%), and sands (10–20%).

The index properties of all investigated mineral and organic soils are presented in Table 1.

Table 1. Index properties of organic soils at the Antoniny, Koszyce, Nielisz, Stegny, and SGGW campus test sites.

Site	Type of Soil	Organic Content I_{om} (%)	$CaCO_3$ Content (%)	Water Content w_n (%)	Liquid Limit w_L (%)	Density: Unit Weight of Soil ρ (t/m^3)	Density: Specific Weight of Soil ρ_s (t/m^3)
Antoniny	Amorphous Peat	65–75	10–15	310–340	305–450	1.05–1.10	1.45–1.50
	Calcareous Gyttja	5–20	65–90	105–140	80–110	1.25–1.40	2.2–2.30
Koszyce	Amorphous Peat	70–85	5–15	400–550	450	1.05–1.1	1.45–1.50
	Calcareous Gyttja (G_y)	10–20	65–80	120–160	80–110	1.20–1.35	2.1–2.25
	Calcareous Gyttja (G_y)	15–20	65–75	180–220	100–110	1.25–1.30	2.2
Nielisz	Organic Mud (M_{or})	20–30	-	120–150	130–150	1.25–1.30	2.25–2.3
	Organic Mud (M_{or})	10–20	-	105–120	110–130	1.30–1.45	2.30–2.40
Stegny	Pliocene Clays	-	-	19.20–28.50	67.6–88.0	2.1–2.2	2.68–2.73
SGGW Campus	Boulder Clay	-	-	5.20–20.10	21.9–26.6	2.0–2.2	2.68–2.73

2. Literature Review

2.1. Methodology and Interpretation of Dilatometer Test Results

Over 46 years ago, Prof. Silvano Marchetti designed and built the first dilatometer at L'Aquila University in Italy; the design and principles of soil research were presented by him in 1975 during the American Society of Civil Engineers (ASCE) conference in Raleigh [7]. DMT tests consist of measuring the gas pressure acting on the diaphragm of a dilatometer blade at selected subsoil depths (Figure 1). In soil tests, two pressures are usually measured (A and B); they force the center of the membrane to move 0.05 mm to the ground (reading A) and deflect the center of the membrane towards the ground by approximately 1.05 mm (reading B). To extend the dilatometer testing, pressure measurements are sometimes taken as the membrane returns to ground contact (C reading). Readings A, B, and C are corrected for the inertia of the diaphragm and marked as p_0, p_1, and p_2, respectively. Pressures p_0 and p_1 and the value of the vertical effective stress σ'_{v0} are used to determine the following dilatometer indexes: material index I_D, horizontal stress index K_D, and dilatometer modulus E_D [8–10].

$$p_0 = 1.05(A - Z_M + \Delta A) - 0.05(B - Z_M - \Delta B),\ (\text{MPa}), \tag{1}$$

- The 1.10 mm corrected pressure reading in DMT p_1:

$$p_1 = B - Z_M - \Delta B,\ (\text{MPa}), \tag{2}$$

- Corrected third reading in DMT p_2:

$$p_2 = C - Z_M - \Delta A\ (\text{MPa}), \tag{3}$$

- Material index I_D:

$$I_D(-) = f(A,\ B,\ u_0) = \frac{P_1 - P_0}{P_0 - u_0}, \tag{4}$$

- Horizontal stress index K_D:

$$K_D(-) = f(A,\ u_0, \sigma'_{v0}, B) = \frac{p_0 - u_0}{\sigma'_{v0}}, \tag{5}$$

- Dilatometer modulus E_D:

$$E_D(\text{MPa}) = f(A,\ B) = 34.7 \cdot (P_1 - P_0), \tag{6}$$

- Pore pressure index U_D:

$$U_D(-) = f(A,\ C,\ u_0, B) = \frac{P_2 - u_0}{P_0 - u_0}, \tag{7}$$

where p_0—pressure reading A corrected for Z_m and ΔA membrane stiffness at 0.05 mm expansion, and 0.05 mm expansion itself, to estimate the total soil stress acting normal to the membrane immediately before its expansion into the soil (0.00 mm expansion);

p_1—pressure reading B corrected for Z_m and ΔB membrane stiffness at 1.10 mm expansion to give the total soil stress acting normal to the membrane at 1.10 mm membrane expansion;

p_2—pressure reading C corrected for Z_m and ΔA membrane stiffness at 0.05 mm expansion and used to estimate pore water pressure;

σ'_{v0}—pre-insertion in situ overburden stress;

u_0—pore water pressure acting in the center of the membrane before insertion of the DMT blade (often assumed as hydrostatic below the groundwater table);

Z_m—gage pressure deviation from zero when vented to atmospheric pressure (offset used to correct pressure readings to the true gage pressure).

2.2. *Existing In Situ Methods for Determining Relative Density and Degree of Saturation in Non-Cohesive Soils*

By definition, relative density is a parameter that characterizes non-cohesive soils. It is the ratio of soil compaction in the natural state to the highest possible compaction of a specific soil. There are two types of dependence for determining relative density (D_r) on the basis of DMT. The first is the relationship presented by Reyna and Chameau (1991) [11], and Mayne (2001) [12], and it depends on the horizontal pressure index (K_D) from DMT tests. The second relationship was described by Marchetti (1992) [13]; it is a function of the dilatometer blade resistance (q_D) and effective vertical stress (σ'_{v0}). In the literature, there are many formulas to determine relative density from in situ studies [11–13]. These dependencies are presented below:

$$D_r = -1.082 + 0.204 \cdot \left(\frac{q_D}{\sigma'_{v0}}\right)^{0.4}, \tag{8}$$

where D_r is relative density (as decimal); σ'_{v0} is effective geostatic stress (kPa); and q_D is wedge resistance.

$$D_r = \left[\frac{1}{40 \cdot (K_D - 1)} + \frac{1}{120}\right]^{-1.0}, \tag{9}$$

For normally consolidated (NC) uncemented sands, the recommended equation for the relative density (D_r) of non-cohesive soils is shown in Formula (8) [11], where parameter D_r is related to K_D from DMT research. This correlation is influenced by the additional $K_D - D_r$ data points (also in Formula (9)) obtained by Tanaka and Tanaka (1998) [14] from the Ohgishima and Kemigawa sites, where parameter D_r was established on the basis of high-quality samples collected using the freezing method. In fractured sands in the range of preconsolidated stress (OC) (Formula (10)), parameter D_r will be overestimated because part of the K_D value is due to the influence of preconsolidation and cementation. At present, it is difficult to clearly assess the value of parameter D_r.

$$D_r = 100 \cdot \left(\frac{K_D - 1}{7}\right)^{0.5}, \tag{10}$$

Formulas (8)–(10) do not sufficiently describe a satisfying (small cyclic shear stress factor) state of non-cohesive soils in a wide range of interpretations of the saturation state with the shear wave velocity propagation for a given soil medium, and their influence on the compressibility of the air–water mixture filling the pores and on the compressibility of the soil skeleton. Therefore, later in this article, a decision is made to develop a new relationship for determining the degree of relative density of non-cohesive soils based on vs. obtained from SDMT tests.

3. Materials and Methods

3.1. Material

This paper contains the test results of sands located in the subsoil and in embankments on the test sites that were presented in Section 1, where a laboratory and field testing program was carried out under and outside of the main dam embankment [15–18]. This research was carried out according to the Casagrande method modified by Prószyński. Soil grains in the range from 0.001 to 0.1 mm were collected using the areometric method, and the remaining fractions in the range above 0.1 mm were collected by the sieve method. The results of fraction testing from 0.001 to 0.1 mm were read together with smaller percent passing, and above 0.1 mm with larger percent retaining. After dividing the range of a particular fraction into clay, silt, sand, gravel, and cobble (the sum of these fractions amounts to 100%), the soil type was determined according to PN-86 B-02480. For the embankment layer at the Antoniny site, the clay fraction content was 0%, the clay fraction ranged from 2 to 3%, the sand fraction ranged from 88 to 91%, and the gravel fraction was about 7%. At the Koszyce site, the clay fraction content was 0%, the clay fraction ranged from 2 to 3%, the sand fraction ranged from 89 to 92%, and the gravel fraction was about 5%. At the Nielisz site, the clay fraction content was 0%, the clay fraction ranged from 3 to 5%, the sand fraction ranged from 87 to 92%, and the gravel fraction was about 2%. At the SGGW campus site, the clay fraction content was 1%, the clay fraction was 4%, the sand fraction ranged from 88 to 95%, and the gravel fraction was 0%. At the Stegny site, the clay fraction content was 0%, the clay fraction was 3%, the sand fraction was 92%, and the gravel fraction was about 5%. The grain size distribution curve obtained from laboratory tests for mineral soils from the described sites is presented in Figure 2. The index properties of mineral soils in the Antoniny, Koszyce, Nielisz, Stegny, and WULS-SGGW campus test sites are presented in Table 2. Figure 3 presents the diagram chart proposed by Marchetti and Crapps for the analyzed sites [19].

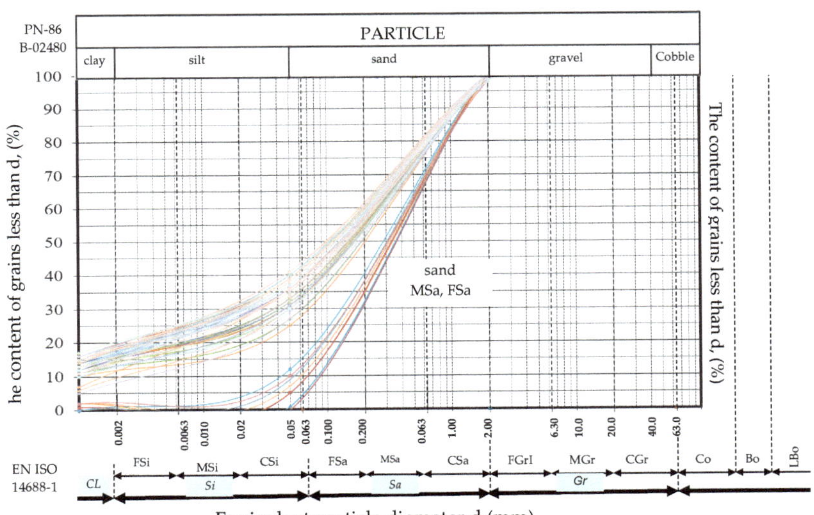

Figure 2. Grain size distribution curve obtained from laboratory tests for the sandy soils from the described sites.

Table 2. Index properties of sandy soils in the Antoniny, Koszyce, and Nielisz embankment test sites, and Stegny and WULS-SGGW campus mineral layer test sites.

Sites	Type of Soil	$CaCO_3$ Content (%)	Water Content w_n (%)	Density	
				Unit Weight of Soil ρ (t/m^3)	Specific Weight of Soil ρ_s (t/m^3)
Antoniny	sand	<1	6.5	1.7	2.65–2.67
Koszyce		<1	6.2	1.85	2.65–2.67
Nielisz		1–3	6.1	1.85	2.65–2.67
Stegny		<1	5.4	1.7	2.68–2.66
WULS-SGGW Campus		1–3	6.7	1.85	2.68–2.66

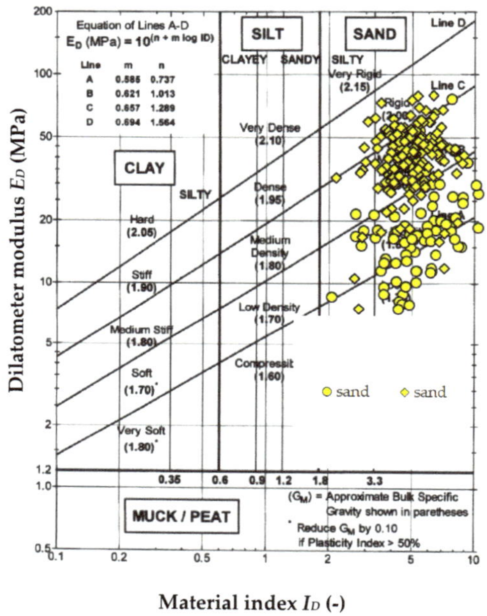

Figure 3. Chart for estimating unit weight (normalized to γ_w) of sands in the Antoniny, Koszyce, Nielisz, Stegny, and WULS-SGGW campus test sites [19].

3.2. Laboratory Test with TDR/MUX/mpts Meter

The laboratory TDR/MUX/mpts meter [20] was adapted to measure the following parameters: volumetric moisture expressed as percentage value using the LP/MS probe; soil salinity with the LP/MS probe, expressed in S/m; parent pressure inside the soil sample with the LP/p probe, expressed in mbar; and temperature inside the soil sample with the LP/t probe, expressed in °C. All tests were performed in the Laboratory of the Department of Geotechnics of the WULS–SGGW (Figure 4).

Each of the sensors was placed in a previously prepared soil sample. The tests were carried out on 2 soil samples: silty sand and silty clay. Before testing, the soil samples were compacted and placed in a steel cylinder with holes for sensor placement. Then, the Lp probe was prepared, which required specialized calibration. Preparation of the meter included the following steps: 1. Inserting the LP/t sensor (temperature sensor) into distilled water, while the LP/p sensor (pressure sensor) should be deaerated and put into a vessel with distilled water. 2. Turning on the readout logging program and defining the names of the saved files. In the program start window, the appropriate channel in which the program will work should be selected. 3. Calibrating the sensors, inserting the moisture probe into the calibrator, and then into distilled water, so that the water reaches the sensor cap, waiting for 2 min, and then following the instructions on the computer monitor. 4. Starting the readings. The probes should be placed in the ground in order not to damage the ground structure.

The first to be tested was a silty sand sample, which was pressed into a steel cylinder that was previously weighed on a laboratory scale. Then, the sensors of the LP probe were pressed, and everything was placed on the scale. The following results were obtained: cylinder weight = 344.50 g; weight of cylinder + silty sand = 801.40 g; total weight = 858.85 g; weight of analyzed silty sand = 456.90 g. The prepared soil sample with the inserted sensors is shown in Figure 5.

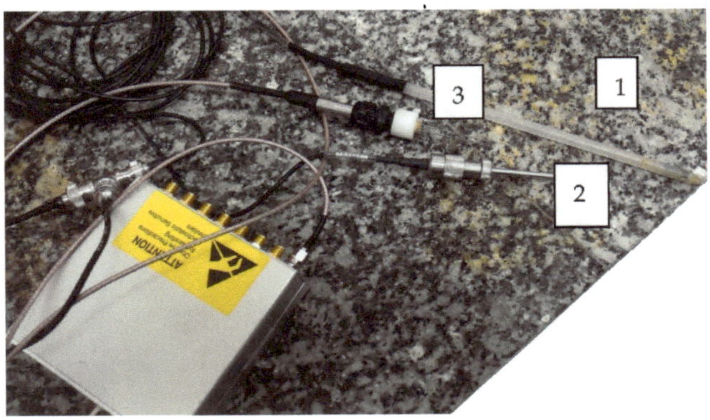

Figure 4. TDR/MUX/mpts meter with the following sensors: 1. temperature, 2. suction pressure, and 3. moisture and salinity (Laboratory of the Department of Geotechnics, WULS-SGGW).

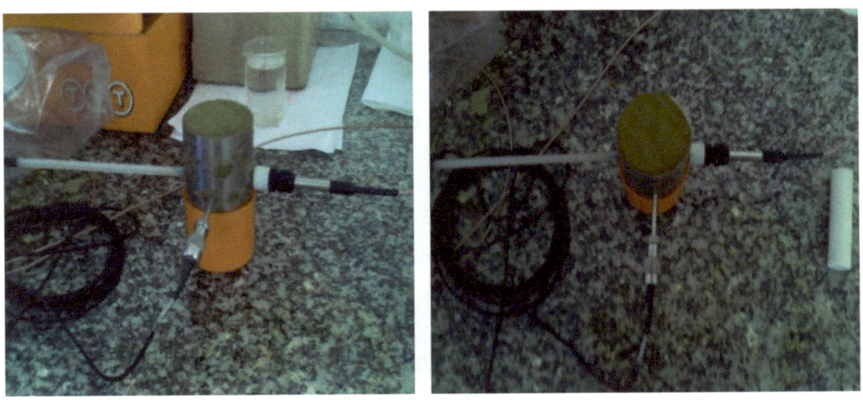

Figure 5. A sample of silty sand with inserted sensors (Laboratory of the Department of Geotechnics, WULS–SGGW).

The volume of the test sample was also determined (237.6 cm^3), and the sample humidity was calculated. The reading was set to 60 s. After 20 days of the analysis, the "drying" of the sample was completed, and the sample was placed on the laboratory scale: weight of total sample after "drying" = 771.85 g; weight of silty sand after drying = 369.90 g. After weighing, 100 g of water was added to the rehydrate, and the sample returned to its original condition. After 8 days of contact of the sample with water, the sandy clay analysis was completed and the soil was reweighed: final weight of soil with sensors and cylinder = 819.30 g; weight of hydrated clay = 474.80 g; weight of the remaining water = 16.80 g. After drying the sample, cracks in the sample were noticed, indicating shrinkage of the tested soils. Soil shrinkage is shown in Figure 6.

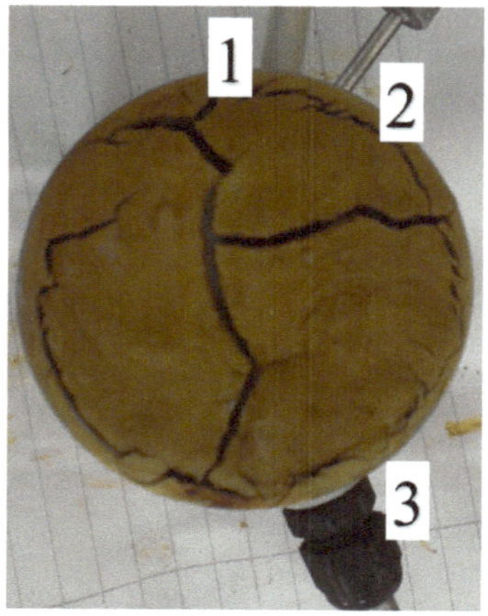

Figure 6. Soil sample with cracks and probe sensors: 1. temperature, 2. suction pressure, and 3. moisture and salinity (Laboratory of the Department of Geotechnics, WULS–SGGW).

Finally, the following dependencies were plotted: soil sample temperature vs. experiment duration; sample moisture vs. time; salinity vs. time; soil sample swelling due to humidity. After the test, the test soil was removed, the probe tips and the metal cylinder were cleaned, and then a clay sample was placed in the cylinder. The clay tests were carried out in the same way as the silty sand tests.

3.3. Estimation Theory

For the determination of parameters D_r and S_r based on DMT tests, parameters p_0, p_1, u_0, and σ'_{v0} were taken into account in the statistical analysis. These quantities can be treated as random variables. A total of 48 DMT tests from the sand layer were used, with a count of 10 to 35 measurements (depth profiles), with a total of 255 measurements. The distribution of the tested random variables was checked. For example, at the Stegny test site, a set of investigations was performed, which consisted of a borehole (BH), 10 DMT soundings (including 3 SDMT), 5 RCPT soundings, and 5 dynamic DP soundings. Surveys were located in the immediate vicinity of the borehole (within two meters): SDMT—3 units, RCPT—5 units, DP—5 units, and the results of these surveys were used to create a database (see attached Figure 7).

Measurements in the SDMT survey were performed every 20 cm, which gave a total of 17 measurements in the profile from 1 to 4.2 m. Each of these measurements was paired with a result from the DP dynamic tests or RCPT—this returned a total number of 3 (SDMT) $\times$ 17 $\times$ 5 (DP), i.e., 255, measurements.

The remaining 7 DMTs from the Stegny test site and studies from the Antoniny, Koszyce, Nielisz, and SGGW campus sites were used to verify the proposed relationships (D_r and S_r). In the case of most studies, i.e., random samples in a statistical sense, there was no reason to reject the hypothesis of normality (Shapiro–Wilk tests suitable for small samples were applied; they are available, e.g., in the Statgraphics statistical software package). No other distribution was found to which the tested quantities complied [21–23].

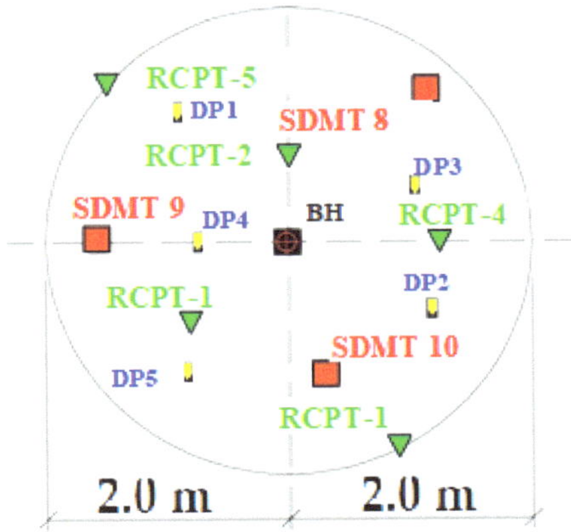

Figure 7. Location of DH, SDMT, and RCPT test point positions—the Stegny site.

The purpose of determining new formulas for calculating relative density (D_r) is to derive the dependencies closely related to the parameters of the measurements with the dynamic DP probe. Apart from the parameters of the DMT tests, the results of RCPT were also used. Several stages to determine the authors' dependencies calculating the subsoil state differed for the given parameters. This is due to the fact that we had different data for specific sites. For determining the relative density (D_r) on the basis of the dependence of DP and RCPT results with the transition to DMT, consisting of performing all the calculations carried out in this work, the solver module was used. It is an addition that extends the functionality of MS Office after it is imported. This function is most often used for linear programming including the modeling and optimization of any type of decision problem. This linear programming should be based on the creation of a reality model, in which the objective function is an important element, the value of which is subordinated to a specific profitability criterion (max/min). The solution of the function is assisted by the use of variable coefficients (for relative density determined in this paper, α_0, α_1, and α_2 were assumed), due to which maximum or minimum values are achieved.

Observation of the obtained results shows that there is a strong relationship between parameter D_r and dilatometer pressures, i.e., p_0 and p_1, and shear wave velocity, as well as u_0 and σ'_{v0}: $D_r = f\,(p_0, p_1, u_0, \sigma'_{v0})$. For non-cohesive soils in the studied sites, a formula was proposed on their basis, and the lowest value of the mean square relative deviation was calculated for them. The formula that was used for the calculations is our original proposal. The summarized test results, i.e., p_0 and p_1 vs. values from DMT and u_0 tests, allowed determining the formula for relative density (D_r) in the following form:

$$D_r = f(\alpha_0, \alpha_1, \alpha_2, p_0, p_1, V_s, u_0, \sigma'_{v0}), \tag{11}$$

$$D_r = \alpha_0 \cdot \left(\frac{p_1 - u_0}{\sigma'_{v0}}\right)^{\alpha_1} \cdot \left(\frac{V_s}{100}\right)^{\alpha_2}, \tag{12}$$

where α_0, α_1, and α_2 are the coefficients. The formula was created dependent on variables p_0, p_1, u_0, σ'_{v0}, which belong to variable cells in the solver function. An additional column was added, the formula of which determines the percentage deviation between the result of the calculated relative density (D_r) for a given depth and the value of the result obtained from DP tests. The given depth is the corresponding level of the measuring point at which the readings from both SDMT (DMT) and DP are recorded. These are the readings every 20 cm. The purpose of the calculations was to obtain the lowest possible mean square relative deviation *MSRD*. The values for the statistical compilation were calculated from the following formulas:

Maximal relative deviation:

$$MRD \max_{i=1,2,\ldots,m}\left|\frac{y_i-\widetilde{y}_i}{y_i}\right|\cdot 100\%, \tag{13}$$

Mean square relative deviation:

$$MSRD = \left[\sqrt{\frac{1}{m}\sum_{i=1}^{m}\left(\frac{y_i-\widetilde{y}_i}{y_i}\right)^2}\right]\cdot 100\%, \tag{14}$$

The function was programmed in a way that allowed achieving the lowest possible target cell value. The results of the calculations are presented in Section 4 below. For non-cohesive soils located in the study area, i.e., in the "Stegny" and WULS-SGGW sites, and the data obtained from the dilatometer test, a formula was proposed, and the relative density was calculated.

As for the second parameter, which is the degree of saturation (S_r), the results of dilatometer tests (DMT) were also used to determine it. Following the observations, there was a strong relationship between parameter S_r and pressures p_0 and p_1, and parameters u_0 and σ'_{v0}: $S_r = f(p_0, p_1, u_0, \sigma'_{v0})$. For non-cohesive soils located in the studied sites, a formula was proposed on their basis, and the lowest value of the mean square relative deviation was calculated. The formula on the basis of which the calculations were performed is also an original proposal of our team and attains the following form:

$$S_r = f(\beta_0,\ \beta_1,\ p_0,\ p_1,\ u_0,\ \sigma'_{v0}), \tag{15}$$

$$S_r = \beta_0\cdot\left[\frac{(P_0-u_0)(P_1-u_0)}{\sigma'_{v0}}\right]^{\beta_1}, \tag{16}$$

4. Test Results

The DMT test results obtained for the Antoniny, Koszyce, Nielisz, Stegny, and WULS-SGGW campus sites are presented in Figure 8. They were taken into account in the construction of a new correlation for mineral and organic soils presented in the following chapter.

4.1. Estimation of Relative Density

Using the solver module, the values of the coefficients (α_0, α_1, α_2) were determined, where there is the lowest value of the mean square relative deviation for pressure, expressed in kPa. For the proposed formula for relative density (D_r), the following statistical values were obtained (summarized in Figure 9 below). The obtained pattern is as follows:

$$D_r = \alpha_0\cdot\left(\frac{p_1-u_0}{\sigma'_{v0}}\right)^{\alpha_1}\cdot\left(\frac{V_s}{100}\right)^{\alpha_2}, \tag{17}$$

where $\alpha_0 = 0.35$, $\alpha_1 = 0.14$, and $\alpha_2 = 0.16$.

The values obtained from the DP test related to the values calculated based on Formulas (8)–(10) using q_D and K_D from DMT tests show a large range of the mean error results. The correlation between the values calculated (from Formulas (8)–(10)) and measured is presented in Figure 10.

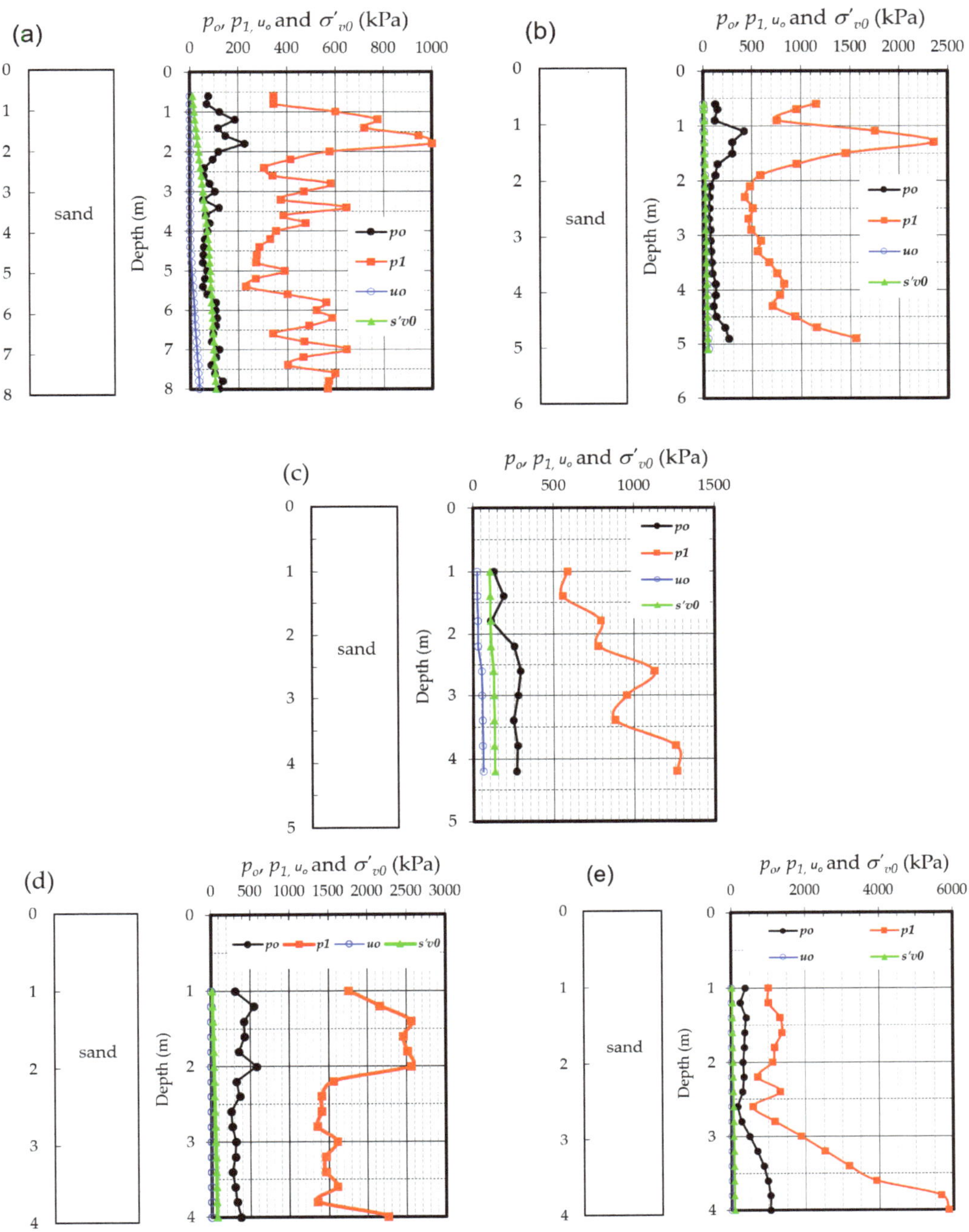

Figure 8. Profiles of p_0, p_1, and p_2 from dilatometer (DMT) and hydrostatic pressure u_0 tests for the: (**a**) Antoniny site; (**b**) Koszyce site; (**c**) Nielisz site; (**d**) Stegny site; and (**e**) WULS-SGGW campus site.

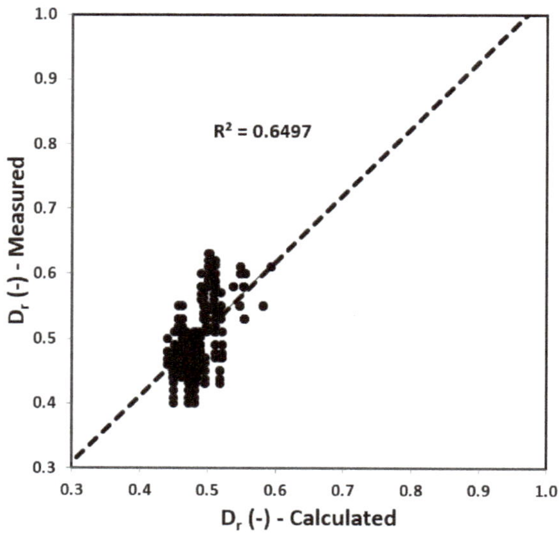

Figure 9. Summary of the values obtained from the DP tests and calculated using Formula (17) from the analyzed Stegny site.

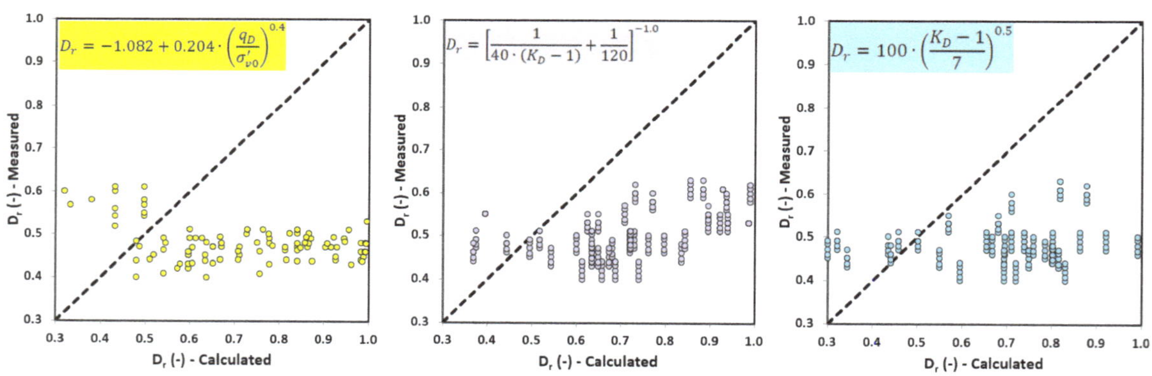

Figure 10. Values obtained from TDR tests and RCPT related to the values calculated from formulas (Equations (8)–(10)) based on DMT tests.

From the chart, it can be seen that the values calculated using the formula correspond well with the values obtained from the DP tests. They behave predictably and fit into an overall trend line, while the DP test values have higher amplitudes. The relationship between the results obtained from DP tests and those determined on the basis of the Formula (17) is presented above.

Using the proposed dependencies to determine the relative density (D_r) in the case of fine-grained soils, the compaction index I_s can also be determined as follows using and modifying the formula developed by Pisarczyk (1975, 2015) [24–26] and the dependencies proposed herein:

$$I_S = 0.855 + 0.058 \cdot \left[K_D^{*0.14} \cdot \left(\frac{V_s}{100}\right)^{0.16} \right], \quad (18)$$

where $K_D^*(-) = \frac{p_1 - u_0}{\sigma'_{v0}}$.

4.2. Estimation of the Degree of Saturation (S_r)

The results of the TDR/MUX/mpt tests and RCPT were used to determine parameter S_r. The following graphs were obtained during the tests: temperature dependence on time; soil swelling dependence on time; time dependence of humidity; and dependence of soil salinity on time. Graphs for a silty sand sample are presented below (Figures 11–13). In these figures, the following observations can be made:

- When examining the dependence of salinity on time, the amount of salinity decreases with soil drying but increases with the addition of water. The minimum salinity value is 0.044, while the maximum value is 0.126.
- During the analysis of the time dependence of humidity, the minimum value was obtained with complete drying of the soil; this value was 17.6%, while the highest was 38.2%.
- When analyzing the dependence of temperature on time, daily temperature fluctuations may be noticed; the lowest was 20.4 °C, while the highest was 25.1 °C.

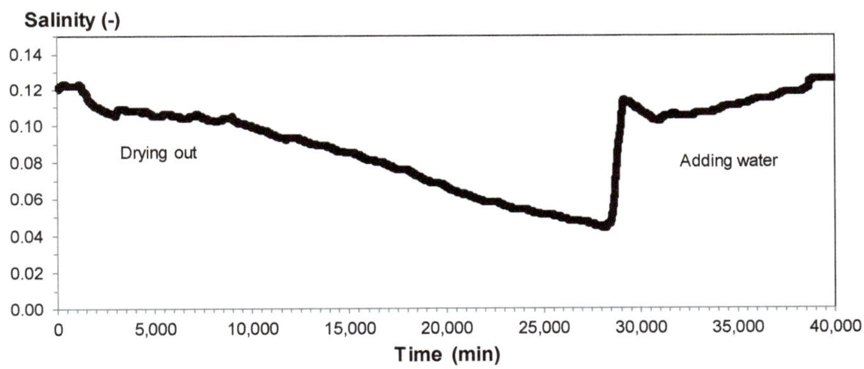

Figure 11. Time dependence of salinity for silty sand from depth level of 3.5 m.

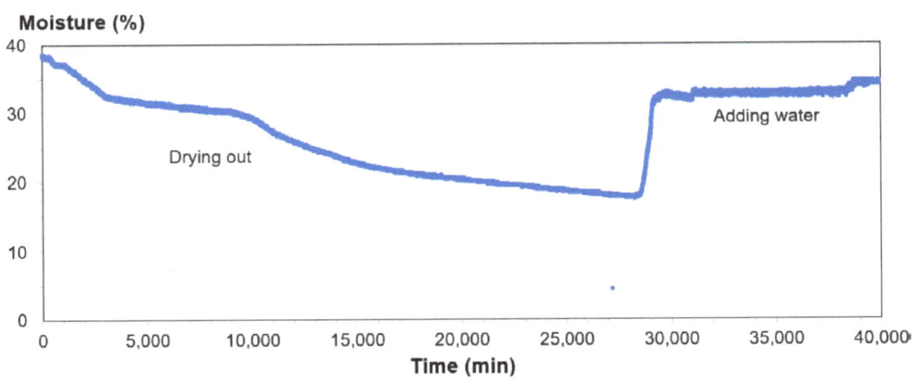

Figure 12. Time dependence of humidity for silty sand from depth level of 3.5 m.

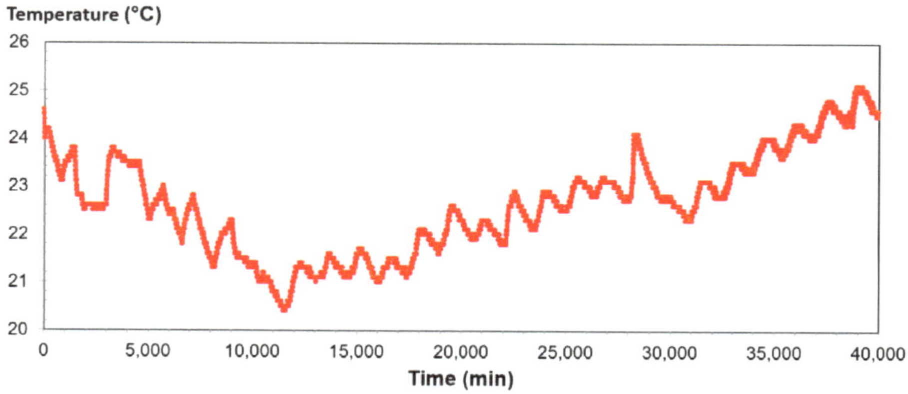

Figure 13. Relationship between temperature and time for silty sand from depth level of 3.5 m.

According to the diagram (Figure 14), the concordance between the parameter calculated on the basis of DMT and the parameter obtained from the TDR and RCPT results is high (*MSRD* = 12%, *MRD* = 22%). These values provide a good prediction for determining the degree of saturation in both the aeration and saturation zones based on DMT tests. On this basis, the following relationship is proposed:

$$S_r = 1.50 \cdot \left[\frac{(P_0 - u_0) \cdot (P_1 - u_0)}{\sigma'_{v0}} \right]^{-0.33}, \quad (19)$$

$$S_r = 1.50 \cdot \left[\frac{B_D}{\sigma'_{v0}} \right]^{-0.33}, \quad (20)$$

where the dilatometer pressure number $B_D = [(P_0 - u_0) \cdot (P_1 - u_0)]^{-0.33}$.

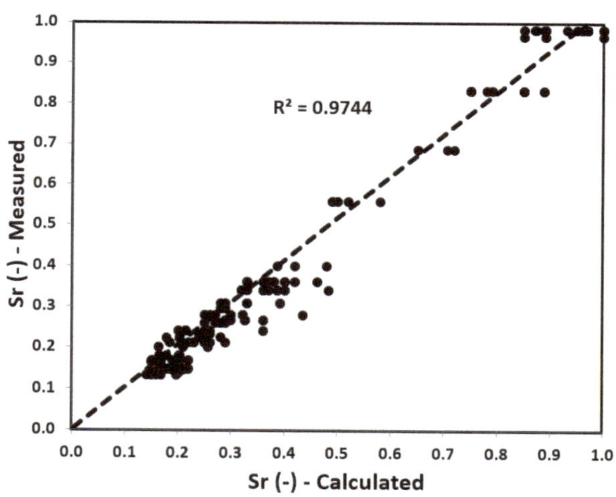

Figure 14. S_r values obtained from TDR tests and RCPT related to the calculated S_r values from the proposed formula based on DMT tests.

5. Discussion of the Results

The obtained values of the relative density and degree of saturation for soils from the proposed two new nomogram charts are comparable to those obtained directly after probing from another device or from the calculation methods. The proposed nomograms are convenient and user-friendly, very helpful for both students and designers.

The nomogram can be used in several ways depending on the data available to the user. With dilatometer pressures p_0 and p_1, u_0, and σ_{v0} from SDMT values of the tested soil, the approximate relative density and degree of saturation can be read (*procedure: on the nomogram moving from point A to B, then to C, the value at point C is the value of the parameter we are looking for*). The nomograms were used to determine the state of sands in the studied sites, and soils with a granular material (such as sand). The proposed nomogram chart was used to compare the values of the relative density and degree of saturation obtained from the tests by several methods (Figure 15).

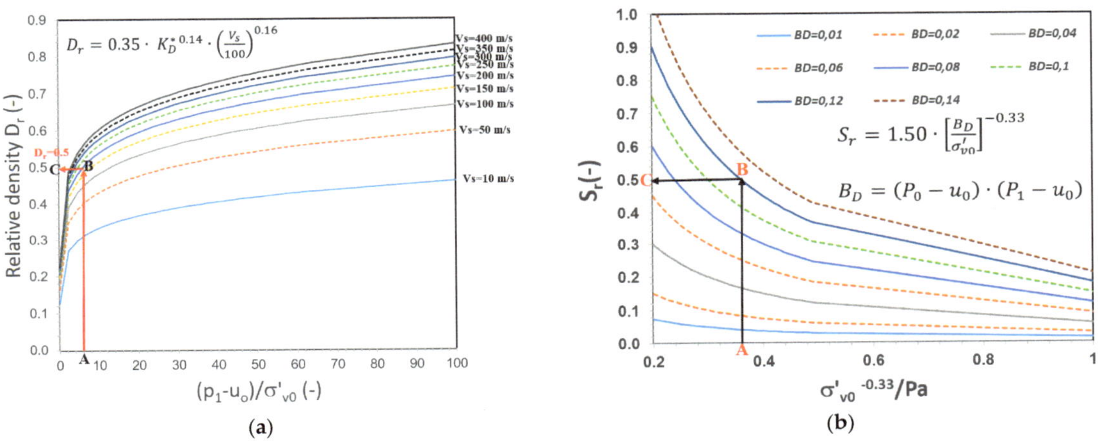

Figure 15. Proposed nomogram charts for determining: (**a**) relative density (D_r) and (**b**) degree of saturation (S_r) based on Marchetti's dilatometer (DMT).

The soil type is important for selecting the method of calculating both the degree of compaction (D_r) and the degree of saturation (S_r). Therefore, it is important to become acquainted with the conditions in the course of preparations, e.g., geology, stratigraphy, and hydrology of the study area.

This study was able to propose a formula and a new nomogram for determining parameter D_r, which are regional and allow for obtaining results similar to those from DP tests without distinguishing whether the ground is above or below the groundwater table.

The deviation values obtained in the course of the calculations were acceptable; their values were as follows: a mean square relative deviation (*MSRD*) = 8.0%, and a maximal relative deviation (*MRD*) = 22%.

The obtained formula is as follows: $D_r = \alpha_0 \cdot \left(\frac{p_1 - u_0}{\sigma'_{v0}}\right)^{\alpha_1} \cdot \left(\frac{V_s}{100}\right)^{\alpha_2}$, where $\alpha_0 = 0.35$, $\alpha_1 = 0.14$, and $\alpha_2 = 0.16$.

The performed calculations show that the deviation values were satisfactory for the degree of humidity. The lowest value of 0.02% was obtained for the measurement both below the groundwater table and above the groundwater table; these values were, respectively, a mean square relative deviation (*MSRD*) = 12.0%, and a maximal relative deviation (*MRD*) = 22%. The formula for calculating the degree of saturation that was finally obtained is as follows: $S_r = 1.50 \cdot \left[\frac{B_p}{\sigma'_{v0}}\right]^{-0.33}$, where the dilatometer pressure number $B_p = (P_0 - u_0) \cdot (P_1 - u_0)$.

6. Conclusions

This study presented the results of dynamic probing (DP), time-domain reflectometry (TDR/MUX/MPTS), resistivity cone penetration tests (RCPT), Marchetti dilatometer tests (DMT), and seismic dilatometer tests (SDMT), from which it is possible to develop a relationship to calculate the relative density (D_r) and degree of saturation (S_r) of selected sandy soils. Probing was conducted at five sites (Antoniny, Koszyce, Nielisz, Stegny, and SGGW campus) in Poland.

Based on the results obtained, two relationships were proposed for determining the relative density (D_r) and the degree of saturation (S_r) in mineral soils based on SDMT (DMT). In addition, this paper proposed a new nomogram chart for determining the relative density (D_r) and saturation degree (S_r) from DMT and SDMT tests. The proposed formula and the new nomogram for determining parameter D_r are local in nature. In the future, our study will continue to be devoted to checking the proposed formulas and nomogram charts on other sites at home and abroad.

The proposed nomogram charts may be limited to a greater extent for additional laboratory tests that must be performed to obtain these values. Thanks to the empirical method established, we may reduce the time needed to assess the ground state for non-cohesive soils.

Author Contributions: Conceptualization, S.R. and M.L.; methodology, S.R., M.L. and M.B.; formal analysis, S.R.; investigation, S.R., M.L. and M.B.; writing—original draft preparation, S.R.; writing—review and editing, S.R., M.L. and M.B.; visualization, M.L.; supervision, S.R.; project administration, S.R. and M.L.; funding acquisition, S.R. All authors have read and agreed to the published version of the manuscript.

Funding: This research received no external funding.

Institutional Review Board Statement: Not applicable.

Informed Consent Statement: Not applicable.

Data Availability Statement: The data cannot be shared because it also belongs to an ongoing research.

Conflicts of Interest: The authors declare no conflict of interest.

References

1. Jamiolkowski, M.; Leroueil, S.; Lo Presti, D.C.F. Design parameters from theory to practice. Theme lecture. In Proceedings of the International Conference on Geotechnical Engineering for Coastal Development, Yokohama, Japan, 3–6 September 1991.
2. Fretti, C.; Lo Presti, D.; Salgado, R. The Research Dilatometer: In situ and Calibration Chamber Test Results. *Riv. Ital. Geotec.* **1992**, *26*, 237–243.
3. Młynarek, Z. Laboratory tests and in situ subsoil tests. General lecture. In Proceedings of the Materiały na XI Krajową Konferencję Mechaniki Gruntów i Fundamentowania, Gdańsk, Poland, 18–20 June 1997; Volume 3, pp. 113–126. (In Polish).
4. Zabielska-Adamska, K. Water Content–Density Criteria for Determining Geomembrane–Fly Ash Interface Shear Strength. In *MATEC Web of Conferences*; Édition Diffusion Presse Sciences: Les Ulis, France, 2019; Volume 262, p. 04005.
5. Zabielska-Adamska, K. Characteristics of compacted fly ash as a transitional soil. *Materials* **2020**, *13*, 1387. [CrossRef] [PubMed]
6. Marchetti, S. In Situ Tests by Flat Dilatometer. *J. Geotech. Geoenviron. Eng.* **1980**, *106*, 299–321. [CrossRef]
7. Marchetti, S. A new in situ test for the measurement of horizontal soil deformability. In *Proceedings of the Conference on In Situ Measurement of Soil Properties*; ASCE: Raleigh, NC, USA, 1975; Volume 2, pp. 255–259.
8. Lutenegger, A.J.; Kabir, M.G. Dilatometer C-reading to help determine stratigraphy. In Proceedings of the International Symposium on Penetration Testing, ISOPT-1, Orlando, FL, USA, 20–24 March 1988; Volume 1, pp. 549–553.
9. Marchetti, S.; Totani, G. Ch evaluations from DMTA dissipation curves. In *Proceedings of the 9th International Conference on Soil Mechanic and Foundation Engineering*; American Society Engineers: Washington, DC, USA, 1989; Volume 1, pp. 281–286.
10. Totani, G.; Calabrese, M.; Marchetti, S.; Monaco, P. Use of in situ flat dilatometer (DMT) for ground characterization in the stability analysis of slopes. In Proceedings of the 14th International Conference on Soil Mechanics and Foundation Engineering, Hamburg, Germany, 6–12 September 1997; Volume 1, pp. 607–610.
11. Reyna, F.; Chameau, J.L. Dilatometer based liquefaction potential of sites in the Imperial Valley. In Proceedings of the International Conferences on Recent Advances in Geotechnical Earthquake Engineering and Soil Dynamics, St. Louis, MI, USA, 11–15 March 1991.
12. Marchetti, S. *The Flat Dilatometer Test*; Publication No FHWA-SA-91-044; Department of Transportation: Washington, DC, USA, 1992.

13. Mayne, P.W.; Christopher, B.R.; DeJong, J. *Manual on Subsurface Investigations*; National Highway Institute; Publication No. FHWA NHI-01-031; Federal Highway Administration: Washington, DC, USA, 2001.
14. Tanaka, H.; Tanaka, M. Characterization of Sandy Soils using CPT and DMT. *Soils Found. Jpn. Geot. Soc.* **1998**, *38*, 55–65. [CrossRef]
15. Wolski, W.; Szymański, A.; Mirecki, J.; Lechowicz, Z.; Larsson, R.; Hartlen, J.; Garbulewski, K.; Bergdahl, U. *Two Stage-Constructed Embankments on Organic Soils*; Report 32; Swedish Geotechnical Institute: Linköping, Sweden, 1988.
16. Wolski, W.; Szymański, A.; Lechowicz, Z.; Larsson, R.; Hartlen, J.; Bergdahl, U. *Full-Scale Failure Test on Stage-Constructed Test Fill on Organic Soil*; Report 36; Swedish Geotechnical Institute: Linköping, Sweden, 1989.
17. Lechowicz, Z.; Rabarijoely, S. Wykorzystanie Badań In Situ w Ocenie Wzmocnienia Słabonośnego Podłoża Zapory Budowanej Etapowo. The Use In Situ Tests in the Evaluation of Subsoil Strengthening at Stages in Dam Construction. Available online: http://www.architectura.actapol.net/pub/18_4_27.pdf (accessed on 1 November 2021). (In Polish).
18. Wierzbicki, G.; Ostrowski, P.; Bartold, P.; Bujakowski, F.; Falkowski, T.; Osiński, P. Urban geomorphology of the Vistula River valley in Warsaw. *J. Maps* **2021**, *17*, 170–185. [CrossRef]
19. Marchetti, S.; Crapps, D.K. Flat dilatometer manual. In *Internal Report of GPE*; GPE Inc.: Gainesville, FL, USA, 1981.
20. Skierucha, W.; Wilczek, A.; Walczak, R.T. Recent software improvements in moisture [TDR method], matric pressure, electrical conductivity and temperature meters of porous media. *Int. Agrophysics* **2006**, *20*, 229–235.
21. Bleistein, N.; Handelsman, R.A. *Asymptotic Expansions of Integrals*; Courier Corporation: Chelmsford, MA, USA, 1975.
22. Lechowicz, Z.; Rabarijoely, S.; Szczypiński, P. The use of dilatometer test to determine the type and state of organic soils. In *Scientific Review Engineering and Environmental Sciences*; SGGW: Warszawa, Poland, 2004; pp. 191–201. (In Polish)
23. Lechowicz, Z.; Fukue, M.; Rabarijoely, S.; Sulewska, M.J. Evaluation of the Undrained Shear Strength of Organic Soils from a Dilatometer Test Using Artificial Neural Networks. *Appl. Sci.* **2018**, *8*, 1395. [CrossRef]
24. Pisarczyk, S. Relationship between the relative density and the degree of compaction index of non-cohesive soils. *Inżynieria Bud.* **1975**, *5*, 213–216. (In Polish)
25. Pisarczyk, S. Bulk soils. In *Geotechnical Properties and Methods of Their Investigation*; Oficyna Wydawnicza Politechniki Warszawskiej: Warszawa, Poland, 2015. (In Polish)
26. Straż, G.; Borowiec, A. Evaluation of the unit weight of organic soils from a CPTM using an Artificial Neural Networks. *Arch. Civ. Eng.* **2021**, *67*, 259–281.

materials

MDPI

Review

From Light-Powered Motors, to Micro-Grippers, to Crawling Caterpillars, Snails and Beyond—Light-Responsive Oriented Polymers in Action

Mikołaj Rogóż , Zofia Dziekan, Klaudia Dradrach, Michał Zmyślony, Paweł Nałęcz-Jawecki, Przemysław Grabowski, Bartosz Fabjanowicz, Magdalena Podgórska, Anna Kudzia and Piotr Wasylczyk *

Photonic Nanostructure Facility, Faculty of Physics, University of Warsaw, Pasteura 5, 02-093 Warsaw, Poland
* Correspondence: pwasylcz@fuw.edu.pl

Abstract: "How would you build a robot, the size of a bacteria, powered by light, that would swim towards the light source, escape from it, or could be controlled by means of different light colors, intensities or polarizations?" This was the question that Professor Diederik Wiersma asked PW on a sunny spring day in 2012, when they first met at LENS—the European Laboratory of Nonlinear Spectroscopy—in Sesto Fiorentino, just outside Florence in northern Italy. It was not just a vague question, as Prof. Wiersma, then the LENS director and leader of one of its research groups, already had an idea (and an ERC grant) about how to actually make such micro-robots, using a class of light-responsive oriented polymers, liquid crystal elastomers (LCEs), combined with the most advanced fabrication technique—two-photon 3D laser photolithography. Indeed, over the next few years, the LCE technology, successfully married with the so-called direct laser writing at LENS, resulted in a 60 micrometer long walker developed in Prof. Wiersma's group (as, surprisingly, walking at that stage proved to be easier than swimming). After completing his post-doc at LENS, PW returned to his home Faculty of Physics at the University of Warsaw, and started experimenting with LCE, both in micrometer and millimeter scales, in his newly established Photonic Nanostructure Facility. This paper is a review of how the ideas of using light-powered soft actuators in micromechanics and micro-robotics have been evolving in Warsaw over the last decade and what the outcomes have been so far.

Keywords: light-responsive materials; liquid crystal elastomers; micro-robotics; soft robotics; actuators

Citation: Rogóż, M.; Dziekan, Z.; Dradrach, K.; Zmyślony, M.; Nałęcz-Jawecki, P.; Grabowski, P.; Fabjanowicz, B.; Podgórska, M.; Kudzia, A.; Wasylczyk, P. From Light-Powered Motors, to Micro-Grippers, to Crawling Caterpillars, Snails and Beyond—Light-Responsive Oriented Polymers in Action. *Materials* **2022**, *15*, 8214. https://doi.org/10.3390/ma15228214

Academic Editor: Gerard Lligadas

Received: 7 October 2022
Accepted: 15 November 2022
Published: 18 November 2022

1. Introduction

A one-millimeter spider that can walk, jump, fly (on a gossamer), spin a super-strong web, catch prey and replicate for millions of generations (adapting along the way to the changing environment, if necessary), not only inspires awe, but also makes us realize how disappointing and, in most cases, futile are our attempts to mimic nature in the small scale. There is a glaring gap between the scale of meters or centimeters, where industrial mechanisms, machines and robots have a well-established footing, and the molecular scale, where we have gained enough understanding to make molecules dance at our will [1]. The smallest practical electrical motors available are approximately 1 mm in diameter, and given drivetrains of similar scale, a centimeter-long solar-powered walker might ultimately be possible [2]. With piezo drives, some spectacular results have been demonstrated, including a several-second, self-sustained flight [3] on the scale of a small dragonfly (albeit with very limited control).

The challenges in building small autonomous mechanisms are threefold:

- How to design the mechanics to overcome all the problems with friction, van der Waals forces, fluid viscosity and moving with very low Reynolds numbers [4]? The laws governing mechanics and motion in the micro-world are different from those we are used to: gravity and the electromagnetic forces are governed by the same constants, but their relations change with scale. As a result, new challenges as well as new opportunities arise: crawling on an upside-down glass ceiling is easy for a millimeter-scale snail robot [5], while a micron-size walker may struggle to lift its legs, glued to the surface by van der Waals forces [6,7];

- How to fabricate motors, gears and other elements on the sub-millimeter and smaller scales, reliably and cheaply?
- How to supply the mechanisms with energy, either remotely or from an onboard source?

For around ten years now, our group at the Faculty of Physics at the University of Warsaw have been trying, with varying success, to approach microscale mechanics and micro-robotics with soft, light-responsive materials: liquid crystal elastomers. In what follows, we present our achievements to date and sketch out possible avenues of future work.

1.1. Liquid Crystal Elastomers as Light-Powered Actuators

Apart from traditional engines and gears, soft actuators capable of powering simple mechanisms and machines, driven by various stimuli, have been demonstrated in small scales [8]. In dielectric elastomer actuators (DEAs), a compressible membrane is sandwiched between flexible electrodes that interact via electrostatic forces [9], resulting in large strains, although with very high voltages needed [10]. Another approach to electrically driven actuators uses conjugated, conductive polymers [11], and asymmetric swelling of gels in ionic solutions can generate shape changes as well [12]. Magnetically driven actuators have also been developed, mostly based on soft polymers with suspended magnetic particles and spatiotemporally varying magnetic fields to control them [13,14]. Material deformation can also be induced by chemical reactions, e.g., selective modification of chemical bonds in polymers [15]. Pressure-driven soft actuators rely on structures with spatially varying stiffness, deforming upon expansion of gas- or liquid-filled chambers [16]—they can be scaled down [17], but still need substantial tubing to deliver the pressurized medium.

It was probably De Gennes who, around 1975, first came up with the idea of anisotropic crosslinked polymer networks with embedded orientational order of nematogenic monomers [18]. This, in turn, triggered various attempts towards fabricating, understanding and using stimuli-responsive-oriented polymers, in particular liquid crystal elastomers (LCEs). LCEs are solid, elastic polymers with a well-defined alignment of molecules (this alignment orientation is often described by the so-called director), just as in the familiar liquid crystals. If cleverly designed, both at the molecular (chemistry) and mesogenic (orientation) level, LCEs may exhibit rather unique mechanical properties, in particular very large, reversible and fast deformations in response to an external stimulus: heat [19]; light triggering photochemical reactions [20,21]; light inducing photo-thermal heating [19,22,23]; electric field [24,25] or the presence of a solvent [26].

Due to the tight packing of rod-shaped molecules, arranged in cross-linked polymer chains, the order (anisotropic) <-> disorder (isotropic) transition reduces the effective chain length along the alignment direction (the director), at the same time increasing the spacing between the molecules in the perpendicular directions (Figure 1A). As a result, the material deforms in a manner determined by: (a) the spatial distribution of the director; (b) the spatio-temporal distribution of the stimulus, e.g., light intensity; (c) the properties of the material itself (stiffness, elasticity, thermal conductivity) and (d) the surrounding environment (e.g., in our LCEs, the response will usually be different in the air (good thermal insulator) and under water (good thermal conductor)). If stimulated by light, additional degrees of freedom in the mechanical response may involve the light spectrum [27,28] or polarization [29,30]. Thus, LCE elements of various sizes and forms [31–33] can be used as micro-actuators in various configurations as well as remotely powered "muscles" for robots (Figure 1B, cf. [34] for a review of LCE actuators and [7] of LCE soft robots).

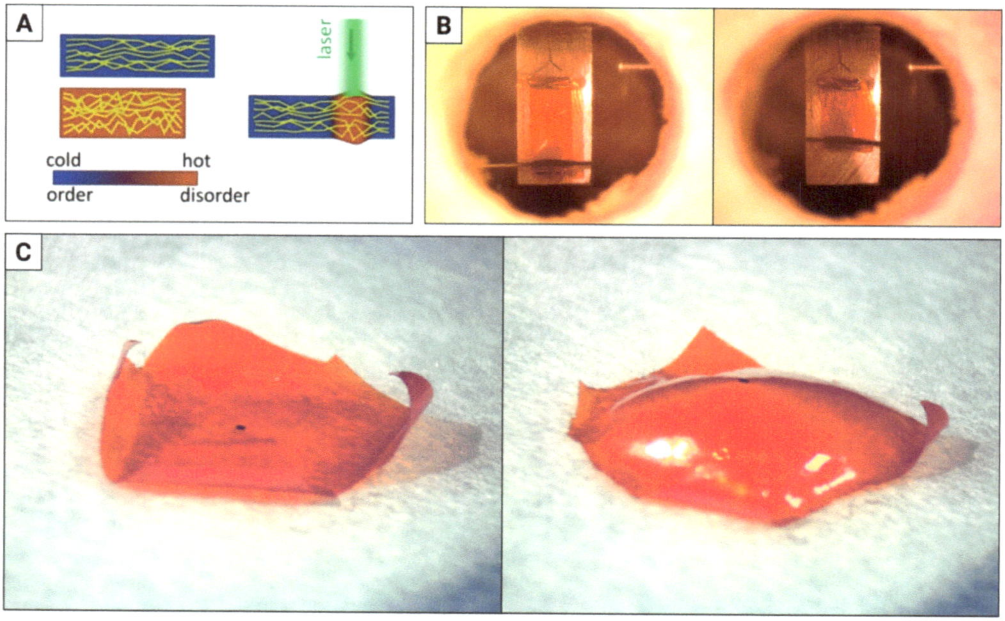

Figure 1. Light-responsive LCE—from a simple contracting strip to films with arbitrary molecular alignment. (**A**) A piece of material with cross-linked polymer chains (olive) arranged in one direction undergoes a phase change in response to temperature increase, either in the entire volume (left) or locally, when illuminated with a green laser beam (right): the molecular order decreases, the polymer chains effectively shorten along the director and macroscopic deformation results. (**B**) A 10 mm long strip of LCE film heated in an oven from 29 °C to 150 °C shortens by approximately 35%. (**C**) A 22 × 22 mm piece of flat LCE film with azimuthal director orientation (the black dot marks the center of rotational symmetry) buckles when heated on a hot plate from 22 °C to 80 °C.

From the material science and chemistry perspective, LCEs can be categorized by the synthesis pathways—the monomer may be a precursor and a synthesis from low-molar-mass LC is carried out, or a polymer chain (macromer) is grown first and then the material is crosslinked. In terms of the crosslinks' density, they can be loosely crosslinked liquid crystal polymers, called elastomers, or a highly crosslinked polymer network. As for the location of the mesogenic moieties, they can be built into the polymer backbone in the so-called main-chain LCE, or covalently bonded with the polymer backbone as side groups in the side-chain LCEs.

Following the first experimental demonstration by Finkelmann et al. [35,36], both main-chain and side-chain LCEs were synthesized via different synthesis routes. In the so-called "Finkelmann method", elastomers with Si-O in the polymer backbone are obtained, capable of high strains, but hydrolyzation reaction is time-consuming and sensitive to impurities and the presence of oxygen. Much later, the "click-chemistry" found its way into the LCE synthesis. These reactions are not sensitive to oxygen and allow for the control of the LCE properties by tuning the structure of the crosslinked liquid crystal polymer [37,38]. The most notable examples are the Michael addition and thiol-ene reactions, leading to LCEs with mechanically programmable alignment, either by shear thinning (in 3D printing) or by surface anchoring [39]. It is also possible to quickly fabricate LCE's thin films—in a matter of hours—from low-molar-mass LCs, through acrylate homopolymerization. A side-chain LCE is formed with a relatively high number of crosslinks and the alignment in thin films can be induced by rubbing or photopatterning. Additionally, in these materials a sensitizer (e.g., dyes) can be introduced into the matrix before as well as after polymerization and crosslinking.

There are two approaches to actuate LCEs with light. Additional molecules (e.g., dyes) can be embedded into the matrix that absorbs light of a specific wavelength. Most often azobenzene moieties are used, likely due to the extensive studies related to the photoisomerization of azobenzene itself and azobenzene derivatives since the 1960s. An azo-moiety can be built into the polymer backbone (as a monomer or a crosslinker [40–42]) and when exposed to light (typically in the UV band), isomerization occurs (trans molecules, supporting the nematic phase, bend and the order parameter in the polymer matrix decreases). After the irradiation stops, the bent molecules go back to their previous state, yet this reversal is slow (it can be sped up with exposure to visible light). Because of the slow response of these reactions (seconds to minutes), some light-responsive LCEs make use of molecules or nano-particles that absorb light and generate heat (which, in turn, reduces the order parameter and induces conformational changes) [43]. Usually, they are dispersed in the LCE matrix: dye molecules, carbon nanotubes [44] or gold nanoparticles [45,46], but coating an LCE actuator surface with NIR-absorbing polymers also works [47]. A combination of photochemical and photothermal responses in light-responsive liquid crystal polymer networks for soft actuators was also presented [48].

1.2. LCE Fabrication in Different Scales

LCE elements and structures are often fabricated by UV light-induced polymerization of the liquid monomer, crosslinker and photoinitiator mixture, typically in a glass cell, tens of micrometers thick. Properly prepared orienting surfaces—either rubbed mechanically or coated with photo-orienting layers—guarantee the director orientation within the LCE film in 2 or 2.5 dimensions, if the opposing layers have different properties (Figure 1C). Different elements, typically of the order of a few millimeters, can then be cut with a blade or a laser beam [49]. More complex director patterns—and thus film deformations—are available with patterning of the orienting layers [50–52]. Another fabrication method involves shaped molds with the director orientation induced by a magnetic field [53]. In extrusion 3D printing, the molecules are oriented by rheological effects while they flow through a small nozzle [54,55]. The most advanced technology so far—3D laser photolithography (direct laser writing)—uses a focused infrared laser beam that is scanned across a drop of liquid monomer, triggering polymerization via two-photon absorption in a volume (voxel) as small as a fraction of a cubic micron [56].

1.3. Problems Yet to Be Solved

While several simple devices have been demonstrated with contracting, bending, twisting or even more complex photo-responsive LCE actuators, typically only their basic response to the light stimulus has been studied. At the same time, even monochromatic light (e.g., in the form of a laser beam(s)) offers a number of degrees of freedom that can be used to control the mechanisms: average power; pulse duration; pulse energy and the light intensity spatial distribution. If the development of mechanisms and machines based on photo-responsive elastomers is to extend beyond single contracting or bending strips of a stimulus-responsive material, a better understanding of the photo-mechanical response of LCEs will be needed, including the interplay of the energy flow and heat dissipation with different light pulse durations and the influence of the light spatial intensity distribution.

Our studies of the millimeter-scale LCE actuators [57] hint that: (a) if used in the transient regime, the actuator response time is independent of laser power; (b) for pulsed laser actuation, long laser pulses result in a smaller actuator response, compared to short laser pulses delivering the same energy; this is true down to a certain pulse duration, determined by the time constants of heat transfer to the environment; (c) when part of the actuator is illuminated with a laser beam of constant total power, but of varying size, the photo-mechanical response is independent of the illuminated area.

Two of the biggest problems we have discovered so far are as follows:

- light absorption in the LCE films decreasing over time, mainly due to the light-absorbing dye bleaching; this can be potentially bypassed by using quantum dots or other, more robust absorbers, well known from fluorescence microscopy, where dye bleaching has been studied and addressed for some time;
- the wear and tear of the actuator after many cycles of operation; this calls for further insights into the mechanical properties of the light-responsive elements, probably at the level of polymer chemistry.

2. Micro-Motors—Direct Conversion of Light Energy into Mechanical Work

"Can light drive a motor?" was the opening question asked by Ikeda et al. [58] in 2008. Most energy-harvesting systems convert (solar) light energy either to heat (in solar thermal collectors) or to electricity (in solar cells). Direct conversion of light energy into mechanical energy has been demonstrated, from the molecular to macroscopic scales [59–61]. Ultimately, this may enable mechanical devices remotely powered with light (delivered either via free space or through optical fibers), where using electrical cables is not possible.

Yamada et al., in [44], demonstrated a light-driven motor using the contraction of an LCE/polyethylene laminated film irradiated with UV and visible light. The motor presented by Geng et al. [62] used a looped strip of hydroxypropyl cellulose film that deformed under humid air to drive a rotating element. The LCE rod described in [63] can roll with either a light beam or a heated surface as the energy source, but it remains an open question if this deserves to be called a "motor". In a similar fashion, various LCE tubes and helical ribbons have been demonstrated as drives for centimeter-scale light-powered rolling vehicles [64].

Historically, we first developed micro-robots—the water strider, the caterpillar [50] and the snail [5]—only later to realize that the very same caterpillar, if held in place, can make a rotary motor [65] (if shaped into a ring—"eating its tail") or a linear inchworm motor (if working in a team of two) [52].

2.1. Rotary Motor

Many high-end camera lenses use so-called ultrasonic motors (USMs) to drive the auto-focus mechanism. Typically, piezoelectric USMs are composed of a ring-shaped rotor and stator (Figure 2A). The voltage applied sequentially to the rotor segments induces travelling wave deformations that couple by friction to the stator, thus setting the former in motion [66–68].

Inspired by this design, we built a light-driven micromotor, where the travelling wave deformation results from the photo-mechanical response in a 5.5 mm diameter LCE ring, illuminated with a laser beam moving around the disc circumference (Figure 2B) [65]. In our experiments, we tested LCE discs with different director distributions, in particular, one with the azimuthal orientation on both sides (A-A) and one with the azimuthal and radial orientations (A-R). The A-R disc proved to be 10 times faster (ω = 5.88 rad/min) and 14 times more efficient at converting laser beam revolutions to disc (rotor) revolutions—see [69] for the video of the rotating motor. We also studied these effects by performing finite element numerical simulation, where the dynamic photomechanical response of the LCE discs was modelled as a local strain tensor (Figure 2D).

Despite low speed and efficiency, the LCE micro-rotor has some advantages, such as the ability to be scaled down and to be powered remotely with light energy.

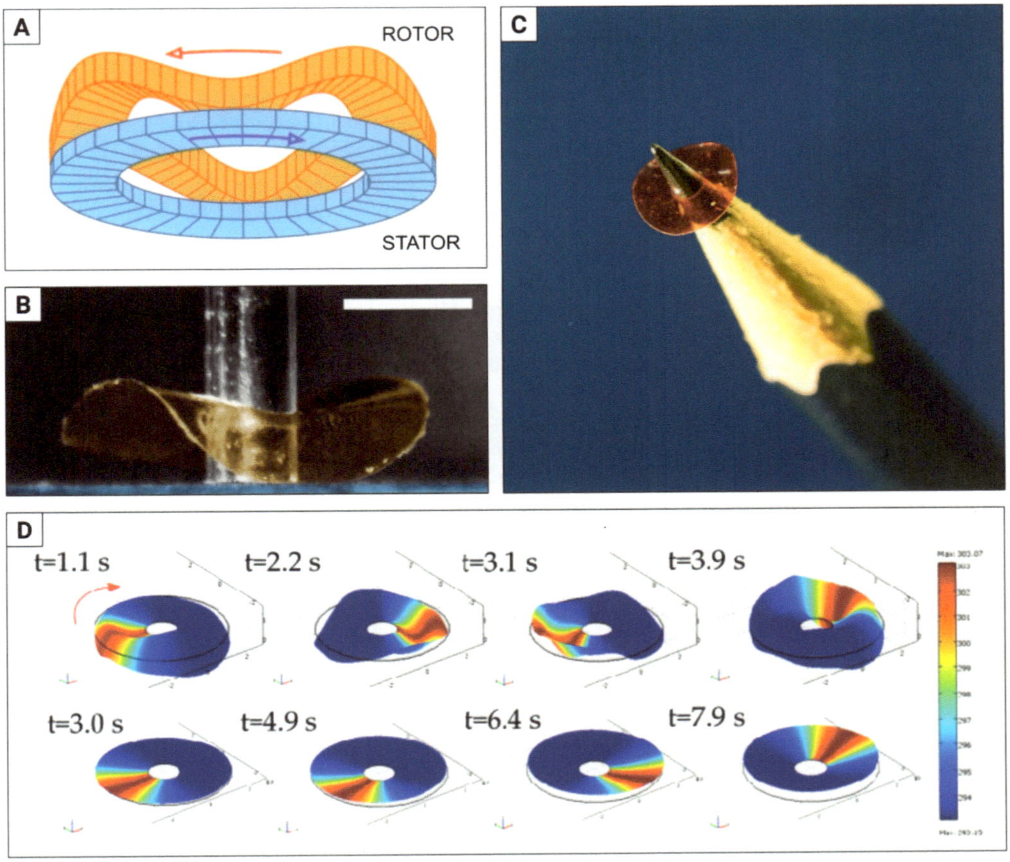

Figure 2. Rotary motor directly powered with light. (**A**) Schematic of the rotary piezoelectric (ultrasonic) motor. Travelling wave deformation generated in the rotor (orange) interacts via friction with the stator (blue) and sets the former in motion. (**B**) LCE rotary micro-motor, seen from the side. The LCE disc (rotor, orange) rotates with respect to the stator (rough solid surface, blue), around a steel axis. The white scale bar is 2 mm long. (**C**) The LCE rotor on a pencil tip for the scale demonstration (although, sadly, we must admit it cannot yet be used to sharpen pencils). (**D**) Numerical simulations of the LCE deformation upon illumination with a spatially modulated (rotating) laser beam. The top row is for azimuthal–radial, and the bottom row for azimuthal–azimuthal director orientation (see text for details). Adapted from [65].

2.2. Linear Stepping Inchworm Motor

Even basic understanding of the mechanics of soft materials reveals that it is much easier to build an actuator that generates a pulling force compared to the one that would push using light-responsive LCEs. To extend the portfolio of devices beyond simple LCE strips (often, rather bombastically, pitched as "motors"), we designed and built a linear stepping inchworm motor with two LCE accordion-like actuators [52]. Linear displacement is often a desired mode of operation and various linear motors are in use, either powered from rotary drives or directly from pressure, electromagnetic forces or shape change in different materials. The first inchworm motors had a rotor moved by a sequential action of piezo actuator(s) [70,71]. In a linear inchworm motor, two or more actuators are operated sequentially to push/pull (rather than rotate) a shaft along its axis.

When fabricating the actuators for our motor, we tested a new LCE-orientation technique that we had conceived a long time before—rubbing overwriting. Mechanical rubbing is commonly used for orienting liquid crystal molecules [72,73]. Our first approach to fabricating LCE films with patterned alignment with rubbing through masks was used to make crawling caterpillars [50], but it required at least two masks and their precise alignment. Here we used a different approach: we rubbed the poly-vinyl alcohol (PVA)-coated glass surface in one direction, then covered certain areas with a mask, and rubbed everything in the perpendicular direction. Accordion-like 50-micron-thick LCE actuators fabricated with this procedure can contract by up to 80% upon heating [50] and were used in the miniature linear motor. A laser beam reflected from a mirror mounted on a galvo scanner illuminates (and thus heats up) two of these actuators, that, in turn, set a small, heavy gripper into an orbital motion (Figure 3B). Each scanning cycle consists of grip–move–release sequence (Figure 3C) and, as a result, the shaft moves (in either direction, defined by the laser scanning direction) at speeds of up to 25 mm/s (see [74] for the video).

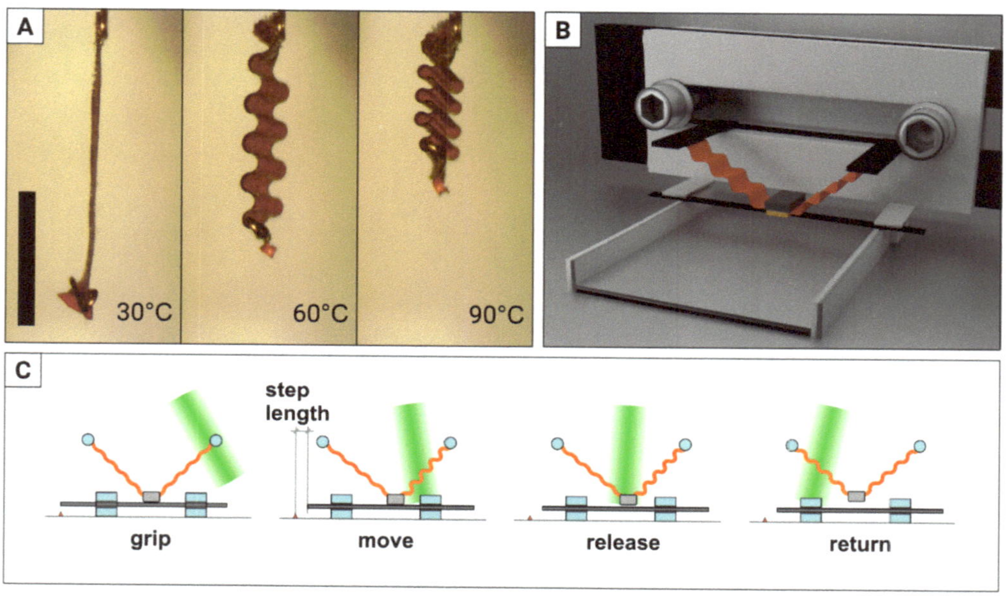

Figure 3. Light-powered linear inchworm motor. (**A**) Accordion-like actuator heated in an oven exhibits very large contraction, with up to 80% stroke. (**B**) CAD rendering of a linear stepping motor with two actuators (orange), gripper (dark grey/yellow) and a sliding shaft (black, under the gripper). (**C**) The sequential action of the actuators, powered by a scanned green laser beam, results in orbital motion of the gripper that, in turn, moves the shaft. The black scale bar in (**A**) is 5 mm long. Adapted from [52].

The motor design can be straightforwardly adapted to a rotary configuration by replacing the linear shaft with a circular one. If equipped with a position sensor, e.g., optical, the stepping motor can operate in a closed loop configuration, where the length of each step is not relevant, as the distance from the target position is continuously measured and adjusted, including with sub-single-step accuracy.

3. Bio-Inspired Millimeter-Scale Robots

For centuries, scientists and engineers have been fascinated by the movement of various animals at different scales and the possibility of copying this movement in man-made machines [75]. Nevertheless, after several decades and huge resources invested in robotics, soft robotics and safe interactions of robots with humans in particular, the results are disappointing: most robots remain awkward automata, at best capable of performing basic repeatable tasks in well-controlled environments (perhaps the latest autonomous cars may soon change this rather gloomy picture). Robotics in the microscale is at an even worse stage—we are nowhere near scaling down robots to millimeters and below, not to mention their reliable fabrication and operation in real-life applications.

3.1. The Caterpillar

Conventional robots, made of multiple rigid parts connected by joints, typically have few degrees of freedom and poor adaptability to the environment. One attractive area of research is the realm of soft-bodied animals—segmented worms, mollusks, cephalopods and insects at some development stages (caterpillars)—where using elastic materials could allow their continuous movement to be mimicked, offering the bio-inspired robots the ability to move in confined spaces and to adjust to topologically complex environments [76]. To date, some attempts have been made to replicate them in real scale, with limited success [77].

Caterpillar locomotion consists of cycles of inching and crawling: the animal lifts and steps forward every pair of its legs—starting from the tail, towards the head. To detach the legs from the ground, it deforms parts of its body, generating a travelling wave of deformation along the body. Such deformations can be induced with spatially varying laser beam in the accordion-like actuators, such as those used in our stepping inchworm motor (compare Figure 3A). Our caterpillar robot [50] was fabricated with a 14.8 × 3.8 mm strip of 50 μm thick LCE film (Figure 4A). When illuminated locally with a scanned green laser beam steered by a galvo scanner (Figure 4B) the film deforms, becomes curved and lifts from the ground, thus generating a wave of deformation. When placed on a rough surface (e.g., sandpaper), this results in crawling locomotion, with a typical step length of 0.3 mm and a maximum speed of 30 mm/min (see Figure 4C for the snapshots and [78] for the video). This is about six times slower than the caterpillar Cucullia verbasci, commonly found in Europe and North Africa [79]. The robot is also able to squeeze through narrow slits, climb a sloped surface and push loads several times its own mass.

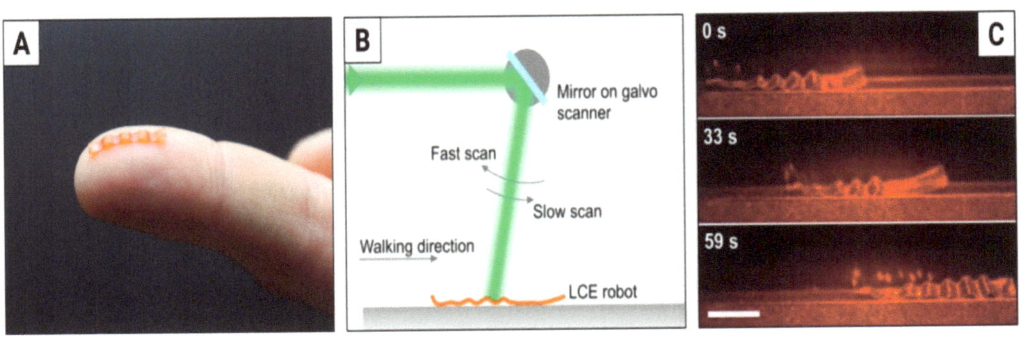

Figure 4. Natural scale crawling caterpillar robot. (**A**) The caterpillar robot on the fingertip of one of its creators. (**B**) Schematic of the experimental setup—the green laser beam is scanned along the robot's body with a galvo mirror driven by an asymmetric sawtooth signal. The beam was scanned at 0.4 Hz and had 2.5 W of power. (**C**) Snapshots of the video with the light-driven caterpillar crawling on a level surface. The white scale is 5 mm long. The laser light is filtered out with an orange optical filter. Adapted from [50].

With a maximum speed of 30 mm/min, our caterpillar robot may be compared with a few similar demonstrations. In [80], a 29 mm-long robot made of LCE performs inchworm locomotion, powered through cables and reaching 1.91 mm/min. A composite tensegrity robot made of carbon nanotube-doped LCE was able to navigate a labyrinth with the average speed of 6.87 cm/min [81]. Liquid crystal elastomer–carbon nanotube composite was also used in [82] in a crawling and jumping untethered robot, reaching 42 cm/min (half the body length) per minute.

3.2. The Snail

Gastropods—snails and slugs—have a single ventral foot, in which pedal waves propagate, propelling the animal (Figure 5A,B). To further increase the interaction between the foot and the surface, the former is covered with slippery mucus. The (apparently) low complexity of this design and its versatility made snail locomotion a promising target for implementations in robotics.

Figure 5. Light-powered snail mimicking the adhesive locomotion of terrestrial gastropods. (**A**) A garden banded-snail Cepea hortensis meets the 10 mm long light-powered snail robot (having no shell, though). (**B**) In snails and slugs, pedal waves propagate along the ventral foot contact surface with a velocity V_S, propelling the animal with an average speed V_{CM}. In a similar way, the light-induced elastomer deformation moves along the robot's soft body (yellow) covered with an artificial mucus layer (purple). In both cases, the deformations are in fact much smaller. (**C**) Snapshots from a video with the snail robot crawling on a horizontal glass plate topped with glycerine as an artificial mucus. The average speed is 1 mm/min, the black scale bar is 5 mm long. The material contraction was of the order of 0.1 mm. Adapted from [5].

A 50 µm thick strip of LCE with planar nematic alignment film was placed on a glass substrate covered with a layer of glycerine (which proved to perform best as an artificial mucus) [5]. As with the caterpillar robot, local light-induced deformation was driven by a spatially scanned laser beam, but this time the elastomer contracted along its length, remaining in contact with the glass via the mucus layer. As the laser beam moved, the contraction propagated with a typical speed vs. on the order of several cm/s, resulting in an average robot speed V_{CM} of a few mm/s (see [83] for the video and Figure 5C for snapshots). The robot could move on various surfaces—in the same way as snails—from coarse sandpaper to PTFE (Teflon) [84,85], crawl over a glass tube obstacle, move horizontally upwards (V_{CM} = 0.8 mm/min), downwards (V_{CM} = 1 mm/min) and upside down (V_{CM} = 3 mm/min). Unlike snails that, interestingly, have no reverse gear [86], it could also move backward by reversing the direction of the laser beam scan. In this case, we are not aware of any similar demonstrations of robots performing the mucus-assisted locomotion in natural scale, even though it may be an interesting alternative for robots operating in challenging environments, and it also provides good security margins due to large contact area with the surface.

3.3. The Water Strider

Not all our LCE-related projects have resulted in spectacular success. If they had, this would be a clear indication that the challenges were easy to overcome. In fact, a number of the projects have been abandoned, while others were put on hold, even for many years in some cases.

One example of the latter group is the light-powered, natural-scale water strider robot. There has been some substantial effort invested in understanding the dynamics of strider locomotion on the water surface, at some point leading to the so-called Denny paradox (baby striders seemingly move their legs "too slowly" to propel themselves, and yet they still manage to do so) [87], and later to its (apparent) resolution [88]. Striders offer a very good example of animal locomotion to be mimicked in the lab, as they do not need to struggle with friction, capillary or van der Waals forces while moving gracefully on the water's surface, supported by surface tension (Figure 6A). Our first natural-scale (about 2.5 cm long) robo-strider was built (twisted) from 0.13 mm diameter copper wire, had the simplest muscle-actuator made of a bending strip and had to be guided by a thread, placed just above the water's surface, to keep on track and remain within the green laser beam powering the LCE actuator (Figure 6B).

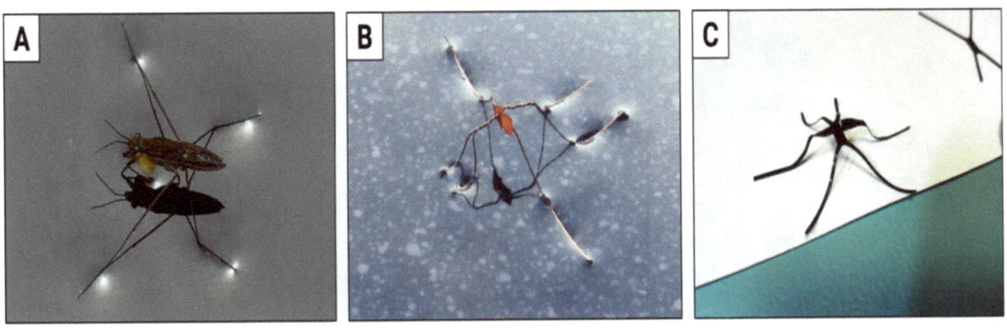

Figure 6. From the pond to the lab (and back)—a light-powered natural-scale water strider robot with LCE muscle. (**A**) Water striders are common inhabitants of reservoirs and rivers in many climate zones. (**B**) First prototype of the light-powered robo-strider with the orange LCE bending actuator powering a pair of legs. (**C**) The latest generation of the robot body (here without the actuator installed) was 2D printed with a 3D extrusion printer and then hot-shaped to take the final form. Water strider photo by Schnobby licensed under CC BY-SA 3.0.

Surprisingly enough, it did manage to move, with an average speed of around 180 mm/min—6000 times slower than some members of the Gerridae family. The project has been on hold for a number of years, but has recently re-emerged with a new approach to the strider body—this time with 3D printing (Figure 6C). We hope this will provide a much lighter frame, at the same time offering higher flexibility in the design and repeatability in manufacturing, should we decide to go for a swarm of light-driven water striders one day. As for increasing the speed, there is a class of so-called snap actuators [89], where a piece of material accumulates energy until it reaches a point of nearly instantaneous shape transition, releasing it to the mechanism to be driven (strider legs)—we hope this may be a way towards faster swimming speeds, perhaps one day comparable with wild-born animals.

Other approaches to making robo-striders included a jumping robot with nickel titanium shape memory alloy actuator [90] and a large scale (10 cm long) robot with as many as 12 legs and piezoelectric actuator, reaching the maximum speed of 180 cm/min [91].

As for now, we are not aware of any attempts to involve light-responsive materials in similar constructions.

3.4. The Ant

The second example of a project awaiting its turn is a millimeter-scale walking ant, powered and controlled by two colors of light, corresponding to the two degrees of freedom in the movement of its six legs. Two separate stimuli, applied in sequence, are the minimum for true non-reciprocal motion [4]. Insects (as opposed to arthropods, for instance) walk on six legs—this seems to be optimized for the number of points of contact with the ground, as they form two tripods that are lifted sequentially and shifted with respect to each other. To better understand this mode of locomotion, we have built several models (see Figure 7A for an example). The next step was to develop two LCE films, responding independently to two different colors of light (Figure 7B). Importantly, the illumination for the millimeter-scale walker must be provided by LEDs, not lasers, as the illuminated area must be large enough to let the walker make at least several steps, before walking into the dark [92]. Ultimately, the mechanical design must take into account many limitations of LCEs, e.g., illumination from one (or two) sides only and the available deformations of the actuators—one concept is presented in Figure 7C.

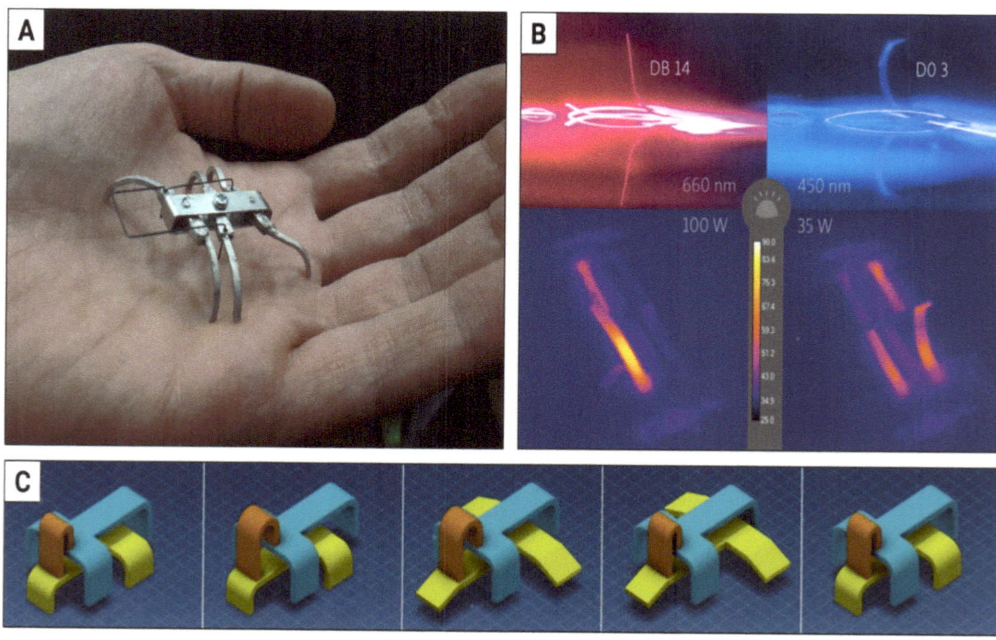

Figure 7. The ultimate walker: LCE ant with two degrees of freedom. (**A**) One of the models built to better understand six-legged locomotion. (**B**) Stripes of LCE with two different dyes, respond by bending to two high-power LEDs with 660 nm and 450 nm centered spectra, respectively. White light (top) and IR (bottom) images. (**C**) One of the concepts of the mechanical design of a six-legged walker with two tripods: one solid (blue) and one with lifting legs (yellow), shifted relative to each other with a bending strip (orange), performing a sequence of deformations that result in a step forward.

4. Micrometer-Scale Light-Power Tools

Perhaps our failures in mimicking nature, in particular when small scale and/or large quantities of mechanisms are in play, stem from the way we approach fabrication. In the lab, workshop or factory, we would start with materials (a plank of wood, a sheet of metal, a piece of wire, a stretch of foil) and then cut, drill, mill or grind to make individual parts first, and then glue, weld, solder or rivet, to join them together. Obviously, this is not the way things are made in the natural world—there, they grow. This, then, is a question we asked ourselves: can we perhaps grow micro-mechanisms, instead of making them in the traditional way [93]?

To this end, we developed a method of fabricating micrometer-scale elastomer structures with a photo-mechanical response by sending UV light via an optical fiber immersed in an oriented liquid monomer, so that the polymerization occurs at the fiber tip—see Figure 8A–C for the snapshots of the fascinating growth process and [94] for the time-lapse video. In our experiments, a cone-shaped structure grows that bends when visible (green) light is delivered via the same fiber (Figure 8D and [94] for the video).

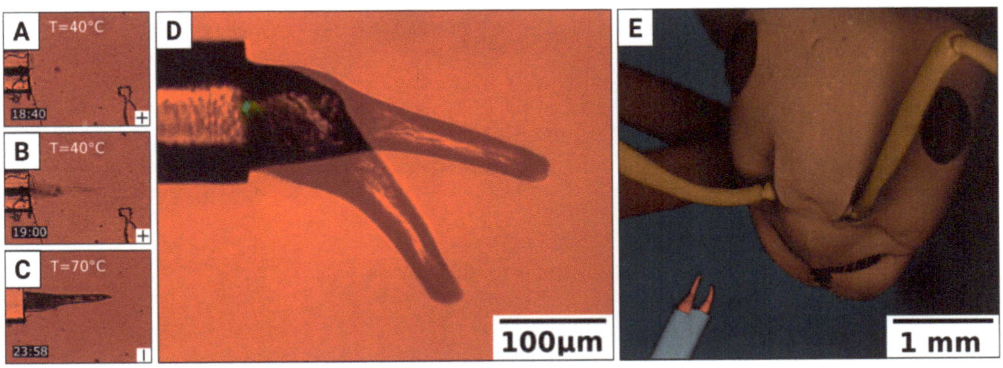

Figure 8. Fiber-grown microscale gripper. (**A–C**) Snapshots from the video show the bending actuator growth inside an LC-mixture-filled cell. Time in min: sec is shown in the bottom left corner of each frame. (**D**) Bending actuator in action, powered with green light delivered through the fiber (two photographs have been superimposed to show the magnitude of the deformation). (**E**) Two fibers with bending actuators make a light-powered gripper, here next to a Formica polyctena ant with its mandibles about ten times the size of our tool. Adapted from [93].

Gripping objects is fundamental for living organisms and in many machines. Mechanical grippers are typically powered by electric, pneumatic, hydraulic or piezoelectric servos and work well at larger scales, but their complexity and need for transmitting force from a distant servo to the gripper elements prevent their miniaturization and remote control. By joining two fibers with LCE-bending structures, we built a micrometer-scale gripper, powered and operated remotely with light energy delivered through the fibers (Figure 8E).

The gripper, nicknamed "optical pliers" (to avoid confusion with "optical tweezers" [95]), can deliver a gripping force of the order of 10^{-7} N that may be compared, for instance, to a single bending structure weight of approximately 3×10^{-9} N (see [94] for the video of the experiment where this force was determined). They do not require any displacement transmission—only energy is transmitted—which makes them very simple and potentially very reliable. In addition, they demonstrate that, with clever technologies, opto-mechanical micro-structures can be fabricated without resorting to any complex (and expensive) microfabrication technology, such as laser photolithography.

Will we one day be able to program and control the growth, so that we can make actuators, or even more complex tools and mechanisms, on fibers, at will? Where would the "genetic information" be stored? In the material, in the growth conditions (molecular orientation, temperature, flow) or in the light that initiates the polymerization—its wavelength(s), polarization, or spatial or temporal modulation? Perhaps in the structure of the optical fiber itself? The concept of "building by growth" involves many more questions than answers and opens up a new, fascinating avenue of research.

5. Conclusions and Outlook

Ten years of working and playing with soft, light-responsive materials have taught us many lessons. The most important one is perhaps the lesson of humility, when comparing our results in micromechanics and micro-robotics with the wonders of the natural world that we try to mimic.

At the moment we have several ongoing projects, including the following:

- Orienting LCE with the electric field during laser photolithography [96]. If successful, this technology will open up the ultimate realm of 5D photo-mechanical microstructures: the 3D-printed shape (with sub-micron resolution) with two angles of director orientation, programmable over the entire volume;
- Orienting LCE molecules with writing direction in laser photolithography. This method is somehow similar to orienting by squeezing through a small nozzle in 3D printing of LCEs [97], but on a much smaller scale. As it does not require any photo aligning layers or electrodes, it can be used with minute substrates, such as the end face of an optical fiber;
- Exploring the theme of "materials as machines" [98], we have developed a conveyor belt with LCE-sorting mechanisms, where small objects can be sorted into a number of buckets with LCE stripes (levers) that respond to their different colors and push them from the belt, without any sensors, data processing or separate actuators;
- Since our lab has its origins in photonics and optics, we are constantly playing with ideas of combining LCE structures with optical elements, e.g., optical fibers. We have tested several approaches to fiber switches, either with self-standing light-responsive actuators or with LCE micro-structures grown on the fiber tip.

Liquid crystal elastomers have been pitched as "promising materials" [99] that "open up new horizons in micro actuation and complex, remotely powered and controlled soft-robotics" [50]. Over the last decade, we have learned a lot about their many problems, not least related to repeatable fabrication, reliability and durability. If ever they may find their way into practical applications, the route will be a long one and not without hurdles and pitfalls. However, the journey into the world of light-responsive materials has given us a lot of joy and we hope that some of this is visible in our publications, including this one.

Author Contributions: M.R. was involved in most of the projects: he designed and made the caterpillar robot, with K.D. developed the snail robot, performed the earliest experiments with growing the (contracting) LCE micro-actuators on optical fibers, and worked on photo-orienting layers for RoboLEC with M.P. He is also involved, with Z.D., in all of the ongoing projects, where Z.D. focuses her efforts on LCE reorientation in laser photolithography; M.Z. led the optical pliers project, with contributions from K.D. and P.N.-J., who developed the early version of the tilting mirror mount with three contracting actuators grown on three optical fibers; P.G., accompanied by K.D., conducted many experiments designed to better understand the LCE response dynamics to various light stimuli. He also conceived the idea of robotic LCE film fabrication, which he later turned into RoboLEC with B.F.; P.W. was the group leader over the years and, to a greater or lesser extent, supervised all projects and made sure the lab worked smoothly from day to day. In the latter, he was supported by A.K., whose main project was the development of the two-color LCE actuators for the robotic ant. The manuscript was written by M.R., Z.D., K.D. and P.W. All authors have read and agreed to the published version of the manuscript.

Funding: The studies on LCE technology and applications were generously funded between 2019–2022 by the National Science Centre (NCN, Poland) with grant No. 2018/29/B/ST7/00192 "Micro-scale actuators based on photo-responsive polymers". They were also funded in part by the Polish Ministry of Science and Higher Education (MNiSW) "Diamentowy Grant" project No. DI2016 015046, awarded to Mikołaj Rogóż and by the NCN grant No. 2019/03/X/ST7/02065 "Liquid crystal crosslinked polymers as actuators in microfluidic devices" awarded to Klaudia Dradrach.

Institutional Review Board Statement: Not applicable.

Informed Consent Statement: Not applicable.

Data Availability Statement: The datasets used and analyzed during the studies presented are available on reasonable request from the corresponding author.

Acknowledgments: PW dedicates this article to Diederik Wiersma. Not only did he introduce me to the wonderful world of liquid crystal elastomers and micro-robots, but also, perhaps more importantly, he taught me that, despite all hurdles and problems, moonshot projects are indeed possible in the academic environment.Our early work was conducted in close collaboration with Wiersma's group in the European Laboratory of Nonlinear Spectroscopy (LENS) in Florence, Italy, in particular with Hao Zeng, Camilla Parmeggiani and Daniele Martella. Hao Zeng, who was instrumental in these developments, after receiving his PhD moved to Tampere, Finland, where he joined the group of Arri Priimagi and thus our Nordic link was established and was maintained until the present. After receiving his MSc in Warsaw, Michał Zmyślony joined the group of John Biggins in the Department of Engineering at the University of Cambridge, soon followed by Klaudia Dradrach, who is currently a postdoc there, working (with some spectacular results) on applications of light-driven mechanisms in microfluidics. As for numerical simulations, we often relied on the invaluable support of Jakub Haberko from the AGH University of Science and Technology in Kraków, Poland and Chen Xuan from the Jiaotong-Liverpool University in Suzhou, China. We also appreciate the patience of Piotr Tykarski from the Faculty of Biology, University of Warsaw, during invaluable consultations on different grippers in the animal kingdom, the generosity of Piotr Ślipiński from the Museum and Institute of Zoology, Polish Academy of Sciences in Warsaw, who donated several ant specimens from his collection. Throughout our journey in the fantastic world of light-responsive mechanics and robotics, we were very lucky to have a range of Polish experts who cooperated with us on our endeavors. Wiktor Piecek and Rafał Węgłowski from the Military University of Technology in Warsaw were (and are) always ready to share their extensive expertise in liquid crystal technology and have supplied the glass slides and cells with photoalignment layers for several experiments. Some of these layers were prepared with unique chemical compounds synthesized by Joanna Konieczkowska from the group of Ewa Schab-Balcerzak from the Centre of Polymer and Carbon Materials, Polish Academy of Sciences in Zabrze.

Conflicts of Interest: The authors declare no competing interest.

References

1. Kistemaker, J.C.M.; Lubbe, A.S.; Feringa, B.L. Exploring Molecular Motors. *Mater. Chem. Front.* **2021**, *5*, 2900–2906. [CrossRef]
2. Challenge for the World's Smallest | History of Namiki DC Coreless Motor | Adamant Namiki Precision Jewel Co., Ltd. Available online: https://www.ad-na.com/en/product/dccorelessmotor/micromotor.html (accessed on 26 August 2022).
3. Jafferis, N.T.; Helbling, E.F.; Karpelson, M.; Wood, R.J. Untethered Flight of an Insect-Sized Flapping-Wing Microscale Aerial Vehicle. *Nature* **2019**, *570*, 491–495. [CrossRef]
4. Purcell, E.M. Life at Low Reynolds Number. *Am. J. Phys.* **1977**, *45*, 3–11. [CrossRef]
5. Rogóż, M.; Dradrach, K.; Xuan, C.; Wasylczyk, P. A Millimeter-Scale Snail Robot Based on a Light-Powered Liquid Crystal Elastomer Continuous Actuator. *Macromol. Rapid Commun.* **2019**, *40*, 1900279. [CrossRef]
6. Zeng, H.; Wasylczyk, P.; Parmeggiani, C.; Martella, D.; Burresi, M.; Wiersma, D.S. Light-Fueled Microscopic Walkers. *Adv. Mater.* **2015**, *27*, 3883–3887. [CrossRef]
7. Zeng, H.; Wasylczyk, P.; Wiersma, D.S.; Priimagi, A. Light Robots: Bridging the Gap between Microrobotics and Photomechanics in Soft Materials. *Adv. Mater.* **2018**, *30*, 1703554. [CrossRef]
8. Hines, L.; Petersen, K.; Lum, G.Z.; Sitti, M. Soft Actuators for Small-Scale Robotics. *Adv. Mater.* **2017**, *29*, 1603483. [CrossRef]
9. Rosset, S.; Niklaus, M.; Dubois, P.; Shea, H.R. Mechanical Characterization of a Dielectric Elastomer Microactuator with Ion-Implanted Electrodes. *Sens. Actuators A Phys.* **2008**, *144*, 185–193. [CrossRef]
10. Brochu, P.; Pei, Q. Advances in Dielectric Elastomers for Actuators and Artificial Muscles. *Macromol. Rapid Commun.* **2010**, *31*, 10–36. [CrossRef]
11. Bay, L.; West, K.; Sommer-Larsen, P.; Skaarup, S.; Benslimane, M. A Conducting Polymer Artificial Muscle with 12% Linear Strain. *Adv. Mater.* **2003**, *15*, 310–313. [CrossRef]
12. Must, I.; Kaasik, F.; Põldsalu, I.; Mihkels, L.; Johanson, U.; Punning, A.; Aabloo, A. Ionic and Capacitive Artificial Muscle for Biomimetic Soft Robotics: Ionic and Capacitive Artificial Muscle for Biomimetic Soft Robotics. *Adv. Eng. Mater.* **2015**, *17*, 84–94. [CrossRef]
13. Lum, G.Z.; Ye, Z.; Dong, X.; Marvi, H.; Erin, O.; Hu, W.; Sitti, M. Shape-Programmable Magnetic Soft Matter. *Proc. Natl. Acad. Sci. USA* **2016**, *113*, E6007–E6015. [CrossRef]
14. Diller, E.; Zhuang, J.; Zhan Lum, G.; Edwards, M.R.; Sitti, M. Continuously Distributed Magnetization Profile for Millimeter-Scale Elastomeric Undulatory Swimming. *Appl. Phys. Lett.* **2014**, *104*, 174101. [CrossRef]
15. Grinthal, A.; Aizenberg, J. Adaptive All the Way down: Building Responsive Materials from Hierarchies of Chemomechanical Feedback. *Chem. Soc. Rev.* **2013**, *42*, 7072. [CrossRef] [PubMed]

16. De Volder, M.; Reynaerts, D. Pneumatic and Hydraulic Microactuators: A Review. *J. Micromech. Microeng.* **2010**, *20*, 043001. [CrossRef]
17. Paek, J.; Cho, I.; Kim, J. Microrobotic Tentacles with Spiral Bending Capability Based on Shape-Engineered Elastomeric Microtubes. *Sci. Rep.* **2015**, *5*, 10768. [CrossRef]
18. De Gennes, P.-G. One Type of Nematic Polymers. *C. R. Seances Acad. Sci. Ser. B* **1975**, *281*, 101–103.
19. Sawa, Y.; Urayama, K.; Takigawa, T.; DeSimone, A.; Teresi, L. Thermally Driven Giant Bending of Liquid Crystal Elastomer Films with Hybrid Alignment. *Macromolecules* **2010**, *43*, 4362–4369. [CrossRef]
20. Yu, Y.; Nakano, M.; Ikeda, T. Directed Bending of a Polymer Film by Light. *Nature* **2003**, *425*, 145. [CrossRef]
21. Braun, L.; Linder, T.; Hessberger, T.; Zentel, R. Influence of a Crosslinker Containing an Azo Group on the Actuation Properties of a Photoactuating LCE System. *Polymers* **2016**, *8*, 435. [CrossRef]
22. Pevnyi, M.; Moreira-Fontana, M.; Richards, G.; Zheng, X.; Palffy-Muhoray, P. Studies of Photo-Thermal Deformations of Liquid Crystal Elastomers under Local Illumination. *Mol. Cryst. Liquid Cryst.* **2017**, *647*, 228–234. [CrossRef]
23. Yang, H.; Buguin, A.; Taulemesse, J.-M.; Kaneko, K.; Méry, S.; Bergeret, A.; Keller, P. Micron-Sized Main-Chain Liquid Crystalline Elastomer Actuators with Ultralarge Amplitude Contractions. *J. Am. Chem. Soc.* **2009**, *131*, 15000–15004. [CrossRef] [PubMed]
24. Urayama, K.; Honda, S.; Takigawa, T. Electrooptical Effects with Anisotropic Deformation in Nematic Gels. *Macromolecules* **2005**, *38*, 3574–3576. [CrossRef]
25. Spillmann, C.M.; Ratna, B.R.; Naciri, J. Anisotropic Actuation in Electroclinic Liquid Crystal Elastomers. *Appl. Phys. Lett.* **2007**, *90*, 021911. [CrossRef]
26. Urayama, K. Selected Issues in Liquid Crystal Elastomers and Gels. *Macromolecules* **2007**, *40*, 2277–2288. [CrossRef]
27. Kumar, K.; Knie, C.; Bléger, D.; Peletier, M.A.; Friedrich, H.; Hecht, S.; Broer, D.J.; Debije, M.G.; Schenning, A.P.H.J. A Chaotic Self-Oscillating Sunlight-Driven Polymer Actuator. *Nat. Commun.* **2016**, *7*, 11975. [CrossRef]
28. Cheng, Z.; Wang, T.; Li, X.; Zhang, Y.; Yu, H. NIR–Vis–UV Light-Responsive Actuator Films of Polymer-Dispersed Liquid Crystal/Graphene Oxide Nanocomposites. *ACS Appl. Mater. Interfaces* **2015**, *7*, 27494–27501. [CrossRef]
29. Tabiryan, N.; Serak, S.; Dai, X.-M.; Bunning, T. Polymer Film with Optically Controlled Form and Actuation. *Opt. Express* **2005**, *13*, 7442. [CrossRef]
30. Martella, D.; Nocentini, S.; Micheletti, F.; Wiersma, D.S.; Parmeggiani, C. Polarization-Dependent Deformation in Light Responsive Polymers Doped by Dichroic Dyes. *Soft Matter* **2019**, *15*, 1312–1318. [CrossRef]
31. Gelebart, A.H.; Jan Mulder, D.; Varga, M.; Konya, A.; Vantomme, G.; Meijer, E.W.; Selinger, R.L.B.; Broer, D.J. Making Waves in a Photoactive Polymer Film. *Nature* **2017**, *546*, 632–636. [CrossRef]
32. Liu, X.; Kim, S.-K.; Wang, X. Thermomechanical Liquid Crystalline Elastomer Capillaries with Biomimetic Peristaltic Crawling Function. *J. Mater. Chem. B* **2016**, *4*, 7293–7302. [CrossRef] [PubMed]
33. Cheng, Z.; Ma, S.; Zhang, Y.; Huang, S.; Chen, Y.; Yu, H. Photomechanical Motion of Liquid-Crystalline Fibers Bending Away from a Light Source. *Macromolecules* **2017**, *50*, 8317–8324. [CrossRef]
34. Ge, F.; Zhao, Y. Microstructured Actuation of Liquid Crystal Polymer Networks. *Adv. Funct. Mater.* **2020**, *30*, 1901890. [CrossRef]
35. Finkelmann, H.; Kock, H.-J.; Rehage, G. Investigations on Liquid Crystalline Polysiloxanes Liquid Crystalline Elastomers—A New Type of Liquid Crystalline Material. *Makromol. Chem. Rapid Commun.* **1981**, *2*, 317–322. [CrossRef]
36. Küpfer, J.; Finkelmann, H. Nematic Liquid Single Crystal Elastomers. *Makromol. Chem. Rapid Commun.* **1991**, *12*, 717–726. [CrossRef]
37. Yoon, H.-H.; Kim, D.-Y.; Jeong, K.-U.; Ahn, S. Surface Aligned Main-Chain Liquid Crystalline Elastomers: Tailored Properties by the Choice of Amine Chain Extenders. *Macromolecules* **2018**, *51*, 1141–1149. [CrossRef]
38. Barnes, M.; Cetinkaya, S.; Ajnsztajn, A.; Verduzco, R. Understanding the Effect of Liquid Crystal Content on the Phase Behavior and Mechanical Properties of Liquid Crystal Elastomers. *Soft Matter* **2022**, *18*, 5074–5081. [CrossRef]
39. Herbert, K.M.; Fowler, H.E.; McCracken, J.M.; Schlafmann, K.R.; Koch, J.A.; White, T.J. Synthesis and Alignment of Liquid Crystalline Elastomers. *Nat. Rev. Mater.* **2022**, *7*, 23–38. [CrossRef]
40. Finkelmann, H.; Nishikawa, E.; Pereira, G.G.; Warner, M. A New Opto-Mechanical Effect in Solids. *Phys. Rev. Lett.* **2001**, *87*, 015501. [CrossRef]
41. Sánchez-Ferrer, A.; Merekalov, A.; Finkelmann, H. Opto-Mechanical Effect in Photoactive Nematic Side-Chain Liquid-Crystalline Elastomers: Opto-Mechanical Effect in Photoactive. *Macromol. Rapid Commun.* **2011**, *32*, 671–678. [CrossRef]
42. Ahn, S.; Ware, T.H.; Lee, K.M.; Tondiglia, V.P.; White, T.J. Photoinduced Topographical Feature Development in Blueprinted Azobenzene-Functionalized Liquid Crystalline Elastomers. *Adv. Funct. Mater.* **2016**, *26*, 5819–5826. [CrossRef]
43. Mehta, K.; Peeketi, A.R.; Liu, L.; Broer, D.; Onck, P.; Annabattula, R.K. Design and Applications of Light Responsive Liquid Crystal Polymer Thin Films. *Appl. Phys. Rev.* **2020**, *7*, 041306. [CrossRef]
44. Marshall, J.E.; Ji, Y.; Torras, N.; Zinoviev, K.; Terentjev, E.M. Carbon-Nanotube Sensitized Nematic Elastomer Composites for IR-Visible Photo-Actuation. *Soft Matter* **2012**, *8*, 1570–1574. [CrossRef]
45. Liu, X.; Wei, R.; Hoang, P.T.; Wang, X.; Liu, T.; Keller, P. Reversible and Rapid Laser Actuation of Liquid Crystalline Elastomer Micropillars with Inclusion of Gold Nanoparticles. *Adv. Funct. Mater.* **2015**, *25*, 3022–3032. [CrossRef]
46. Hauser, A.W.; Liu, D.; Bryson, K.C.; Hayward, R.C.; Broer, D.J. Reconfiguring Nanocomposite Liquid Crystal Polymer Films with Visible Light. *Macromolecules* **2016**, *49*, 1575–1581. [CrossRef]

47. He, Q.; Wang, Z.; Wang, Y.; Wang, Z.; Li, C.; Annapooranan, R.; Zeng, J.; Chen, R.; Cai, S. Electrospun Liquid Crystal Elastomer Microfiber Actuator. *Sci. Robot.* **2021**, *6*, eabi9704. [CrossRef] [PubMed]
48. Lahikainen, M.; Zeng, H.; Priimagi, A. Reconfigurable Photoactuator through Synergistic Use of Photochemical and Photothermal Effects. *Nat. Commun.* **2018**, *9*, 4148. [CrossRef]
49. Cheng, Y.; Lu, H.; Lee, X.; Zeng, H.; Priimagi, A. Soft Actuators: Kirigami-Based Light-Induced Shape-Morphing and Locomotion (Adv. Mater. 7/2020). *Adv. Mater.* **2020**, *32*, 2070047. [CrossRef]
50. Rogóż, M.; Zeng, H.; Xuan, C.; Wiersma, D.S.; Wasylczyk, P. Light-Driven Soft Robot Mimics Caterpillar Locomotion in Natural Scale. *Adv. Opt. Mater.* **2016**, *4*, 1689–1694. [CrossRef]
51. Ware, T.H.; McConney, M.E.; Wie, J.J.; Tondiglia, V.P.; White, T.J. Voxelated Liquid Crystal Elastomers. *Science* **2015**, *347*, 982–984. [CrossRef]
52. Rogóż, M.; Haberko, J.; Wasylczyk, P. Light-Driven Linear Inchworm Motor Based on Liquid Crystal Elastomer Actuators Fabricated with Rubbing Overwriting. *Materials* **2021**, *14*, 6688. [CrossRef] [PubMed]
53. Yao, Y.; Waters, J.T.; Shneidman, A.V.; Cui, J.; Wang, X.; Mandsberg, N.K.; Li, S.; Balazs, A.C.; Aizenberg, J. Multiresponsive Polymeric Microstructures with Encoded Predetermined and Self-Regulated Deformability. *Proc. Natl. Acad. Sci. USA* **2018**, *115*, 12950–12955. [CrossRef] [PubMed]
54. Wehner, M.; Truby, R.L.; Fitzgerald, D.J.; Mosadegh, B.; Whitesides, G.M.; Lewis, J.A.; Wood, R.J. An Integrated Design and Fabrication Strategy for Entirely Soft, Autonomous Robots. *Nature* **2016**, *536*, 451–455. [CrossRef]
55. Truby, R.L.; Lewis, J.A. Printing Soft Matter in Three Dimensions. *Nature* **2016**, *540*, 371–378. [CrossRef]
56. Zeng, H.; Wasylczyk, P.; Cerretti, G.; Martella, D.; Parmeggiani, C.; Wiersma, D.S. Alignment Engineering in Liquid Crystalline Elastomers: Free-Form Microstructures with Multiple Functionalities. *Appl. Phys. Lett.* **2015**, *106*, 111902. [CrossRef]
57. Grabowski, P.; Haberko, J.; Wasylczyk, P. Photo-Mechanical Response Dynamics of Liquid Crystal Elastomer Linear Actuators. *Materials* **2020**, *13*, 2933. [CrossRef]
58. Yamada, M.; Kondo, M.; Mamiya, J.; Yu, Y.; Kinoshita, M.; Barrett, C.J.; Ikeda, T. Photomobile Polymer Materials: Towards Light-Driven Plastic Motors. *Angew. Chem. Int. Ed.* **2008**, *47*, 4986–4988. [CrossRef]
59. Wang, H.; Pumera, M. Fabrication of Micro/Nanoscale Motors. *Chem. Rev.* **2015**, *115*, 8704–8735. [CrossRef]
60. Xu, L.; Mou, F.; Gong, H.; Luo, M.; Guan, J. Light-Driven Micro/Nanomotors: From Fundamentals to Applications. *Chem. Soc. Rev.* **2017**, *46*, 6905–6926. [CrossRef]
61. Chen, H.; Zhao, Q.; Du, X. Light-Powered Micro/Nanomotors. *Micromachines* **2018**, *9*, 41. [CrossRef]
62. Bisoyi, H.K.; Li, Q. Light-Driven Liquid Crystalline Materials: From Photo-Induced Phase Transitions and Property Modulations to Applications. *Chem. Rev.* **2016**, *116*, 15089–15166. [CrossRef] [PubMed]
63. Ahn, C.; Li, K.; Cai, S. Light or Thermally Powered Autonomous Rolling of an Elastomer Rod. *ACS Appl. Mater. Interfaces* **2018**, *10*, 25689–25696. [CrossRef] [PubMed]
64. Lu, X.; Guo, S.; Tong, X.; Xia, H.; Zhao, Y. Tunable Photocontrolled Motions Using Stored Strain Energy in Malleable Azobenzene Liquid Crystalline Polymer Actuators. *Adv. Mater.* **2017**, *29*, 1606467. [CrossRef]
65. Dradrach, K.; Rogóż, M.; Grabowski, P.; Xuan, C.; Węgłowski, R.; Konieczkowska, J.; Schab-Balcerzak, E.; Piecek, W.; Wasylczyk, P. Traveling Wave Rotary Micromotor Based on a Photomechanical Response in Liquid Crystal Polymer Networks. *ACS Appl. Mater. Interfaces* **2020**, *12*, 8681–8686. [CrossRef] [PubMed]
66. Morita, T. Miniature Piezoelectric Motors. *Sens. Actuators A Phys.* **2003**, *103*, 291–300. [CrossRef]
67. Oh, J.-H.; Jung, H.-E.; Lee, J.; Lim, K.-J.; Kim, H.-H.; Ryu, B.-H.; Park, D.-H. Design and Performances of High Torque Ultrasonic Motor for Application of Automobile. *J. Electroceram.* **2009**, *22*, 150–155. [CrossRef]
68. Spanner, K.; Koc, B. Piezoelectric Motors, an Overview. *Actuators* **2016**, *5*, 6. [CrossRef]
69. Traveling Wave Rotary Micromotor. Available online: https://www.youtube.com/watch?v=2w5vf_kfO5Y&ab_channel=PNaF (accessed on 1 October 2022).
70. Smits, J.G. Design Considerations of a Piezoelectric-on-Silicon Microrobot. *Sens. Actuators A Phys.* **1992**, *35*, 129–135. [CrossRef]
71. Bexell, M.; Tiensuu, A.-L.; Schweitz, J.-Å.; Söderkvist, J.; Johansson, S. Characterization of an Inchworm Prototype Motor. *Sens. Actuators A Phys.* **1994**, *43*, 322–329. [CrossRef]
72. Montazami, R.; Spillmann, C.M.; Naciri, J.; Ratna, B.R. Enhanced Thermomechanical Properties of a Nematic Liquid Crystal Elastomer Doped with Gold Nanoparticles. *Sens. Actuators A Phys.* **2012**, *178*, 175–178. [CrossRef]
73. Han, W.C.; Sim, G.W.; Kim, Y.B.; Kim, D.S. Reversible Curvature Reversal of Monolithic Liquid Crystal Elastomer Film and Its Smart Valve Application. *Macromol. Rapid Commun.* **2021**, *42*, 2100404. [CrossRef]
74. Inchworm Motor Based on Liquid Crystal Elastomer Actuators Fabricated with Rubbing Overwriting. Available online: https://www.youtube.com/watch?v=PhxzyFMsWJU&ab_channel=PNaF (accessed on 1 October 2022).
75. Lombardero, M.; Yllera, M. del M. Leonardo Da Vinci's Animal Anatomy: Bear and Horse Drawings Revisited. *Animals* **2019**, *9*, 435. [CrossRef]
76. Kim, S.; Laschi, C.; Trimmer, B. Soft Robotics: A Bioinspired Evolution in Robotics. *Trends Biotechnol.* **2013**, *31*, 287–294. [CrossRef]
77. Pilz da Cunha, M.; Debije, M.G.; Schenning, A.P.H.J. Bioinspired Light-Driven Soft Robots Based on Liquid Crystal Polymers. *Chem. Soc. Rev.* **2020**, *49*, 6568–6578. [CrossRef] [PubMed]
78. Light-Driven Soft Robot Mimics Caterpillar Locomotion in Natural Scale. Available online: https://www.youtube.com/watch?v=mAGK8JG0gVY&ab_channel=PNaF (accessed on 1 October 2022).

79. Brackenbury, J. Fast Locomotion in Caterpillars. *J. Insect Physiol.* **1999**, *45*, 525–533. [CrossRef]
80. Wang, C.; Sim, K.; Chen, J.; Kim, H.; Rao, Z.; Li, Y.; Chen, W.; Song, J.; Verduzco, R.; Yu, C. Soft Ultrathin Electronics Innervated Adaptive Fully Soft Robots. *Adv. Mater.* **2018**, *30*, 1706695. [CrossRef] [PubMed]
81. Wang, Z.; Li, K.; He, Q.; Cai, S. A Light-Powered Ultralight Tensegrity Robot with High Deformability and Load Capacity. *Adv. Mater.* **2019**, *31*, 1806849. [CrossRef] [PubMed]
82. Ahn, C.; Liang, X.; Cai, S. Bioinspired Design of Light-Powered Crawling, Squeezing, and Jumping Untethered Soft Robot. *Adv. Mater. Technol.* **2019**, *4*, 1900185. [CrossRef]
83. Snail Soft Robot, Light-Powered, Made of LCE. Available online: https://www.youtube.com/watch?v=u3z3TdB4BT0&ab_channel=PNaF (accessed on 1 October 2022).
84. Denny, M. The Role of Gastropod Pedal Mucus in Locomotion. *Nature* **1980**, *285*, 160–161. [CrossRef]
85. Denny, M.W. A Quantitative Model for the Adhesive Locomotion of the Terrestrial Slug, *Ariolimax columbianus*. *J. Exp. Biol.* **1981**, *91*, 195–217. [CrossRef]
86. Parker, G.H. The Mechanism of Locomotion in Gastropods. *J. Morphol.* **1911**, *22*, 155–170. [CrossRef]
87. Suter, R.; Rosenberg, O.; Loeb, S.; Wildman, H.; Long, J. Locomotion on the Water Surface: Propulsive Mechanisms of the Fisher Spider. *J. Exp. Biol.* **1997**, *200*, 2523–2538. [CrossRef] [PubMed]
88. Hu, D.L.; Chan, B.; Bush, J.W.M. The Hydrodynamics of Water Strider Locomotion. *Nature* **2003**, *424*, 663–666. [CrossRef] [PubMed]
89. Kim, Y.; van den Berg, J.; Crosby, A.J. Autonomous Snapping and Jumping Polymer Gels. *Nat. Mater.* **2021**, *20*, 1695–1701. [CrossRef] [PubMed]
90. Koh, J.-S.; Yang, E.; Jung, G.-P.; Jung, S.-P.; Son, J.H.; Lee, S.-I.; Jablonski, P.G.; Wood, R.J.; Kim, H.-Y.; Cho, K.-J. Jumping on Water: Surface Tension–Dominated Jumping of Water Striders and Robotic Insects. *Science* **2015**, *349*, 517–521. [CrossRef]
91. Song, Y.S.; Sitti, M. Surface-Tension-Driven Biologically Inspired Water Strider Robots: Theory and Experiments. *IEEE Trans. Robot.* **2007**, *23*, 578–589. [CrossRef]
92. Kudzia, A. Liquid Crystal Elastomer Films as Light-Driven Actuators with the Spectral Degree of Freedom. Bachelor's Thesis, University of Warsaw, Warsaw, Poland, 2021.
93. Zmyślony, M.; Dradrach, K.; Haberko, J.; Nałęcz-Jawecki, P.; Rogóż, M.; Wasylczyk, P. Optical Pliers: Micrometer-Scale, Light-Driven Tools Grown on Optical Fibers. *Adv. Mater.* **2020**, *32*, 2002779. [CrossRef]
94. Optical Pliers: Micrometer-Scale, Light-Driven Tools Grown on Optical Fibers. Available online: https://www.youtube.com/watch?v=j-R-SB6EBqs&ab_channel=PNaF (accessed on 1 October 2022).
95. Ashkin, A.; Dziedzic, J.M. Optical Trapping and Manipulation of Viruses and Bacteria. *Science* **1987**, *235*, 1517–1520. [CrossRef]
96. Carlotti, M.; Tricinci, O.; den Hoed, F.; Palagi, S.; Mattoli, V. Direct Laser Writing of Liquid Crystal Elastomers Oriented by a Horizontal Electric Field. *Open Res. Europe* **2021**, *1*, 129. [CrossRef]
97. Zhang, C.; Lu, X.; Fei, G.; Wang, Z.; Xia, H.; Zhao, Y. 4D Printing of a Liquid Crystal Elastomer with a Controllable Orientation Gradient. *ACS Appl. Mater. Interfaces* **2019**, *11*, 44774–44782. [CrossRef]
98. McCracken, J.M.; Donovan, B.R.; White, T.J. Materials as Machines. *Adv. Mater.* **2020**, *32*, 1906564. [CrossRef] [PubMed]
99. Wani, O.M.; Zeng, H.; Wasylczyk, P.; Priimagi, A. Programming Photoresponse in Liquid Crystal Polymer Actuators with Laser Projector. *Adv. Opt. Mater.* **2018**, *6*, 1700949. [CrossRef]

materials

MDPI

Review

Review on Polymer, Ceramic and Composite Materials for CAD/CAM Indirect Restorations in Dentistry—Application, Mechanical Characteristics and Comparison

Aleksandra Skorulska [1,*], Paweł Piszko [2,*], Zbigniew Rybak [3], Maria Szymonowicz [3] and Maciej Dobrzyński [1]

1 Department of Pediatric Dentistry and Preclinical Dentistry, Wroclaw Medical University, Krakowska 26, 50-425 Wrocław, Poland; maciej.dobrzynski@umed.wroc.pl
2 Polymer Engineering and Technology Division, Wrocław University of Science and Technology, Wyb. Wyspiańskiego 27, 50-370 Wrocław, Poland
3 Department of Experimental Surgery and Biomaterials Research, Wroclaw Medical University, Ulica Bujwida 44, 50-368 Wrocław, Poland; zbigniew.rybak@umed.wroc.pl (Z.R.); maria.szymonowicz@umed.wroc.pl (M.S.)
* Correspondence: aleksandraskorulska@gmail.com (A.S.); pawel.piszko@pwr.edu.pl (P.P.)

Abstract: The aim of this review article is to present various material groups, including ceramics, composites and hybrid materials, currently utilized in the field of CAD/CAM. The described technology is amongst the most important in modern prosthetics. Materials that are applicable in this technique are constantly tested, evaluated and improved. Nowadays, research on dental materials is carried out in order to meet the increasing demand on highly aesthetic and functional indirect restorations. Recent studies present the long-term clinical success of restorations made with the help of both ceramic and composite materials in the CAD/CAM method. However, new materials are developed and introduced that do not have long-term in vivo observations. We can outline a monolithic polymer-infiltrated ceramic network and zirconia teeth support that show promising results to date but require further assessment. The materials will be compared with regard to their mechanical and clinical properties, purpose, advantages and limitations.

Keywords: CAD/CAM; mechanical properties; dental materials; dental ceramics; resin composites; biocompatibility

Citation: Skorulska, A.; Piszko, P.; Rybak, Z.; Szymonowicz, M.; Dobrzyński, M. Review on Polymer, Ceramic and Composite Materials for CAD/CAM Indirect Restorations in Dentistry—Application, Mechanical Characteristics and Comparison. *Materials* **2021**, *14*, 1592. https://doi.org/10.3390/ma14071592

Academic Editors: Jakub Zdarta and Francesco Inchingolo

Received: 3 February 2021
Accepted: 18 March 2021
Published: 24 March 2021

1. Introduction

CAD/CAM stands for computer-aided design and computer-aided manufacturing. It is applied in different branches of engineering, science or even art as a rapid prototyping method for accelerating the design process and smoothening its transition into manufacturing [1]. In this review article, we focus on its constantly growing role in dentistry, more precisely, in dental prosthetics. This field of dentistry serves for the manufacturing of either polymer, ceramics or composite teeth restoration. CAD/CAM allows us to provide a patient with dental prosthesis, such as crowns, veneers, bridges, inlays, onlays and dental implant supported restorations [2–4]. In many cases, due to the technology used in CAD/CAM, it is possible to scan a patient's oral cavity, design and create restoration and finally bond it in the patient's mouth in the course of one day. There are both chairside systems and systems which further outsource the prosthesis manufacturing available on the market. To obtain the required properties of restorations, such as high aesthetics, biocompatibility, durability and functionality, a wide range of constantly evolving materials are used in the system described below [5].

2. Materials and Methods

An electronic search of the English-language and Polish-language literature published between 1987 and August 31, 2020, was conducted within PubMed, Scopus, Web of Science and Google Scholar databases. A combination of free-text words was used: CAD/CAM, ceramics, resin composites, dental material, methacrylates, biocompatibility, cytotoxicity and chairside systems. A group of articles was outlined for further verification with respect to their contribution to the topic. Authors focused on articles regarding polymer, ceramic and composite dental CAD/CAM materials along with their mechanical characterization, wearing, cytotoxicity and clinical assessment. Afterwards, the reference list was selected and narrowed down in the scope of the relevance in the field of intraoral CAD/CAM materials utilized in dentistry and their characteristics.

3. Historical Outlook on CAD/CAM

The first versions of the system were invented in the 1980s [6]. The idea of the digital assisted prosthetic system was first developed and introduced as a result of the cooperation of three research centres, the first established by the University of Zurich and Brains and Brandestini Instruments of Switzerland, the second by Hennson International of France and the third group by the University of Minnesota [7]. The aim was to provide the patient with a prosthetic restoration in a fast and impressionless process. Moreover, the authors intended to make durable posterior teeth restorations in a natural colour [6]. Recent years have shown that it is an innovative, developing and forward-thinking method of designing and forming dental prostheses.

4. Advantages of CAD/CAM

The individual properties of CAD/CAM restorations depend, among others, on material which is used for manufacturing. There is wide range of benefits shared by all used materials, which make described technology very attractive both for the patient and dentist. Among them, we can list the following: shorter time of prosthetic treatment; patient's satisfaction; and avoiding the traditional method of making impressions, which has been replaced with user-friendly intraoral scanners [8,9]. There are randomised clinical trials that confirm higher efficiency and comfort using digital scanning compared to conventional impression [10,11]. Moreover, it provides the opportunity to combine high aesthetics, durability and functionality in one restoration [12,13]. The aspect of accuracy is very important as well. It was confirmed that the CAD/CAM restorations, such as single crowns, fixed dental prostheses and implant-retained fixed dental prostheses, are characterized by sufficient marginal adaptation [14,15]. This fact is significant for further plaque accumulation. Potential caries development is less likely to occur when the marginal adaptation is within the clinically acceptable marginal discrepancy range. Replacing the classical procedure of taking impressions by the digital technique helps not only to reduce procedure time [16] and increase the patient's positive feelings but also to maintain an adequate level of precision (4 to 80 μm for scans with a limited area) [17]. Furthermore, scanning oral cavity by an intraoral scanner provides an image of prosthetic substrate on the computer screen almost immediately and under magnification, which helps to control the preparation process as well as plan further restoration. This is an important benefit for many operators. The technology provides the opportunity to use new materials for prosthetic reconstruction and maintain the quality control of the process [18]. All these positive aspects of CAD/CAM technology are reflected in patients' satisfaction and long-term restoration success considering both ceramic [19,20] and composite restorations [21,22].

5. Limitations and Handling of the System

The technology of CAD/CAM, without a doubt, is very innovative and provides a broad range of opportunities. However, this technique is still considered expensive, and despite the development of new systems and increasing competition on the market, the prices remain high [2,23]. The whole process of creating a restoration using CAD/CAM comprises many steps, such as scanning the oral cavity by an intraoral scanner, computer designing using specific software and modelling a restoration either from a solid block of restorative material or using an additive technique [23]. All of this requires highly trained personnel [2], and the technique learning curve can range from a few days to several months [24]. Moreover, as opposed to the traditional way of planning prosthetic reconstruction, in the CAD/CAM system, the involvement of the patient is minimalised [25]. After scanning the oral cavity, the dentist decides the shade, shape and occlusal relation of the prosthetic restoration. Considering clinical cases regarding patients with maxilla-mandibular disorders and occlusion distortions, the CAD/CAM system itself may not be sufficient to obtain correct teeth relation [26]. Moreover, the size of the blocks limits designing and milling restorations exceeding their sizes. This indicates clinical problems including inaccurate occlusal vertical dimension and incorrect centric relation [27].

The accuracy of digital scans depends on the length of the arch included in the impression and is favourable for short distances [28]. The survival rate of CAD/CAM restorations may vary for different types of materials. It is mostly presented in short- and medium-term studies, which makes it more difficult to evaluate and compare to conventional prosthetic restorations. For example, ceramic material Vita Mark II (VITA Zahnfabrik, Bad Säckingen, Germany) inlays showed survival rates of 90.6% after 8 years and 85.7–89% after 10 years [13,29–31]. Therefore, we observe that survival rate decreases over time. The mechanical aspects, such as the flexural strength or mean modulus of resilience, differ for various types of utilised materials [32]. It can be assumed that not all kinds of materials are suitable for all clinical applications. The prosthetic restorations made using CAD/CAM are not free of defects. The main reported complications are framework fractures and recurrent periodontal disease for reinforced glass ceramics and glass infiltrated alumina [16]. Thus, there is still room for improvements in the described technique.

Operators should be aware of certain limitations regarding patients with CAD/CAM restorations. For example, applying lasers for periodontal or conservative reasons among patients with zirconia-based restorations can be performed. It should be considered that the surface of the restorative material can be affected by the laser [33]. Moreover, mechanical limitations are of significant importance, which should not be omitted while describing CAD/CAM restorations. The study by Romanyk et al. shows that subtractive machining results in strength-limiting, surface and subsurface damage in the restorations, which may be clinically relevant [34].

6. Currently Used Dental CAD/CAM Systems

The list of producers offering CAD/CAM software (e.g. CEREC SW 5.1.3, Dentsply Sirona, York, Pennsylvania, United States) and manufacturing systems is broad and has rapidly grown in recent years. CAD/CAM systems can be classified as either in-office or laboratory systems [35]. The two most popular systems currently available on the market are CEREC by Dentsply Sirona (York, PA, USA) and Planmeca by Planmeca Oy (Helsinki, Finland) [36]. Both of them are complex and consist of numerous elements. For example, Sirona offers the CEREC Omnicam scanner, software for CAD/design and for CAM and also the milling unit, which is the CEREC MC, X and XL 4-axis milling machine [8]. Other recognized and used total CAD/CAM systems are Carestream Dental (Atlanta, GA, USA), Dental Wings (Montreal, QC, Canada) and Zfx (Dachau, Germany) [36]. There is also the possibility to buy parts included in a CAD/CAM system, which are offered separately by different companies. The choice of adequate system depends on the prosthetic experience of operators and the equipment of the dental office, but should be also dictated by the patient's therapeutic needs [37].

7. Computer-Aided Design

After scanning the oral cavity using an intraoral scanner, such as powder-free CS 3600 by Carestream Dental [8], we are able to obtain a digital image of the oral cavity, which is the field for further prosthetic restoration. We can divide scanners into two groups: those that require powder and powder-free. Powder scanners require an opaque reflective coating, such as titanium dioxide powder, on the teeth before scanning in order to eliminate reflection and to create an equal surface. On the other hand, powder coating may reduce the scan precision and marginal adaptation of definitive restoration. [38]. Nowadays, mostly powder-free scanners, such as CS 3600 by Carestream Dental (Atlanta, GA, USA), are used [8]. In this group of scanners, there is no risk of mixing intraoral liquids and powder, and thus blurring the boundaries of the preparation. The digital impression is required in multiple steps of the restorative process, including preliminary scanning for the clean-up process followed by the margin setup by the technician before the fabrication takes place. Therefore, they are not considered to save time in comparison to the conventional impressions [39]. Moreover, the resolution of full arch digital impressions is limited in comparison to the conventional impressions [40].

Appropriate software helps to design a restoration in the most optimal way, allowing the operator to make changes simultaneously if needed. In the case of Carestream Dental, a dedicated software for the system is CS Restore. Plenty of different software is available on the market, such as DWOS Chairside (Dental Wings), Cerec SW 4.5. and Cerec Premium SW 4.5 (Dentsplay Sirona), Zfx CAD software (Zfx), Planmeca PlanCAD Easy, integrated in Planmeca Romexis (Planmeca) or MyCrown Design (Fona Dental) [8].

8. Materials for CAD/CAM

The spectrum of materials utilized in computer-aided manufacturing is very broad. It includes not only acrylics polymers but multiple ceramic materials and resin composites [41]. Moreover, the conventionally utilized restoration materials for dental prosthetics include metals. However, due to the necessity of ceramic veneering, we are no longer in the scope of chairside CAD/CAM due to the required postprocessing [42]. Thus, in this review, metals are not covered, and the main focus is put on polymer, ceramic and composite materials.

Each material has different processing parameters, and, therefore, the whole system needs to be adjusted for a specific material. The success of prosthetic treatment using CAD/CAM technology depends, to a certain extent, on material selection but also on all steps of a treatment: from case classification, to correct preparation, precise scanning, planning and designing, resulting in the manufacturing and cementing of a restoration. Along with the desirable characteristics of materials used for restorations in the chairside procedure, the efficiency lies in the possibility of high-speed milling without damage and a short time of preparation of the restoration after milling [43]. The examples of CAD/CAM blocks before processing are depicted in Figure 1 (below).

Figure 1. Examples of commercial CAD/CAM blocks. From the left: resin matrix composite Vita Enamic, resin matrix composite CERASMART and composite Brilliant Crios, reprinted with permission from ref. [44] (Copyright 2020 Inżynier i Fizyk Medyczny).

8.1. Dental Ceramics

There are various types of dental ceramics with respect to their chemical composition, method of obtaining and structure. They can be classified into 3 groups: resin matrix ceramics (RMCs), silicate ceramics and oxide ceramics (see Figure 2) [43]. In general, ceramics can be characterized by strength, brittleness, transparency and hardness [17,45,46]. All of the ceramic restorations made using the CAD/CAM system can be used both in posterior and aesthetic segments and are becoming increasingly popular every year. Their main advantages include biocompatibility, low plaque adherence susceptibility and colour stability [17].

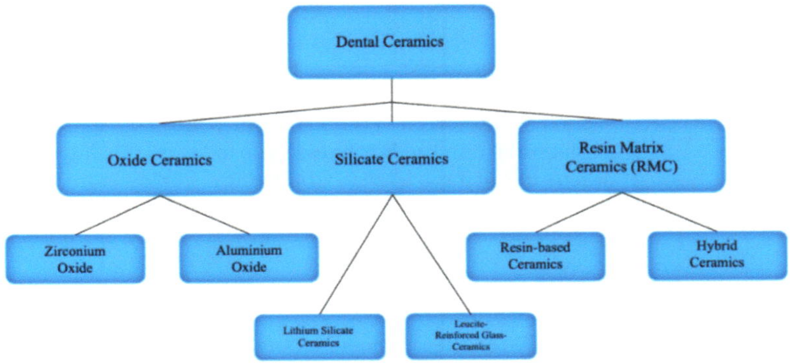

Figure 2. Dental ceramics classification.

Ceramics are crystalline and non-metallic materials, while glass ceramics are composite-type materials in which the glassy phase is the matrix and the ceramic is the reinforcing filler [43]. All-ceramic CAReviD/CAM restorations demand a rounded shoulder or a heavy chamfer around the circumference of the prepared tooth. In most cases, luting using adhesive resin cements is indicated for all-ceramic crowns. This helps to increase fracture resistance [47]. The outline of the ceramics division is presented in Figure 1. Each of the enlisted ceramic type has a different clinical application due to its properties (see Table 1).

Table 1. Examples of commercial ceramic CAD/CAM material with its application.

Type of Ceramics	Brand, Manufacturer	Clinical Application
Resin Matrix Ceramics	Lava Ultimate, 3M-ESPE VITA Enamic, VITA-Zahnfabrik Cerasmart, GC	Onlays, inlays, veneers, single crowns, implant crowns
Lithium Silicate Ceramics	IPS e.max CAD, Ivoclar Vivadent VITA Suprinity PC, VITA Zahnfabrik Celtra Duo, Dentsply Sirona	Inlays, onlays, veneers, crowns
Leucite-Reinforced Glass Ceramics	IPS Empress CAD, Ivoclar Vivadent	Veneers, inlays, onlays, crowns
Zirconium Oxide Ceramics	NobelProcera Zirconia, Nobel Biocare	Single crowns, bridges, prosthetic restorations covering the entire dental arches, mainly posterior segment
Aluminium Oxide Ceramics	InCeram Alumina, VITA Zahnfabrik	Single crowns, bridges

8.1.1. Resin Matrix Ceramics

This is a relatively new material to the market, but it is claimed that it shows some favourable characteristics for dental prosthetics. Resin matrix ceramics are characterized by lucrative milling properties and, compared with silica-based ceramics, have a higher load capacity and better modulus of elasticity [36]. The aesthetic aspect is also satisfying for resin matrix ceramics, which shows optical properties similar to natural teeth.

We can distinguish resin-based ceramics (e.g., Lava Ultimate by 3M-ESPE, Seefeld, Germany) that contain a polymer matrix with at least 80% nanosized ceramic filler particles and hybrid ceramics (e.g., VITA Enamic by VITA-Zahnfabrik, Bad Säckingen, Germany and Cerasmart by GC, Leuven, Belgium) made of a ceramic network infiltrated with a polymer using polymer-infiltrated ceramic network (PICN) technology [36,44]. Recent studies report that resin-based ceramics show flexural strengths up to 230 MPa, characteristic strength (σ_0: 300 MPa) and relatively low Young's modulus [48]. The manufacture of Lava Ultimate characterizes this material as less brittle than glass ceramic and resistant to chipping and cracking when milled [49]. VITA Hybrid ceramics combine the properties of both composites and ceramics, which leads to sufficient flexibility, optimal distribution of chewing forces and high resistance to loads [17]. The manufacture process of VITA Enamic ensures that the tendency of fracturing is lower in comparison with pure ceramics, and CAD/CAM processing is superior [50]. Moreover, their optical properties are similar to natural teeth, and they are characterized by lower abrasion for opposing teeth compared to ceramics [44].

8.1.2. Silicate Ceramics

These are non-metallic inorganic ceramic materials that contain a glass phase. Among silica-based ceramics, we can distinguish feldspathic and lithium silicate ceramics. Examples are Vitablocs TriLuxe by Vita and IPS Empress CAD Multi by IvoclarVivadent. They can be characterized by favourable optical aspects, such as high translucency and natural appearance. However, the presence of glass in their compositions contributes to the brittleness and low fracture strength [36]. Silicate ceramics require hydrofluoric (HF) acid etching to enhance micromechanical retention [51] and adhesive bonding. After acid etching, the glassy matrix is dissolved and crystalline phase is exposed, and thus, the surface of the ceramic becomes available for the micromechanical interlocking of resin cement [52]. The study evaluating tensile bond strength for lithium disilicate ceramic confirms that etching the bonding surface of restorations with hydrofluoric acid is still a "gold standard" [53].

8.1.3. Leucite-Reinforced Glass Ceramics

The clinical long-term valuation of leucite-reinforced glass restoration ceramics (e.g., Duraceram and Dentsply Degussa) was discussed. Leucite-reinforced ceramics are not recommended for crowns in the posterior segment due to their lower mechanical properties compared to other glass ceramics [54]. However, their aesthetic qualities are sufficient, and the wear resistance of the enamel antagonist is similar to other glass ceramic materials [55]. In recent years, they have been replaced by lithium silicate ceramics, which have better physical properties and sufficient optical properties.

8.1.4. Lithium Silicate Ceramics

Some sources claim that lithium silicate ceramics (e.g., IPS e.max CAD by Ivoclar Vivadent, Schaan, Liechtenstein, VITA Suprinity PC by VITA Zahnfabrik and Celtra Duo by Dentsply Sirona) are the strongest of all the available silicate ceramics with a flexural strength of around 407 MPa [36]. First, lithium disilicate ceramic was introduced to the market in 1998 (IPS Empress 2) [56]. Its chemical composition—a crystalline phase consisting of lithium disilicate and lithium orthophosphate—indicates higher fracture resistance without a negative influence on the translucency of the material [36,57,58].

It shows good clinical results in follow-ups with a failure-free rate at the level of 93% [59]. There is also a study which indicates its superior colour stability in different staining solutions, such as coffee or cola, compared to high-translucency zirconia, nanoceramic or hybrid ceramic [60].

8.1.5. Oxide Ceramics

Oxide ceramics exhibit highly favourable mechanical properties, while their aesthetic qualities are slightly worse than silicate ceramics due to their low translucency. They can be divided into aluminium-oxide- and zirconium-oxide-based ceramics.

The oxide ceramics are divided into two major groups:

- Aluminium oxide ceramics

Described as glass-infiltrated aluminium oxide core ceramics (InCeram Alumina, VITA Zahnfabrik, Bad Säckingen, Germany), they are characterised by a flexural strength of 500 Mpa. They show satisfying results in long-term follow-ups, notwithstanding in recent years, when they have mostly been replaced by more popular zirconia ceramics characterized by superior physical properties [61].

- Zirconium oxide ceramics

Commercial CAD/CAM zirconium oxide ceramics are present in the form of yttria-stabilized tetragonal zirconia polycrystal (Y-TZP) [62,63] in products such as Lava Plus (3M, ESPE) or Kavo Everest (KaVo Dental). In the product range of IPS e.max ZirCAD (Ivoclar Vivadent), we can find Y-TZP stabiliized by a 3,4 or 5% addition of yttrium oxide (3Y-TZP, 4Y-TZP or 5Y-TZP). The mechanical properties are dependent on the chemical composition. For example, the flexural strength of ZirCad products decreases with the addition of yttrium oxide [64] (see Figure 3). For a long-term durable bond, a complex protocol is recommended by a producer [36]. Tooth-Supported Zirconia Single Crowns show good results in long-term in vivo observations [65] and general biocompatibility [66].

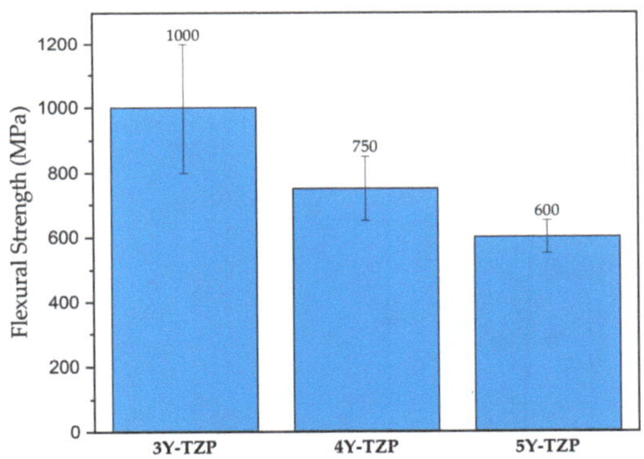

Figure 3. Flexural strength IPS e.max ZirCAD zirconium oxide ceramics varying in yttrium oxide content: 3%-3Y-TZP, 4%-4Y-TZP and 5%-5Y-TZP Adapted from ref. [64].

8.2. Polymer-Based Materials

Due to its mechanical properties and biocompatibility, poly(methyl methacrylate) (PMMA) was introduced as a CAD/CAM material (see Figure 4) for manufacturing prostheses [67,68]. Moreover, PMMA resin is among the oldest acrylic materials used in dentistry. The PMMA in CAD/CAM blocks occurs in cross-linked form (eg. Telio CAD, Ivoclar Vivadent), unlike conventional dental application where it is subjected to photocuring. The highly cross-linked nature of those materials puts them before conventionally polymerized interim resins in terms of durability and processing ease [69].

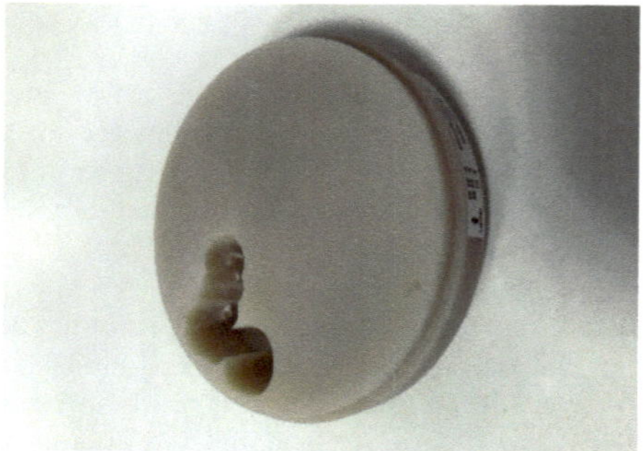

Figure 4. PMMA disk, reprinted with permission from ref. [44] (Copyright 2020 Inżynier i Fizyk Medyczny).

As stated in Alp et al. (2019), the CAD/CAM PMMA-based polymers exhibited greater flexural strength than bis-acrylate derivatives as well as conventional PMMA interim resin material [68]. Figure 5 (below) clearly shows that three CAD/CAM PMMA-based materials (M-PM-DISC, Polident PMMA and Telio CAD) exhibit greater flexural strength than Protemp 4 (Bis-acrylate composite resin) or Art Concept ArtDentine, which is conventional PMMA. During the storage of the prepolymerized PMMA-based polymer, the relaxation of the material occurs.

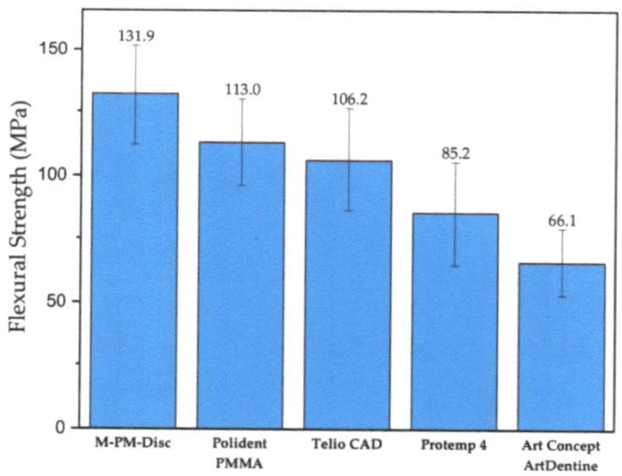

Figure 5. Flexural strength of different commercial interim PMMA-based resins. Adapted from ref. [68].

Among the other high-performance polymers utilized in the CAD/CAM systems for manufacturing we poly(ether ether ketone), commonly known as PEEK. Like many other polymeric materials used in dentistry, it has thermoplastic characteristics and increased biocompatibility. It is worth mentioning that PEEK has an elasticity modulus at the level of 3–4 GPa [70], which is very close to the modulus of Type 3 Spongeous bone [71]. This polymer was concluded to be applicable for removable prosthetics due to its mechanical properties, low discoloration and minimal monomer content when produced by means of CAD/CAM [72,73].

PMMA-based materials have been suggested as a long-term interim prosthesis [74–76]. By definition, this material should have adequate mechanical resistance to withstand the mechanical forces of the jaw during teeth collision.

8.3. Composites

With the development of CAD/CAM materials, and increasing doctor and patient expectations, hybrid structures, such as composite resins, hybrid ceramics or conventional materials with additives altering their physical properties, have been introduced to the market. Composites are known to be materials made of at least two substances, exhibiting properties from each of them. It is no surprise that regarding CAD/CAM, an additive for polymeric or ceramic materials is used in order to enhance the tribological or mechanical properties or simply the aesthetics of the prostheses. For example, 0.3–0.5 μm ceramic particles are added to PEEK for property optimization [77]. The division of such composites is based on their structure and manufacturing process. We can outline polymer-infiltrated ceramic networks (PICN) and resin composite blocks (RCB). The former consists of two phases: a ceramic scaffold and an interpenetrating polymeric network [42]. The latter are formed by transferring the filler into a monomer mixture [78].

A study on the resistance of resin composite crowns was performed using the example of the maxillary first molar, and lithium disilicate glass ceramic (IPS e.max CAD by Ivoclar vivadent) was utilized as a control reference [79]. In the same work, an evaluation of fracture strength and biaxial flexural strength showed satisfactory results. All CAD/CAM composite blocks that were tested (among them Shofu block HC by Shofu Inc., KZR-CAD HR by Yamamoto precious metal Co. and Katana avencia by Kuraray noritake dental) exhibited about 3–4 times higher fracture strength than the average maximum bite force of the molar tooth (700–900 N). An example of a composite CAD/CAM disk is presented in Figure 6 (below).

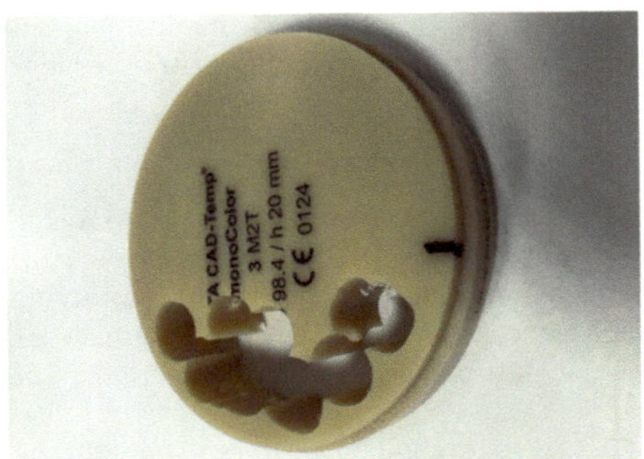

Figure 6. Vita CAD-Temp disk, reprinted with permission from ref. [44] (Copyright 2020 Inżynier i Fizyk Medyczny).

It is crucial to mention that dental composites vary in properties depending on composition. A comprehensive study conducted by Stawarczyk et al. [76] evaluated the mechanical as well as optical behaviour of commercial composites. The research concluded that regarding flexural strength, CAD/CAM composites generally exhibit higher flexural strength over leucite ceramic material, but lower than lithium silicate ceramic. The same study stated that glass ceramics showed a lower discoloration rate in comparison with CAD/CAM composites.

9. Nanoscale Aspect of CAD/CAM Materials

Nanotechnology has a wide range of applications in medicine and dentistry. By incorporating nanomaterials, one can alter the properties of different materials. It is possible to enhance the optical, chemical, electrical and mechanical properties of materials by adding suitable nanoparticles [80,81]. Nanosized filler particles are added to the resin-based composites to fill the space between larger filler particles and at the same time reduce the content of the resin [82].

When considering nanoparticles, safety is of vital importance. The intraoral CAD/CAM restorations as well as other dental materials are subjected to processes of wearing and degradation over time. Therefore, the issue of the release of nanoparticles over time and their possible negative effect on the organism needs to be addressed [83]. When scaling down to nanosized particles, we observe an increase in the surface area, which enhances their chemical activity in the scope of potential interactions inducing an adverse cellular response [84].

Among the materials utilized in CAD/CAM technology, we can outline resin matrix ceramics, also called nanoceramics [48]. This material was invented in order to meet the need of combining the ceramics' high aesthetic and mechanical advantages of composites. The examples are Cerasmart, GC Dental Products and Lava Ultimate, 3M ESPE, which owes its beneficial properties to nanotechnology used to bind nanoceramic particles into a resin matrix [49]. The producer describes its composition as 80% nanoceramic and 20% resin matrix [85] (average particles size: 20 nm for silica particles and 4–11 nm for zirconia particles) [60], while Ceramsmart is composed of 71 wt.% silica and barium glass nanoparticles [52]. Nanoceramics are considered optimal materials for restorations in aesthetic segments. Their high translucency is an effect of nanosized zirconia and silica particles that decrease light scattering [52,86]. Moreover, the addition of nanoparticles to composite resins improves the tensile and compressive strength, which contributes to reducing secondary caries by the elimination of microleakage [80].

Whereas nanotechnology in dentistry provides a wide array of possibilities in modifying the properties of the material (enhancing tribological and mechanical properties, lowering cytotoxicity), it is necessary that researchers and clinicians take into consideration its long-term effect and potential toxicity. Although nanoparticle addition is generally considered not to cause a negative toxicological response, both beneficial and adverse implications of nanotechnology in dentistry are thoroughly covered by R.N. Al Kahtani et al. [87].

10. Comparison of CAD/CAM Materials

Both resin composite and ceramic CAD/CAM materials have advantages and drawbacks for intraoral application. It is of vital importance to select the correct material depending on the patient's personal needs and considering mechanical and visual material characteristics. Resin composites are attractive because of their machinability and intraoral reparability, while glass ceramics/ceramics may offer superior mechanical and aesthetic properties [43].

The final dental restoration or prosthesis should resist occlusal forces that appear during the clinical service [68]. Moreover, the mechanical parameters of conventional interim materials change over time and are affected by wearing and their chemical characteristics (e.g., susceptible for water sorption [88]). We can outline multiple mechanical properties characterizing CAD/CAM materials: flexural strength, fatigue stress resistance, hardness and elastic modulus in addition to optical properties such as colour and translucency. Every major CAD/CAM material group is presented in the Table 2, along with their flexural strength, hardness, elastic modulus and composition.

Flexural strength is an important property giving an idea of the general mechanical strength and rigidity of the presented material with respect to dental application prosthesis [89]. This might lead to the conclusion that the higher the value is, the better. High flexural strength is essential to successful clinical procedure, but one must take into consideration other mechanical properties depending on the final application, i.e., onlay, inlay, crown, bridge or arch. Among those with the highest flexural strength, we can outline IPS e.max zirCAD (zirconium oxide ceramics, Ivoclar Vivadent, Schaan, Liechtenstein) with 850 MPa or VITA In-Ceram ALUMINA (aluminium oxide ceramics, VITA Zahnfabrik, Bad Säckingen, Germany) with 419 MPa. On the other hand, there are materials that exhibit very low flexural strength, such as VITA CAD-Temp (composite, VITA Zahnfabrik, Bad Säckingen, Germany) with 80 MPa or VITA ENAMIC (hybrid ceramic, VITA Zahnfabrik, Bad Säckingen, Germany) with 150–160 MPa.

The listed parameters clearly show the fluctuations in their hardness values for every type, e.g., aluminium oxide ceramics have 2035 HV hardness, while PMMA has just 26 HV—approximately two orders of magnitude more. Hardness is essential with regard to material wearing and scratching resistance. Hence, the ideal enamel replacement material would have similar hardness to the human teeth tissues, which is approximately 274.8 ± 18.1 HV [90]. We can see that the VITA ENAMIC hybrid ceramic exhibits a value of 200 HV, which indicates that it is a potential candidate for enamel substitution.

Materials are subjected to deformation after load. Those with a low elasticity modulus will be more strongly deformed than materials with a higher modulus [91]. This, yet again, indicates the necessity for clinical application. When mechanical load bearing or teeth mechanical collision is present, a restoration wears at a faster rate. A dentist might consider a material with a higher elastic modulus, such as VITA In-Ceram ALUMINA (aluminium oxide ceramics, VITA Zahnfabrik, Bad Säckingen, Germany) with 410 GPa [92] or IPS e.max CAD (lithium silicate ceramic, Ivoclar Vivadent, Schaan, Liechtenstein) with 103 GPa [93] for a longer lifespan of the restoration. As it is shown, every material has its weak points and advantages, and, therefore, the dentistry specialist needs to bear this in mind while selecting the right material for the purpose.

Fatigue stress has a significant influence on degradation and the material's fracture response. Mechanical degradation and water-assisted corrosion lead to a reduction in the stress intensity threshold for fracture initiation, as concluded in a comprehensive study on 8 commercial CAD/CAM materials. Moreover, the study gave an insight into their lifetime predictions, showcasing the maximum applied stress as the percentage of characteristic strength [94].

The advantage of CAD/CAM materials over the direct fabrication technique was confirmed in a study where fixed partial dentures made from three interim resin materials were stored in different conditions for 5000 thermocycles. The study concluded that the flexural strength of acrylate-based CAD/CAM-fabricated fixed partial dentures (Luxatemp AM Plus, Carcon Base PMMA, Trim) was greater than that of those manufactured from the same materials in a direct manner [67]. Although the fabrication method affected the maximum force at fracture values of CAD/CAM and directly fabricated fixed partial dentures, the scanning electron microscopy analysis did not show any porosities or voids that may affect the overall strength among specimens. Therefore, this could be explained by the higher load bearing capacity of the polymeric phase manufactured by means of CAD/CAM. Moreover, as we observe a major development in the field of nanomaterials and complexity in material composition (see Table 2), in the near future, the technology will allow for manufacturing additives for dental materials which will affect the mechanical properties in a selective manner for a tailored-fit CAD/CAM block.

Table 2. Mechanical properties and chemical composition of selected CAD/CAM materials.

Material Type	Product, Manufacturer	Flexural Strength (MPa)	Hardness (HV)	Elastic Modulus (GPa)	Composition	References
Composites	VITA CAD-Temp, Vita Zahnfabrik	80	n/a	2.8	Acrylate polymer with microparticle filler	[44,95]
Aluminium oxide ceramics	VITA In-Ceram ALUMINA, Vita Zahnfabrik	419	2035	410	Al_2O_3 (82 wt.%), La_2O_3 (12 wt.%), SiO_2 (4.5 wt.%), CaO (0.8 wt.%), other oxides (0.7 wt.%)	[96]
Zirconium oxide ceramics	IPS e.max zirCAD, Ivoclar Vivadent	1200	n/a	206.3	3 mol% Yttria-stabilized tetragonal zirconia polycrystals (3Y-TZP)	[93,97]
Lithium silicate ceramics	IPS e.max CAD, Ivoclar Vivadent	353.1	617	102.7	SiO_2, Li_2O, K_2O, P_2O_5, SiO_2, ZnO	[85,91,98,99]

Table 2. *Cont.*

Material Type	Product, Manufacturer	Flexural Strength (MPa)	Hardness (HV)	Elastic Modulus (GPa)	Composition	References
Leucite-Reinforced Glass Ceramics	IPS Empress CAD, Ivoclar Vivadent	160	632.2	62	SiO_2 (60–65 wt.%), Al_2O_3 (16–20 wt.%) K_2O (10–14 wt.%) Na_2O (3.5–6.5 wt.%), other oxides (0.5 wt.%), pigments	[100]
Resin-based ceramics	Lava Ultimate, 3M	200	96	12	Polymerizable resin, dispersed nanometric colloidal silica, ZrO_2 spherical particles	[48,85,93,101]
PMMA	Polident PMMA, Polident	114	26	2.77	PMMA, pigment	[102]
Hybrid cermics	VITA ENAMIC, Vita Zahnfabrik	150–160	200	30	SiO_2, Al_2O_3, Na_2O, K_2O, B_2O_3, ZrO_2, CaO, urethane dimethylacrylate, triethylene glycol dimethylacrylate	[50,85]
PEEK	PEEK-OPTIMA™, Invibio	165	n/a	3.70	PEEK	[103–105]

11. Adhesion–Bonding of CAD/CAM Restoration

Effective and stable bonding contributes to long-term high clinical success rates. Resin bonding and self-adhesive resin cements are vastly recommended for CAD/CAM restorations. A. Mine et al. created a review providing a broad outlook on the bonding procedures of CAD/CAM materials [106]. The study states that the bonding procedure should be preceded by generating microretentive surfaces by hydrofluoric acid etching. The next step of the procedure is silanization, the aim of which is ensuring chemical adhesion [107]. The study considers the presented bonding procedure regarding indirect resin composite materials (including Lava Ultimet, KATANA AVENCIA block, Gradia Block, Ceras- mart, Paradigm and Block HC) and CAD/CAM polymer-infiltrated ceramics (such as Vita Enamic) [107]. Authors of the aforementioned review also notice that it is possible to improve bonding to CAD/CAM PMMA resin materials by using materials containing MMA. These are general recommendations on bonding for most CAD/CAM materials, which may vary depending on the producer's recommendations and personal experience of clinical operators. As presented in another study from 2018 [108], resin bonding has long been the gold standard for the retention and reinforcement of silica-based ceramics. However, due to the complexity of the bonding procedure compared to conventional cementation, many dentists seek a simplified bonding protocol. The bonding procedure from the technical and scientific documentation of Vita Enamic can be presented as an example [50]. Authors propose the following protocol: Etching for 60 s with VITA CERAMICS ETCH (5% hydrofluoric acid gel) then silanizing with VITASIL, VITA or Monobond Plus, Ivoclar Vivadent. The next step is using the bonding composites RelyX Unicem (3M, Seefeld, Germany) and Variolink II (Ivoclar Vivadent, Schaan, Lichtenstein) in accordance with the manufacturer's instructions. This procedure ensures optimal bonding with compressive shear strength of approximately 20 MPa.

12. User Experience of CAD/CAM Hybrid Materials

In fact, VITA Enamic (VITA Zahnfabrik) seems to be an adequate material not only for crowns in the posterior segment but also in aesthetic segment or restorations with reduced wall thickness. This provides the possibility to create restorations based on implant frameworks. The structure of VITA Enamic is based on two interpenetrating networks—dominant ceramic mesh (86%) is reinforced by a polymer (14%). As mentioned, the structure of the hybrid material allows us to combine the beneficial properties of ceramics and composites. VITA Enamic has optimal flexibility, stress resistance and light conductivity, which provide better visual adaptation. To adjust the optical aspect, it is available in mono- and multichromatic types in three translucency levels. All these factors influence not only patient satisfaction but also make dentist work more comfortable. After designing, as an effect of milling, we can obtain anatomically precise restorations. It is worth mentioning that milling, which in our case was performed by the CEREC System, is faster (about 6.5 min for molar crown) for this material compared to others. According to the manufacturer, milling performed on Sirona MC XL (normal mode, Dentsply Sirona, New York City, NY, USA) for VITA Enamic blocks takes 7:56 min for the inlay, 7:10 min for the anterior crown and 9:07 for the posterior crown [50]. The next step after milling is the cutting of the sprue, polishing, preparing the inside surface by etching with hydrofluoric acid applying a silane and bonding agent and cementing using resin cement. To support the information provided by the manufacturer and our clinical experience, the study of flexural strength, flexural modulus, fracture strength and microhardness of different CAD/CAM materials may be referenced. Researchers compared IPS e.max CAD (lithium disilicate), VITA Suprinity (zirconia- reinforced lithium silicate), GC Cerasmart (hybrid high-performance polymer composite resin) and VITA Enamic. As a result of that study, all tested materials were considered as suitable for posterior full-crown restorations, but hybrid materials showed lower hardness and stiffness compared to glass ceramics [109]. Similar properties were presented in another study, which showed that the microhardness, flexural strength and fracture toughness of VITA Enamic are similar compared to IPS e.max CAD (Ivoclar Vivadent), IPS Empress CAD (Ivoclar Vivadent) and VITA Mark II (VITA Zahnfabrik) [110].

13. Surface Finishing and Disguise of CAD/CAM Restorations.

Polishing and glazing of CAD/CAM restorations as the finishing elements is crucial for smoothing the surface of the restoration and preserving the optical properties of the material. After milling the ceramic and composite resin blocks by diamond burs, the surface becomes rough. The roughness of the surface not only increases the level of biofilm accumulation [111,112], which may negatively affect gum tissues, but also facilitates discoloration [113,114]. There is also a number of mechanical consequences of surface roughness, such as decreasing mechanical resistance [115] or causing wearing of the opposing dentition. To obtain a smoother surface of the restoration, after milling, a polishing is necessary. This can be performed using different disks, polishing kits and pastes depending on the material or the operator's preference. Finopol Diamond Polisher rubber wheels (Finopol, Praha, Czech Republic) or Hatho Habbras Discs (Hatho, Eschbach, Germany) are examples of utilized instruments. Glaze application, according to the manufacturer's instructions, is a recommended procedure for ceramic CAD/CAM restorations. Afterwards, it is fired in a ceramic oven. Examples of glaze products are Vita Akzent Fluid and Vita Glaze AKZ 25, Empress Universal Glaze and IPS Empress Glaze Paste [116]. Glazing is also used to obtain the effect of advanced characterization. After the restoration is cleaned, dried and fixed, the operator can apply small amounts of shades and stains at the surface of the restoration. To obtain the desired colours and workable consistency, it is necessary to mix glaze liquid and different shades and stains on the pad. Glaze should be applied by a dedicated brush in a very subtle way, considering that the firing process results in a more intense appearance. It is recommended to apply a small amount of incisal shade to the incisal region and cusps in order to increase translucency. To give a restoration a more natural look, it is possible to apply warm colours to the area of the central fossa. After this characterization, short bursts of glaze spray in two or more series on the axial walls following the occlusal wall should be applied. Once a restoration is glazed and dried, the operator may proceed to the firing procedure using an appropriate programme. The subsequent steps are the cooling, cleaning and cementation of the restoration. The glazing procedure was summarised in the example of IPS e.max CAD Glaze given by the manufacturer Ivoclar Vivadent [117].

A study investigating the colour stability of zirconia-reinforced lithium silicate ceramic and lithium disilicate glass ceramic in beverages after two months showed that the glaze procedure led to enhanced colour stability. Moreover, authors noticed that due to the polishing, all changes in colour were clinically acceptable [113]. Another study on CAD/CAM materials' surface roughness proves that mechanical polishing is capable of reducing the surface roughness, while glazing is facultative for fully crystallized/polymerized materials and desirable for partially crystallized materials, such as lithium-based ceramics to reduce the effect of roughening by milling [116]. The study by Tekçe N et al. on the surface of selected CAD/CAM resin restorative materials also results in the conclusion that glazing the surface contributes to its smoothing [118]. Similar conclusions were made by Vichi et al. in a study, where the roughness and gloss of the surface of VITA Suprinity and IPS e.max CAD (silica-based glass ceramics) were assessed after finishing and polishing. Authors claim that the most effective procedures for lowering the roughness and yielding the highest gloss of tested CAD-CAM materials are manual finishing and polishing for 60 s and applying glazing paste [119]. On the other hand, when glaze is not applied or the restoration requires additional adjustment in a patient's oral cavity after the finishing procedure, faster plaque accumulation and discoloration may occur [119]. There is also a theory that finishing and polishing the surface of hybrid ceramic materials may negatively affect the physical properties of the restoration [120]. Marrelli et al. devote a lot of attention to surface roughness, which may influence the mechanical strength of zirconia-based CAD/CAM crowns and bridges. In their study, it is also proved that colouring using a commercial colouring liquid (Zirkon Zahn) has no significant effects on the mechanical strength of the zirconium ceramic-based specimens (the example of Kavo Everest Bio ZS Blank) [121].

14. Biocompatibility and Cytotoxicity of CAD/CAM Materials

It is extremely important for the new dental restoration to adapt to the conditions in the oral cavity, not only regarding their shape but with mechanical and physical properties. The crucial quality in the biological aspect is the compatibility with the surrounding tissues. Biocompatibility is an interdisciplinary phenomenon which covers biological, chemical and physical interactions and is highly connected with the concept of cytotoxicity—mainly in terms of the cellular response.

In addition to possessing the mechanical properties, as well as chemical and thermal resistance comparable to human bone, a CAD/CAM material needs to be biocompatible with the surrounding tissues. A material is expected not to cause any irritation, swelling or any kind of intolerance in the oral cavity. Therefore, a potential material needs to be evaluated in the scope of biocompatibility.

Human biofilm contains about 1000 species of bacteria [122] that adhere not only to the surface of the teeth but also to a prosthetic restoration. This adhesion depends on the type and roughness of the material's surface. The adhesion and development of microorganisms being part of biofilm on different materials used for CAD/CAM was studied. As a result, it can be claimed that tested sintered materials, such as IPS e.max and polished IPS e.max, showed the best "anti-adhesive properties" with respect to *Streptococcus mutans* and *Lactobacillus rhamnosus* [99]. In another study, materials such as VITA CAD-, Celtra Duo, IPS e.max CAD and VITA YZ were tested to determine the cytotoxic effects and collagen type I secretions on human gingival fibroblasts. Results show that after 72 h, all groups reached biologically acceptable levels of cytotoxic potential. Moreover, it is concluded that the ceramic materials (lithium disilicate) present a better cell response than the polymers [123].

In terms of polymeric materials, the biocompatibility assessment of PEEK has been conducted since the early 1990s, and in vivo biocompatibility has been assessed as positive [124].

Acrylates are another vast branch of materials used in dentistry and specifically in CAD/CAM technology. They are known for their allergy potential. However, it is shown that due to the material processing, this effect can be minimized, and the general biocompatibility is sufficient for dental application. The aforementioned biocompatibility is confirmed by a very small and practically undetectable level of residual toxic monomer in the samples evaluated in vivo and in vitro tests [77].

Considering composites as a material for CAD/CAM blocks, there is a vital aspect of monomer release which depends on the degree of polymerization and further degradation. Depending on the degree of conversion and monomer composition, bisphenol dental composites can release bisphenol A (BPA; low weight monomers, such as HEMA and TEGDMA; high weight monomers, such as Bis-GMA and UDMA; and additives, such as free radicals and photoinitiator molecules) [98]. Composite blocks currently available for CAD/CAM technology are characterized by more applicable biocompatibility properties. They exhibit a higher degree of conversion, utilize less toxic monomers in their composition and lack in photoinitiators [98]. There are studies performed on RCB, Lava Ultimate (LAVA) and Vita Enamic (VITA) and Paradigm MZ100 (MZ100) apparatus, which compare resin blocks for CAD/CAM and conventional composites, which prove that no monomer elution is seen from the RCB. However, the authors stipulate that resin blocks for CAD/CAM show some worrisome results regarding cytotoxicity, and they require more studies [125].

Another evaluation of resin material was performed in regard to human gingival fibroblast response. Materials were divided into several groups with different chemical compositions and fabricating methods. CAD/CAM materials in the study were represented by Yamahachi dental materials containing poly(methyl methacrylate) polymer and a prefabricated hybrid ceramic block Vita Enamic, which is a polymer-infiltrated ceramic and was used as a control group. The results show that poly(methyl methacrylate) and bis-acryl have lower cytotoxicity to human gingival fibroblasts than poly(ethyl methacrylate). Moreover, CAD/CAM restorations, as they are prefabricated from resin blocks, prevent residual monomer and achieve high cell attachment [126]. Another study conducted on gingival cells proved no significant difference in CAD/CAM block cytotoxicity [78].

Studies on the biocompatibility of lithium metasilicate glass ceramics utilized in the CAD/CAM technique were performed. The material was not assessed as cytotoxic with the usage of methyl tetrazolium salt and Alizarin Red. It showed the best cellular adhesion and proliferation after 21 days [127].

15. Current Demand for the CAD/CAM Restorations

As already mentioned, CAD/CAM technology is constantly improving and gains popularity among dental offices and their patients. However, as an expensive and innovative technology, it is not available for everyone. Moreover, if the dentist does not perform a large volume of restorations, the CAD/CAM system investment will not pay off [39]. Most dental practices offering restorations made using CAD/CAM are located in high-income areas, such as Western Europe and the United States of America [41]. In 2016, more than 30,000 dentists around the world owned scanning and milling machines. They are most popular in the United States and Canada where almost one-third of all CAD/CAM devices are used. Moreover, all over the world, more than 15 million CEREC restorations alone have been completed [32]. In Poland, there are around 30 dental practices that use the CAD/CAM Cerec System [128]. At this point, it is worth mentioning that in Poland, the country of origin of this review, dental services provided using CAD/CAM technology are not part of the benefits reimbursed by the National Health Found. Remaining only at the field of private dentistry, they are available for the part of the society who can afford them.

In many cases, it is not considered that the final effect depending on long-term clinical success and patients' quality of life with this type of prosthetic restoration may outperform other clinical approaches. There are many clinical studies proving the long-term success of CAD/CAM restorations [108,129,130], which can confirm this thesis.

16. Conclusions

There is a wide range of materials used in CAD/CAM, including polymers, ceramics and composites, which are becoming more accessible and easier to handle. This review presented modern materials for CAD/CAM along with their characterization and highlighted their mechanical and clinical properties enabling satisfactory long-term restorations. Furthermore, high biocompatibility and aesthetic properties are subsequent advantages of the described material group.

New generations of powder-free intraoral scanners with greater resolution are being developed; hence, the restorations are designed faster and more accurately for a higher clinical success rate and prolonged longevity. The technology is gaining a strong position in dentistry, especially in the field of fixed partial dentures and crowns. Undeniably, the technology has its downsides, including the cost (of both materials and equipment) as well as the need for highly qualified personnel. Moreover, the correct selection of a material requires an experienced clinician. The technique of producing prosthetic restorations must always be adapted to every patient personally. For example, in the case of maxillo-mandibular relation disorders, the inability to define an occlusal plan may occur. Another limitation is the inaccurate horizontal and occlusal vertical dimension. Each clinical case has to be considered individually, and material and methods have to be chosen in order to meet the patient's personal needs.

We can observe an increased number of composite materials on the market due to the possibility of mixing the properties of both the polymer, ceramic matrix and other filler particles. The comparison and selection of the correct type of material allows us to provide the patient with restorations characterized by acceptable and satisfying durability, biocompatibility and aesthetics. There are multiple mechanical and clinical parameters describing the materials, many of which are given by the manufacturer or are possible to find in the literature (e.g., provided in this article).

In the field of in situ restoration materials, there is still a necessity for advancement in terms of developing materials with superior properties to the contemporary used materials, as well as conducting long-term studies of the biocompatibility and wear of multiple materials in vivo. The aspect of surface finishing is of particular interest in the industry, as it is proved to affect the mechanical properties of the restoration. The application of CAD/CAM in dentistry provides state-of-art dental care. Hence, it is vital for the CAD/CAM framework in dentistry to be developed for further benefit of the patients.

Author Contributions: Conceptualization, M.D. and Z.R.; methodology, P.P.; validation, M.D., Z.R. and M.S.; formal analysis, M.D. and Z.R.; investigation, A.S.; resources, A.S. and P.P.; data curation, P.P.; writing—original draft preparation, A.S. and P.P.; writing—review and editing, M.D. and M.S.; visualization, P.P.; supervision, M.D., Z.R. and M.S.; project administration, M.D., Z.R. All authors have read and agreed to the published version of the manuscript.

Funding: This research received no external funding.

Institutional Review Board Statement: Not applicable.

Informed Consent Statement: Not applicable.

Data Availability Statement: The data presented in this study are available in the article.

Conflicts of Interest: The authors declare no conflict of interest. The funders had no role in the design of the study; in the collection, analyses, or interpretation of data; in the writing of the manuscript, or in the decision to publish the results.

References

1. Boboulos, M.A. *CAD-CAM & Rapid Prototyping Application Evaluation*; Bookboon: London, UK, 2010; ISBN 9788776816766.
2. Davidowitz, G.; Kotick, P.G. The Use of CAD/CAM in Dentistry. *Dent. Clin. N. Am.* **2011**, *55*, 559–570. [CrossRef]
3. Okoński, P.; Lasek, K.; Mierzwińska-Nastalska, E. Kliniczne zastosowanie wybranych materiałów ceramicznych. *Protet. Stomatol.* **2012**, *62*, 181–189.
4. Duret, F.; Blouin, J.L.; Duret, B. CAD-CAM in dentistry. *J. Am. Dent. Assoc.* **1988**, *117*, 715–720. [CrossRef]
5. Fasbinder, D.J. Materials for chairside CAD/CAM restorations. *Compend. Contin. Educ. Dent.* **2010**, *31*, 702–710. [PubMed]
6. Mörmann, W.H. The evolution of the CEREC system. *J. Am. Dent. Assoc.* **2006**, *137*, 7–13. [CrossRef]
7. Rekow, D. Computer-aided design and manufacturing in dentistry: A review of the state of the art. *J. Prosthet. Dent.* **1987**, *58*, 512–516. [CrossRef]
8. Zaruba, M.; Mehl, A. Chairside systems: A current review Chairside-Systeme: Eine aktuelle Übersicht. *Int. J. Comput. Dent.* **2017**, *20*, 123–149.
9. Joda, T.; Brägger, U. Patient-centered outcomes comparing digital and conventional implant impression procedures: A randomized crossover trial. *Clin. Oral Implant. Res.* **2016**, *27*, e185–e189. [CrossRef] [PubMed]
10. Gjelvold, B.; Chrcanovic, B.R.; Korduner, E.K.; Collin-Bagewitz, I.; Kisch, J. Intraoral Digital Impression Technique Compared to Conventional Impression Technique. A Randomized Clinical Trial. *J. Prosthodont.* **2016**, *25*, 282–287. [CrossRef]
11. Yuzbasioglu, E.; Kurt, H.; Turunc, R.; Bilir, H. Comparison of digital and conventional impression techniques: Evaluation of patients' perception, treatment comfort, effectiveness and clinical outcomes. *BMC Oral Health* **2014**, *14*, 1–7. [CrossRef]
12. Giordano, R. Materials for chairside CAD/CAM-produced restorations. *J. Am. Dent. Assoc.* **2006**, *137*, 14S–21S. [CrossRef]
13. Li, R.W.K.; Chow, T.W.; Matinlinna, J.P. Ceramic dental biomaterials and CAD/CAM technology: State of the art. *J. Prosthodont. Res.* **2014**, *58*, 208–216. [CrossRef] [PubMed]
14. Papadiochou, S.; Pissiotis, A.L. Marginal adaptation and CAD-CAM technology: A systematic review of restorative material and fabrication techniques. *J. Prosthet. Dent.* **2018**, *119*, 545–551. [CrossRef] [PubMed]
15. Abdullah, A.O.; Pollington, S.; Liu, Y. Comparison between direct chairside and digitally fabricated temporary crowns. *Dent. Mater. J.* **2018**, *37*, 957–963. [CrossRef]
16. Ahmed, K.E. We're Going Digital: The Current State of CAD/CAM Dentistry in Prosthodontics. *Prim. Dent. J.* **2018**, *7*, 30–35. [CrossRef] [PubMed]
17. Kuźniar-Folwarczny, A.; Sulewski, M.; Błaszczyk, A.; Sulewska, A.; Kosior, P.; Dobrzyński, M. Charakterystyka ceramicznych materiałów stomatologicznych stosowanych w gabinetowych systemach CAD/CAM, 2019, 8, 479–487. *Inż. Fiz. Med.* **2019**, *8*, 479–487.
18. Miyazaki, T.; Hotta, Y.; Kunii, J.; Kuriyama, S.; Tamaki, Y. A review of dental CAD/CAM: Current status and future perspectives from 20 years of experience. *Dent. Mater. J.* **2009**, *28*, 44–56. [CrossRef] [PubMed]
19. Otto, T.; Dent, M. Case series Clinical Results from a Long Term Case Series using Chairside CEREC CAD-CAM Inlays and Onlays. *Int. J. Prosthodont.* **2008**, *21*, 53–59.
20. Sjögren, G.; Molin, M.; Van Dijken, J.W.V. A 5-year clinical evaluation of ceramic inlays (Cerec) cemented with a dual-cured or chemically cured resin composite luting agent. *Acta Odontol. Scand.* **1998**, *56*, 263–267. [CrossRef]
21. Zimmermann, M.; Koller, C.; Reymus, M.; Mehl, A.; Hickel, R. Clinical Evaluation of Indirect Particle-Filled Composite Resin CAD/CAM Partial Crowns after 24 Months. *J. Prosthodont.* **2018**, *27*, 694–699. [CrossRef]
22. Tunac, A.T.; Celik, E.U.; Yasa, B. Two-year performance of CAD/CAM fabricated resin composite inlay restorations: A randomized controlled clinical trial. *J. Esthet. Restor. Dent.* **2019**, *31*, 627–638. [CrossRef] [PubMed]
23. Alghazzawi, T.F. Advancements in CAD/CAM technology: Options for practical implementation. *J. Prosthodont. Res.* **2016**, *60*, 72–84. [CrossRef] [PubMed]
24. Rahim Hussain, D. CAD CAM In Restorative Dentistry, College of Dentistry. Maser's Thesis, University of Baghdad, Baghdad, Iraq, 2018.
25. Jia-Mahasap, W. Comparative Assessment of Conventional vs. CAD/CAM Complete Denture Fabrication Techniques on Patient Satisfaction, Quality of Life and Prosthesis Biofilm. Master's Thesis, University of Iowa, Iowa City, IA, USA, 2017.
26. Bilgin, M.S.; Baytaroğlu, E.N.; Erdem, A.; Dilber, E. A review of computer-aided design/computer-aided manufacture techniques for removable denture fabrication. *Eur. J. Dent.* **2016**, *10*, 286–291. [CrossRef] [PubMed]
27. Saponaro, P.C.; Yilmaz, B.; Heshmati, R.H.; McGlumphy, E.A. Clinical performance of CAD-CAM-fabricated complete dentures: A cross-sectional study. *J. Prosthet. Dent.* **2016**, *116*, 431–435. [CrossRef] [PubMed]
28. Vecsei, B.; Joós-Kovács, G.; Borbély, J.; Hermann, P. Comparison of the accuracy of direct and indirect three-dimensional digitizing processes for CAD/CAM systems—An In Vitro study. *J. Prosthodont. Res.* **2017**, *61*, 177–184. [CrossRef]
29. Pallesen, U.; Van Dijken, J.W.V. An 8-year evaluation of sintered ceramic and glass ceramic inlays processed by the Cerec CAD/CAM system. *Eur. J. Oral Sci.* **2000**, *108*, 239–246. [CrossRef] [PubMed]
30. Sjögren, G.; Molin, M.; van Dijken, J.W. A 10-year prospective evaluation of CAD/CAM-manufactured (Cerec) ceramic inlays cemented with a chemically cured or dual-cured resin composite. *Int. J. Prosthodont.* **2004**, *17*, 241–246.
31. Zimmer, S.; Göhlich, O.; Rüttermann, S.; Lang, H.; Raab, W.H.M.; Barthel, C.R. Long-term survival of cerec restorations: A 10-year study. *Oper. Dent.* **2008**, *33*, 484–487. [CrossRef]

32. Awada, A.; Nathanson, D. Mechanical properties of resin-ceramic CAD/CAM restorative materials Presented at the American Association of Dental Research/Canadian Association of Dental Research Annual Meeting, Charlotte, NC, March 2014. *J. Prosthet. Dent.* **2015**, *114*, 587–593. [CrossRef]
33. Popa, D.; Bordea, I.R.; Burde, A.V.; Crişan, B.; Câmpian, R.S.; Constantiniuc, M. Surface modification of zirconia after laser irradiation. *Optoelectron. Adv. Mater. Rapid Commun.* **2016**, *10*, 785–788.
34. Romanyk, D.L.; Martinez, Y.T.; Veldhuis, S.; Rae, N.; Guo, Y.; Sirovica, S.; Fleming, G.J.P.; Addison, O. Strength-limiting damage in lithium silicate glass-ceramics associated with CAD–CAM. *Dent. Mater.* **2019**, *35*, 98–104. [CrossRef]
35. Uzun, G. An overview of dental cad/cam systems. *Biotechnol. Biotechnol. Equip.* **2008**, *22*, 530–535. [CrossRef]
36. Blatz, M.B.; Conejo, J. The Current State of Chairside Digital Dentistry and Materials. *Dent Clin. N. Am.* **2019**, *63*, 175–197. [CrossRef] [PubMed]
37. Steinmassl, P.-A.; Klaunzer, F.; Steinmassl, O.; Dumfahrt, H.; Grunert, I. Evaluation of Currently Available CAD/CAM Denture Systems. *Int. J. Prosthodont.* **2017**, *30*, 116–122. [CrossRef]
38. Hategan, S.I.; Ionel, T.F.; Goguta, L.; Gavrilovici, A.; Negrutiu, M.L.; Jivanescu, A. Powder and Powder-Free Intra-Oral Scanners: Digital Impression Accuracy. *Prim. Dent. J.* **2018**, *7*, 40–43. [CrossRef] [PubMed]
39. Abdullah, A.; Muhammed, F.; Zheng, B.; Liu, Y. An Overview of Computer Aided Design/Computer Aided Manufacturing (CAD/CAM) in Restorative Dentistry. *J. Dent. Mater. Tech.* **2018**, *7*, 1–10. [CrossRef]
40. Ender, A.; Mehl, A. In-Vitro evaluation of the accuracy of conventional and digital methods of obtaining full-arch dental impressions. *Quintessence Int.* **2015**, *46*, 9–17. [CrossRef]
41. Beuer, F.; Schweiger, J.; Edelhoff, D. Digital dentistry: An overview of recent developments for CAD/CAM generated restorations. *Br. Dent. J.* **2008**, *204*, 505–511. [CrossRef] [PubMed]
42. Lambert, H.; Durand, J.C.; Jacquot, B.; Fages, M. Dental biomaterials for chairside CAD/CAM: State of the art. *J. Adv. Prosthodont.* **2017**, *9*, 486–495. [CrossRef]
43. Ruse, N.D.; Sadoun, M.J. Resin-composite blocks for dental CAD/CAM applications. *J. Dent. Res.* **2014**, *93*, 1232–1234. [CrossRef] [PubMed]
44. Rogula, J.; Kuźniar-Folwarczny, A.; Sulewski, M.; Błaszczyk, A. Charakterystyka kompozytowych materiałów stomatologicznych stosowanych w gabinetowych systemach CAD/CAM. *Inżynier i Fizyk Medyczny* **2020**, *9*, 57–61.
45. Charlton, D.G.; Roberts, H.W.; Tiba, A. Measurement of select physical and mechanical properties of 3 machinable ceramic materials. *Quintessence Int.* **2008**, *39*, 573–579.
46. Kelly, J.R.; Nishimura, I.; Campbell, S.D. Ceramics in dentistry: Historical roots and current perspectives. *J. Prosthet. Dent.* **1996**, *75*, 18–32. [CrossRef]
47. Datla, S.R.; Alla, R.K.; Alluri, V.R.; Babu, J.P.; Konakanchi, A. Dental Ceramics: Part II–Recent Advances in Dental Ceramics. *Am. J. Mater. Eng. Technol.* **2015**, *3*, 19–26. [CrossRef]
48. Spitznagel, F.A.; Boldt, J.; Gierthmuehlen, P.C. CAD/CAM Ceramic Restorative Materials for Natural Teeth. *J. Dent. Res.* **2018**, *97*, 1082–1091. [CrossRef]
49. 3M. LavaTM Ultimate CAD/CAM Restorative for CEREC. Available online: .https://www.3m.com/3M/en_US/company-us/all-3m-products/~{}/lava-ultimate-cerec-3M-Lava-Ultimate-CAD-CAM-Restorative-for-CEREC-/?N=5002385+3294768924&rt=rud (accessed on 14 March 2021).
50. Vita Zahnfabrik. *Vita Enamic -Technical and Scientific Documentation*. 2019. Available online: https://www.vita-zahnfabrik.com/en/VITA-ENAMIC-17427.html (accessed on 14 March 2021).
51. Thurmond, J.W.; Barkmeier, W.W.; Wilwerding, T.M. Effect of porcelain surface treatments on bond strengths of composite resin bonded to porcelain. *J. Prosthet. Dent.* **1994**, *72*, 355–359. [CrossRef]
52. Kurtulmus-Yilmaz, S.; Cengiz, E.; Ongun, S.; Karakaya, I. The Effect of Surface Treatments on the Mechanical and Optical Behaviors of CAD/CAM Restorative Materials. *J. Prosthodont.* **2019**, *28*, e496–e503. [CrossRef] [PubMed]
53. Klosa, K.; Boesch, I.; Kern, M. Long-term bond of glass ceramic and resin cement: Evaluation of titanium tetrafluoride as an alternative etching agent for lithium disilicate ceramics. *J. Adhes. Dent.* **2013**, *15*, 377–37783. [CrossRef]
54. El-Mowafy, O.; Brochu, J.F. Longevity and clinical performance of IPS-Empress ceramic restorations—A literature review. *J. Can. Dent. Assoc.* **2002**, *68*, 233–237. [PubMed]
55. Mayinger, F.; Lümkemann, N.; Musik, M.; Eichberger, M.; Stawarczyk, B. Comparison of mechanical properties of different reinforced glass-ceramics. *J. Prosthet. Dent.* **2020**, 1–8. [CrossRef]
56. Tysowsky, G.W. The science behind lithium disilicate: A metal-free alternative. *Dent. Today* **2009**, *28*, 112–113. [PubMed]
57. Pieger, S.; Salman, A.; Bidra, A.S. Clinical outcomes of lithium disilicate single crowns and partial fixed dental prostheses: A systematic review. *J. Prosthet. Dent.* **2014**, *112*, 22–30. [CrossRef] [PubMed]
58. Schultheis, S.; Strub, J.R.; Gerds, T.A.; Guess, P.C. Monolithic and bi-layer CAD/CAM lithium-disilicate versus metal-ceramic fixed dental prostheses: Comparison of fracture loads and failure modes after fatigue. *Clin. Oral Investig.* **2013**, *17*, 1407–1413. [CrossRef]
59. Reich, S.; Endres, L.; Weber, C.; Wiedhahn, K.; Neumann, P.; Schneider, O.; Rafai, N.; Wolfart, S. Three-unit CAD/CAM-generated lithium disilicate FDPs after a mean observation time of 46 months. *Clin. Oral Investig.* **2014**, *18*, 2171–2178. [CrossRef] [PubMed]
60. Eldwakhly, E.; Ahmed, D.R.M.; Soliman, M.; Abbas, M.M.; Badrawy, W. Color and translucency stability of novel restorative CAD/CAM materials. *Dent. Med. Probl.* **2019**, *56*, 349–356. [CrossRef]

61. Conejo, J.; Nueesch, R.; Vonderheide, M.; Blatz, M.B. Clinical Performance of All-Ceramic Dental Restorations. *Curr. Oral Heal. Reports* **2017**, *4*, 112–123. [CrossRef]
62. Raigrodski, A.J. Clinical and laboratory considerations for the use of CAD/CAM Y-TZP-based restorations. *Pract. Proced. Aesthet. Dent.* **2003**, *15*, 469–476; quiz 477.
63. Kelly, P.M.; Rose, L.R.F. The martensitic transformation in ceramics - Its role in transformation toughening. *Prog. Mater. Sci.* **2002**, *47*, 463–557. [CrossRef]
64. Gebhardr, A.; Todd, J.C. IPS e.max ZirCAD - Scientific Documentation. Available online: https://asia.ivoclarvivadent.com/zoolu-website/media/document/42081/IPS+e-max+ZirCAD (accessed on 5 November 2020).
65. Rojas Vizcaya, F. Retrospective 2- to 7-Year Follow-Up Study of 20 Double Full-Arch Implant-Supported Monolithic Zirconia Fixed Prostheses: Measurements and Recommendations for Optimal Design. *J. Prosthodont.* **2018**, *27*, 501–508. [CrossRef] [PubMed]
66. Mallineni, S.K.; Nuvvula, S.; Matinlinna, J.P.; Yiu, C.K.; King, N.M. Biocompatibility of various dental materials in contemporary dentistry: A narrative insight. *J. Investig. Clin. Dent.* **2013**, *4*, 9–19. [CrossRef]
67. Alt, V.; Hannig, M.; Wöstmann, B.; Balkenhol, M. Fracture strength of temporary fixed partial dentures: CAD/CAM versus directly fabricated restorations. *Dent. Mater.* **2011**, *27*, 339–347. [CrossRef] [PubMed]
68. Alp, G.; Murat, S.; Yilmaz, B. Comparison of Flexural Strength of Different CAD/CAM PMMA-Based Polymers. *J. Prosthodont.* **2019**, *28*, e491–e495. [CrossRef]
69. Güth, J.F.; e Silva, J.S.A.; Beuer, F.; Edelhoff, D. Enhancing the predictability of complex rehabilitation with a removable CAD/CAM-fabricated long-term provisional prosthesis: A clinical report. *J. Prosthet. Dent.* **2012**, *107*, 1–6. [CrossRef]
70. Rodriguez, F.; Cohen, C.; Ober, C.K.; Archer, L. *Principles of Polymer Systems*; CRC Press: Boca Raton, FL, USA, 2014.
71. Matarasso, S.; Rasperini, G.; Iorio Siciliano, V.; Salvi, G.E.; Lang, N.P.; Aglietta, M. A 10-year retrospective analysis of radiographic bone-level changes of implants supporting single-unit crowns in periodontally compromised vs. periodontally healthy patients. *Clin. Oral Implant. Res.* **2010**, *21*, 898–903. [CrossRef]
72. Taufall, S.; Eichberger, M.; Schmidlin, P.R.; Stawarczyk, B. Fracture load and failure types of different veneered polyetheretherketone fixed dental prostheses. *Clin. Oral Investig.* **2016**, *20*, 2493–2500. [CrossRef]
73. Tekin, S.; Cangül, S.; Adıgüzel, Ö.; Değer, Y. Areas for use of PEEK material in dentistry. *Int. Dent. Res.* **2018**, *8*, 84–92. [CrossRef]
74. Güth, J.F.; Almeida E Silva, J.S.; Ramberger, M.; Beuer, F.; Edelhoff, D. Treatment concept with CAD/CAM-fabricated high-density polymer temporary restorations. *J. Esthet. Restor. Dent.* **2012**, *24*, 310–318. [CrossRef] [PubMed]
75. Bähr, N.; Keul, C.; Edelhoff, D.; Eichberger, M.; Roos, M.; Gernet, W.; Stawarczyk, B. Effect of different adhesives combined with two resin composite cements on shear bond strength to polymeric CAD/CAM materials. *Dent. Mater. J.* **2013**, *32*, 492–501. [CrossRef]
76. Stawarczyk, B.; Liebermann, A.; Eichberger, M.; Güth, J.F. Evaluation of mechanical and optical behavior of current esthetic dental restorative CAD/CAM composites. *J. Mech. Behav. Biomed. Mater.* **2016**, *55*, 1–11. [CrossRef]
77. Ardelean, L.C.; Bortun, C.M.; Podariu, A.C.; Rusu, L.C. Acrylates and Their Alternatives in Dental Applications. *Acrylic Polym. Healthc.* **2017**. [CrossRef]
78. Alamoush, R.A.; Kushnerev, E.; Yates, J.M.; Satterthwaite, J.D.; Silikas, N. Response of two gingival cell lines to CAD/CAM composite blocks. *Dent. Mater.* **2020**, *36*, 1214–1225. [CrossRef]
79. Okada, R.; Asakura, M.; Ando, A.; Kumano, H.; Ban, S.; Kawai, T.; Takebe, J. Fracture strength testing of crowns made of CAD/CAM composite resins. *J. Prosthodont. Res.* **2018**, *62*, 287–292. [CrossRef]
80. Bapat, R.A.; Joshi, C.P.; Bapat, P.; Chaubal, T.V.; Pandurangappa, R.; Jnanendrappa, N.; Gorain, B.; Khurana, S.; Kesharwani, P. The use of nanoparticles as biomaterials in dentistry. *Drug Discov. Today* **2019**, *24*, 85–98. [CrossRef] [PubMed]
81. Mitra, S.B.; Wu, D.; Holmes, B.N. An application of nanotechnology in advanced dental materials. *J. Am. Dent. Assoc.* **2003**, *134*, 1382–1390. [CrossRef] [PubMed]
82. Schmalz, G.; Hickel, R.; van Landuyt, K.L.; Reichl, F.X. Nanoparticles in dentistry. *Dent. Mater.* **2017**, *33*, 1298–1314. [CrossRef] [PubMed]
83. Suh, W.H.; Suslick, K.S.; Stucky, G.D.; Suh, Y.H. Nanotechnology, nanotoxicology, and neuroscience. *Prog. Neurobiol.* **2009**, *87*, 133–170. [CrossRef]
84. Ai, J.; Biazar, E.; Jafarpour, M.; Montazeri, M.; Majdi, A.; Aminifard, S.; Zafari, M.; Akbari, H.R.; Rad, H.G. Nanotoxicology and nanoparticle safety in biomedical designs. *Int. J. Nanomed.* **2011**, *6*, 1117. [CrossRef]
85. Ludovichetti, F.S.; Trindade, F.Z.; Werner, A.; Kleverlaan, C.J.; Fonseca, R.G. Wear resistance and abrasiveness of CAD-CAM monolithic materials. *J. Prosthet. Dent.* **2018**, *120*, 318–e1. [CrossRef]
86. Lee, Y.K. Influence of scattering/absorption characteristics on the color of resin composites. *Dent. Mater.* **2007**, *23*, 124–131. [CrossRef]
87. AlKahtani, R.N. The implications and applications of nanotechnology in dentistry: A review. *Saudi Dent. J.* **2018**, *30*, 107–116. [CrossRef] [PubMed]
88. Thompson, G.A.; Luo, Q. Contribution of postpolymerization conditioning and storage environments to the mechanical properties of three interim restorative materials. *J. Prosthet. Dent.* **2014**, *112*, 638–648. [CrossRef]

89. Naveen, K.S.; Singh, J.P.; Viswambaran, C.M.; Dhiman, R.K. Evaluation of flexural strength of resin interim restorations impregnated with various types of silane treated and untreated glass fibres. *Med. J. Armed Forces India* **2015**, *71*, S293–S298. [CrossRef] [PubMed]
90. Chun, K.J.; Lee, J.Y. Comparative study of mechanical properties of dental restorative materials and dental hard tissues in compressive loads. *J. Dent. Biomech.* **2014**, *5*. [CrossRef] [PubMed]
91. Wendler, M.; Belli, R.; Petschelt, A.; Mevec, D.; Harrer, W.; Lube, T.; Danzer, R.; Lohbauer, U. Chairside CAD/CAM materials. Part 2: Flexural strength testing. *Dent. Mater.* **2017**, *33*, 99–109. [CrossRef] [PubMed]
92. VITA IN-CERAM. Available online: http://www.cerec.co.il/downloads/vita_in_ceram.pdf (accessed on 5 November 2020).
93. Belli, R.; Wendler, M.; de Ligny, D.; Cicconi, M.R.; Petschelt, A.; Peterlik, H.; Lohbauer, U. Chairside CAD/CAM materials. Part 1: Measurement of elastic constants and microstructural characterization. *Dent. Mater.* **2017**, *33*, 84–98. [CrossRef] [PubMed]
94. Wendler, M.; Belli, R.; Valladares, D.; Petschelt, A.; Lohbauer, U. Chairside CAD/CAM materials. Part 3: Cyclic fatigue parameters and lifetime predictions. *Dent. Mater.* **2018**, *34*, 910–921. [CrossRef]
95. VITA CAD-Temp. Available online: https://erp.somuden.es/instrucciones/804-6.pdf (accessed on 5 November 2020).
96. Osorio, E.; Toledano, M.; da Silveira, B.L.; Osorio, R. Effect of different surface treatments on In-Ceram Alumina roughness. An AFM study. *J. Dent.* **2010**, *38*, 118–122. [CrossRef]
97. ZirCAD. The Innovative Zirconium Oxide. Available online: https://www.ivoclarvivadent.com/en/p/all/products/all-ceramics/ips-emax-technicians/ips-emax-zircad (accessed on 5 November 2020).
98. Mainjot, A.K.; Dupont, N.M.; Oudkerk, J.C.; Dewael, T.Y.; Sadoun, M.J. From Artisanal to CAD-CAM Blocks: State of the Art of Indirect Composites. *J. Dent. Res.* **2016**, *95*, 487–495. [CrossRef] [PubMed]
99. Dobrzynski, M.; Pajaczkowska, M.; Nowicka, J.; Jaworski, A.; Kosior, P.; Szymonowicz, M.; Kuropka, P.; Rybak, Z.; Bogucki, Z.A.; Filipiak, J.; et al. Study of Surface Structure Changes for Selected Ceramics Used in the CAD/CAM System on the Degree of Microbial Colonization, In Vitro Tests. *Biomed Res. Int.* **2019**, *2019*. [CrossRef]
100. IPS Empress CAD. Available online: https://www.ivoclarvivadent.com/en/productcategories/ips-empress-system-technicians/ips-empress-cad (accessed on 5 November 2020).
101. Lava Ultimate: Technical Product Profile. Available online: http://www.d-way.cz/data/product/13/23/files/Lava_Ult_TPP.pdf (accessed on 4 November 2020).
102. Polident CAD/CAM Discs. Available online: https://www.polident.si/en/products/cad-cam-discs/2019071510431758 (accessed on 5 November 2020).
103. Toth, J.M.; Wang, M.; Estes, B.T.; Scifert, J.L.; Seim, H.B.; Turner, A.S. Polyetheretherketone as a biomaterial for spinal applications. *Biomaterials* **2006**, *27*, 324–334. [CrossRef]
104. Ferguson, S.J.; Visser, J.M.A.; Polikeit, A. The long-term mechanical integrity of non-reinforced PEEK-OPTIMA polymer for demanding spinal applications: Experimental and finite-element analysis. *Eur. Spine J.* **2006**, *15*, 149–156. [CrossRef]
105. PEEK OPTIMA Natural–Invibio. Available online: https://invibio.com/materials/peek-optima-natural (accessed on 4 November 2020).
106. Mine, A.; Kabetani, T.; Kawaguchi-Uemura, A.; Higashi, M.; Tajiri, Y.; Hagino, R.; Imai, D.; Yumitate, M.; Ban, S.; Matsumoto, M.; et al. Effectiveness of current adhesive systems when bonding to CAD/CAM indirect resin materials: A review of 32 publications. *Jpn. Dent. Sci. Rev.* **2019**, *55*, 41–50. [CrossRef]
107. El Zohairy, A.A.; De Gee, A.J.; Mohsen, M.M.; Feilzer, A.J. Microtensile bond strength testing of luting cements to prefabricated CAD/CAM ceramic and composite blocks. *Dent. Mater.* **2003**, *19*, 575–583. [CrossRef]
108. Blatz, M.B.; Vonderheide, M.; Conejo, J. The Effect of Resin Bonding on Long-Term Success of High-Strength Ceramics. *J. Dent. Res.* **2018**, *97*, 132–139. [CrossRef]
109. Furtado de Mendonca, A.; Shahmoradi, M.; de Gouvêa, C.V.D.; De Souza, G.M.; Ellakwa, A. Microstructural and Mechanical Characterization of CAD/CAM Materials for Monolithic Dental Restorations. *J. Prosthodont.* **2019**, *28*, e587–e594. [CrossRef] [PubMed]
110. Sonmez, N.; Gultekin, P.; Turp, V.; Akgungor, G.; Sen, D.; Mijiritsky, E. Evaluation of five CAD/CAM materials by microstructural characterization and mechanical tests: A comparative In Vitro study. *BMC Oral Health* **2018**, *18*, 1–13. [CrossRef] [PubMed]
111. Queiroz, J.R.C.; Fissmer, S.F.; Koga-Ito, C.Y.; Salvia, A.C.R.D.; Massi, M.; Sobrinho, A.S.d.S.; Júnior, L.N. Effect of diamond-like carbon thin film coated acrylic resin on candida albicans biofilm formation. *J. Prosthodont.* **2013**, *22*, 451–455. [CrossRef] [PubMed]
112. Haralur, S.B. Evaluation of efficiency of manual polishing over autoglazed and overglazed porcelain and its effect on plaque accumulation. *J. Adv. Prosthodont.* **2012**, *4*, 179–186. [CrossRef]
113. Kanat-Ertürk, B. Color Stability of CAD/CAM Ceramics Prepared with Different Surface Finishing Procedures. *J. Prosthodont.* **2020**, *29*, 166–172. [CrossRef] [PubMed]
114. Stawarczyk, B.; Sener, B.; Trottmann, A.; Roos, M.; Özcan, M.; Hämmerle, C.H.F. Discoloration of manually fabricated resins and industrially fabricated CAD/CAM blocks versus glass-ceramic: Effect of storage media, duration, and subsequent polishing. *Dent. Mater. J.* **2012**, *31*, 377–383. [CrossRef]
115. De Jager, N.; Feilzer, A.J.; Davidson, C.L. The influence of surface roughness on porcelain strength. *Dent. Mater.* **2000**, *16*, 381–388. [CrossRef]
116. Mota, E.G.; Smidt, L.N.; Fracasso, L.M.; Burnett, L.H.; Spohr, A.M. The effect of milling and postmilling procedures on the surface roughness of CAD/CAM materials. *J. Esthet. Restor. Dent.* **2017**, *29*, 450–458. [CrossRef]

117. Ivoclar Vivadent AG IPS, e.max CAD Glaze—Advanced Characterization. Available online: https://www.youtube.com/watch?v=_pMPUyfnZNA (accessed on 14 October 2020).
118. Tekçe, N.; Fidan, S.; Tuncer, S.; Kara, D.; Demirci, M. The effect of glazing and aging on the surface properties of CAD/CAM resin blocks. *J. Adv. Prosthodont.* **2018**, *10*, 50–57. [CrossRef] [PubMed]
119. Vichi, A.; Fabian Fonzar, R.; Goracci, C.; Carrabba, M.; Ferrari, M. Effect of finishing and polishing on roughness and gloss of lithium disilicate and lithium silicate zirconia reinforced glass ceramic for CAD/CAM systems. *Oper. Dent.* **2018**, *43*, 90–100. [CrossRef]
120. Ekici, M.A.; Egilmez, F.; Cekic-Nagas, I.; Ergun, G. Physical characteristics of ceramic/glasspolymer based CAD/CAM materials: Effect of finishing and polishing techniques. *J. Adv. Prosthodont.* **2019**, *11*, 128–137. [CrossRef]
121. Marrelli, M.; Maletta, C.; Inchingolo, F.; Alfano, M.; Tatullo, M. Three-point bending tests of zirconia core/veneer ceramics for dental restorations. *Int. J. Dent.* **2013**, *2013*. [CrossRef] [PubMed]
122. D'Argenio, V.; Salvatore, F. The role of the gut microbiome in the healthy adult status. *Clin. Chim. Acta* **2015**, *451*, 97–102. [CrossRef] [PubMed]
123. Rizo-Gorrita, M.; Herráez-Galindo, C.; Torres-Lagares, D.; Serrera-Figallo, M.Á.; Gutiérre-Pérez, J.L. Biocompatibility of Polymer and Ceramic CAD/CAM Materials with Human Gingival Fibroblasts (HGFs). *Polymers* **2019**, *11*, 1446. [CrossRef] [PubMed]
124. Wenz, L.M.; Merritt, K.; Brown, S.A.; Moet, A.; Steffee, A.D. In Vitro biocompatibility of polyetheretherketone and polysulfone composites. *J. Biomed. Mater. Res.* **1990**, *24*, 207–215. [CrossRef]
125. Hussain, B.; Le Thieu, M.K.; Johnsen, G.F.; Reseland, J.E.; Haugen, H.J. Can CAD/CAM resin blocks be considered as substitute for conventional resins? *Dent. Mater.* **2017**, *33*, 1362–1370. [CrossRef]
126. Shim, J.; Kim, H.; Park, S.; Yun, H.; Ryu, J. Comparison of Various Implant Provisional Resin Materials for Cytotoxicity and Attachment to Human Gingival Fibroblasts. *Int. J. Oral Maxillofac. Implants* **2019**, *34*, 390–396. [CrossRef] [PubMed]
127. Daguano, J.K.M.B.; Milesi, M.T.B.; Rodas, A.C.D.; Weber, A.F.; Sarkis, J.E.S.; Hortellani, M.A.; Zanotto, E.D. In Vitro biocompatibility of new bioactive lithia-silica glass-ceramics. *Mater. Sci. Eng. C* **2019**, *94*, 117–125. [CrossRef] [PubMed]
128. Reference Offices. Available online: https://www.amadar.pl/sprzet/system-cadcam/poznaj-system-cerec/gabinety-referencyjne/ (accessed on 18 November 2020).
129. Joda, T.; Bragger, U.; Zitzmann, N.U. CAD/CAM implant crowns in a digital workflow: Five-year follow-up of a prospective clinical trial. *Clin. Implant Dent. Relat. Res.* **2019**, *21*, 169–174. [CrossRef] [PubMed]
130. Aslan, Y.U.; Coskun, E.; Ozkan, Y.; Dard, M. Clinical Evaluation of Three Types of CAD/CAM Inlay/Onlay Materials after 1-Year Clinical Follow Up. *Eur. J. Prosthodont. Restor. Dent.* **2019**, *27*, 131–140. [CrossRef] [PubMed]

materials

MDPI

Article

Pristine and Poly(Dimethylsiloxane) Modified Multi-Walled Carbon Nanotubes as Supports for Lipase Immobilization

Iryna Sulym [1,*], Jakub Zdarta [2,*], Filip Ciesielczyk [2], Dariusz Sternik [3], Anna Derylo-Marczewska [3] and Teofil Jesionowski [2]

[1] Chuiko Institute of Surface Chemistry of NASU, 17 General Naumov Str., 03164 Kyiv, Ukraine
[2] Institute of Chemical Technology and Engineering, Faculty of Chemical Technology, Poznan University of Technology, Berdychowo 4, 60965 Poznan, Poland; Filip.Ciesielczyk@put.poznan.pl (F.C.); teofil.jesionowski@put.poznan.pl (T.J.)
[3] Institute of Chemical Sciences, Maria Curie-Sklodowska University in Lublin, M.C. Sklodowska Sq.3, 20031 Lublin, Poland; dsternik@poczta.umcs.lublin.pl (D.S.); annad@hektor.umcs.lublin.pl (A.D.-M.)
* Correspondence: irynasulym@ukr.net (I.S.); jakub.zdarta@put.poznan.pl (J.Z.); Tel.: +38-044-422-96-72 (I.S.); +48-61-665-3747 (J.Z.)

Abstract: The presented study deals with the fabrication of highly stable and active nanobiocatalysts based on *Candida antarctica* lipase B (CALB) immobilization onto pristine and poly(dimethylsiloxane) modified MWCNTs. The MWCNTs/PDMS nanocomposites, containing 40 wt.% of the polymer with two molecular weights, were successfully synthesized via adsorption modification. The effect of PDMS chains length on the textural/structural properties of produced materials was studied by means of the nitrogen adsorption–desorption technique, Raman spectroscopy, and attenuated total reflectance Fourier transform infrared spectroscopy. P-MWCNTs and MWCNTs/PDMS nanocomposites were tested as supports for lipase immobilization. Successful deposition of the enzyme onto the surface of P-MWCNTs and MWCNTs/PDMS nanocomposite materials was confirmed mainly using ATR-FTIR spectroscopy. The immobilization efficiency, stability, and catalytic activity of the immobilized enzyme were studied, and the reusability of the produced biocatalytic systems was examined. The presented results demonstrate that the produced novel biocatalysts might be considered as promising materials for biocatalytic applications.

Keywords: multi-walled carbon nanotubes; poly(dimethylsiloxane); polymer nanocomposites; *Candida antarctica* lipase B; lipase immobilization; enzymes stability and reusability

Citation: Sulym, I.; Zdarta, J.; Ciesielczyk, F.; Sternik, D.; Derylo-Marczewska, A.; Jesionowski, T. Pristine and Poly(Dimethylsiloxane) Modified Multi-Walled Carbon Nanotubes as Supports for Lipase Immobilization. *Materials* **2021**, *14*, 2874. https://doi.org/10.3390/ma14112874

Academic Editor: Klára Hernádi

Received: 11 May 2021
Accepted: 25 May 2021
Published: 27 May 2021

1. Introduction

Carbon occurs in many forms, and the properties of each form with respect to their special structure make carbon a truly unique building block for nanomaterials [1]. Owing to their interesting electrical, magnetic, mechanical, and thermal properties, carbon nanotubes (CNTs) have become the most promising materials in many scientific and technological fields [2–7]. CNTs are made of one or more graphene sheets rolled-up to form tubes. Single-walled CNTs (SWCNTs) comprise a single graphene layer seamlessly wrapped into a cylindrical tube. Multi-walled carbon nanotubes (MWCNTs) comprise an array of concentric cylinders coaxially arranged around a central hollow core with van der Waals forces between adjacent layer [8,9]. CNTs exhibit extremely high surface area, large aspect ratio, low density, remarkably high mechanical strength, and electrical as well thermal conductivities [8,10]. Functionalization of carbon nanotube surfaces can be proceeded via covalent or non-covalent modification. Non-covalent surface modificationof CNTs, which includes adsorption of surfactants, polymers, or biological macromolecules, is a method that does not affect their intrinsic structure [11]. Among many methods, non-covalent attachment (polymer wrapping and absorption) is of key importance when CNTs functionalize with polymer molecules [11–13]. Polymeric materials includingsilicon rubbers, in particular poly(dimethylsiloxane) (PDMS), are used in everyday materials. It is related to their excellent durability and mechanical properties that results from their high cross-linking density and degree of polymerization. PDMS is hydrophobic, chemically inert, and electrically resistant, and exhibits dielectric strength, high elasticity, and easy processing and isenvironmentally-friendly [14–17]. The addition of PDMS to MWCNTs improves the mechanical, electrical, and thermal properties of the resulting nanocomposites [18–20]. Based on our previous studies [21–25] concerning materials prepared in a similar way but containing chemically different highly dispersed nano-oxides as well as PDMS modified MWCNTs, it was proved thatpolymer–filler interactions depend on nanocomposite compositions and inorganic particle types (oxides). On the other hand, those interactions are responsible for surface properties of the resulting nanocomposites such as roughness, textural porosity, or hydrophobicity, which furthermore affect their potential applications as catalysts supports, adsorbents, fillers, etc.

This current research may be a part of work on preparation of active biocatalyst composed with the functional support and biomolecules—e.g., enzymes.

Recently, many studies have been devoted to the immobilization of enzymatically active substances on inorganic, e.g., silica-based, supports towards heterogeneous biocatalyst preparation. This is especially related to the nature of such systems—their availability and relatively low cost as well as high mechanical strength and durability of biocatalyst granules in the reaction media [26,27]. In particular, lipases are the most widely used biomolecules in enzyme technology because of their widespecificity for some substrates, low production cost, wide pH activity profiles, as well as ability to catalyze various different reactions, such as hydrolysis, esterification, amination, transesterification, alcoholysis, etc. [28–30]. Graphene oxide (GO), having a large surface area (2630 m^2/g) and abundant functional groups (such as epoxide, hydroxyl, and carboxylic groups), provides a great substrate for enzyme immobilization without any surface modification or any coupling agents [31]. Nevertheless, among the various nanostructured materials that might be used as novel supports for enzyme immobilization and stabilization, CNTs are of great interest to many research centers worldwide due to their stability, high adsorption capacity, improvedretention of catalytic activity, and biocompatibility [32–35]. Both SWCNTs and MWCNTs have been used to immobilize enzymes [36]. MWCNTs are structurally similar to SWCNTs, but their diameters can range from a few nanometers to dozens of nanometers [37]. SWCNTs are attractive due to their larger surface area for protein interaction, excellent biocompatibility, antifouling properties, and high conductivity, but MWNTs are desirable because they are easier to prepare, exhibit better dispersibility, and are commercially available at a relative lower price, which makes them more feasible for industrial applications. Therefore, MWNTs are suitable materials as enzyme supports [38]. Enzymes can be immobilized on the surface of MWCNTs by adsorption or covalent binding, whichresultsin enhanced catalytic performance and stability. Moreover, lipases are well-known interfacially active catalysts and exhibit their catalytic abilities at the interface between the organic phase containing hydrophobic substrates and aqueous phase, so they can be activated at the hydrophobic–hydrophilic interface [34,39].

Herein, in the present work, for the first time, the methodology to design and characterize an alternative, highly stable, and active nanobiocatalyst based on *Candida antarctica* lipase B (CALB) immobilized onto pristine and poly(dimethylsiloxane)modified MWCNTsis presented and discussed. The idea was to combine the textural properties as well as functionality of PDMS modified with lipase activity, and to obtain a novel type of biocatalyst dedicated to biotechnological applications. As a result, high enzyme loading, its improved stability and reusability, as well as activity of the biocatalyst produced, were expected. The innovative nature of the presented study is based not only on application of a novel, previously undescribed support material for lipase immobilization, but also on the possibility to use MWCNT surface modifying agents at various molecular weights to examine their effects on enzyme loading and catalytic activity.

2. Materials and Methods

2.1. Chemicals and Materials

Commercial poly(dimethylsiloxane) fluids of two molecular weights (Wacker Chemie AG, linear, –CH_3 terminated, code names: PDMS-100 and PDMS-12500 with MW ≈ 3410 and 39,500 g/mol, respectively) and multi-walled carbon nanotubes (MWCNTs, obtained by catalytic chemical vapor deposition (CCVD) [40] using pyrolysis of propylene on complex metal oxide catalysts) [41] were used as initialmaterials forpolymer nanocompositepreparation.

Candida antarctica lipase B (CALB) (EC 3.1.1.3, ~200 U/g), para-nitrophenyl palmitate (p-NPP), para-nitrophenol (p-NP), gum arabic and Triton X-100 (laboratory grade), sodium hydroxide, sodium chloride, 50 mM acetate buffer (pH 3–5), 50 mM phosphate buffer (pH 6–8), and 50 mM Tris-HCl (pH 8 and 9) were supplied from Sigma-Aldrich (St. Louis, MO, USA). All of the reagents were of analytical grade and were used directly without any further purification.

2.2. Preparation of MWCNTs/PDMS Nanocomposites

PDMS-100 and PDMS-12500 fluids were physically adsorbed onto pristine multi-walled carbon nanotubes (P-MWCNTs) in the amount of 40 wt.%. Before adsorption, the samples were dried at 110 °C for 2 h. A hexane solution of PDMS (1 wt.% PDMS) was prepared, and its estimated amount wasadded to a fixed amount of dry carbon powder material. The suspension was mechanically stirred and finally dried at room temperature for 48 h and then at 80 °C for 3 h. All samples modified with PDMS in the amount of 40 wt.% were in the form of powder similar to P-MWCNTs, while neat PDMS-100 and PDMS-12500 were liquids. The prepared polymer nanocomposites were marked as MWCNTs/PDMS-100(40) and MWCNTs/PDMS-12500(40), respectively.

2.3. Lipase Immobilization

The pristine MWCNTs and MWCNTs/PDMS nanocomposites were used as supports for the immobilization of *Candida antarctica* lipase B (CALB). In all experiments, 100 mg of P-MWCNTs or modified MWCNTs (MWCNTs/PDMS-100(40) and MWCNTs/PDMS-12500(40)) were added to 5 mL of lipase solution at concentration of 5 mg/mL in 50 mM phosphate buffer solution at pH 7. The samples were then shaken for 24 h using a KS 4000i Control incubator (IKA Werke GmbH, Staufen im Breisgau, Germany) at ambient temperature. Next, samples were centrifuged (4000 rpm at 4 °C) using an Eppendorf 5810 R centrifuge (Hamburg, Germany) and furthermore washed several times with 50 mM phosphate buffer in order to remove unbounded lipase. The samples were labelled as CALB@P-MWCNTs, CALB@MWCNTs/PDMS-100(40), and CALB@MWCNTs/PDMS-12500(40), respectively.

2.4. Analysis of Nanocomposites before Lipase Immobilization

2.4.1. Textural Characterization

To analyze the textural characteristics of P-MWCNTs and MWCNTs/PDMS nanocomposites, low-temperature (77.4 K) nitrogen adsorption–desorption isotherms were recorded using an automatic gas adsorption analyzer ASAP 2420 (Micromeritics Instrument Corp., Norcross, GA, USA). Beforehand, the measurement samples were degassed at 110 °C for 2 h in a vacuum chamber. The values of surface area (S_{BET}) were calculated according to the standard BET method (using Micromeritics software). The total pore volume, V_p, was evaluated from the nitrogen adsorption at p/p_0 = 0.98–0.99 (p and p_0 denote the equilibrium and saturation pressure of nitrogen at 77.4 K, respectively). The nitrogen desorption data were used to compute the pore size distributions (PSD, differential $f_V(R) \sim dVp/dR$ and $f_S(R) \sim dS/dR$)), using a model with slit-shaped and cylindrical pores and voids between spherical nanoparticles (SCV) with a self-consistent regularization (SCR) procedure for MWCNTs/PDMS samples and slit-shaped pores for P-MWCNTs [42,43]. The differential PSD with respect to pore volume $f_V \sim dV/dR$, $\int f_V dR \sim V_p$ were re-calculated to incremental PSD (IPSD) at $\phi_V(R_i) = (f_V(R_{i+1}) + f_V(R_i))(R_{i+1} - R_i)/2$ at $\sum\phi_V(R_i) = V_p$). The f_V and f_S functions were also used to calculate contributions of micropores (V_{micro} and S_{micro} at radius $R \leq 1$ nm), mesopores (V_{meso} and S_{meso} at 1 nm < R < 25 nm), and macropores (V_{macro} and S_{macro} at 25 nm < R < 100 nm) to the total pore volume and surface area.

2.4.2. Spectral Analysis

The Raman spectra were recorded over the 3200–500 cm^{-1} range using the in Via Reflex Microscope DMLM Leica Research Grade, Reflex (Renishaw, Wotton-under-Edge, UK) with Ar^+ ion laser excitation at λ_0 = 514.5 nm. For each sample, the spectra were recorded at several points in order to ascertain the homogeneity of the sample, and the average results were plotted. The surface functional groups of the pristine MWCNTs and MWCNTs/PDMS nanocomposites, before and after lipase immobilization, were investigated using Fourier transform infrared spectroscopy (FTIR) in attenuated total reflectance (ATR) mode (Vertex 70 spectrometer, Bruker, Germany). The FTIR spectra were recordedat a wavenumber range of 4000–500 cm^{-1}, and at a resolution of 1 cm^{-1}.

2.5. Characterization of Free and Immobilized Lipase

The activity of free and immobilized lipase was estimated based on the model reaction of p-nitrophenyl palmitate hydrolysis to p-nitrophenol. The spectrophotometric measurements at 410 nm, using a Jasco V-750 UV–Vis spectrophotometer (Jasco, Tokyo, Japan), were used for this purpose. In the reaction, 10 mg of free lipase and a corresponding amount of the biocatalytic system produced (CALB@MWCNTs, CALB@MWCNTs/PDMS-100(40) and CALB@MWCNTs/PDMS-12500(40)), containing 10 mg of the enzyme, were used. Reactions were carried out for 2 min under continuous stirring (800 rpm). After the assumedtime, the reaction was terminated, and absorbance was measured. The activity of the free and immobilized lipase was calculated using a standard calibration curve for p-NP. The amount of biocatalyst that hydrolyzed 1 µmol of p-NPP per minute was defined as the one unit of lipase activity. The highest measured activity was defined as 100% relative activity. The effect of pH on the activity of the free and immobilized lipase was determined based on the above-mentioned reaction at a temperature of 30 °C, in the pH range 3–11, using buffer solution at the desired pH. The effect of temperature on the activity of the free and immobilized lipase was determined according to the above-presented methodology at pH 7 (50 mM phosphate buffer), over a temperature rangefrom 10 to 80 °C. All measurements were made in triplicate; error bars are presented as means ± standard deviation.

2.6. Stability and Reusability of Free and Immobilized Lipase

Thermal stability over time was determined during incubation of free and immobilized enzyme for 180 min under optimal process conditions (30 °C and pH 7). The relative activity of free and immobilized lipase was further determined based on the model hydrolysis reaction of p-nitrophenyl palmitateat every specified period of time, applying spectrophotometric measurements at 410 nm. The initial lipase activity was defined as 100% relative activity. The inactivation constant (k_D) and enzyme half-life ($t_{1/2}$) were determined based on the linear regression slope.

Storage stability of free lipase and products after immobilization were examined based on the above-mentioned model reaction of p-NPP hydrolysis over 30 days of storage at 4 °C in 50 mM phosphate buffer at pH 7. The relative activity was measured every 2 days.

The reusability of the produced biocatalytic systems was also determined using the same model hydrolysis reaction carried out under optimal process conditions, over 10 repeated catalytic cycles. Immobilized lipase was separated from the reaction mixture by centrifugation, washed several timeswith 50 mM phosphate buffer at pH 7, and placed in the fresh substrate solution.

The effect of 5% Triton X-100 and 0.5 M NaCl on relative activity of the immobilized enzyme was examined over 24 h of incubation in a proper solution. After a specified period of time, the relative activity of the immobilized lipase was determined based on the model reaction of p-NPP hydrolysis. All measurements were made in triplicate; error bars are presented as means ± standard deviation.

3. Results

3.1. Analysis of Nanocomposites before Lipase Immobilization

3.1.1. Parameters of the Porous Structure

The structural characteristic of P-MWCNTs and MWCNTs/PDMS nanocomposites wasstudied using low-temperature nitrogen adsorption–desorption isotherms (Figure 1a and Table 1). All of the materials were characterized withthe nitrogen adsorption isotherms of type II (H3 type of hysteresis loops) according to the IUPAC classification [44,45]. Capillary condensation occurred at pressure $p/p_0 > 0.85$ (due to adsorption in broad mesopores and macropores).

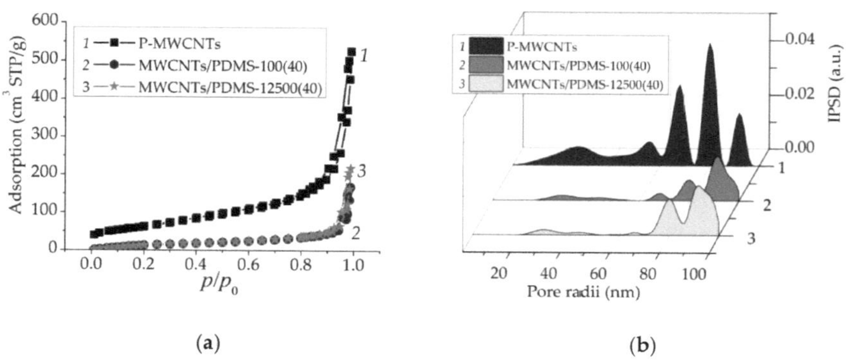

Figure 1. (**a**) Nitrogen adsorption–desorption isotherms and (**b**) incremental pore size distributions for P-MWCNTs (curve *1*) and MWCNTs/PDMS nanocomposites (curves *2*, *3*).

Table 1. Textural characteristics of P-MWCNTs and MWCNTs/PDMS nanocomposites.

Sample	S_{BET} (m²/g)	S_{micro} (m²/g)	S_{meso} (m²/g)	S_{macro} (m²/g)	V_{micro} (cm³/g)	V_{meso} (cm³/g)	V_{macro} (cm³/g)	V_p (cm³/g)	$R_{p,V}$ (nm)
P-MWCNTs	222	74	134	14	0.039	0.418	0.357	0.814	23
MWCNTs/PDMS-100(40)	76	0	56	20	0	0.056	0.203	0.259	62
MWCNTs/PDMS-12500(40)	77	0	51	26	0	0.054	0.283	0.337	65

Surface area values (Table 1, S_{BET}) demonstrated a significant reduction after adsorption of both types of PDMS onto carbon nanotube surfaces. Moreover, the total pore volume (V_p) decreased for the MWCNTs/PDMS-100 and MWCNTs/PDMS-12500 nanocomposites by 68 and 59%, respectively, as compared to the P-MWCNTs. Moreover, it was observed that the pore average radii in MWCNTs/PDMS (−100, −12,500) samples was three times greater than that of P-MWCNTs.

The incremental pore size distribution IPSD functions (Figure 1b) show that the textural characteristics ofMWCNTschanged after the modification with polymer. The textural porosity of the pristine MWCNT resulted from mesopores and secondly due to micropore presence. MWCNTs/PDMS nanocomposites were characterized by bimodal porous structures (Figure 1b). In addition, MWCNTs/PDMS samples were characterized with a significant decrease in mesopore contributions to the total porosity with a simultaneous increase in contributions of macropores as compared to P-MWCNTs.

3.1.2. Raman Spectroscopy

Raman spectroscopy is a very valuable tool for the characterization of carbon-based nanostructures. This technique is used to analyze the presence of amorphous and crystalline phases corresponding to differences in graphitization. The spectra were collected in the most informative range for carbon materials of 3200–500 cm^{-1} (Figure 2). Three major peaks at 1341 cm^{-1} as the D-band (sp^3 carbons in non-graphitic structures), at 1570 cm^{-1} as the G-band (sp^2 carbons in graphitic structures), and its second-order harmonic at 2672 cm^{-1} as the G′-band, were noted [46]. The ratio between the integral intensities of the G and D bands (A_G/A_D ratio as a measure of the graphitization degree) is an indicator of the crystallinity degree [47]. The value of A_G/A_D was calculated by deconvolution of the spectra using the Lorentzian function. After adsorption of polymer, relative intensity ratio A_G/A_D tended to decrease from 1.1 for P-MWCNTs to 0.95 for MWCNTs/PDMS nanocomposites, respectively.

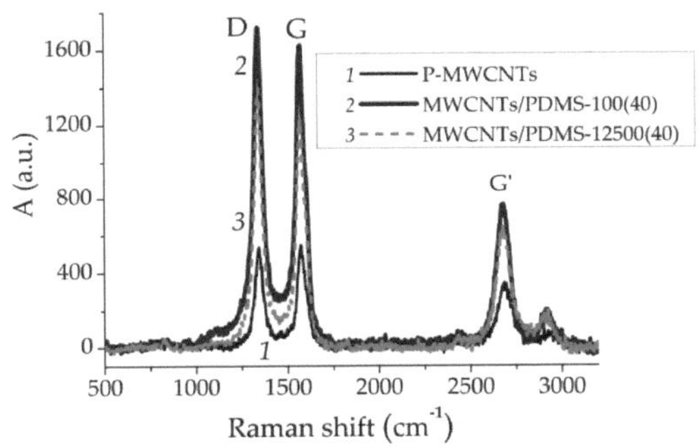

Figure 2. Raman spectra for P-MWCNTs (curve *1*) and MWCNTs/PDMS nanocomposites (curves *2*, *3*).

3.1.3. ATR-FTIR Spectroscopy

Fourier transform infrared spectroscopy was used to determine the nature of chemical groups present on the surface of analyzed materials as well as to indirectly confirm nanotube modification and enzyme immobilization (Figure 3).

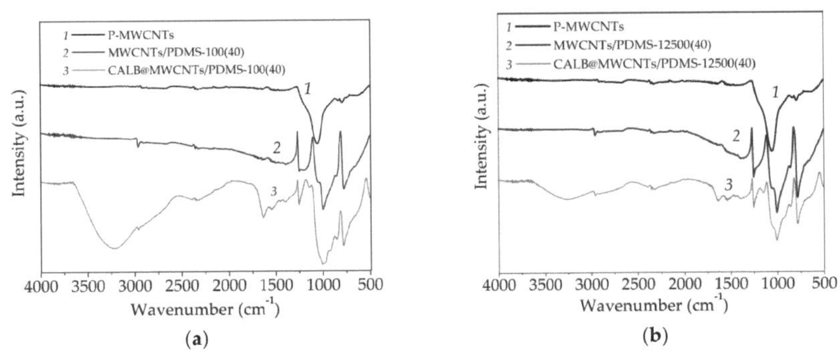

Figure 3. FTIR spectra of (**a**) P-MWCNTs, MWCNTs/PDMS-100, and (**b**) MWCNTs/PDMS-12500 nanocomposites, before and after lipase immobilization.

The FTIR spectrum of the P-MWCNTs showeda broad peak with a maximum at 1060 cm^{-1} that corresponded to the stretching vibrations of C–O bonds in carboxylic groups present onto the surface of MWCNTs. Upon modification by PDMS, irrespectively of the molecular mass of the modifying agent used, additional signals could be observed. The small signal at 2950 cm^{-1} was related to the presence of C–H stretching vibrations, peaks at 1250 and 780 cm^{-1} corresponded to the stretching and bending vibrations of Si-CH_3 groups, whereas peaks at 1010 and 1055 cm^{-1} were characteristic for the stretching vibrations of Si–O–Si bonds [48]. After enzyme immobilization onto both materials, the presence of additional signals, characteristic for the enzyme structure, was clearly seen. Among them, the most important was a peak at 3400 cm^{-1}, assigned to the stretching vibrations of –OH groups, and two signals at 1655 and 1545 cm^{-1}, ascribed to the stretching vibrations of amide I and amide II bands, respectively. Further, it could be seen that the intensity of signals characteristic for enzyme was higher in the CALB@MWCNTs/PDMS-100(40) spectrum, as compared to the CALB@MWCNTs/PDMS-12500(40) spectrum.

3.2. Characterization of Free and Immobilized Lipase

The next stage included tests of obtained materials (MWCNTs modified with 40 wt.% of PDMS-100 and PDMS-12500) as supports for enzyme immobilization. Lipase was selected as a model enzyme, as itexhibitsimproved catalytic activity in a hydrophobic microenvironment. The effect of various process conditions on thestability and activity of the immobilized enzyme was determined, and the reusability of the produced biocatalytic systems was examined.

3.2.1. pH Profiles of Free and Immobilized Lipase

Free lipase and biocatalytic systems produced showed maximum activity at pH 7 (Figure 4). Further, their pH profiles weresimilar. In the tested pH range (beside pH 7), free enzyme exhibited relative activity not higher than 60% and even less than 30% at pH ranges from 3 to 5 and from 9 to 11. By contrast, enzyme immobilized on both pristine and modified MWCNTs showed about 10–30% higher relative activity over whole analyzed pH range. Further, lipase deposited onto MWCNTs/PDMS nanocomposites retained over 80% of its relative activity over a wide pH rangefrom 6 to 9 and more than 30% relative activity at pH 3 to 10. It should also be highlighted that lipase immobilized onto MWCNTs/PDMS-100(40) material exhibited around 5–10% higher activity than enzyme immobilized onto carbon nanotubes functionalized by PDMS withhigher molecular weight.

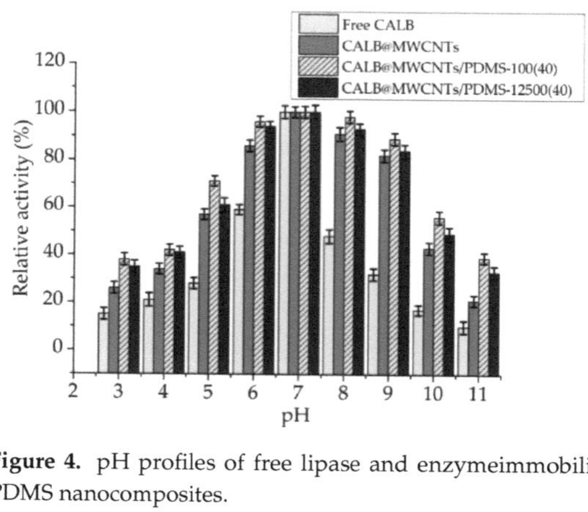

Figure 4. pH profiles of free lipase and enzymeimmobilized onto P-MWCNTs and MWCNTs/PDMS nanocomposites.

3.2.2. Temperature Profiles of Free and Immobilized Lipase

Temperature profiles of free and immobilized lipase were determined over a temperature range of 10–80 °C (Figure 5). The optimal temperature for all analyzed samples was found to be 30 °C. Even a slight change in temperature conditions resulted in a sharp decrease of catalytic activity of free enzyme. Only at temperatures ranging from 20 to 40 °C did free lipase show over 60% of relative activity. Although temperature profiles of immobilized biomolecules exhibited similarity in shape to those of free enzyme, significantly higher relative activityof those systems was observed over the whole analyzed temperature range. At temperatures ranging from 20 to 70 °C, over 60% of relative activity was noticed for all analyzed biocatalytic systems. Finally, similarly as analyzing pH effect on relative activity of immobilized lipase, slightly higher activity was noticed when MWCNTs/PDMS-100(40) material was used as support for enzyme immobilization.

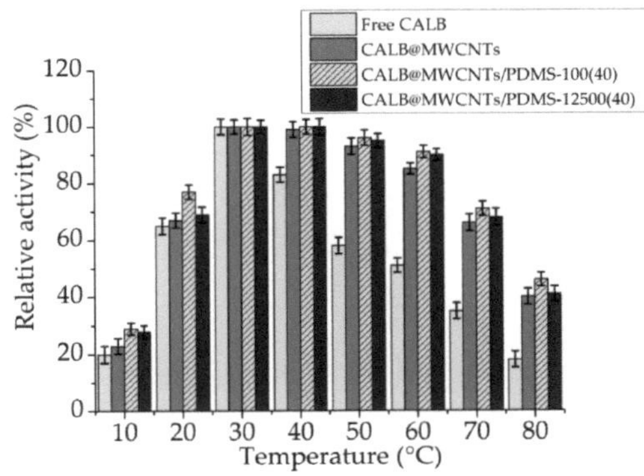

Figure 5. Temperature profiles of free lipase and enzyme immobilized onto P-MWCNTs and MWCNTs/PDMS nanocomposites.

3.2.3. Thermal Stability of Free and Immobilized Lipase

Thermal stability of free and immobilized lipase was determined via samples incubation for 3 h at a temperature of 30 °C and atpH 7 (Figure 6). A relative activity decrease over incubation time for free and immobilized lipase was observed. However, the drop of catalytic properties was more pronounced for free enzyme, which retained less than 20% of its relative activity after 3 h of incubation. Significantly higher values of relative activity were noticed for biocatalytic systems produced. The decrease of catalytic activity of immobilized lipase was much slower as compared to the free counterpart; after 1 h and 3 h of incubation, immobilized enzyme showed around 40% and 50% higher values of relative activity, respectively. Finally, both biocatalytic systems obtained using MWCNTs/PDMS nanocomposites showed relative activity exceeding 80% after specific incubation periods.

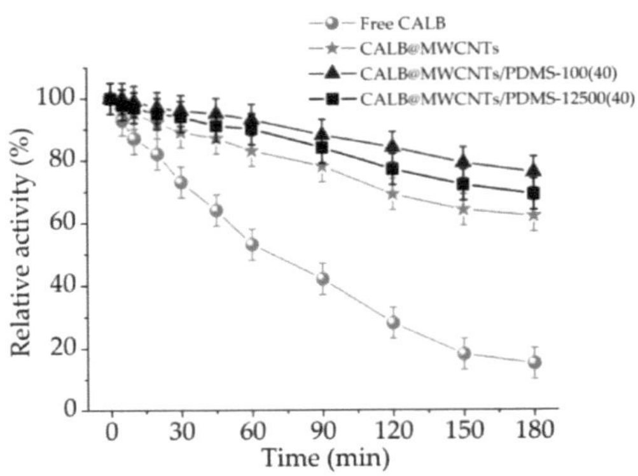

Figure 6. Thermal stability of free lipase and enzyme immobilized onto P-MWCNTs and MWCNTs/PDMS nanocomposites.

In order to clearly present improvement of lipase stability and activity upon immobilization, enzyme inactivation constant (k_D) and enzyme half-life ($t_{1/2}$) were determined

(Table 2). These parameters were calculated based on the linear regression slope from the above-presented Figure 6. Free lipase was characterized by k_D = 0.01075 min^{-1} and a half-life of 64.74 min. Inactivation constant and enzyme half-life of immobilized lipase were improved. The most predominant increase of enzyme stability was noticed for lipase immobilized onto MWCNT/PDMS-100(40) material. Over a seven-fold lower inactivation constant (0.00126 min^{-1}) and consequently over a seven-fold higher enzyme half-life (446.15 min) were observed for this particular biocatalytic system.

Table 2. Inactivation constant and half-life of free lipase and enzyme immobilized onto P-MWCNTs and MWCNTs/PDMS nanocomposites.

Parameter	Free CALB	CALB@P-MWCNTs	CALB@MWCNTs/ PDMS-100(40)	CALB@MWCNTs/ PDMS-12500(40)
k_D (min^{-1})	0.01075	0.00268	0.00156	0.00208
$t_{1/2}$ (min)	64.74	259.70	446.15	334.61

3.2.4. Storage Stability and Reusability of Free and Immobilized Lipase

From apractical application point of view, storage stability and reusability are the crucial properties determining possible large-scale use of the immobilized enzymes. In this study storage stability of free and immobilized lipase was followed over 30 days, and reusability potential was examined over ten repeated reaction cycles (Figure 7).

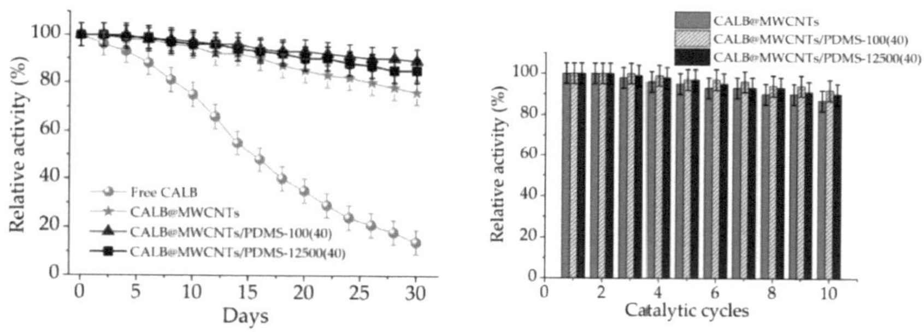

(a) (b)

Figure 7. (**a**) Storage stability and (**b**) reusability of free lipase and enzyme immobilized onto P-MWCNTs and MWCNTs/PDMS nanocomposites.

It can be seen (Figure 7a) that storage stability of the lipase gradually decreased from the first day of storage; after 30 days it reached less than 20%. By contrast, storage stability of all tested biocatalytic systems with immobilized enzyme was improved significantly. Obtained biocatalytic systems showed over 90% relative activity after 6 days of storage and over 80% after 30 days. Further, lipase immobilized onto MWCNTs/PDMS nanocomposites showed over 90% activity retention at the end of the test.

Results of the reusability study (Figure 7b) showed that relative activity of the lipase immobilized onto MWCNTs/PDMS nanocomposites and P-MWCNTs remained almost unaltered for the first three reaction steps. Over the next experimental steps, catalytic activity decreased slightly. After ten cycles, relative activity of lipase immobilized onto P-MWCNTs reached 85%, whereas activity of lipase immobilized onto MWCNTs/PDMS nanocompositesattained over 90%.

3.2.5. Effect of Solvents on the Immobilized Lipase

The effects of surfactant (Triton X-100) and salt solution (0.5 M NaCl) on catalytic activity of immobilized enzymes and stability of enzyme binding were determined by incubation of produced biocatalytic systems in the presence of the mentioned solutions over time. In Figure 8, it can be seen that relative activity of immobilized enzymes decreased gradually over first 6 h of incubation. After that time, all biocatalytic systems showed less than 50% and less than 30% of relative activity in the presence of Triton X-100 and NaCl, respectively. Further treatment of the biocatalysts withthe solvents did not result in such a pronounced drop in relative activity. After 24 h of incubation of lipase immobilized onto MWCNTs/PDMS nanocompositesin Triton X-100 and NaCl solution, analyzed samples retained over 40% and over 20% of relative activity, respectively.

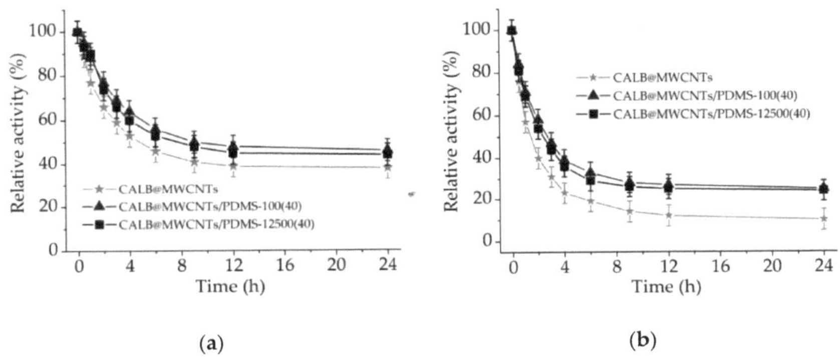

Figure 8. Effect of (**a**) 5% Triton X-100 solution and (**b**) 0.5 M NaCl solution on the relative activity of lipase immobilized onto P-MWCNTs and MWCNTs/PDMS nanocomposites.

4. Discussion

4.1. Analysis of Nanocomposites before Lipase Immobilization

The presented results demonstrate the changes in the textural/structural properties of MWCNTs after modification with polymer. The obtained data can be discussed also in terms of the absolute values of the surface area, S_{BET}, which sharply decreased from 222 m^2/g to 76–77 m^2/g after PDMS (−100, −12,500) grafting (in the amount of 40 wt.%) onto carbon nanotube surfaces. That can be explained by two factors: reducing the content of MWCNTs in the resulting polymer composites after PDMS modification, as carbon nanotubes are responsible forsurface area, as well as increasing the size (diameter) of MWCNTs due to the polymer grafting onto carbon nanotubes. It is known that surface area is inversely proportional to the particle size of the dispersed phase. In general, the polymer adsorption leads to suppression of the values of V_p, V_{meso}, and V_{macro} because each long PDMS macromolecule can bind carbon nanotubes and aggregate themin more compacted structures, whichleads to adecrease in the volume of voids between MWCNTs [49].

It was found that the prepared nanocomposites are characterized by different graphitization degrees according to the data of Raman spectroscopy. A relatively lower A_G/A_D ratio (about 0.9) for polymer nanocomposites indicates a low graphitization degree and shows that graphitic layers are semi-crystalline and possess many defects related to the D band due to introduction of new functional groups to carbon nanotube surfaces. The G′ peak appears at 2672 cm^{-1} as an overtone of the D band and is believed to originate from finite-size disordered structures of graphite (i.e., with the sp^2C atoms) in the surface layers of the nanocomposites [50]. It would be interesting to check in futurework the results obtained here with respect to other polymers and other carbon-based fillers.

The results of FTIR analysis confirmed the carbon structure of the P-MWCNTs and indicatedthe presence of carboxylic groups on their surface that facilitate further MWC-

NTs modification. Upon PDMS adsorption onto MWCNTs, new signals are observed in the FTIR spectra of both modified samples, whichsuggests effective modification using poly(dimethylsiloxane) at various molecular weights. Nevertheless, the most important findings were made based on analysis of FTIR spectra of samples after lipase immobilization. The presence of signals characteristic for vibrations of amide I, amide II, and hydroxyl groups clearly indicate effective deposition of the enzyme onto the surface of both modified materials [51]. Moreover, the higher intensity of the signals ascribed for lipase structure, observed in FTIR spectrum of the system formed using MWCNTs modified with PDMS-100, indicate that a greater amount of the enzyme was immobilized, and immobilized biocatalysts retained higher catalytic properties [52].

4.2. Immobilized Lipase Characterization

Obtained data clearly showed that although pH and temperature profiles of free and immobilized lipase are similar, enzymes attached to P-MWCNTs and MWCNTs/PDMS nanocompositesshowed significantly higher relative activity over wider pH and temperature ranges as compared to free counterparts. Moreover, significant improvement of thermal stability and enzyme half-life of the lipase after immobilization was observed. A drop of the catalytic properties of the lipase in conditions different than optimal is related to the electrostatic repulsion of positively and negatively charged ionic groups in the enzyme structure and is also caused by thermal denaturation of the enzyme at harsh temperature conditions [53,54]. By contrast, immobilized lipase showed over 80% relative activity over wide pH (6–9) and temperature (30–60 °C) ranges. This might be explained by the fact that upon immobilization, an external backbone for the enzyme structure is provided due the formation of stable enzyme–support interactions, which stabilize enzyme structure and protect against biocatalyst denaturation at harsh reaction conditions [55]. Nevertheless, similarity in the pH and temperature profiles, and the presence of optima at the same conditions for free and immobilized lipases, as well asretention of high catalytic activity by produced systems indicate that immobilization did not significantly interfere with enzyme structure and its active site. It should be highlighted that among tested samples, the highest activity and tolerance to reaction conditions are ascribed to the lipase immobilized onto MWCNTs/PDMS-100(40) material. This is directly related to the fact that PDMS provides the hydrophobic nature of the surface and consequently the hydrophobic microenvironment for the immobilized lipase. In these conditions, lipase might undergo a phenomenon called interfacial activation, whichis based on opening of the polypeptide lid of the enzyme active site, leading to improvement of the activity of the immobilized enzyme [56,57]. Finally, significant enhancement of enzyme thermal stability (up to 50% higher relative activity, as compared to free enzyme) and reduction ofinactivation constant of immobilized enzymes are related to the fact that immobilization providesa protective environment for the enzyme molecules, whichreduces conformational changes of the enzyme structure in the presence of long-heat exposure. The advancement of using PDMS modified support for lipase immobilization was recently proved. Li et al. [58] modified silk fabric by amino-functional poly(dimethylsiloxane) (PDMS) and used it as a support for lipase from *Candida* sp. immobilization. It was shown that lipase activity and stability increased upon immobilization onto the hydrophobic surface. However, it was emphasized that the amount of the PDMS used affected catalytic properties of the immobilized enzyme. In another study, macroporous ZIF-8 was modified with PDMS in order to builda hydrophobic pore space for lipase from *Aspergillus niger* immobilization. Immobilized lipase showed improved stability and was found as an effective biocatalyst in the transesterification process in biodiesel production [59].

Determination of the storage stability and reusability of immobilized lipases is of crucial importance, as these parameters are key, determining possible practical application of the biocatalytic systemsproduced. All produced biocatalysts showed over 80% relative activity, even after 30 days of storage and 10 repeated uses, clearly indicating possible large-scale application. Such improvement of long-term stability and reusability is mainly related to the stabilization of the enzyme structure upon immobilization as well as the protective effects of both the MWCNTs support and the PDMS layer on the enzyme structure against inactivation over storage and reuse [60]. In the next part of the study, in order to determine the stability of the formed enzyme–support interactionsand the effects of various solvents on relative activity of immobilized enzymes, produced systems were incubated for 24 h in Triton X-100 and NaCl solutions. A significant drop in relative activity of immobilized lipase in the presence of both solvents might be explained mainly by two factors. Firstly, enzyme support interactions are based mainly on hydrogen and van der Waals interactions, which in the presence of Triton X-100 and NaCl lead to the partial elution of the enzyme from the support and decreased catalytic properties. Further, ionic strength might also affect catalytic properties of the immobilized enzyme by disturbing ionic interactions in the structure of the enzyme [61]. Partial elution of the enzyme and its inhibition by the reaction products are also explained by a slight decrease in the relative activity of the enzyme under repeated use. All of the above-mentioned facts negatively affect storage stability and reusability of the biocatalytic systems produced. Nevertheless, retention of over 80% activity after 30 days of storage, and 10 repeated catalytic cycles by the designed system, suggest that further study ofapplication of MWCNTs and PDMS modified MWCNTs as supportsfor enzyme immobilization are still required. In another study, Jamie et al. [62] immobilized lipase by covalent binding onto MWCNTs modified by n-2-hydroxysuccinimide/(1-ethyl-3-(3-dimethylaminopropyl)carbodiimide) (NHS/EDC) approach. Significant improvement of enzyme activity and operational stability was noticed, which is in agreement with the findings presented in this manuscript. Further, Khan et al. [63] immobilized lipase by adsorption onto MWCNTs treated with NHO_3 and H_2SO_4. In this study, a protective effect of the support material was confirmed; however, it was highlighted that the initial concentration of the enzyme solution plays an important role in the final activity of immobilized lipase.

Recently, lipases of various origin were immobilized using a wide range of support materials, including sol–gel derived silica, zeolites, as well as synthetic polymers and biopolymers [64–68] (Table 3). In the presented studies, usually adsorption immobilization was applied resulting in production of biocatalytic systems characterized by retention of high catalytic activity and significant long-term stability and reusability. Further, application of obtained systems in hydrolysis reactions results in the attaining of over 90% process efficiency. In this context, lipase immobilized using biopolymers (modified chitin, spongin scaffolds), the application of which results in 100% transesterification efficiency, seems to be of particular importance [30,68]. Presented in this study approach, where MWCNTs modified with PDMS were used, results in production of a highly active biocatalytic system that retained 94% of its catalytic activity and over 90% of activity after 20 days of storage and 10 repeated catalytic cycles. High long-term stability and recycle potential of the obtained systems facilitates their potential in real condition applications, for instance in biodiesel production or in the pharmaceutical industry.

Table 3. Comparison of the most important parameters of lipase immobilized using various support materials. n.a.—not available.

Enzyme	Support	Type of Immobilization	Reusability	Storage Stability	Activity Retention	Process Efficiency	Ref.
Lipase from *Rhizomucor miehei*	Pure silica zeolites	Adsorption	60% after 4 catalytic cycles	n.a.	68%	93% of methyl myristate conversion	[64]
Fusarium solanipisi recombinant cutinase with high lipolytic activity	Zeolite	Adsorption	n.a.	89% after 45 days	74%	91% of trycaprylin transformation	[65]
Commercial lipases from *Rhizomucor miehei*	Polypropylene	Adsorption	85% after 8 catalytic cycles	n.a.	over 70%	90% of sunflower oil methanolysis	[66]
Lipase from *Rhizornucor rniehei*	Sol–gel silica	Entrapment	n.a.	75% after 20 days	86%	n.a.	[67]
Lipase B from *Candida antarctica*	*Hippospongiacommunis* spongin scaffolds	Adsorption	82% after 20 catalytic cycles	85% after 20 days	91%	100% of rapeseed oil methanolysis	[30]
Lipase B from *Candida antarctica*	Chitin modified by POSS * compounds	Adsorption	87% after 15 catalytic cycles	90% after 20 days	87%	100% of rapeseed oil methanolysis	[68]
Lipase B from *Candida antarctica*	MWCNTs modified by PDMS	Adsorption	91% after 10 catalytic cycles	90% after 20 days	94%	n.a.	this study

* POSS—polyoctahedralsilsesquioxanes.

5. Conclusions

In the presented study, the fabrication of highly stable and active biocatalysts based on *Candida antarctica* lipase B (CALB) immobilized onto pristine and modified MWCNTs by poly(dimethylsiloxane) was reported. During material characterization, it was proved that the textural characteristics ofMWCNTs change after the modification with polymer, and that the prepared nanocomposites are characterized by a different graphitization degree, which results from, e.g., surface modification of carbon nanotubes with polymer—a lower graphitization degree; graphitic layers are semi-crystalline and possess many defects related to the introduction of new functional groups to carbon nanotube surfaces. Effective MWCNTs with poly(dimethylsiloxane) as well as enzyme loading were confirmed by bands present on FTIR spectra, characteristic for both modifier and biomolecule structures, which all together confirmed relative high potential of synthesized MWCNTs-based materials as a support for lipase immobilization. Enzyme loaded onto P-MWCNTs and MWCNTs/PDMS nanocomposites showed significantly higher relative activity over wider pH and temperature ranges as compared to free counterparts. Moreover, significant improvement of thermal stability and enzyme half-life of the lipase after immobilization was observed. It was confirmed that after immobilization, the external backbone for the enzyme structure was provided due the formation of stable enzyme–support interactions, which stabilize the enzyme structure and protect against biocatalyst denaturation under harsh reaction conditions. This fact suggests wide application potential of designed novel types of biocatalytic systems in various technological/biotechnological applications.

Author Contributions: I.S. and J.Z. conceived and designed the experiments, analyzed and interpreted the data, and wrote the original draft of the manuscript. I.S. calculated the textural parameters for P-MWCNTs and nanocomposites using a self-consistent regularization procedure. D.S. and A.D.-M. participated in the measurement of Raman spectroscopy and the low-temperature nitrogen adsorption–desorption technique for nanocomposites. F.C. and T.J. supervised the work and performed the final reviewing and editing of the manuscript. All authors have read and agreed to the published version of the manuscript.

Funding: The authors acknowledge financial support by the Centre for East European Studies (University of Warsaw) within the framework of "NAGRODA im. I. WYHOWSKIEGO" supporting the scientific internship of I. Sulym at Poznan University of Technology and Maria Curie-Skłodowska University in Lublin (November 2018–May 2019). The financial support of the Visegrad Fund (Contract number 51910525) is greatly appreciated. This research was also partially supported by Ministry of Education and Science Poland.

Institutional Review Board Statement: Not applicable.

Informed Consent Statement: Not applicable.

Data Availability Statement: The datasets supporting the conclusions of this work are included within the article. Any raw data generated and/or analyzed in the present study are available from the corresponding author upon request.

Acknowledgments: The authors would like to express their gratitude to Yuri I. Sementsov (Chuiko Institute of Surface Chemistry, National Academy of Sciences of Ukraine, Kyiv, Ukraine) for providing the pristine MWCNTs.

Conflicts of Interest: The authors declare no conflict of interest.

References

1. Dresselhaus, M.S.; Jorio, A.; Hofmann, M.; Dresselhaus, G.; Saito, R. Perspectives on Carbon Nanotubes and Graphene Raman Spectroscopy. *Nano Lett.* **2010**, *10*, 751–758. [CrossRef]
2. Xu, T.; Yang, J.; Liu, J.; Fu, Q. Surface modification of multi-walled carbon nanotubes by O_2 plasma. *Appl. Surf. Sci.* **2007**, *253*, 8945–8951. [CrossRef]
3. Yudianti, R. Analysis of Functional Group Sited on Multi-Wall Carbon Nanotube Surface. *Open Mater. Sci. J.* **2011**, *5*, 242–247. [CrossRef]
4. Karousis, N.; Tsotsou, G.-E.; Evangelista, F.; Rudolf, P.; Ragoussis, N.; Tagmatarchis, N. Carbon Nanotubes Decorated with Palladium Nanoparticles: Synthesis, Characterization, and Catalytic Activity. *J. Phys. Chem. C* **2008**, *112*, 13463–13469. [CrossRef]

5. Ciraci, S.; Dag, S.; Yildirim, T.; Gülseren, O.; Senger, R.T. Functionalized carbon nanotubes and device applications. *J. Phys. Condens. Matter* **2004**, *16*, R901–R960. [CrossRef]
6. Holger, F.; Bettinger, D. *Carbon Nanotubes: Basic Concepts and Physical Properties*; WILEY-VCH Verlag GmbH & Co. KGaA: Weinheim, Germany, 2004.
7. Ibrahim, K.S. Carbon nanotubes-properties and applications: A review. *Carbon Lett.* **2013**, *14*, 131–144. [CrossRef]
8. Wepasnick, K.A.; Smith, B.A.; Bitter, J.L.; Fairbrother, D.H. Chemical and structural characterization of carbon nanotube surfaces. *Anal. Bioanal. Chem.* **2010**, *396*, 1003–1014. [CrossRef] [PubMed]
9. He, X.; Kitipornchai, S.; Wang, C.; Liew, K. Modeling of van der Waals force for infinitesimal deformation of multi-walled carbon nanotubes treated as cylindrical shells. *Int. J. Solids Struct.* **2005**, *42*, 6032–6047. [CrossRef]
10. Ebbesen, T.W.; Lezec, H.J.; Hiura, H.; Bennett, J.W.; Ghaemi, H.F.; Thio, T. Electrical conductivity of individual carbon nanotubes. *Nature* **1996**, *382*, 54–56. [CrossRef]
11. Mallakpour, S.; Soltanian, S. Surface functionalization of carbon nanotubes: Fabrication and applications. *RSC Adv.* **2016**, *6*, 109916–109935. [CrossRef]
12. Moniruzzaman, M.; Winey, K.I. Polymer Nanocomposites Containing Carbon Nanotubes. *Macromolecules* **2006**, *39*, 5194–5205. [CrossRef]
13. Fujigaya, T.; Nakashima, N. Non-covalent polymer wrapping of carbon nanotubes and the role of wrapped polymers as functional dispersants. *Sci. Technol. Adv. Mater.* **2015**, *16*, 024802. [CrossRef]
14. Jin, M.; Feng, X.; Xi, J.; Zhai, J.; Cho, K.; Feng, L.; Jiang, L. Super-Hydrophobic PDMS Surface with Ultra-Low Adhesive Force. *Macromol. Rapid Commun.* **2005**, *26*, 1805–1809. [CrossRef]
15. Wang, Y.; Huang, Z.; Gurney, R.S.; Liu, D. Superhydrophobic and photocatalytic PDMS/TiO_2 coatings with environmental stability and multifunctionality. *Colloids Surf. Physicochem. Eng. Asp.* **2019**, *561*, 101–108. [CrossRef]
16. Shena, C.; Wang, H.; Zhang, T.; Zeng, Y.J. Silica coating onto graphene for improving thermal conductivity and electrical insulation of graphene/polydimethylsiloxane nanocomposites. *Mater. Res. Technol.* **2019**, *35*, 36–43. [CrossRef]
17. Deng, W.; Lei, Y.; Lin, Y.; Zhou, S.; Zhang, A. Absorptive supramolecular elastomer wound dressing based on polydimethylsiloxane–(polyethylene glycol)–polydimethylsiloxane copolymer: Preparation and characterization. *RSC Adv.* **2016**, *6*, 51694–51702. [CrossRef]
18. Kong, K.; Mariatti, M.; Rashid, A.; Busfield, J. Enhanced conductivity behavior of polydimethylsiloxane (PDMS) hybrid composites containing exfoliated graphite nanoplatelets and carbon nanotubes. *Compos. Part B Eng.* **2014**, *58*, 457–462. [CrossRef]
19. Guo, Y.; Pan, L.; Yang, X.; Ruan, K.; Han, Y.; Kong, J.; Gu, J. Simultaneous improvement of thermal conductivities and electromagnetic interference shielding performances in polystyrene composites via constructing interconnection oriented networks based on electrospinning technology. *Compos. Part A Appl. Sci. Manuf.* **2019**, *124*, 840–848. [CrossRef]
20. Meng, L.-Y.; Park, S.-J. Superhydrophobic carbon-based materials: A review of synthesis, structure, and applications. *Carbon Lett.* **2014**, *15*, 89–104. [CrossRef]
21. Klonos, P.; Sulym, I.; Borysenko, M.; Kripotou, S.; Kyritsis, A.; Pissis, P.; Gun'Ko, V. Interfacial interactions and complex segmental dynamics in systems based on silica-polydimethylsiloxane core–shell nanoparticles: Dielectric and thermal study. *Polymers* **2015**, *58*, 9–21. [CrossRef]
22. Klonos, P.; Sulym, I.; Kyriakos, K.; Vangelidis, I.; Zidropoulos, S.; Sternik, D.; Borysenko, M.; Kyritsis, A.; Deryło-Marczewska, A.; Gun'Ko, V. Interfacial phenomena in core–shell nanocomposites of PDMS adsorbed onto low specific surface area fumed silica nanooxides: Effects of surface modification. *Polymers* **2015**, *68*, 158–167. [CrossRef]
23. Klonos, P.; Dapei, G.; Sulym, I.Y.; Zidropoulos, S.; Sternik, D.; Deryło-Marczewska, A.; Borysenko, M.V.; Gun'Ko, V.M.; Kyritsis, A.; Pissis, P. Morphology and molecular dynamics investigation of PDMS adsorbed on titania nanoparticles: Effects of polymer molecular weight. *Eur. Polym. J.* **2016**, *74*, 64–80. [CrossRef]
24. Klonos, P.; Sulym, I.Y.; Sternik, D.; Konstantinou, P.; Goncharuk, O.V.; Deryło–Marczewska, A.; Gun'Ko, V.M.; Kyritsis, A.; Pissis, P. Morphology, crystallization and rigid amorphous fraction in PDMS adsorbed onto carbon nanotubes and graphite. *Polymers* **2018**, *139*, 130–144. [CrossRef]
25. Sulym, I.; Kubiak, A.; Jankowska, K.; Sternik, D.; Terpilowski, K.; Sementsov, Y.; Borysenko, M.; Derylo-Marczewska, A.; Jesionowski, T. Superhydrophobic MWCNTs/PDMS-nanocomposite materials: Preparation and characterization. *Physicochem. Probl. Miner. Process.* **2019**, *55*, 1394–1400. [CrossRef]
26. Rodgers, L.; Knott, R.; Holden, P.; Pike, K.; Hanna, J.; Foster, L.; Bartlett, J. Structural evolution and stability of sol–gel biocatalysts. *Phys. B Condens. Matter.* **2006**, *385–386*, 508–510. [CrossRef]
27. Kovalenko, G.A.; Perminova, L.V.; Beklemishev, A.B.; Parmon, V.N. Heterogeneous Biocatalysts Prepared by Immuring Enzymatic Active Components inside Silica Xerogel and Nanocarbons-In-Silica Composites. *Catalysts* **2018**, *8*, 177. [CrossRef]
28. Sarmah, N.; Revathi, D.; Sheelu, G.; Rani, K.Y.; Sridhar, S.; Mehtab, V.; Sumana, C. Recent advances on sources and industrial applications of lipases. *Biotechnol. Prog.* **2018**, *34*, 5–28. [CrossRef]
29. Rodríguez, K.; Martinez, R.; Bernal, C. Selective immobilization of *Bacillus subtilis* lipase A from cell culture supernatant: Improving catalytic performance and thermal resistance. *Process. Biochem.* **2020**, *92*, 214–223. [CrossRef]
30. Zdarta, J.; Norman, M.; Smułek, W.; Moszyński, D.; Kaczorek, E.; Stelling, A.L.; Ehrlich, H.; Jesionowski, T. Spongin-Based Scaffolds from *Hippospongia communis* Demosponge as an Effective Support for Lipase Immobilization. *Catalysts* **2017**, *7*, 147. [CrossRef]

31. Zhang, J.; Zhang, F.; Yang, H.; Huang, X.; Liu, H.; Zhang, J.; Guo, S. Graphene Oxide as a Matrix for Enzyme Immobilization. *Langmuir* **2010**, *26*, 6083–6085. [CrossRef]
32. Asmat, S.; Anwer, A.H.; Husain, Q. Immobilization of lipase onto novel constructed polydopamine grafted multiwalled carbon nanotube impregnated with magnetic cobalt and its application in synthesis of fruit flavours. *Int. J. Biol. Macromol.* **2019**, *140*, 484–495. [CrossRef]
33. Yan, Y.; Miao, J.; Yang, Z.; Xiao, F.-X.; Yang, H.B.; Liu, B.; Yang, Y. Carbon nanotube catalysts: Recent advances in synthesis, characterization and applications. *Chem. Soc. Rev.* **2015**, *44*, 3295–3346. [CrossRef]
34. Wang, L.; Liu, X.; Jiang, Y.; Zhou, L.; Ma, L.; He, Y.; Gao, J. Biocatalytic Pickering Emulsions Stabilized by Lipase-Immobilized Carbon Nanotubes for Biodiesel Production. *Catalysts* **2018**, *8*, 587. [CrossRef]
35. Zhang, C.; Luo, S.; Chen, W. Activity of catalase adsorbed to carbon nanotubes: Effects of carbon nanotube surface properties. *Talanta* **2013**, *113*, 142–147. [CrossRef]
36. Pavlidis, I.V.; Tsoufis, T.; Enotiadis, A.; Gournis, D.; Stamatis, H. Functionalized Multi-Wall Carbon Nanotubes for Lipase Immobilization. *Adv. Eng. Mater.* **2010**, *12*, B179–B183. [CrossRef]
37. Wang, L.; Xu, R.; Chen, Y.; Jiang, R. Activity and stability comparison of immobilized NADH oxidase on multi-walled carbon nanotubes, carbon nanospheres, and single-walled carbon nanotubes. *J. Mol. Catal. B Enzym.* **2011**, *69*, 120–126. [CrossRef]
38. Feng, W.; Ji, P. Enzymes immobilized on carbon nanotubes. *Biotechnol. Adv.* **2011**, *29*, 889–895. [CrossRef] [PubMed]
39. Reis, P.; Holmberg, K.; Watzke, H.; Leser, M.; Miller, R. Lipases at interfaces: A review. *Adv. Colloid Interface Sci.* **2009**, *147–148*, 237–250. [CrossRef]
40. Kartel, M.; Sementsov, Y.; Mahno, S.; Trachevskiy, V.; Bo, W. Polymer Composites Filled with Multiwall Carbon Nanotubes. *Univ. J. Mater. Sci.* **2016**, *4*, 23–31. [CrossRef]
41. Melezhyk, A.V.; Sementsov, Y.I.; Yanchenko, V.V. Synthesis of thin carbon nanotubes on co-precipitated metaloxide catalysts. *Russ. J. Appl. Chem.* **2005**, *78*, 938–946.
42. Gun'ko, V.M. Composite materials: Textural characteristics. *Appl. Surf. Sci.* **2014**, *307*, 444–454. [CrossRef]
43. Gun'ko, V.M.; Mikhalovsky, S.V. Evaluation of slitlike porosity of carbon adsorbents. *Carbon* **2004**, *42*, 843–849. [CrossRef]
44. Gregg, S.J.; Sing, K.S.W. *Adsorption, Surface Area and Porosity*, 2nd ed.; Academic Press: London, UK, 1982.
45. Kruk, M.; Jaroniec, M. Gas Adsorption Characterization of Ordered Organic−Inorganic Nanocomposite Materials. *Chem. Mater.* **2001**, *13*, 3169–3183. [CrossRef]
46. Dresselhaus, M.; Dresselhaus, G.; Jorio, A.; Filho, A.S.; Saito, R. Raman spectroscopy on isolated single wall carbon nanotubes. *Carbon* **2002**, *40*, 2043–2061. [CrossRef]
47. Dresselhaus, M.; Dresselhaus, G.; Saito, R.; Jorio, A. Raman spectroscopy of carbon nanotubes. *Phys. Rep.* **2005**, *409*, 47–99. [CrossRef]
48. Johnson, L.M.; Gao, L.; Iv, C.W.S.; Smith, M.; Efimenko, K.; Cushing, K.; Genzer, J.; López, G.P. Elastomeric microparticles for acoustic mediated bioseparations. *J. Nanobiotechnol.* **2013**, *11*, 22. [CrossRef]
49. Sulym, I.; Goncharuk, O.; Sternik, D.; Terpilowski, K.; Derylo-Marczewska, A.; Borysenko, M.V.; Gun'Ko, V.M. Nanooxide/Polymer Composites with Silica@PDMS and Ceria–Zirconia–Silica@PDMS: Textural, Morphological, and Hydrophilic/Hydrophobic Features. *Nanoscale Res. Lett.* **2017**, *12*, 1–10. [CrossRef]
50. Wang, S.; Li, T.; Wu, L.; Zhang, L.; Dong, L.; Hu, X.; Li, C.-Z. Second-order Raman spectroscopy of char during gasification. *Fuel Process. Technol.* **2015**, *135*, 105–111. [CrossRef]
51. Wong, P.T.; Wong, R.K.; Caputo, T.A.; Godwin, T.A.; Rigas, B. Infrared spectroscopy of exfoliated human cervical cells: Evidence of extensive structural changes during carcinogenesis. *Proc. Natl. Acad. Sci. USA* **1991**, *88*, 10988–10992. [CrossRef]
52. Portaccio, M.B.E.; Della Ventura, B.; Mita, D.G.; Manolova, N.; Stoilova, O.; Rashkov, I.; Lepore, M. FT-IR microscopy characterization of sol–gel layers prior and after glucose oxidase immobilization for biosensing applications. *J. SolGel Sci. Technol.* **2011**, *57*, 204–211. [CrossRef]
53. Colla, L.M.; Ficanha, A.M.M.; Rizzardi, J.; Bertolin, T.E.; Reinehr, C.O.; Costa, J.A.V. Production and Characterization of Lipases by Two New Isolates of Aspergillus through Solid-State and Submerged Fermentation. *BioMed Res. Int.* **2015**, *2015*, 725959. [CrossRef]
54. Zhu, Y.-T.; Ren, X.-Y.; Liu, Y.-M.; Wei, Y.; Qing, L.-S.; Liao, X. Covalent Immobilization of Porcine Pancreatic Lipase on Carboxyl-Activated Magnetic Nanoparticles: Characterization and Application For Enzymatic Inhibition Assays. *Mater. Sci. Eng. C* **2014**, *38*, 278–285. [CrossRef]
55. Narwal, S.K.; Saun, N.K.; Gupta, R. Characterization and Catalytic Properties of Free and Silica-Bound Lipase: A Comparative Study. *J. Oleo Sci.* **2014**, *63*, 599–605. [CrossRef]
56. Peña, S.A.; Rios, N.S.; Carballares, D.; Gonçalves, L.R.; Fernandez-Lafuente, R. Immobilization of lipases via interfacial activation on hydrophobic supports: Production of biocatalysts libraries by altering the immobilization conditions. *Catal. Today* **2021**, *362*, 130–140. [CrossRef]
57. Rodrigues, R.C.; Virgen-Ortíz, J.J.; dos Santos, J.C.; Berenguer-Murcia, Á.; Alcantara, A.R.; Barbosa, O.; Ortiz, C.; Fernandez-Lafuente, R. Immobilization of lipases on hydrophobic supports: Immobilization mechanism, advantages, problems, and solutions. *Biotechnol. Adv.* **2019**, *37*, 746–770. [CrossRef] [PubMed]
58. Li, W.; Shen, H.; Tao, Y.; Chen, B.; Tan, T. Amino silicones finished fabrics for lipase immobilization: Fabrics finishing and catalytic performance of immobilized lipase. *Process. Biochem.* **2014**, *49*, 1488–1496. [CrossRef]

59. Hu, Y.; Dai, L.; Liu, D.; Du, W. Hydrophobic pore space constituted in macroporous ZIF-8 for lipase immobilization greatly improving lipase catalytic performance in biodiesel preparation. *Biotechnol. Biofuels* **2020**, *13*, 1–9. [CrossRef]
60. Dong, L.; Ge, C.; Qin, P.; Chen, Y.; Xu, Q. Immobilization and catalytic properties of candida lipolytic lipase on surface of organic intercalated and modified MgAl-LDHs. *Solid State Sci.* **2014**, *31*, 8–15. [CrossRef]
61. Lee, J.H.; Kim, S.B.; Park, C.; Kim, S.W. Effect of a buffer mixture system on the activity of lipases during immobilization process. *Bioresour. Technol.* **2010**, *101*, S66–S70. [CrossRef]
62. Jamie, A.; Alshami, A.S.; Maliabari, Z.O.; Ateih, M.A.; Al Hamouz, O.C.S. Immobilization and enhanced catalytic activity of lipase on modified MWCNT for oily wastewater treatment. *Environ. Prog. Sustain. Energy* **2016**, *35*, 1441–1449. [CrossRef]
63. Khan, A.K.; Mubarak, N.M.; Abdullah, E.; Khalid, M.; Nizamuddin, S.; Baloch, H.A.; Siddiqui, M. Immobilization of Lipase Enzyme Carbon Nanotubes via Adsorption. *IOP Conf. Ser. Mater. Sci. Eng.* **2019**, *495*, 012055. [CrossRef]
64. Macario, A.; Giordano, G.; Frontera, P.; Crea, F.; Setti, L. Hydrolysis of Alkyl Ester on Lipase/Silicalite-1 Catalyst. *Catal. Lett.* **2007**, *122*, 43–52. [CrossRef]
65. Gonqalves, A.P.V.; Lopes, J.M.; Lemos, F.; Ram, F.; Ramoa Ribeiro, F.; Prazeres, D.M.F.; Cabral, J.M.S.; Aires-Barros, M. Effect of the immobilization support on the hydrolytic activity of a cutinase from *Fusarium solanipisi*. *J. Mol. Catal. B Enzym.* **1996**, *1*, 53–60.
66. Soumanou, M.M.; Bornscheuer, U.T. Improvement in lipase-catalyzed synthesis of fatty acid methyl esters from sunflower oil. *Enzym. Microb. Technol.* **2003**, *33*, 97–103. [CrossRef]
67. Reetz, M.T.; Zonta, A.; Simpelkamp, J. Efficient immobilization of lipases by entrapment in hydrophobic sol-gel materials. *Biotechnol. Bioeng.* **1996**, *49*, 527–534. [CrossRef]
68. Zdarta, J.; Wysokowski, M.; Norman, M.; Kołodziejczak-Radzimska, A.; Moszyński, D.; Maciejewski, H.; Ehrlich, H.; Jesionowski, T. Candida antarctica Lipase B Immobilized onto Chitin Conjugated with POSS® Compounds: Useful Tool for Rapeseed Oil Conversion. *Int. J. Mol. Sci.* **2016**, *17*, 1581. [CrossRef] [PubMed]

materials

MDPI

Review

Review on Preformed Crowns in Pediatric Dentistry—The Composition and Application

Klaudia Sztyler [1,*], Rafal J. Wiglusz [2,*] and Maciej Dobrzynski [1,*]

1 Department of Pediatric Dentistry and Preclinical Dentistry, Wroclaw Medical University, Krakowska 26, 50-425 Wroclaw, Poland

2 Institute of Low Temperature and Structure Research, Polish Academy of Sciences, Okolna 2, 50-422 Wroclaw, Poland

* Correspondence: klaudia.sztyler@umw.edu.pl (K.S.); r.wiglusz@intibs.pl (R.J.W.); maciej.dobrzynski@umw.edu.pl (M.D.); Tel.: +48-71-7840-361 (K.S. & M.D.); +48-71-3954-159 (R.J.W.)

Abstract: The purpose of this review is to compare and contrast the various types of preformed crowns that can be used to restore the primary teeth in children. Historically, preformed crowns have been widely available for the past 50 years. The clinical performance of preformed crowns has evolved to meet higher functional, mechanical, and aesthetic demands. Preformed crowns are available in a range of prefabricated sizes and shapes. Preformed crowns can vary depending on their properties, compounds, methods of preparation, and biocompatibility.

Keywords: pediatric crowns; primary teeth; material composition; teeth restoration; repair procedures

Citation: Sztyler, K.; Wiglusz, R.J.; Dobrzynski, M. Review on Preformed Crowns in Pediatric Dentistry—The Composition and Application. *Materials* **2022**, *15*, 2081. https://doi.org/10.3390/ma15062081

Academic Editors: Jakub Zdarta and Agata Zdarta

Received: 31 December 2021
Accepted: 3 March 2022
Published: 11 March 2022

Publisher's Note: MDPI stays neutral with regard to jurisdictional claims in published maps and institutional affiliations.

1. Introduction

Dental caries is one of the most widespread medical conditions both in adults and children [1]. According to WHO Oral Health facts, more than 530 million children suffer from dental caries of primary teeth [2]. A survey from 2019 conducted by Public Health England shows that one in four five-year-olds have had dental caries [3]. The National Dental Inspection Program in Scotland found that 15% of Scottish children had at least one tooth extracted due to caries before the age of five [4]. This number increases to 42% for eight-year-olds [5]. The consequences of primary tooth decay include local and systemic problems. Local potential problems include pain due to pulp or periapical tissue inflammation and infection of permanent tooth buds, which can interfere with odontogenesis and cause a defect called Turner's tooth. Premature loss of primary teeth can potentially cause malocclusion, tongue movement disorders, chewing disorders, change of facial features, and behavioral difficulties. The presence of caries in primary teeth increases the risk of caries in permanent teeth. Systemic implications include symptoms of infection such as high temperature and apathy. Long-term pain can lead to eating difficulties, which can cause weight loss and growth and development disorders. This illustrates how crucial it is to keep primary teeth healthy and prevent the development of caries [6]. If caries develops, it is highly recommended to implement treatment as soon as possible. The treatment options for irreversible caries contain nonrestorative cavity control [7] including the Hall technique [8] and techniques associated with the removal of caries. Removal can be performed selectively, either through selective caries removal or stepwise caries removal [9,10], or non-selectively by removing all demineralized dentin. The use of the latter-mentioned procedure has not been recommended [11]. After preparation, the dental tissue must be restored. Choosing the best method of restoration is another important step to provide the best treatment results. The most chosen restoration material is glass ionomers, composite resin, compomer, and amalgam, or prefabricated crowns [9]. Many studies compare these materials to each other in terms of durability, secondary caries, endodontic complications, or restoration loss. According to many studies, the best outcome is achieved by using prefabricated crowns [12–14]. A study by Isabel Cristina et al. analyzed the survival rate of stainless-steel crowns compared to composite restorations in primary teeth undergoing endodontic treatment. The study showed that the overall success rate was higher in the SSC group (88%) compared to composite restorations (75%), but more importantly, failure of composite restoration resulted in endodontic treatment failure, whereas endodontic failure in SSCs group was not associated with restoration failure [15]. Stainless steel crowns present the highest success rate, with a performance level of 96.1%. In comparison, resin modified glass-ionomer shows a 93.6% success rate, compomer 91.2%, and metal-reinforced glass ionomer cement 57.4% [16]. Roberts et al. compared the survival of resin-modified glass ionomer cement (RMGIC) to PMC. Their study shows that PMCs present a very high survival rate, especially for large cavities and teeth that have undergone endodontic treatment [13]. It is also worth noting that prefabricated crowns can be used to restore non-carious lesions or developmental defects such as hypoplasia, hypomineralization, etc. [17].

2. Materials and Methods

Pediatric crowns are a prefabricated solution for full crown restoration of deciduous teeth. They are available in sets containing different sizes and shapes dedicated to primary teeth. The first commercially available primary crowns were made of steel and contained large amounts of nickel. Subsequent generations of crowns had improved the composition of metal crown material, as well as enhancement of their prefabricated shape. Today, most pediatric steel crowns are made of stainless steel, however, their compositions can vary. Due to the poor aesthetics of grey stainless-steel crowns, patients prefer white crowns to match the color of other natural teeth. Most crowns include pre-veneered stainless-steel crowns (PVSSCs), crowns made of polymers, pre-veneered aluminum crowns, or prefabricated zirconia crowns. These types of crowns are intended to mimic the natural color of the teeth, which is highly desired by parents, especially in the anterior segment of the dental arch. The use of crowns allows the reconstruction of teeth with severe damage, both caused by caries and processes associated with the disruption of hard tissue development. They are often the best, or one of the few solutions determining clinical success and maintenance of the tooth in the oral cavity until physiological tooth replacement.

This review will focus on basic types of prefabricated pediatric crowns by comparing their characteristics, indications, contraindications, advantages, and disadvantages.

It should be noted that the quantitative compositions of individual brands of preformed pediatric crowns can vary. The information on composition was gathered from product safety data sheets. The selected publications contain a thorough description of each group of crowns, their properties, advantages, disadvantages, indications, contraindications, preparation protocols, and treatment results (Figure 1) [18].

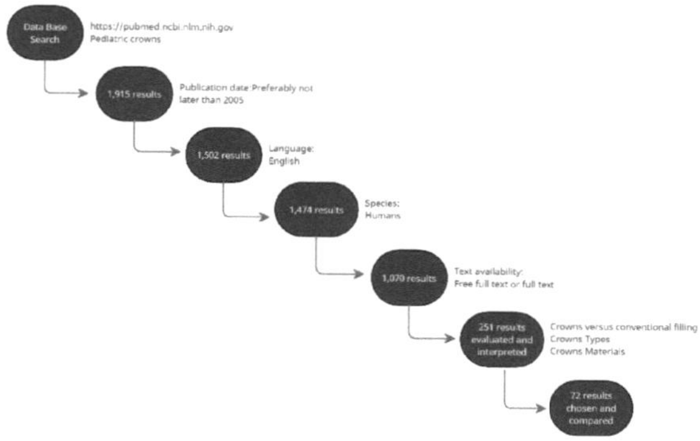

Figure 1. A flowchart of the database search strategy (accessed on 1 August 2021).

3. Crowns

Prefabricated crowns have been widely used in pediatric dentistry for the last 50 years [19]. The three recently most used ones are preformed metal crowns, resin veneered stainless steel crowns, and strip crowns [20]. More aesthetic solutions include pre-veneered crowns and zirconia crowns. The use of pediatric crowns involves the proper preparation of the tooth crown for it to fit well, or it may involve no preparation and the use of the Hall technique. In the Hall technique, the reconstruction of the tooth is performed without local anesthesia by placing the crown on the remaining tooth tissue and pressing it to the correct position using finger pressure or the patient's occlusion force. The use of crowns is especially recommended for teeth after pulp treatment or with advanced decay damage. They are also a good solution in the case of developmental disorders of dental hard tissue. They can also be used as a method of reconstruction of deciduous teeth during procedures under general anesthesia. Their primary purpose is to allow tight restoration, with a long-term positive outcome and without major failures. The overall procedure should cause as little pain and as little trauma for the young patient as possible. Innes et al. in their systematic review conclude that the use of crowns is associated with a reduced risk of major failure, pain, and formation of abscess in the long term compared to conventional restorations. The use of crowns may be associated with a higher risk of gingival bleeding [17]. There are a small number of studies comparing different types of crowns. Therefore, we cannot, in the current state of knowledge, say which crowns are best. This refers to the use of zirconia crowns as a replacement for stainless steel crowns [21]. Pediatric crowns should be easy to adapt, the bonding strength should be high enough to withstand masticatory forces. Crown materials should be safe for antagonist teeth. They should not hamper oral hygiene maintenance and should be biocompatible with the surrounding tissue [22–24]. Sahana S et al. divide crowns into two groups (see Figure 2) [25].

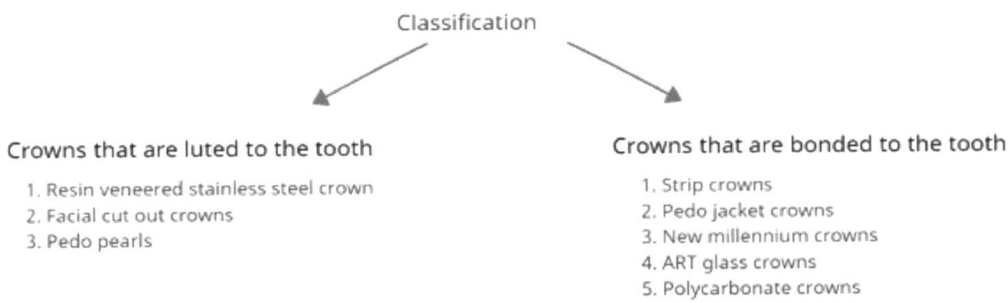

Figure 2. Classification of crowns used in pediatric dentistry.

3.1. Preformed Metal Crowns

Preformed metal crowns (PMCs) can be divided into two groups depending on the composition. The first group is represented by stainless steel crowns (SSCs), while the second group include nickel-chromium crowns [19]. SSCs were introduced to dentistry in 1950. Their first prototypes were usually too large, with straight and very long sides. Proper adaptation to teeth required many steps such as trimming, contouring, crimping, and finishing [26,27]. The next generation of SCCs was focused on improving the imitation of natural tooth anatomy, which would help simplify the adaptation process. Introducing nickel-chromium crowns helped eliminate many disadvantages found in the first SSCs. Firstly, Cr-Ni crowns are fully shaped and resistant to defects [27]. Secondly, thanks to improved anatomical accuracy, they rarely require trimming [27]. It is worth noting that they also need to be modified to improve adaptation, but they usually need fewer steps to achieve it. Figure 3 presents an example of modern preformed metal crowns available on the market (below).

Figure 3. Preformed metal crowns.

Preformed metal crowns (PMCs) are represented on the market by three main types of crowns. The first is untrimmed crowns. They require long adaptation due to a lack of trimming and contouring in the production process. An example of these crowns is Rocky Mountain Crowns. The second type are pre-trimmed crowns. The sides of these crowns are straight and festooned to follow the gingival crest line. They still require contouring and sometimes need to be trimmed [28]. The third type are pre-contoured crowns. The sides of these crowns are pre-contoured and festooned. They show the best imitation of anatomical geometry, yet occasionally minimal trimming and recontouring are required in the adaptation process. An example of these crowns is Unitek Stainless Steel Primary Crowns by 3M [28]. PMC usually contains 67% iron, 10–13% nickel, 17–19% chromium, and 4% of minor elements, although each brand available on the market has a slightly different composition in Table 1.

Table 1. The composition of preformed metal crowns.

Brand	Composition	Toxicity	Ref
DENOVO Stainless Steel Crowns DENOVO DENTAL	Iron—69.22% Chromium—18.12% Nickel—9.58% Manganese—1.59% Silicon—0.52% Molybdenum—0.39% Copper—0.35% Cobalt—0.18% Carbon—0.05% Phosphorus—Trace Amounts Titanium—Trace Amounts Sulphur—Trace Amounts Aluminum—Trace Amounts Oxygen—Trace Amounts	Iron—30,000 mg/kg Oral—Rat Chromium—>9000 mg/kg Oral—Rat Nickel—n/a Manganese—9000 mg/kg Oral—Rat Silicon—n/a Molybdenum—n/a Copper—3160 mg/kg Cobalt—n/a Carbon—6171 mg/kg Oral—Rat Boric Acid—n/a Petroleum distillates—n/a Potassium Fluoborate—n/a Potassium Fluoride—n/a	[29]

Table 1. *Cont.*

Brand	Composition	Toxicity	Ref
Hu-Friedy PEDO CROWNS Hu-Friedy	Carbon—0.03% Sulphur—0.03% Silicon—0.75% Molybdenum—2.00% Phosphorus—0.045% Copper—0.22% Molybdenum—n/a Nickel—8.0–12% Chromium—18–20% Cobalt—n/a Iron—69.00%	n/a	[30]
Primary Stainless Steel Crowns 3M ESPE	Stainless steel 12597-68-1—100% Iron—65–74% Chromium—17–19% Nickel—9–13%	n/a calculated acute toxicity estimate >5000 mg/kg	[31] [32]
Unitek Primary SSC 3M ESPE	Stainless steel 12597-68-1—100% Iron—65–74% Chromium—17–19% Nickel—9–13%	n/a calculated acute toxicity estimate >5000 mg/kg	[31] [32]
KTR DentalCrown KTR	Iron—70–90% Chromium—15–35% Nickel—5–10% Manganese—2.5% Silicon—2.5% Copper—2.5%	n/a	[33]
Acero Stainless Steel Crown Acero Crowns	n/a	n/a	

Indications for use in primary molar teeth [19,27,28,34]:

1. Decay affecting two or more tooth surfaces;
2. Inability to place an amalgam filling;
3. Restoration after pulp treatment procedures;
4. Restoration in non-carious lesions or developmental defects;
5. Restoration of fractured primary molars;
6. Severe bruxism;
7. Restoration in children who require general anesthesia for treatment;
8. In children with a high risk and high susceptibility to caries;
9. An abutment for a space maintainer.

Indications for use in permanent molar teeth [19]:

1. Temporary restoration after tooth fracture;
2. Temporary restoration until orthodontic opinion and treatment plan;
3. Temporary restoration before final prosthetic restoration;
4. Economic reasons;
5. Restoration in non-carious lesions or developmental defects;
6. Restoration of a permanent molar that must cover the entire crown.

Contraindications for the use of PMC [28]:

1. Allergy or vulnerability to nickel;
2. Uncooperative patient;

3. A primary tooth near its exfoliation time;
4. A radiograph showing resorption of more than half of the tooth root.

The main advantages of PMCs are their capability of tight and long-term survival restoration even when the risk of caries is high. The application procedure is simple, which lowers the possibility of errors during the treatment. Lastly, they have the advantage of multiple clinical indications [13,14,19,35]. Their main disadvantage is low aesthetics. They also cannot be used on partially erupted teeth [19]. Possible complications during the preparation and application of PMCs are crown draft, usually towards a massively destructed wall, interproximal ledge, or poor margins when crowns are incorrectly adapted [36–40].

Preparation [27,28] (Figure 4):

1. Prior to placing the PMC, the dentist should discuss the treatment with the child and the parents/guardians and obtain their consent;
2. The dentist should estimate the crown size, to enable it to click into place. When choosing the appropriate crown size, it is recommended to measure the mesial-distal width between the contact points of the adjacent teeth with calipers. If this cannot be achieved, the mesial width of the contralateral tooth in the opposite arch can be measured. It is advised that the smallest matching crown should be selected;
3. Local anesthesia and tooth isolation;
4. If necessary, caries removal, pulp therapy;
5. Tooth restoration with glass-ionomer cement or compomer;
6. Occlusal reduction of about 1.5 mm;
7. Mesial and distal reduction, so that the probe can pass through, maximally 1 mm;
8. No buccal and lingual reduction or minimal reduction;
9. Try in of the crown; the crown should go maximally 1 mm subgingival, if it goes deeper, it requires adaptation. Trimming is performed with special crown scissors or an abrasive wheel. After trimming, the crown needs to be crimped with crimping pliers. Finally, the margins should be thinned with white stone and finely polished;
10. Cementation with the use of resin-modified glass ionomer, polycarboxylate phosphate cements, or RelyX™ Luting Plus Cement. Placing is usually performed from the lingual side and rolled during the preparation to the buccal margin.

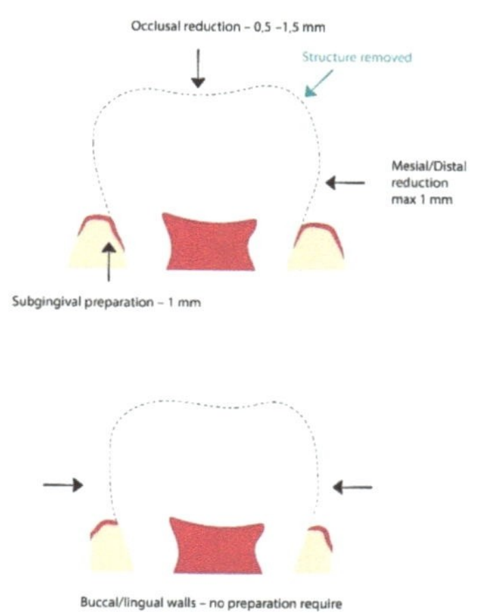

Figure 4. Proper preparation for preformed metal crowns. The dashed line shows the range of hard tissue removal, which will later be used for proper crown fit and restoration [25].

Sometimes crown adaptation involves various difficulties. If the smallest size of PMC is too large to be used for restoration, metal edges can be sized down by cutting and overlapping, which results in the reduction in crown circumference. In the next step, overlapped margins are welded together [27]. On the other hand, if the largest PMC available is too small, the best fitting crown can be cut, and filled with an additional piece of orthodontic stainless steel band material welded over the space [41]. When multiple crowns are fitted and adjusted, it is recommended to reduce proximal surfaces more than usual [42]. Another option for adjusting PMC is the Hall technique.

3.2. Hall Technique

The Hall Technique is a method of adjusting and establishing PMCs which belongs to nonrestorative cavity control techniques [43]. The first step of this technique is to separate the tooth by using dental floss or separating pliers [44,45]. Separators are left in for approximately five days. After that, the separators are removed, and the proper size of the crown is estimated and placed on a tooth with glass ionomer cement [8]. The crown is fitted on a tooth with the pressure exerted by the dentist with his or her fingers or with the child's bite force [46]. The last step is the removal of excess cement. This procedure does not require local anesthesia, caries removal, or tooth preparation. The idea of this technique is to arrest caries and force them to change into less cariogenic flora, which will stop or at least slow down the progression of caries [47].

Indications for the use of the Hall technique in primary molars [44]:

1. Used in occlusal caries or non-cavitated teeth, if the patient is unable to tolerate fissure sealant, partial caries removal, or conventional restoration;
2. Proximal caries or non-cavitated teeth if the patient is unable to tolerate partial caries removal or conventional restoration.

Contradictions for the use of the Hall technique in primary molars [7]:

1. Pulp infection;
2. Irreversible pulpitis;
3. Pulp exposure;
4. Lack of clear band of dentine on the radiograph;
5. Clinical or radiological signs of peri-radicular pathology;
6. Extremely damaged crowns.

This technique presents many advantages. Firstly, it is very non-invasive, because it does not require the injection of local anesthesia, caries removal, or tooth preparation [48]. The procedure is quick and less traumatic, which can improve the child's future cooperation [48]. Some authors show that it is also more cost-effective than conventional restorations [49,50]. On the other hand, this technique is controversial and causes some concerns. Firstly, there is no tooth preparation which leaves the crown without additional space, which in turn leads to premature occlusal contact after cementation. In the days following the procedure, biting forces resolve premature contact and usually after one or two days the juvenile patient does not feel discomfort [43]. It is worth noting that further studies are needed to analyze the effect of the Hall technique on occlusion and the temporomandibular joint [46]. Another important disadvantage is the use of non-aesthetic PMCs. Overall, the Hall technique shows promising results. Data show that it can provide a lack of pain and infection and general high effectiveness [51–53].

3.3. Open-Faced Stainless Steel Crowns

This is a form of the use of SSCs in the anterior section of the dental arch. The procedure includes adapting proper SSC. If needed, the crown is trimmed, crimped, and polished. After the crown is cemented and the cement sets, the labial wall of the crown is cut out and the luting cement is partially removed to create undercuts. In the following step, the space is filled with a more aesthetic material such as composite [19,54].

Indications to use open-faced SSC [54]:

1. Crown fracture;
2. Pulp protection.

Contraindications to use open-faced SSC [54]:

1. Allergy or vulnerability to nickel;
2. Uncooperative patient;
3. A primary tooth near its exfoliation time;
4. A radiograph showing resorption of more than half of the tooth root;
5. Tooth fracture level below gingival margin.

Their main advantage is better aesthetics compared to traditional SSCs, however, the procedure is time-consuming and requires a dry restoration area. The restoration may have poor color stability and the metal margins of the crown might still be visible [55].

3.4. Pre-Veneered Stainless Steel Crowns

PVSSCs combine the mechanical properties of SSCs with the additional aesthetic factor of composite resin or thermoplastic resin [19]. The aesthetic part is either chemically or mechanically bonded to the crown [56]. At first, the restoration of anterior primary teeth were introduced to their indications; later on, they were also developed to restore primary molars [19,54,56]. The examples of PVSSCs available on the market are Nusmile Primary Crowns, Kinder Krowns, Cheng Crowns, Flex Crowns, Dura Crowns, and Whiter Biter [56,57]. The exemplary composition of these crowns presented in Nu Smile Pediatric Crowns safety data sheet contains composite paste, iron, copper, silver, 2-hydroxyethyl methacrylate, chromium, nickel, zinc, manganese, silicon, molybdenum, cobalt, and carbon [58]. A study by Sean Beattie et al. compared three pre-veneered stainless steel crown manufacturers for their fracture resistance. Their study involved EC crowns, Kinder Krowns, and NuSmile Primary Crowns. The crowns were subjected to uniaxial force. The results showed no significant differences in fracture resistance between the crowns tested, and furthermore, the forces required for fracture in each case exceeded the control child's occlusal force in the 6- to 10-year age range [59]. Their advantages are long durability and a good aesthetic. PVSSCs allow restoration when the treatment area cannot be perfectly dry [19,60]. On the other hand, they require more aggressive tooth preparation compared to SSCs. They come with some limitations such as prefabricated resin shade, which can look artificial [19]. They are also wide mesio-distally, which can cause problems with placing them in patients with crowding [19,54]. The labial section cannot be crimped, because it might weaken the aesthetic facing and cause premature failure [19,56]. It is also worth noting that clinically try-in crowns that do not meet the proper parameters and require sterilization procedure, which can exert stress on the resin [56]. To reduce the impact of stress, it is recommended to use steam sterilization [56].

Preparation [56] (Figure 5):

1. Discuss the procedure with the parents and child and obtain their consent;
2. The dentist should estimate the crown size;
3. Local anesthesia and tooth isolation;
4. Occlusal reduction around 2 mm or incisal reduction around 2 mm;
5. Circumferential reduction 25–30%;
6. In posterior teeth buccal reduction 1.5–2 mm;
7. Feather-edge subgingival preparation 1.5–2 mm;
8. If necessary, removal of caries and pulp therapy;
9. Try in of the crown;
10. Cementation of the crown. The cement of choice is glass-ionomer.

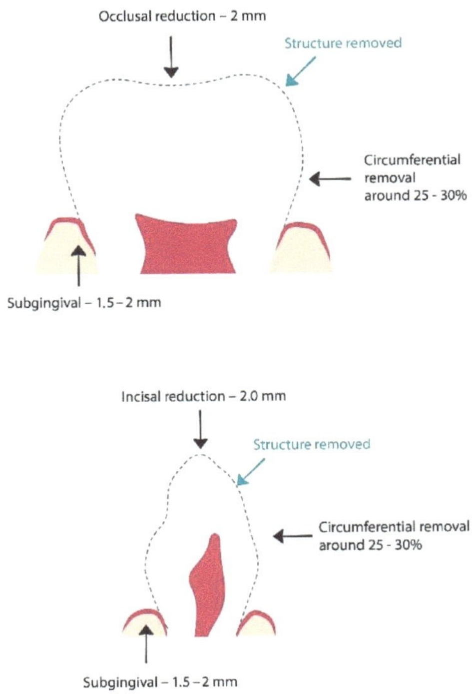

Figure 5. Proper preparation for pre-veneered stainless-steel crowns. The dashed line shows the range of hard tissue removal, which will later be used for proper crown fit and restoration [61].

3.5. Pedo Pearl

These crowns can be included in the group of pre-veneered crowns. The base of these crowns is made of aluminum covered with epoxy paint which gives them an aesthetic tooth color [54]. They are easy to adapt by cutting and crimping [43]. If necessary, they can be covered with composite [62]. Their disadvantages are their soft structure and possible shorter durability [63].

3.6. Polycarbonate Crowns

Anterior primary teeth are usually damaged due to early childhood caries (ECC) caused by bottle feeding without proper hygiene. ECC usually starts on the labial surface of the upper incisors and progresses rapidly. The treatment depends mostly on the co-operation with the patient and includes non-restorative cavity control, preparation, and restoration with conventional materials. Crowns can also be used for restoration and are especially useful when the caries damage is extensive and conventional restoration might be problematic. Polycarbonate crowns are made of aromatic polyesters of carbonic acids [19,54]. They can be described as thermoplastic resins. The use of high temperature (around 130 °C) and pressure makes the material easy to mold and shape into the desired form [19,54]. The material properties are thin structure and flexibility greater than that of acrylic resin crowns [64]. On the other hand, these crowns do not resist high abrasive forces which can cause fracture or premature crown loss [64]. Figure 6 presents an example of polycarbonate crowns available on the market (below).

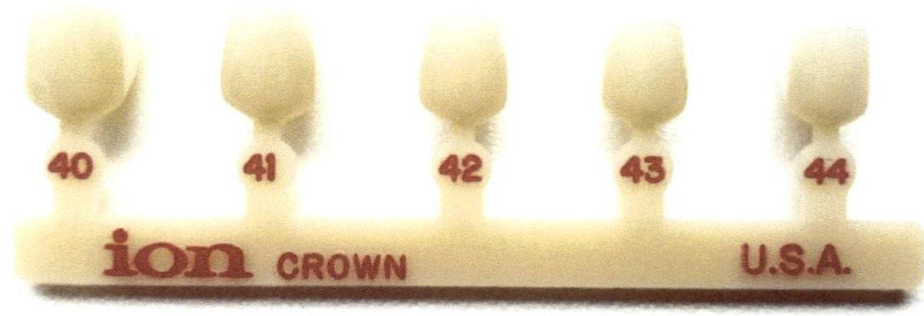

Figure 6. Polycarbonate crowns.

Indications for the use polycarbonate crowns [19,54]:

1. Full restoration of anterior teeth destroyed by caries;
2. ECC as lesion stabilization;
3. Discolored teeth;
4. Restoration after pulp therapy;
5. Restoration in non-carious lesions or developmental defects;
6. Abutment for space maintainers.

Contraindications to use polycarbonate crowns [54]:

1. Too small teeth;
2. Crowded anterior teeth;
3. Excessive tooth damage preventing retention;
4. Bruxism;
5. Excessive abrasion;
6. Overbite;
7. Deep impinging bite.

Preparation [61] (Figure 7):

1. Discuss the procedure with the parents and child and obtain their consent;
2. The dentist should estimate the crown size; the most important part is to properly estimate the mesiodistal dimensions to obtain proper tooth contour;
3. Local anesthesia and tooth isolation;
4. Incisal reduction by about 2 mm;
5. Tiny mesiodistal preparation. The walls should be slightly parallel;
6. Facial/Lingual reduction by about 1 mm;
7. Old protocols suggest performing a chamber 1 mm below gingiva on labial and proximal surfaces;
8. Feather subgingival preparation 1 mm;
9. If pulp procedures had been performed on the tooth, the lingual opening can be used as additional retention;
10. Crown fitting;
11. Cementation;

12. Removal of excess resin cement.

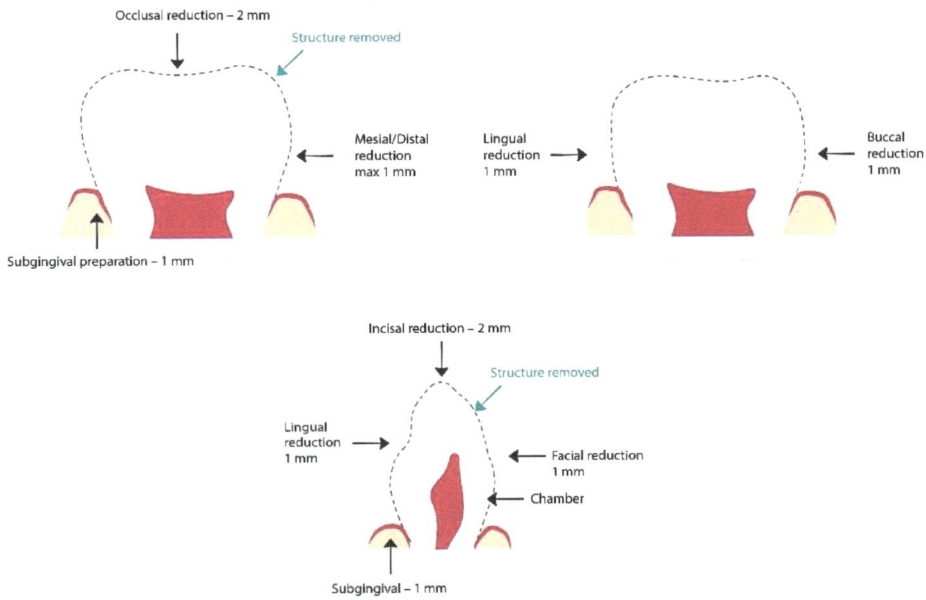

Figure 7. Proper preparation for polycarbonate crowns. The dashed line shows the hard tissue removal range, which will later be used for proper crown fit and restoration [64].

Polycarbonate crowns are at risk of fracture, dislodgement, and discoloration, all of which contribute to a possible unsuccessful treatment outcome [64]. For this reason, many clinicians choose to use a different type of polymer crown type, strip crowns [61].

3.7. Strip Crowns

Strip crowns are transparent plastic forms used to simplify work within upper incisors restoration. They can be filled with both chemical and light curing composite material. Once the material has set, they can be easily removed, leaving a smooth surface. According to Kupietzky et al., their advantages ease of fitting, trimming, and removal [65]. They are also thin and transparent, which makes them easier to match to natural dentition and control composite color. For the best treatment results, patients require proper hygiene instructions and further proper hygienization. Parents must be aware that the lack of proper oral hygiene decreases the chances of successful treatment, which means that they are partially responsible for the procedure's overall outcome [66]. The surrounding soft tissue must be free of inflammation [66]. Strip crowns provide high aesthetics and functionality. They are also cheap and easy to repair [65]. However, their disadvantages include the need to maintain a dry restoration area. Any moisture or blood can interfere with the bonding, and blood can also cause discoloration of the composite material [67]. Their use is also restricted to primary teeth having enough enamel to allow proper bonding after preparation [54]. Minimal reduction is required for proper preparation [61].

Preparation [54]:

1. Discuss the procedure with the parents and child and obtain their consent;
2. Local anaesthesia and tooth isolation;
3. The dentist should estimate the crown size. To facilitate crown size selection, the length of the incisal edge of the tooth being treated or—if the tooth is damaged—of the matching tooth can be used;
4. Reduction in tooth length;
5. Mesial-distal preparation;
6. Knife edge preparation at gingival margin;
7. Choosing composite shade;
8. Preparing vent holes in incisal corners;
9. Firmly seating the crown with composite on the tooth;
10. Curing the composite;
11. Strip crown removal. For safety, the best way is to use a hand piece such as a carver.

Indications to use strip crowns [66]:

1. Extensive decay of the primary anterior teeth;
2. Fractured teeth;
3. Restoration in non-carious lesions or developmental defects;
4. Teeth discoloration;
5. Teeth after pulp therapy.

Contraindications to use strip crowns [66]:

1. Significant teeth tissue loss preventing proper retention;
2. Deep overbite;
3. Periodontal disease.

3.8. *Pedo Jacket Crown*

Similar to strip crowns [19], Pedo Jacket Crowns primarily differ in the material used. They are made from tooth-colored copolyester and filled with resin material. Pedo Jacket Crowns are only available on the market in a single-color shade. Another difference compared to strip crowns is that Pedo Jacket crowns are left on the tooth after polymerization [62]. They cannot be adapted by trimming and reshaping with high-speed finishing bur, as doing so would melt the copolyester [50].

3.9. *New Millennium Crowns*

The New Millennium Crowns are made of laboratory-improved composite resin material [19,54] and are also similar to strip crowns. Their advantages include high aesthetics and parental satisfaction [68]. They can be adapted by reshaping them with a high-speed bur, however their disadvantages include a fragile structure, the need for a dry restoration area, as well as the possible discoloration of the crown by the hemorrhage [19]. They also cannot be crimped [62]. Preparation for New Millennium crowns is similar to strip crown preparation [54].

Indications [19]:

1. Restoration of multi-surface caries;
2. Discolored primary incisors;
3. Anterior teeth fracture;
4. Restoration in non-carious lesions or developmental defects.

Contraindications [19]:

1. Difficulty in keeping the restoration area dry;
2. Overbite;
3. Deep impinging bite;
4. Extensive tooth damage that prevents retention;
5. Periodontal disease.

3.10. Artglass Crowns

Artglass Crowns, also known as Glastech, are made up of polymer glass which forms a three-dimensional molecular network with a cross-linked structure [54]. They contain such fillers as micro-glass and silica, which improve their durability and aesthetics compared to strip crowns [54]. Their longevity is comparable to that of porcelains [62].

3.11. Zirconia Pediatric Crowns

Zirconia has three forms including a monoclinic, tetragonal, and cubic one [69]. These structures are stable in various temperature ranges. The Monoclinic form is stable at room temperature; above 1170 °C, zirconia changes into a tetragonal form, while at 2370 °C, the main form is the cubic one [70]. When zirconia is cooled, the tetragonal phase changes into the monoclinic phase, causing a volumetric expansion of 3–4% [70]. In dentistry, zirconia is used in the form of yttria-stabilized tetragonal polycrystal (Y-TZP), magnesia-partially stabilized zirconia, and zirconia-toughened alumina [69]. Zirconia has many beneficial properties. Firstly, it is very strong, and secondly, it offers good aesthetic properties and good biocompatibility. Zirconia shows high wear and corrosion resistance [69]. It can also resist crack propagation due to a change in the crystalline phase [71]. Zirconia pediatric crowns require minimal preparation; moreover, the whole preparation and restoration process can be completed during a single visit. They are also an alternative for patients with Ni-Cr allergy or sensitivity. Their disadvantage is their high cost. While they cannot be modified, they also show greater thickness than PMCs [70]. According to Sumer et al., zirconia crowns exhibit less plaque accumulation, as evidenced by follow-up visits. They also show nearly zero risk of developing secondary caries and significantly lower restoration loss rate in comparison to strip crowns [71]. A study by Pinar et al. shows that plaque index and gingival index exhibit lower values around zirconia crowns compared to SSCs [72]. This results in better gingival health [73]. Zirconia pediatric crown brands available on the market include Ez-Pedo, NuSmile ZR, and Kinder Crowns Zirconia (see Table 2).

Table 2. The composition of zirconia crowns.

Brand	Composition	Toxicity	Ref
NuSmile Zr NuSmile Pediatric Crowns	Zirconium oxide—88–96% Yttrium oxide—4–6% Hafnium oxide—5% Organic Binder—2–5% Pigment—1–4%	n/a	[74]
Ez-Pedo Ez-Pedo	Zirconium oxide > 85% Hafnium oxide < 5% Yttrium oxide < 6% Organic binder < 5%	Acute oral toxicity: LD50 > 5000 mg/kg (rat) Data for zirconium oxide. Acute dermal toxicity: n/a Acute inhalation toxicity: LC50 > 4.3 mg/L (4 h exposure, rat) Data for zirconium oxide	[75]
Kinder Krowns Zirconia Kinder Krowns	Zirconium dioxide—70–100% Aluminum oxide—0–1% Yttrium oxide—1–5% Iron hydroxideoxid—0–3% Mixture of glycols < 1% Sodium, potassium, boron, and aluminum silicate glass < 1%	n/a	[76]

Indications for the use of zirconia crowns [19]:

1. Decay affecting two or more teeth surfaces;
2. Inability to use amalgam restoration;
3. Restoration after pulp treatment procedures;
4. Restoration in non-carious lesions or development defects;
5. Restoration of fractured primary molars;
6. Restoration of fractured anterior teeth;
7. Bruxism;
8. Restoration in children who require general anesthesia treatment;
9. In children with high caries risk and tendency;
10. An abutment for a space maintainer;
11. Discolored primary incisors.

Preparation [69] (Figure 8):

1. Discuss the procedure with the parents and child and obtain their consent;
2. Local anesthesia and tooth isolation;
3. Reduction in incisal wall of around 1.5–2 mm or occlusal reduction around 2 mm;
4. Buccal reduction around 0.5–1 mm, lingual reduction around 0.75–1.25 mm;
5. Knife edge subgingival preparation 1–2 mm;
6. Checking the occlusion to see if there is adequate clearance from the opposing dentition;
7. Cron selection. This can be achieved by placing the incisal edge of the zirconia crown against the incisal edge of the identical tooth;
8. Cementation with the use of resin-modified glass-ionomer or calcium aluminate cement;
9. Removal of excess cement.

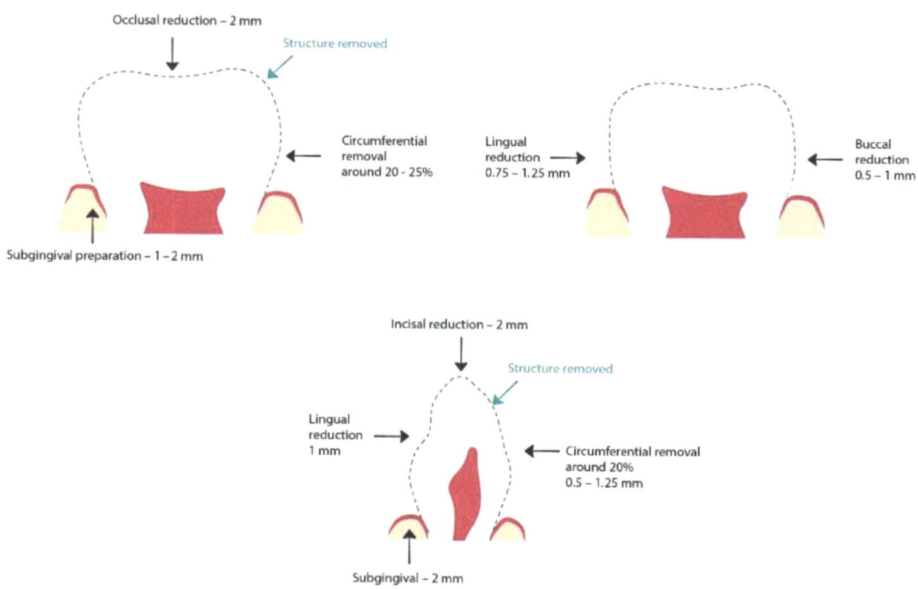

Figure 8. Proper preparation for zirconia crowns. The dashed line shows the hard tissue removal range, which will later be used for proper crown fit and restoration [69].

3.12. Summary of Crowns Used in Pediatric Dentistry

The following Table 3 provides summary of crowns used in pediatric dentistry presenting their advantages and disadvantages.

Table 3. Summary of crowns used in pediatric dentistry [61].

Material	Available Brands	Advantages	Disadvantages
Stainless Steel	Hu-Friedy PEDO CROWNS Hu-Friedy Primary Stainless Steel Crowns 3M ESPE Unitek Primary SSC 3M ESPE Acero Stainless Steel Crowns Acero Crowns DENOVO SSC DENOVO DENTAL	1. Minimal tooth reduction required 2. High strength reliability 3. Good flexibility 4. Easy to contour and crimp 5. To improve the aesthetics, they can be used in open-faced crown technique	1. Low aesthetics 2. Possible nickel allergy and sensitivity
Pre-veneered Stainless Steel	NuSmile Signature NuSmile Pediatric Crowns Cheng Crowns Cheng Crowns Flex Crowns Success Essential Kinder Krowns Next Generation Kinder Krowns	1. High aesthetics 2. Metal edges can be crimped 3. Some brands provide personal customization	1. Pre-veneered material can crack from either crimping or wear-off during usage 2. Requires deeper preparation 3. Possible nickel allergy and sensitivity
Pre-veneered Aluminum	Pedo Pearls Java Crowns	1. High aesthetics 2. Minimal tooth reduction required 3. The tooth-colored coating is flexible which enables contouring and crimping	1. Lower strength 2. May offer decreased longevity 3. The tooth-colored coating is very thin and can wear off during use
Zirconia	Ez-Pedo Ez-Pedo NuSmile Zr NuSmile Pediatric Crowns Kinder Krowns Zirconia	1. Highest strength of all pediatric crowns 2. High aesthetic	1. Expensive 2. Cannot be crimped 3. Require deep preparation 4. Need to have good isolation for effective bonding
Polymer 1. Acrylic 2. Polycarbonate 3. Strip crowns	PedoNatural Crown Strip Crowns Forms 3M ESPE Pedo Jacket Crowns Success Essentials Pediatric Strip Crowns Success Essentials DirectCrown DirectCrown Products	1. Minimal tooth reduction required 2. Some brands offer crowns flexible enough to crimp	1. Due to decreased strength, their suitability for use in posterior location is questionable 2. Some polymerized polymers will not bond to placed resin

4. Risks of Using Pediatric Preformed Crowns

4.1. Periodontal Aspects

Many researchers present a link between restoration using pediatric crowns and plausible periodontal complications. It is worth emphasizing that patients in need of pediatric crowns might lack an oral hygiene routine, which might increase both plaque accumulation and caries risk. This aspect prompts us to properly educate both patients and their parents. Preventive hygiene instructions should be included as a first step in the treatment plan [77,78]. The second aspect that might increase gingivitis risk around pediatric crowns is the inadequate contour of the crown margins [79–83]. Goto found out that posterior crowns presented a higher percentage of gingivitis, which could be caused by more difficult access and the fit of the crown itself [84]. Many studies show that good or moderate fitting of the crown does not significantly increase gingival problems or plaque accumulation [76–80]. The last factor that can be associated with gingivitis is the presence of residual cement left in the gingival pocket [85,86].

4.2. Nickel Allergy and Sensitivity

Nickel percentage content in PMCs changed over the years. Modern PMCs contain around 5–12% nickel, significantly less compared to old formulation nickel-chromium crowns. A study by Feasby et al. shows that a group of children who received old formulation crowns presented an increased nickel-positive patch test, whereas children with modern PMCs showed no statistical difference compared to a control group with no history of nickel appliance use [87]. Nickel hypersensitivity is more frequent in females than males. This relates to ear piercing and the usage of jewelry containing nickel. A study by Keruso et al. and Hoogstraten et al. reported that orthodontic treatment with nickel-containing stainless-steel appliances before ear piercing shows a lower risk of nickel hypersensitivity [88,89]. It is also worth noting that any adjustment to a crown, including cutting or crimping, might increase the risk of corrosion, so margins should be smoothed and polished to a high gloss to minimize this process [90–92]. According to Leila Basir et al., the number of released nickel ions decreased with the trimming of margins. They also noticed increased nickel release as the temperature increased [93]. A study by Dr. Deepak Bhayya et al. shows a difference in nickel ion release from 3M ESPE stainless steel crowns depending on the pH. The results show significant nickel ion release of pH 4.3, 5.5 and 6.3 in artificial saliva, with a maximum nickel release of pH 4.3, followed by 5.5 and 6.3 [94]. According to a survey by the Clinicians Report Foundation, allergy to SSCs is very rare [61].

4.3. Biological Response

Every material introduced to the oral environment can be associated with a biological response. PMCs contain various heavy metals, which might be released due to mechanical, chemical, and thermal intraoral stimulation. While experiments performed on laboratory rats showed that metal pieces containing Fe, Cr, and Ni were found to be cytotoxic to DNA and cultured cells [95], the amounts of metal ions present were not harmful to human health [96,97]. Various human studies show an increase in Ni, Cr, and Fe ions in the saliva; however, although the maximum amounts were always lower than dietary intake and were not capable of causing toxicity, further studies are needed [98–106].

5. Conclusions

In pediatric dentistry, a great deal of effort is required to obtain proper restoration. The first difficulty is to ensure the child's cooperation due to their age; a fear of dentistry is a frequent obstacle in the treatment process. Therefore, it is vital for treatment procedures performed in pediatric dentistry to utilize the simplest and least traumatic procedures that have the best prognosis regarding long-term durability. This review presents one of the approaches for full coverage restoration in pediatric dentistry practice using different types of preformed pediatric crowns. The use of pediatric crowns makes it possible to achieve long-term and positive therapeutic effects. Their use is often easier and faster than manual reconstruction, especially in the case of multi-surface cavities. They are also a good option for restoring teeth after pulp treatment and those with abnormal hard tissue development. PMCs are the most used crowns. Currently, these crowns contain very low levels of nickel and are associated with a low risk of nickel allergy and hypersensitivity. It should be mentioned that there are many types of PMCs available on the market. They differ in terms of composition and prefabrication methods. Therefore, it is important to select solutions corresponding to the dental operator's needs and demands. Over the years, new types of pediatric crowns were introduced to the market to overcome the disadvantages of stainless-steel crowns and respond to the demand for higher aesthetics, such as zirconia crowns. An important factor that requires further investigation is a comparative analysis of the use of crowns from different materials, such as a comparison of the long-term durability of zirconia crowns in comparison with PMCs. Ultimately, it is always up to the dentist and the parents to choose the optimal type of restoration based on the child's cooperation, as well as the parents' aesthetic requirements and economic factors.

Author Contributions: Conceptualization: M.D. and K.S.; methodology: M.D., K.S. and R.J.W.; software: K.S.; validation: R.J.W.; formal analysis: K.S., M.D. and R.J.W.; investigation: K.S.; resource: K.S., M.D. and R.J.W.; data curation: K.S., M.D. and R.J.W.; writing—original draft preparation: K.S. and M.D.; writing—review and editing: K.S., M.D. and R.J.W.; visualization: K.S., M.D. and R.J.W.; supervision: M.D. and R.J.W.; project administration: M.D.; funding acquisition: M.D. All authors have read and agreed to the published version of the manuscript.

Funding: This work was financed by a subsidy from Wroclaw Medical University, number SUBZ. B180.22.091.

Institutional Review Board Statement: Not applicable.

Informed Consent Statement: Not applicable.

Data Availability Statement: Not applicable.

Acknowledgments: The authors would like to thank Edyta Cichos for the graphical support.

Conflicts of Interest: The authors declare no conflict of interest. The funders had no role in the design of the study, data collection, analysis, and interpretation, writing of the manuscript, and the decision to publish the results.

References

1. Pitts, N.B.; Zero, D.T.; Marsh, P.D.; Ekstrand, K.; Weintraub, J.A.; Ramos-Gomez, F.; Tagami, J.; Twetman, S.; Tsakos, G.; Ismail, A. Dental caries. *Nat. Rev. Dis. Primers* **2017**, *3*, 17030. [CrossRef] [PubMed]
2. World Health Organization. Oral Health. Available online: https://www.who.int/news-room/fact-sheets/detail/oral-health (accessed on 25 March 2020).
3. UK. Oral Health Survey of 5-Year-Old Children 2019. 2020. Available online: https://www.gov.uk/government/statistics/oral-health-survey-of-5-year-old-children-2019 (accessed on 20 March 2020).
4. Scotland's National Dental Inspection Programme 2003. Available online: http://www.dundee.ac.uk/ndip/index.htm (accessed on 11 January 2020).
5. O'Brien, M. *Children's Dental Health in the United Kingdom 1993*; H.M. Stationery Office: London, UK, 1994; pp. 5–17.
6. Alkarimi, H.; Watt, R.; Pikhart, H.; Sheiham, A.; Tsakos, G. Dental caries and growth in school-age children. *Pediatrics* **2014**, *133*, e616–e623. [CrossRef] [PubMed]
7. van Strijp, G.; van Loveren, C. No Removal and Inactivation of Carious Tissue: Non-Restorative Cavity Control. In *Caries Excavation: Evolution of Treating Cavitated Carious Lesions*; Karger Publishers: Basel, Switzerland, 2018; Volume 27, pp. 124–136.
8. Innes, N.P.T.; Evans, D.J.P.; Bonifacio, C.C.; Geneser, M.; Hesse, D.; Heimer, M.V.; Kanellis, M.; Machiulskiene, V.; Narbutaite, J.; Olegário, I.C.; et al. The Hall Technique 10 years on: Questions and answers. *Br. Dent. J.* **2017**, *222*, 478–483. [CrossRef] [PubMed]
9. Bjørndal, L. Stepwise Excavation, Caries Excavation: Evolution of Treating Cavitated Carious Lesions. *Monogr. Oral Sci.* **2018**, *27*, 68–81.
10. Ricketts, D.; Lamont, T.; Innes, N.P.T.; Kidd, E.; Clarkson, J.E. Operative caries management in adults and children. *Cochrane Database Syst. Rev.* **2019**, CD003808. [CrossRef]
11. Santamaria, R.; Innes, N.; Machiulskiene, V.; Schmoeckel, J.; Alkilzy, M.; Splieth, C.H. Alternative Caries Management Options for Primary Molars: 2.5-Year Outcomes of a Randomised Clinical Trial. *Caries Res.* **2017**, *51*, 605–614. [CrossRef]
12. Chisini, L.A.; Collares, K.; Cademartori, M.G.; De Oliveira, L.J.C.; Conde, M.C.M.; Demarco, F.F.; Corrêa, M.B. Restorations in primary teeth: A systematic review on survival and reasons for failures. *Int. J. Paediatr. Dent.* **2018**, *28*, 123–139. [CrossRef]
13. Roberts, J.F.; Attari, N.; Sherriff, M. The survival of resin modified glass ionomer and stainless steel crown restorations in primary molars, placed in a specialist paediatric dental practice. *Br. Dent. J.* **2005**, *198*, 427–431. [CrossRef]
14. Hickel, R.; Kaaden, C.; Paschos, E.; Buerkle, V.; Garcia-Godony, F.; Manhart, J. Longevity of occlusally stressed restorations in posterior primary teeth. *Am. J. Dent.* **2005**, *18*, 198–211.

15. Olegário, I.C.; Bresolin, C.R.; Pássaro, A.L.; de Araujo, M.P.; Hesse, D.; Mendes, F.M.; Raggio, D.P. Stainless steel crown vs bulk fill composites for the restoration of primary molars post-pulpectomy: 1-year survival and acceptance results of a randomized clinical trial. *Int. J. Paediatr. Dent.* **2022**, *32*, 11–21. [CrossRef]
16. Schüler, I.M.; Hiller, M.; Roloff, T.; Kühnisch, J.; Heinrich-Weltzien, R. Clinical success of stainless steel crowns placed under general anaesthesia in primary molars: An observational follow up study. *J. Dent.* **2014**, *42*, 1396–1403. [CrossRef] [PubMed]
17. Innes, N.P.; Ricketts, D.; Chong, L.Y.; Keightley, A.J.; Lamont, T.; Santamaria, R.M. Preformed crowns for decayed primary molar teeth. *Cochrane Database Syst. Rev.* **2015**, 2015. [CrossRef] [PubMed]
18. Zbańska, J.; Herman, K.; Kuropka, P.; Dobrzyński, M. Regenerative Endodontics as the Future Treatment of Immature Permanent Teeth. *Appl. Sci.* **2021**, *11*, 6211. [CrossRef]
19. Garg, V.; Panda, A.; Shah, J.; Panchal, P. Crowns in pediatric dentistry: A review. *J. Adv. Med. Dent. Sci. Res.* **2016**, *4*, 41–46.
20. Karthikeyan, G.; Ravindran, V.; Ramamurthy, J. Prevalence of usage of stainless steel crown, strip crown and zirconia in anterior teeth for paediatric dental patients in different age groups. *Int. J. Res. Pharm. Sci.* **2020**, *11*, 1511–1516. [CrossRef]
21. Ludovichetti, F.S.; Stellini, E.; Signoriello, A.G.; Di Fiore, A.; Gracco, A.; Mazzoleni, S. Zirconia vs Stainless steel pediatric crowns: A literature review. *Minerva Dent. Oral Sci.* **2021**, *70*, 112–118. [CrossRef]
22. Kara, N.B.; Yılmaz, Y. Assessment of oral hygiene and periodontal health around posterior primary molars after their restoration with various crown types. *Int. J. Paediatr. Dent.* **2014**, *24*, 303–313. [CrossRef]
23. Zimmerman, J.A.; Feigal, R.J.; Till, M.J.; Hodges, J.S. Parental attitudes on restorative materials as factors influencing current use in pediatric dentistry. *Pediatr. Dent.* **2009**, *31*, 63–70.
24. Walia, T.; Salami, A.A.; Bashiri, R.; Hamoodi, O.M.; Rashid, F. A randomised controlled trial of three aesthetic full-coronal restorations in primary maxillary teeth. *Eur. J. Paediatr. Dent.* **2014**, *15*, 113–118.
25. Sahana, S.; Vasa, A.A.K. Esthetic crowns for primary teeth: A review. *Ann. Essences Dent.* **2010**, *2*, 87–93. [CrossRef]
26. Larry, W. Bybee, Russ Misner, Peldyak, Prefabricated Pediatric Crowns and Method of Producing Prefabricated Pediatric Crowns. U.S. Patent Application No. US 2006/0154211 A1, 13 July 2006.
27. Nash, D.A. The nickel-chromium crown for restoring posterior primary teeth. *JADA* **1981**, *102*, 44–49. [CrossRef] [PubMed]
28. Sajjanshetty, S.; Patil, P.; Hugar, D.; Rajkumar, K. Pediatric Preformed Metal Crowns—An Update. *J. Dent. Allied Sci.* **2013**, *2*, 29–32. [CrossRef]
29. Denovo Stainless Steel Safety Data Sheet.
30. Hu-Friedy's Pedo Crowns, Material Safety Data Sheet. 2011.
31. 3M ESPE Stainless Steel Crowns, Safety Data Sheet. 2017.
32. Kodaira, H.; Ohno, K.; Fukase, N.; Kuroda, M.; Adachi, S.; Kikuchi, M.; Asada, Y. Release and systemic accumulation of heavy metals from preformed crowns used in restoration of primary teeth. *J. Oral Sci.* **2013**, *55*, 161–165. [CrossRef] [PubMed]
33. KTR Dental Crown, Stainless Steel Pediatric Primary Molar Dental Crowns, Safety Data Sheet. 2014.
34. Dimitrova, M.G.; Dimitrov, E.; Andreeva, R. Indications for use of preformed crowns in pediatric dentistry. *J. Med. Dent. Pr.* **2016**, *3*, 439–445. [CrossRef]
35. Santamaría, R.M.; Abudrya, M.H.; Gül, G.; Mourad, M.S.; Gomez, G.F.; Zandona, A.G.F. How to Intervene in the Caries Process: Dentin Caries in Primary Teeth. *Caries Res.* **2020**, *54*, 306–323. [CrossRef]
36. Roberts, J.F.; Sherriff, M. The fate and survival of amalgam and preformed crown molar restorations placed in a specialist paediatric dental practice. *Br. Dent. J.* **1990**, *169*, 237–244. [CrossRef]
37. Curzon, M.E.J.; Roberts, J.F.; Kennedy, D. *Kennedy's Paediatric Operative Dentistry*, 4th ed.; Butterworth-Heinemann: Boston, MA, USA, 1996; pp. 79–87, 114–127.
38. Croll, T.P.; Epstein, D.W.; Castaldi, C.R. Marginal adaptation of stainless steel crowns. *Pediatr. Dent.* **2003**, *25*, 249–252.
39. Spedding, R.H. Two principles for improving the adaptation of stainless steel crowns to primary molars. *Dent. Clin. N. Am.* **1984**, *28*, 157–175.
40. Mcdonald, R.E.; Avery, D.R. Restorative Dentistry. In *Dentistry for the Child and Adolescent*, 11th ed.; Mosby International Ltd.: Indianapolis, IN, USA, 2021; pp. 376–382.
41. Mink, J.R.; Hill, C.J. Modification of the stainless steel crown for primary teeth. *ASDC J. Dent. Child* **1971**, *38*, 197–205.
42. Randall, R.C. Preformed metal crowns for primary and permanent molar teeth: Review of the literature. *Pediatric Dent.* **2002**, *24*, 5.
43. Srinath, S.; Kanthaswamy, A.C. Different Crown Used For Restoring Anterior Primary Teeth: A Review. *J. Pharm. Sci. Res.* **2017**, *9*, 190–193.
44. Mendes, F.M.; De Benedetto, M.S.; Zardetto, C.G.D.C.; Wanderley, M.T.; Correa, M.S.N.P. Resin composite restoration in primary anterior teeth using short-post technique and strip crowns: A case report. *Quintessence Int.* **2004**, *35*, 689–692. [PubMed]
45. Usha, M.; Deepak, V.; Venkat, S.; Gargi, M. Treatment of severely mutilated incisors: A challenge to the pedodontist. *J. Indian Soc. Pedod. Prev. Dent.* **2007**, *25*, 34–36.
46. Wanderley, M.T.; Ferreira, S.L.; Rodrigues, C.R.; Filho, L.E.R. Primary anterior tooth restoration using posts with macroretentive elements. *Quintessence Int.* **1999**, *30*, 432–443.
47. Ramires-Romito, A.C.; Wanderley, M.T.; Oliveira, M.D.; Imparato, J.C.; Corrêa, M.S. Biologic restoration of primary anterior teeth. *Quintessence Int.* **2000**, *31*, 405–411.

48. Waggoner, W.F.; Kupietzky, A. Anterior esthetic fixed appliances for the preschooler: Considerations and a technique for placement. *Pediatr. Dent.* **2001**, *23*, 147–150.
49. Gilchrist, F.; Morgan, A.G.; Farman, M.; Rodd, H.D. Impact of the Hall technique for preformed metal crown placement on undergraduate paediatric dentistry experience. *Eur. J. Dent. Educ.* **2013**, *17*, e10–e15. [CrossRef]
50. Evans, D.; Southwick, C.; Foley, J.; Innes, N.; Pavitt, S. The Hall Technique: A Pilot Trial of a Novel Use of Preformed Metal Crowns for Managing Carious Primary Teeth. Available online: http://www.app.dundee.ac.uk/tuith/Articles/index.htm (accessed on 1 August 2021).
51. Innes, N.P.; Evans, D.J.; Stirrups, D.R. The Hall Technique: A randomized controlled clinical trial of a novel method of managing carious primary molars in general dental practice: Acceptability of the technique and outcomes at 23 months. *BMC Oral Health* **2007**, *7*, 18. [CrossRef]
52. Innes, N.P.; Evans, D.J.; Stirrups, D.R. Sealing caries in primary molars: Randomized control trial, 5-year results. *J. Dent. Res.* **2011**, *90*, 1405–1410. [CrossRef]
53. Santamaria, R.M.; Innes, N.P.T.; Machiulskiene, V.; Evans, D.J.P.; Splieth, C.H. Caries management strategies for primary molars: 1yr randomized control trial results. *J. Dent. Res.* **2014**, *93*, 1062–1069. [CrossRef]
54. Mittal, G.K.; Verma, A.; Pahuja, H.; Agarwal, S.; Tomar, H. Esthetic crowns in pediatric dentistry: A review. *Int. J. Contemp. Med. Res.* **2016**, *3*, 1280–1282.
55. Champagne, C.; Waggoner, W.; Ditmyer, M.; Casamassimo, P.S.; MacLean, J. Parental satisfaction with preveneered stainless steel crowns for primary anterior teeth. *Pediatr. Dent.* **2008**, *29*, 465–469.
56. Cohn, C. *Pre-Veneered Stainless Steel Crowns—An Aesthetic Alternative*; Academy of Dental Therapeutics and Stomatology Supplement to PennWell Publications: Chicago, IL, USA, 2012.
57. Yang, J.N.C.; Mani, G. Crowns for primary anterior teeth. *Int. J. Pedod. Rehabil.* **2016**, *1*, 75–78. [CrossRef]
58. NuSmile Signature Pediatric Crowns, Safety Data Sheet. 2012.
59. Beattie, S.; Taskonak, B.; Jones, J.; Chin, J.; Sanders, B.; Tomlin, A.; Weddell, J. Fracture resistance of 3 types of primary esthetic stainless steel crowns. *J. Can. Dent. Assoc.* **2011**, *77*, b90. [PubMed]
60. Leith, R.; O'Connell, A.C. A clinical study evaluating success of 2 commercially available preveneered primary molar stainless steel crowns. *Pediatr. Dent.* **2011**, *33*, 300–306. [PubMed]
61. Christensen, G.J. Pediatric Crowns Are Growing Up. *Clin. Rep.* **2012**, *5*, 1, 3–4.
62. Aiem, E.; Smail-Faugeron, V.; Muller-Bolla, M. Aesthetic preformed paediatric crowns: Systematic review. *Int. J. Paediatr. Dent.* **2017**, *27*, 273–282. [CrossRef]
63. Yilmaz, Y.; Guler, C. Evaluation of different sterilization and disinfection methods on commercially made preformed crowns. *J. Indian Soc. Pedod. Prev. Dent.* **2008**, *26*, 162–167. [CrossRef]
64. Venkataraghavan, K.; Chan, J.; Karthik, S. Polycarbonate crowns for primary teeth revisited: Restorative options, technique and case reports. *J. Indian Soc. Pedod. Prev. Dent.* **2014**, *32*, 156–159. [CrossRef]
65. Kupietzky, A. Bonded resin composite strip crowns for primary incisors: Clinical tips for a successful outcome. *Pediatr. Dent.* **2002**, *24*, 145–148.
66. Muhamad, A.H.; Abdulgani, A. Strip crowns Technique for restoration of primary anterior teeth: Case report. *IOSR J. Dent. Med. Sci.* **2015**, *14*, 48–53.
67. Ram, D.; Fuks, A.B.; Eidelman, E. Long-term clinical performance of esthetic primary molar crowns. *Pediatr. Dent.* **2004**, *25*, 582–584.
68. Nuvvula, S.; Kamatham, R. Treatment of anterior crossbite in the primary dentition with esthetic crowns: Report of 3 cases. *Pediatr. Dent.* **2012**, *34*, 339–342.
69. Khatri, A. Esthetic zirconia crown in pedodontics. *Int. J. Pedod. Rehabil.* **2017**, *2*, 31–33. [CrossRef]
70. Bica, C.; Pescaru, P.; Stefanescu, A.; Docan, M.O.; Martha, K.; Esian, D.; Cerghizan, D. Applicability of Zirconia-Prefabricated Crowns in Children with Primary Dentition. *Rev. Chim.* **2017**, *68*, 1940–1943. [CrossRef]
71. Alaki, S.M.; Abdulhadi, B.S.; AbdElBaki, M.A.; Alamoudi, N.M. Comparing zirconia to anterior strip crowns in primary anterior teeth in children: A randomized clinical trial. *BMC Oral Health* **2020**, *20*, 313. [CrossRef]
72. Taran, P.K.; Kaya, M.S. A Comparison of Periodontal Health in Primary Molars Restored with Prefabricated Stainless Steel and Zirconia Crowns. *Int. J. Clin. Pediatr. Dent.* **2018**, *40*, 334–339.
73. Larsson, C. Zirconium dioxide based dental restorations. Studies on clinical performance and fracture behaviour. *Swed. Dent. J. Suppl.* **2011**, *213*, 9–84.
74. NuSmile ZR Pediatric Crowns, Safety Data Sheet. 2012.
75. Tosoh Usa Inc. Zpex, Safety Data Sheet. 2015.
76. Zirconia Kinder Krowns, Safety Data Sheet. 2015.
77. Duggal, M.S.; Curzon, M.E.; Fayle, S.A.; Toumba, K.J.; Robertson, A.J. *Restorative Techniques in Paediatric Dentistry: An Illustrated Guide to the Restoration of Carious Primary Teeth*; CRC Press: London, UK, 2021.
78. Henderson, H.Z. Evaluation of the preformed stainless steel crown. *ASDC J. Dent. Child.* **1973**, *40*, 353–358.
79. Goldberg, N.L. The stainless steel crown in pediatric dentistry. *Dent. Dig.* **1969**, *75*, 352–355.
80. Webber, D.L. Gingival health following placement of stainless steel crowns. *ASDC J. Dent. Child.* **1974**, *41*, 186–189. [PubMed]

81. Myers, D.R. A clinical study of the response of the gingival tissue surrounding stainless steel crowns. *ASDC J. Dent. Child.* **1975**, *42*, 281–284. [PubMed]
82. Durr, D.P.; Ashrafi, M.H.; Duncan, W.K. A study of plaque accumulation and gingival health surrounding stainless steel crowns. *ASDC J. Dent. Child.* **1982**, *49*, 343–346. [PubMed]
83. Einwag, J. Effect of entirely preformed stainless steel crowns on periodontal health in primary, mixed dentitions. *ASDC J. Dent. Child.* **1984**, *51*, 356–359. [PubMed]
84. Goto, G.; Imanishi, T.; Machida, Y. Clinical evaluation of preformed crown for deciduous teeth. *Bull. Tokyo Dent. Coll.* **1970**, *11*, 169–176.
85. More, F.G.; Pink, T.C. The stainless steel crown: A clinical guide. *J. Mich. State Dent. Assoc.* **1973**, *55*, 237–242.
86. Fayle, S. UK National Clinical Guidelines in Paediatric Dentistry. *Int. J. Paediatr. Dent.* **2008**, *9*, 311–314. [CrossRef]
87. Zinelis, S.; Lambrinaki, T.; Kavvadia, K.; Papagiannoulis, L. Morphological and compositional alterations of in vivo aged prefabricated pediatric metal crowns (PMCs). *Dent. Mater.* **2008**, *24*, 216–220. [CrossRef]
88. Kerosuo, H.; Kullaa, A.; Kerosuo, E.; Kanerva, L.; Hensten-Pettersen, A. Nickel allergy in adolescents in relation to orthodontic treatment and piercing of ears. *Am. J. Orthod. Dentofac. Orthop.* **1996**, *109*, 148–154. [CrossRef]
89. Van Hoogstraten, I.M.W.; Andersen, K.E.; Von Blomberg, B.M.E.; Boden, D.; Bruynzeel, D.P.; Burrows, D.; Camarasa, J.G.; Dooms-Goossens, A.; Kraal, G.; Lahti, A.; et al. Reduced frequency of nickel allergy upon oral nickel contact at an early age. *Clin. Exp. Immunol.* **1991**, *85*, 441–445. [CrossRef]
90. Grimsdottir, M.R.; Gjerdet, N.R.; Hensten-Pettersen, A. Composition and in vitro corrosion of orthodontic appliances. *Am. J. Orthod. Dentofac. Orthop.* **1992**, *101*, 525–532. [CrossRef]
91. Bishara, S.E.; Barrett, R.D.; Selim, M.I. Biodegradation of orthodontic appliances. Part II. Changes in the blood level of nickel. *Am. J. Orthod. Dentofac. Orthop.* **1993**, *103*, 115–119. [CrossRef]
92. von Fraunhofer, J.A. Corrosion of orthodontic devices. *Semin. Orthod.* **1997**, *3*, 198–205. [CrossRef]
93. Basir, L.; Shamsaei, M.; Ziaei, S.A. Evaluation of nickel releasing from stainless steel crowns regarding to "trimming": An in vitro study. *J. Indian Soc. Pedod. Prev. Dent.* **2018**, *36*, 58–64. [PubMed]
94. Tiwari, S.; Bhayya, D.; Gupta, S.; Saxena, S.; Kathal, S.; Roy, S. Effect of pH on Nickel ion release from stainless steel crowns: An in vitro study. *IERJ* **2016**, *2*, 47.
95. Einwag, J.; Dünninger, P. Stainless steel crown versus multisurface amalgam restorations: An 8-year longitudinal clinical study. *Quintessence Int.* **1996**, *27*, 321–323.
96. Innes, N.; Ricketts, D.; Evans, D. Preformed metal crowns for decayed primary molar teeth. *Cochrane Database Syst. Rev.* **2015**, *2015*. [CrossRef]
97. Humphery, W. Uses of chrome steel in children's dentistry. *Dent. Surv.* **1950**, *26*, 945–949.
98. Bhaskar, V.; Subba Reddy, V.V. Biodegradation of Ni and chromium from space maintainers: An in vitro study. *J. Indian Soc. Pedod. Prev. Dent.* **2010**, *28*, 6. [CrossRef]
99. Feasby, W.; Ecclestone, E.; Grainger, R. Ni sensitivity in pediatric dental patients. *Pediatr. Dent.* **1988**, *10*, 127–129.
100. Yilmaz, A.; Ozdemir, C.E.; Yilmaz, Y. A delayed hypersensitivity reaction to a stainless steel crown: A case report. *J. Clin. Pediatr. Dent.* **2012**, *36*, 235–238. [CrossRef] [PubMed]
101. Kulkarni, P.; Agrawal, S.; Bansal, A.; Jain, A.; Tiwari, U.; Anand, A. Assessment of Ni release from various dental appliances used routinely in pediatric dentistry. *Indian J. Dent.* **2016**, *7*, 81–85. [CrossRef] [PubMed]
102. Morán-Martínez, J.; Monreal-de Luna, K.D.; Betancourt-Martínez, N.D.; Carranza-Rosales, P.; Contreras-Martínez, J.G.; López-Meza, M.C.; Rodríguez-Villarreal, O. Genotoxicity in oral epithelial cells in children caused by Ni in metal crowns. *Genet. Mol. Res.* **2013**, *12*, 3178–3185. [CrossRef]
103. Menek, N.; Basaran, S.; Karaman, Y.; Ceylan, G.; Tunç, E.S. Investigation of Ni ion release from SSCs by square wave voltammetry. *Int. J. Electrochem. Sci.* **2012**, *7*, 6465–6471.
104. Hernández-Martínez, C.T.; Morales-Luckie, R.A.; Robles-Berme, N.L.; Jiménez-Gayosso, S.I.; Ramírez-Cardona, M.; García-Hernández, V.; Lara-Carrillo, E.; Medina-Solís, C.E. Detection of Ni, Fe, and Cr Released in Saliva after Prefabricated Metal Crown Placement in Children. *Preprints* **2018**, 2018040236. [CrossRef]
105. Mohamed, A.A.; Amel Mahmoud Ahmed, A.M.; Mahmoud, T.T. Comparison between Ni and chromium levels in saliva of children having space maintainers versus SSCs (Comparative Study). *Int. J. Sci. Res.* **2013**, *5*, 663–666.
106. Bhat, S.S.; Amanna, E.N.; Hegde, S.K. An in vitro evaluation of nickel and chromium release from different commercially available stainless steel crowns. *J. Indian Soc. Pedod. Prev. Dent.* **2019**, *37*, 31–38. [CrossRef]

MDPI AG
Grosspeteranlage 5
4052 Basel
Switzerland
Tel.: +41 61 683 77 34

Materials Editorial Office
E-mail: materials@mdpi.com
www.mdpi.com/journal/materials

www.ingramcontent.com/pod-product-compliance
Lightning Source LLC
Chambersburg PA
CBHW041423210926
51850CB00013B/3344
* 9 7 8 3 7 2 5 8 3 4 6 7 9 *